RADIATION IN ASTROPHYSICAL PLASMAS

ASTROPHYSICS AND SPACE SCIENCE LIBRARY

VOLUME 204

RADIATION IN ASTROPHYSICAL PLASMAS

by

V. V. ZHELEZNYAKOV

Institute of Applied Physics,
Russian Academy of Science,
Nizhny Novgorod, Russia
and
Advanced School of General and Applied Physics,
Nizhny Novgorod University,
Nizhny Novgorod, Russia

KLUWER ACADEMIC PUBLISHERS

DORDRECHT / BOSTON / LONDON

A C.I.P. Catalogue record for this book is available from the Library of Congress.

ISBN 0-7923-3907-X

Published by Kluwer Academic Publishers,
P.O. Box 17, 3300 AA Dordrecht, The Netherlands.

Kluwer Academic Publishers incorporates
the publishing programmes of
D. Reidel, Martinus Nijhoff, Dr W. Junk and MTP Press.

Sold and distributed in the U.S.A. and Canada
by Kluwer Academic Publishers,
101 Philip Drive, Norwell, MA 02061, U.S.A.

In all other countries, sold and distributed
by Kluwer Academic Publishers Group,
P.O. Box 322, 3300 AH Dordrecht, The Netherlands.

Printed on acid-free paper

Table of Contents

Preface 1

1 Introduction 5

1.1 Parameters of astrophysical plasmas 5

1.1.1 Basic plasma properties and characteristic frequencies 5

1.1.2 Specific values of cosmic plasma parameters and magnetic field strengths 8

1.1.3 Plasma under extreme astrophysical conditions . . . 18

1.2 Electromagnetic radiation and its main characteristics . . . 21

1.2.1 Energy density and specific intensity 21

1.2.2 Polarization of radiation. Polarization tensor and Stokes parameters 27

1.3 Radiation sources and radiation mechanisms. Definitions and general properties 32

1.3.1 Spectral power and flux density of radiation from stationary and moving sources 33

1.3.2 Radiation mechanisms 41

2 Electromagnetic waves in homogeneous plasma 45

2.1 Dielectric permittivity of plasma. Dispersion equations . . . 45

2.1.1 Electromagnetic field equations in a medium 45

2.1.2 Dispersion relations for normal waves 50

2.1.3 Dielectric tensor of plasma 54

2.2 Electromagnetic waves in homogeneous equilibrium plasma 67

2.2.1 Waves in isotropic plasma 67

2.2.2 Ordinary and extraordinary waves in magnetoactive plasma 71

2.2.3 Plasma waves 81

2.2.4 Bernstein modes 84

2.2.5 Waves in relativistic plasma 88

2.2.6 Electromagnetic waves in a system comprising birefringent vacuum and rarefied plasma 91

3 Electromagnetic waves in inhomogeneous astrophysical plasma **95**
3.1 Geometrical optics approximation in an inhomogeneous medium . . . 95
3.1.1 The form of the geometrical optics solution . . . 95
3.1.2 Geometrical optics approximation in isotropic and magnetoactive plasma . . . 96
3.1.3 The ray approach . . . 102
3.1.4 Effect of the group delay of signals . . . 107
3.2 Linear mode coupling in magnetoactive plasma . . . 111
3.2.1 Equations of quasi-isotropic approximation and basic problems of linear mode coupling theory . . . 112
3.2.2 Linear coupling in an inhomogeneous magnetic field 119
3.2.3 Transition layers and "limiting" polarization of radiation . . . 121
3.2.4 Propagation of waves through a region with transverse magnetic field . . . 124
3.2.5 Wave propagation through a zero magnetic field surface . . . 136
3.2.6 On the role of the linear mode coupling in cosmic plasmas with an inhomogeneous magnetic field . . . 142
3.2.7 Coupling of electromagnetic waves and the problem of escape of radio emission from dense plasma . . . 146
3.2.8 Linear mode coupling in a system "plasma + birefringent vacuum". . . . 151

4 Transfer of radiation in astrophysical plasmas **155**
4.1 Transfer of radiation intensity . . . 156
4.1.1 Radiation transfer equation in isotropic plasma . . . 156
4.1.2 Transfer equation for magnetoactive plasma . . . 163
4.1.3 Solutions of radiation transfer equations . . . 166
4.2 Transfer of polarization . . . 175
4.2.1 Faraday and Cotton–Mouton effects in plasma . . . 175
4.2.2 Equations of polarization transfer in magnetoactive plasma . . . 183
4.2.3 Solutions of polarization transfer equations . . . 191

5 Emission by astrophysical plasmas **199**
5.1 Emission by a charged particle . . . 199
5.1.1 Bremsstrahlung . . . 199
5.1.2 Čerenkov radiation . . . 204
5.1.3 Magnetic bremsstrahlung . . . 207

5.1.4 Cyclotron radiation 215
5.1.5 Synchrotron radiation 221
5.1.6 Curvature radiation 231
5.2 Plasma emissivity 233
5.2.1 Bremsstrahlung emissivity 234
5.2.2 Čerenkov emissivity 237
5.2.3 Cyclotron emissivity 238
5.2.4 Gyrosynchrotron emissivity 242
5.2.5 Synchrotron emissivity. Synchrotron mechanism of cosmic radio emission 249

6 Absorption in thermal astrophysical plasmas 257
6.1 Absorption in isotropic plasma 258
6.1.1 Absorption due to particle collisions. Bremsstrahlung radiation mechanism 258
6.1.2 Čerenkov absorption of plasma waves. Landau damping 268
6.2 Absorption in magnetoactive plasma 269
6.2.1 Absorption of electromagnetic waves due to particle collisions 270
6.2.2 Cyclotron absorption. Mechanism of thermal cyclotron emission 272

7 Reabsorption and amplification of radiation in non-equilibrium plasma 287
7.1 Method of kinetic equation and method of Einstein coefficients 288
7.1.1 Method of kinetic equation. Amplification and convective instability 288
7.1.2 Method of Einstein coefficients 292
7.1.3 Population balance equation 296
7.2 Čerenkov instability 307
7.2.1 Instability and amplification in a system "beam – plasma" 307
7.2.2 Buneman instability 313
7.2.3 Loss-cone instability 315
7.2.4 Quasi-linear theory of beam instability 317
7.2.5 Plasma mechanism of radio wave generation 323
7.3 Cyclotron instability 325
7.3.1 Cyclotron instability of longitudinal waves 325
7.3.2 Kinetic instability of Bernstein modes 327
7.3.3 Kinetic instability in the hybrid band 333

7.3.4 Hydrodynamic instability at double plasma resonance 336
7.4 Synchrotron reabsorption and instability 340
7.4.1 Synchrotron reabsorption 340
7.4.2 Reabsorption coefficient for power-law particle energy spectrum 343
7.4.3 Synchrotron instability 345
7.4.4 Negative reabsorption in a system of relativistic electrons with a "monoenergetic" or power-law energy spectrum 348
7.4.5 Concluding remarks 352

8 Scattering of radiation in plasma 357
8.1 Scattering of waves by particles in isotropic plasma 357
8.1.1 Spontaneous and stimulated scattering 357
8.1.2 Integral kinetic equation for photons and radiation transfer equation 360
8.1.3 Scattering of radiation with one-dimensional spectrum. Differential and integral redistribution 363
8.1.4 Stabilization of beam instability 366
8.1.5 Rayleigh scattering in non-relativistic plasma 371
8.1.6 Scattering by free particles. Inverse Compton effect . 377
8.1.7 Kompaneets differential equation and "comptonization" of radiation 384
8.2 Transfer of radiation at cyclotron frequencies 394
8.2.1 Resonant cyclotron scattering 394
8.2.2 Transfer equation for radiation at cyclotron frequencies in collisionless plasma 396
8.2.3 Solution of radiation transfer equation in a homogeneous and inhomogeneous magnetic field 399
8.2.4 Transfer of radiation at cyclotron frequencies in plasma with particle collisions 407
8.2.5 Cyclotron features in the spectra of magnetic degenerate stars 414
8.3 Radiation pressure force in plasma 421
8.3.1 Radiation pressure due to Thomson scattering . . . 421
8.3.2 Pressure force by radiation at cyclotron frequencies . 423
8.3.3 Radiation pressure effects on magnetic degenerate stars 429
8.4 Raman scattering in plasma 433
8.4.1 Conservation laws for elementary scattering events . 434
8.4.2 Kinetic equation for photons and the radiation transfer equation 438

8.4.3 Coalescences and decays in isotropic plasma 443

References **449**

Index **460**

Preface

The interest in the problem of interaction between radiation and astrophysical plasmas arose decades ago. Initially, this was closely related to the discovery of radio emission from the Sun and Galaxy which alerted theoretical radio astronomers to the problem of the origin of extra-terrestrial radio emission. At the beginning of the seventies this interest grew incredibly due to the outstanding achievements of radio astronomical observations. Extra-terrestrial radio sources of new kinds have been discovered: pulsars, quasars, etc. The origin of their emission at different wavelengths has become one of the main astrophysical problems which form our general ideas on the nature of radiating sources.

It has been found that the observed radio emission from cosmic sources is generated by virtue of the mechanisms which work mainly in plasma (an ionized gas). Thus, on the one hand, theory of generation and propagation of electromagnetic waves in cosmic plasmas provides a theoretical basis for understanding the nature of the observed radiation. On the other hand, this theory allows one to judge physical conditions and electromagnetic processes in astrophysical plasmas.

Recently, the theory of generation and propagation of radiation in astrophysical plasmas has outgrown its parent domain of theoretical radio astronomy and is applied successfully to other fields, for example, high-energy astrophysics. Such an expansion has been favored by the generally adopted conviction that the appearance of X- and gamma-ray emission of a number of astrophysical objects (X-ray pulsars and other X-ray sources, cosmic gamma-ray bursts etc.) cannot be explained without the use of methods and ideas developed by plasma astrophysics. General results obtained in this field may also help to better understand the complicated phenomena in laboratory plasmas on the Earth. At the same time, analysis of interaction between radiation and astrophysical plasmas under extreme conditions (strong magnetic fields of white dwarfs and neutron stars or strong gravitational fields in the vicinity of black holes) stimulates the development of plasma physics as a whole. In fact, physics of plasma under extreme conditions in space is a new branch of fundamental science.

The issues related to electromagnetic radiation in astrophysical plasmas are discussed in hundreds of original papers. Of course, these have also been considered in monographs, for example in [1–9], though in a way restricted by the requirement to provide an interpretation of radiation by specific astrophysical sources: interstellar medium, solar chromosphere and corona etc. As a part of the general theory of electromagnetic wave generation, they may be found in numerous books on plasma physics. Nevertheless, complexity, diversity, and significance of radiation mechanisms and propagation processes which influence electromagnetic waves in astrophysical plasmas justify, to a certain extent, an attempt to write a monograph presenting a unified and consistent description of the general principles of generation and transfer of radiation. The interpretation of the observed radiation by specific sources should not be the main goal of such a book. Rather, these issues need to be mentioned only as illustrations for general statements of the theory. Such a style has been adopted in the monographs [10–12] published long ago. The last one, [12], which the present book is related to, was written by the author nearly two decades ago. According to the demands of the science of that time, this book focused on the problems related to theoretical radio astronomy. In recent years the X-ray astronomy has become an advanced and promising part of astrophysics. It has stimulated the development of plasma physics under extreme astrophysical conditions. Therefore, the present book is radically different from [12], especially concerning the chapters on linear mode coupling in plasma, radiation transfer, scattering of radiation in plasma. New sections on gyrosynchrotron emission, comptonization, resonance cyclotron scattering, radiation pressure in plasma, and vacuum birefringence effects in super-strong magnetic fields enlarge the new book compared to the older one. The present monograph yields the description of physical processes involved in interaction between radiation and astrophysical plasmas. Hopefully, it meets the contemporary needs of radio, X-ray and optical astronomy, as long as these touch on the problems considered. In the author's opinion, this book comprises the reasonable minimum necessary for understanding the emission and propagation of electromagnetic waves in astrophysical plasmas; without this minimum one could not succeed in interpreting the results of a number of astronomical observations. Of course, this does not mean that the contents of this book includes everything one needs for fruitful scientific work in the field of plasma astrophysics, though such a drawback can be charged to any monograph on the subject.

The bibliography included also does not pretend to be exhaustive. As a rule, it comprises either publications of principal importance or the ones in which a particular issue is considered in the best way. The choice has been inevitably influenced by the customs and tastes of the author. The subject index does not contain the terms mentioned in the table of contents.

This monograph should be useful for graduate and post-graduate stu-

ERRATA

RADIATION IN ASTROPHYSICAL PLASMAS

V. V. Zheleznyakov

Astropysics and Space Science Library 204 / ISBN 0-7923-3907-X

Page	*Location*	*For*	*Read*
8	Eq. (1.12)	Hz	B_0 Hz
13	Fig. 1.1	frequecy	frequency
23	Eq. (1.28)	dS	$dSd\Omega$
23	Eq. (1.31)	$d\omega dS$	$d\Omega$
30	Line 18_t	arctan $2p$	arctan p
30	Eq. (1.57)	$-\sin\sigma\cos\chi$	$+\sin\sigma\cos\chi$
36	Line 9_t	v_0 (see Fig. 1.5)	v (see Fig. 1.5)
37	Footnote 11	cylindrical	co-rotating cylindrical
64	Eq. (2.95)	$\xi = \frac{k_\perp v_\perp}{m\Omega_B}$	$\xi = \frac{k_\perp p_\perp}{m\Omega_B}$
64	Eq. (2.95)	$\int_{-\infty}^{-\infty}$	$\int_{-\infty}^{+\infty}$
65	The first Eq. from the top	$\int_{-\infty}^{-\infty}$	$\int_{-\infty}^{+\infty}$
66	Eq. (2.102)	$\sum_{l=-\infty}^{\infty}$	$\sum_{l=1}^{\infty}$
80	Footnote 22	(ω/c)	(ω/k)
85	Line 5_b	$s \leq 3$	$s \geq 3$
91	Eq. (2.177)	$4a$	$-4a$
108	Line 2_t	$\|\nabla\epsilon\| \sim 10^{10}$	$\|\nabla\epsilon\| \sim 10^{-10}$
113	Line 5_t	$\epsilon/$	ϵ^{-1}
127	Fig. 3.9	The figure is placed upside down	
130	Line 5_t	$dS(\xi)$	$dS_0(\xi)$
130	Eq. (3.87)	$\frac{\sigma d\sigma}{1+\sigma^2}$	$\left.\frac{\sigma d\sigma}{1+\sigma^2}\right\}$
130	Eq. (3.87)	$\left.\right]\left.\right\}$	$\left.\right]$
132	Footnote 16	$\tilde{\alpha}=0$	$\tilde{\alpha}=1$
134	Line 1_b	I_i	I_l
141	Eq. (3.105)	B_{sh}	B_{sh}^2
141	6_t	perpendicular	collinear
141	Eq. (3.107)	$\omega_{B_{sh}}^2$	$\omega_{B_{sh}}^3$
146	Eq.	$\tan^{5/4}\beta$	$\tan^{5/4}\gamma$
146	Line 14_t	β	γ
149	2_b	$u\cos^2\alpha > 0$	$u\cos^2\alpha > 1$
151	Eq. (3.116)	$(1 - u\sin^2\alpha)$	$(2 - u\sin^2\alpha)$
164	Eq. (4.25)	$-\left(\frac{\partial\omega}{\partial k_j}\right)$	$\left(\frac{\partial\omega}{\partial k_j}\right)$
168	Eq. (4.42)	$T_b(0)$	$T_b(l_0)$
173	Eq. (4.66)	$(\tau^{sc})^{-2}$	τ^{-2}
186	Eq. (4.110)	$+K_{\alpha\beta\gamma\delta}$	$-K_{\alpha\beta\gamma\delta}$
192	Eq. (4.127)	$+(k_1' - k_2')Q$	$-(k_1' - k_2')Q$
194	Eq. (4.133)	$(1+\zeta_c^0)e^{-(k_1''-k_2'')z}$	$(1+\zeta_c^0)e^{(k_1''-k_2'')z}$
197	Line 5_t	$\sqrt{\zeta_l + \zeta_c}$	$\sqrt{\zeta_l^2 + \zeta_c^2}$
200	Eq. (5.5)	c^2	c^3
215	Line 7_b	$c\omega n_j$	$c/\omega n_j$
217	Eq. (5.58)	P_ω	P_Ω
222	Line 5_b	δt	Δt
223	Eq. (5.74)	$1-\beta$	$1-\beta^2$
232	Line 5_b	$\omega_B m_e c^2$	$\omega_B m_e c$
235	Lines 3_t, 4_t	(2.95)	(2.87)
238	Line 7_b	$2\pi p_\perp^2$	$2\pi p_\perp$
244	Eq. (5.150)	$a(\alpha)$	$a_\omega(\alpha)$
247	Eq. in the middle	$(1+\delta^2)^{-1/2}$	$(1+\delta)^{-1/2}$
267	Line 14_b	T_{eff}	T_b
274	Line 7_t	$\beta_T \cos\alpha$	$\beta_{T\omega}\cos\alpha$
275	Line 14_t	$\epsilon_{\alpha\beta}^{pl}$	$\epsilon_{\alpha\beta}$

Page	*Location*	*For*	*Read*
276	Eq. (6.52)	$[\ldots]^{-s}$	$[\ldots]^{s}$
277	Eq. (6.53)	$s\hbar\omega^2$	$\hbar\omega^2$
277	Lines 5_t, 13_t	(6.49)	(6.52)
277	Line 6_t	(6.47), (6.48)	(6.50), (6.51)
289	Line 15_b	$i\phi_1$	$-I\phi_1$
297	Line 3_b	(2.85)	(2.76)
299	Eq. (7.32)	ω_{qm}	ω_{qm}^3
299	Line 5_t	brightness	effective
300	Eq. (7.38)	$\ln\frac{1+x}{x}$	$\left(\ln\frac{1+x}{x}\right)^{-1}$
308	Line 5_t	(7.7)	(7.17)
309	Eq. (7.71)	$\omega_{\rm LS}$	$\omega_{\rm LS}^2$
310	Line 7_t	$\omega^2 = \omega_{\rm L}^2 + k^2 v_T^2$	$\Omega^2 = \omega_{\rm L}^2 + k^2 v_T^2$
312	Line 6_t	$\operatorname{Im} Z$	$\operatorname{Im} Z_{\rm S}$
316	Line 5_b	$1/\arctan(\sqrt{2})$	$\arctan(\sqrt{2})$
317	Line 6_b	(7.7)	(7.8)
317	Eq. (7.95)	$W_{\mathbf{k}}$	$kW_{\mathbf{k}}$
319	Eq. (7.100)	$m_e v^2$	$m_e v^3$
360	Eq. (8.14)	$(2\pi)^6$	$(2\pi)^3$
361	Eq. (8.16)	$(2\pi)^6$	$(2\pi)^3$
362	Eq. (8.19)	γ_j	$2\gamma_j$
363	Eq. (8.20)	2ω	$2\omega'$
363	Line 11_t	$\gamma_j^{\rm ind} > 0$	$\gamma_j^{\rm ind} < 0$
366	Eq. (8.31)	v_T^4	$v_{T_1}^4$
381	Eq. (8.83)	$d\mathcal{E}_\gamma$	$\mathcal{E}_\gamma d\mathcal{E}_\gamma$
383	Line 11_b	ω/ω_B	ω_B/ω
394	Line 15_t	$kD \ll 1$	$kD \gg 1$
396	Line 1_b	$T_{\rm b}$	$T_{\rm eff}$
406	Line 11_t	$\beta_{T_\parallel}^{2s-2}$	$\beta_{T_\perp}^{2s-2}$
407	Eq. (8.174)	A_m^{m-1}	$\int_{-\infty}^{+\infty} A_m^{m-1}$
409	Line 6_b	$I^{(0)}(p_\parallel)$	$I^{(0)}(T_\parallel)$
412	Eq. (8.201)	μ_{js}	$\mu_{js} dz$
417	Line 14_t	$6 \cdot 10^8$	$6 \cdot 10^{10}$
430	Left picture	$\log T\,(K)$ a	$\log T\,(K)$ b
430	Right picture	$\log T\,(K)$ b	$\log T\,(K)$ a
432	$g = 10^8$ cm s^{-2}	$g = 10^8$ cm s^{-2} at $r = R_*$	
433	Eq. (8.257)	$\mathcal{D} = \frac{r_0}{R_*}\sqrt{\frac{2\kappa T_\parallel}{m_{\rm p} g R_*}}$	$\mathcal{D} \sim \sqrt{\frac{2\kappa T_\parallel}{m_{\rm p} g R_*}\frac{r_0}{R_*}}$
433	Eq. (8.258)	$m_{\rm p} g \frac{R_*^2}{r^2}$	$m_{\rm p} g$
439	Eq. (8.272)	$d^3 k_1$	$d^3 k$
439	Eq. (8.273)	$d^3 k_2$	$d^3 k$
443	Eq. (8.283)	$j_{\rm nl}$	$j'_{\rm nl}$
445	Line 8_t	(8.281)	(8.291)

Lines are counted from the top (subscript t) or bottom (subscript b) of a page, excluding displayed formulas, headers, footers, footnotes, and figure captions.

dents and young scientists as a textbook on plasma astrophysics and the issues of plasma physics dealing with radiation. At the same time, the book can be used by specialists on astrophysics, radio astronomy and plasma physics.

The author is sincerely grateful to C. de Jager who has initiated the publishing of this book, and to A. V. Serber whose understanding of plasma astrophysics has guaranteed the authenticity of the English translation of the manuscript.

V. V. Zheleznyakov

March 1995

Introduction

Plasma regions are main sources of extra-terrestrial radiation. It sounds quite natural, since matter in the Universe (for example, stars and their atmospheres, interstellar and interplanetary medium, planet ionospheres and magnetospheres) is mostly in a plasma state. The rare exceptions among radiation sources, such as planet surfaces and lower atmospheres, are not considered in this book. In this respect, the diversity in profiles and scales of phenomena in astrophysical plasmas is not at all surprising. Among the others, basic effects come from electromagnetic processes which govern plasma dynamics, motion and acceleration of charged particles, generation and transfer of radiation. This book focuses on the problems related to generation and propagation of radiation in, as well as its influence on plasma. Thus, section 1.1, the first one of this introductory chapter, points out those plasma characteristics which finally determine the conditions for generation and propagation of electromagnetic radiation; the next section 1.2 reviews the main characteristics of radiation; in the last one, section 1.3, the properties of astrophysical radiation sources which originate due to their spatial motion are discussed and basic definitions necessary to establish general classification of radiation mechanisms are introduced.

1.1. Parameters of astrophysical plasmas

This section starts with the definition of the term "plasma" and points out key properties distinguishing this medium from other ones. Next, main plasma parameters governing its interaction with radiation are introduced. Finally, the astrophysical plasmas are overviewed and specific parameter values relevant to plasmas in interstellar medium, solar corona, magnetic stars, neutron star magnetospheres etc. are given.

1.1.1. BASIC PLASMA PROPERTIES AND CHARACTERISTIC FREQUENCIES

An ionized gas comprising free charged particles is usually called plasma. The following remarks elucidate this definition. Let $N_{\rm tot}$ be the total density of atoms or molecules in a medium and $N_{\rm a}$ the density of neutrals. The degree of ionization is then defined as the ratio

$$\delta_{\rm i} = \frac{N_{\rm tot} - N_{\rm a}}{N_{\rm tot}} = \frac{N_+}{N_{\rm tot}}, \tag{1.1}$$

where N_+ is the density of positive ions, i.e., the amount of such ions in a unit volume. For the sake of simplicity, the negative ions are not considered. The value $\delta_{\rm i} = 0$ corresponds to a neutral gas, while $\delta_{\rm i} = 1$ to a fully ionized plasma. Plasma can be considered as a perfect gas if the mean kinetic energy of comprising particles is much larger than the energy of interaction between the neighboring ones. If T is the kinetic temperature of a fully ionized plasma with $N_+ = N_{\rm tot}$ and $N_{\rm a} = 0$, the mean kinetic energy $\mathcal{E}_{\rm kin} \sim \kappa T$, where κ is the Boltzmann constant. On the other hand, the energy of Coulomb interaction of two nearest neighbors with charges e can be estimated as $\mathcal{E}_{\rm Coul} \sim e^2/r_0$, where $r_0 \sim n_+^{-1/3}$ is mean distance between plasma particles. Thus, the criterion of perfect gas $\mathcal{E}_{\rm kin} \gg \mathcal{E}_{\rm Coul}$ takes the form

$$\kappa T \gg e^2 N_+^{1/3}. \tag{1.2}$$

Astrophysical plasmas, as a rule, are quasi-neutral: the condition

$$\rho_+ + \rho_{\rm e} \simeq 0 \tag{1.3}$$

is satisfied very well when averaging over a sufficiently large volume. Here $\rho_+ = ZeN_+$ and $\rho_{\rm e} = -eN$ are charge densities due to positive ions and electrons, respectively, e and Ze are the values of electron and ion charges, N is the electron density. If plasma is comprised of single charged ions only ($Z = 1$), the quasi-neutrality condition (1.3) simply means that $N_+ \simeq N$. A possible exception can be magnetospheres of rotating neutron stars (pulsars): some models predict that the condition (1.3) is violated in pulsar magnetospheres, and sharp spatial separation of opposite charges can exist.

The above definition of plasma as an ionized, quasi-neutral gas is not quite perfect: in fact, any gas contains some amount of ionized particles. It is not thus clear under which conditions a gas displays to a full extent the specific plasma properties: (i) Debye screening, (ii) plasma oscillations, and (iii) Landau damping. As we see below, this additional condition requires that the size of the region occupied by an ionized gas is sufficiently large [13]:

$$L \gg D = \left(\frac{\kappa T}{8\pi e^2 N} \right)^{1/2}. \tag{1.4}$$

For given L and T the condition (1.4) determines the values of N at which a gas behaves like a plasma. Tremendous sizes of astrophysical objects usually satisfies this criterion very well.

The value D introduced above is called Debye radius. If the condition (1.3) is satisfied, the electric potential of a static charge in plasma is given by

$$\varphi = q\frac{e^{-r/D}}{r}. \tag{1.5}$$

Hence, a charge q is screened by the surrounding space charge induced in plasma, so that its electric field becomes exponentially small at distances $r \gg D$. It is easy to see that the condition (1.2) is equivalent to the requirement that the number of plasma particles inside the Debye sphere is large. Evidently, Debye screening appears only if the criterion (1.4) is satisfied for the medium considered. This effect ensures an ionized gas to be quasi-neutral.

According to the last criterion, weakly damped plasma oscillations (plasma waves) can exist in an ionized gas. The frequency of these oscillations is close to the plasma, or Langmuir frequency

$$\omega_{\rm L} = \left(\frac{4\pi e^2 N}{m_{\rm e}}\right)^{1/2}, \tag{1.6}$$

where $m_{\rm e}$ is electron rest mass, and $\omega_{\rm L}$ is the so-called Langmuir, or plasma frequency (see section 2.2 for more detail). Plasma waves remain weakly damped as long as their wavelength $\lambda \gg 2\pi D$. Otherwise, if $\lambda \lesssim 2\pi D$, strong collisionless damping (Landau damping) occurs which leads to the absorption of plasma waves in a time $t \lesssim 1/\omega$. Thus, such plasma waves are non-propagating (see the remarks in section 2.2; the detailed analysis of the Landau damping is given in chapter 6).

Up to now, only non-magnetized plasma has been considered. But if plasma is permeated by a magnetic field with the induction $\boldsymbol{B}_0$, charged particle trajectories are drastically changed. A collisionless particle in a homogeneous magnetic field moves along a helical line with the axis parallel to the vector $\boldsymbol{B}_0$. The corresponding rotation frequency of an electron is equal to

$$\omega_B = \frac{eB_0}{m_{\rm e}c}, \tag{1.7}$$

where c is the speed of light. This quantity is called electron gyrofrequency or cyclotron frequency. As for relativistic electron with energy $\mathcal{E}$, it rotates in a magnetic field with relativistic gyrofrequency

$$\Omega_B = eB_0/mc = \omega_B m_{\rm e}c^2/\mathcal{E}. \tag{1.8}$$

Here $\mathcal{E}$ and $m = \mathcal{E}/c^2$ stand for energy and mass of an electron, respectively. Rotation frequency of a non-relativistic ion equals the ion gyrofrequency

$$\omega_{B\rm i} = eB_0/m_{\rm i}c, \tag{1.9}$$

and that of a relativistic one, the relativistic ion gyrofrequency

$$\Omega_{B\mathrm{i}} = \omega_{B\mathrm{i}} m_\mathrm{i} c^2 / \mathcal{E}, \tag{1.10}$$

where m_i is the rest mass of an ion.

For the reader's convenience, the numerical formulas facilitating the transformation from electron density and magnetic field strength to plasma frequency and electron/ion gyrofrequency are listed below:

$$\omega_\mathrm{L} \simeq 5.64 \cdot 10^4 N^{1/2},\ \mathrm{s}^{-1}, \qquad f_\mathrm{L} = \omega_\mathrm{L}/2\pi \simeq 8.97 \cdot 10^3 N^{1/2}\,\mathrm{Hz}; \tag{1.11}$$

$$\omega_B \simeq 1.76 \cdot 10^7 B_0\,\mathrm{s}^{-1}, \qquad f_B = \omega_B/2\pi \simeq 2.81 \cdot 10^6\,\mathrm{Hz}; \tag{1.12}$$

$$\omega_{B\mathrm{p}} \simeq 9.6 \cdot 10^3 B_0\,\mathrm{s}^{-1}, \qquad f_{B\mathrm{p}} = \omega_{B\mathrm{p}}/2\pi \simeq 1.53 \cdot 10^3 B_0\,\mathrm{Hz}. \tag{1.13}$$

Here N is measured in cm^{-3}, the magnetic field B_0 in G; ion gyrofrequencies are written for protons with $m_\mathrm{i} = m_\mathrm{p} \simeq 1.8 \cdot 10^3 m_\mathrm{e}$ and denoted as $\omega_{B\mathrm{p}}$ and $f_{B\mathrm{p}}$.

In addition to the above frequencies, there are some other values, in particular, effective frequency of particle collisions ν_eff (see chapter 6) and mean thermal velocities of electrons,

$$v_T = \sqrt{\kappa T/m_\mathrm{e}} \simeq 3.89 \cdot 10^5 T^{1/2}\,\mathrm{cm/s}, \tag{1.14}$$

and protons,

$$v_{T\mathrm{p}} = \sqrt{\kappa T/m_\mathrm{p}} \simeq 9.08 \cdot 10^3 T^{1/2}\,\mathrm{cm/s}, \tag{1.15}$$

in a non-relativistic, equilibrium plasma, which are essentially used to describe plasma properties. Note that kinetic temperature T in formulas (1.14), (1.15) is measured in K.

1.1.2. SPECIFIC VALUES OF COSMIC PLASMA PARAMETERS AND MAGNETIC FIELD STRENGTHS

The preceding section clearly revealed that electron density N, magnetic field strength B_0 and kinetic temperature T (provided equilibrium velocity distribution of plasma particles is realized) are basic parameters which, to a considerable extent, determine electromagnetic properties of plasma. In cosmic plasmas, even restricted to regions and objects being the immediate sources of the observed radiation, these values vary over enormously wide scales. Electron density N may be as low as $\sim (1-5) \cdot 10^{-2}\,\mathrm{cm}^{-3}$ in an interstellar medium (H I regions) and as high as $10^{12}\,\mathrm{cm}^{-3}$ at the base of solar chromosphere or in pulsar magnetospheres.[1] Recent measurements

[1] According to the Goldreich-Julian model of neutron star magnetosphere [14], the pointed out values of the plasma density N may occur in the nearest vicinity of a rotating, strongly magnetized neutron star.

and estimates of neutron star magnetic fields set the upper limit $B_0 \sim 10^{11} - 10^{13}$ G which is (or can be) reached in an astrophysical object. An opposite limit is posed by interstellar medium in which $B_0 \sim 10^{-6}$ G. As far as kinetic temperature is concerned, the observed range is also rather wide, from 10^2 K in interstellar H I regions up to 10^6 K in solar corona and 10^8 K in X-ray sources. A special case is neutron star magnetospheres in which relativistic plasma with electron energies $\mathcal{E} \gg m_e c^2$ may occur. If such a plasma were equilibrium, the corresponding kinetic temperature $T \sim \mathcal{E}/\kappa$ would have been much greater than 10^{10} K.

Many problems concerning generation and propagation of electromagnetic waves in cosmic plasmas can be successfully treated even if the essential parameters are known by the order of magnitude. Table 1.1 comprises the values of N, B_0 and T typical for various astrophysical objects, as well as corresponding frequencies f_L, f_B, $f_{B\mathrm{p}}$ and thermal velocities v_T, $v_{T\mathrm{p}}$ (see the formulas (1.11)-(1.15)). In some cases, however, even the preliminary analysis of these issues requires more or less detailed information on spatial distribution of electron density, magnetic field and temperature over a radiation source. Such data are either determined using astronomical observations or obtained by *in situ* spacecraft measurements. Sometimes a model of the radiation source is initially developed upon general assumptions of the object's nature, and then the proposed model is examined and corrected to provide the best agreement with the observed characteristics of the source radiation.

For example, analysis of the generation and propagation of electromagnetic waves in the Galaxy is done by assuming that the interstellar medium near the Galactic plane is comprised of regions of ionized or neutral hydrogen with different density and temperature called H II or H I regions, respectively (see Table 1.1). H II regions are formed around hot stars belonging to the disk population, so their fraction is decreased with distance from the Galactic plane. The degree of ionization δ_i in such regions is close to unity. Conversely, $\delta_i \sim 10^{-3}$ in H I regions, so that here the amount of neutrals is 10^3 times greater than that of ionized particles.

As for the interstellar magnetic field, the early models assumed it to be composed of quasi-homogeneous field regions of a size 10-100 pc which are chaotically oriented with respect to each other. Such a field structure, however, cannot explain the polarimetric observations of star radiation and cosmic radio emission. It is generally believed now that magnetic field in spiral arms (or, at least, in the arm in which the solar system is situated) is oriented along them and probably also has a helical component with the axis along the arm. Respectively, inhomogeneities of interstellar medium ("clouds") in the Galactic disk are non-isometric, i.e. are not of the same size along different directions. Rather, they are stretched along the magnetic

TABLE 1.1. Typical parameters of astrophysical plasmas: electron density N in cm^{-3}; magnetic field strength B_0 in G; kinetic temperature T in K; plasma, electron cyclotron and ion cyclotron frequencies (f_L, f_B, and f_{B_p}) in Hz; mean thermal velocity of electrons and protons, v_T and v_{T_p}, in $cm\,s^{-1}$.

Object	N	B_0	T	f_L	f_B	f_{B_p}	v_T	v_{T_p}
Interstellar medium:								
– H I regions	10^{-2}	10^{-6}–10^{-5}	10^2	10^3	3–30	$2\cdot10^{-3}$–$2\cdot10^{-2}$	$4\cdot10^6$	10^5
– H II regions	10^2	10^{-6}	10^4	10^5	3	$2\cdot10^{-3}$	$4\cdot10^7$	10^6
Interplanetary medium	10	10^{-4}	10^4	$3\cdot10^4$	$3\cdot10^2$	$1.5\cdot10^{-1}$	$4\cdot10^7$	10^6
Supernova remnants (Crab nebula)	10^3	10^{-3}	$2\cdot10^4$	$3\cdot10^5$	$3\cdot10^3$	2	$5\cdot10^7$	10^6
Jupiter:								
– ionosphere	$3\cdot10^5$	10	$5\cdot10^6$	$3\cdot10^7$	$1.5\cdot10^4$	10^3	10^7	$3\cdot10^5$
– magnetosphere (Io plasma torus)	10^3	10^{-1}	10^4	$3\cdot10^5$	$3\cdot10^5$	$1.5\cdot10^2$	$3\cdot10^7$	10^6
The Sun:								
– chromosphere	10^9–10^{11}	1–$4\cdot10^3$	10^4	$3\cdot(10^8$–$10^9)$	$3\cdot10^6$–10^{10}	$2\cdot10^3$–$6\cdot10^6$	$4\cdot10^7$	10^6
– inner corona	10^8	1–10^3	10^6	10^8	$3\cdot(10^6$–$10^9)$	$2\cdot(10^3$–$10^6)$	$4\cdot10^8$	10^7
Magnetic white dwarfs	-	10^8	-	-	$3\cdot10^{14}$	$2\cdot10^{11}$	-	-
Pulsar magnetospheres	10^{12}	10^{12}	-	10^{10}	$3\cdot10^{18}$	$2\cdot10^{15}$	-	-

field lines [15–17].

Further information on physical conditions in an interstellar medium can be found in the monograph [2], H I regions are described in detail in [18], and [19] deals with conditions in the Galactic center region. Characteristics of cosmic plasma in expanding supernova envelopes and in the inter-galactic medium are also not considered here. These issues are the subject of the reviews [20,21].

Interplanetary medium within the solar system is formed mainly by the solar wind, i.e., plasma outflow from the upper layers of the solar corona. The wind velocity at the Earth's orbit is about $3 \cdot 10^7 - 10^8\,\mathrm{cm\,s^{-1}}$, the corresponding density is about $5\text{–}20\,\mathrm{cm^{-3}}$, and the electron temperature equals 20–40 eV. The interplanetary magnetic field has spiral structure which originate due to the Sun rotation. The magnetic field strength varies over the interval $5 \cdot 10^{-5}\text{–}2 \cdot 10^{-4}$ G at the Earth's orbit and falls off as approximately $1/r$ with distance r from the Sun. More details on the structure of the interplanetary medium can be found in [22].

According to *in situ* measurements by *Pioneer* and *Voyager* spacecrafts, the magnetic field of Jupiter is generally dipole. The magnetic moment is orientated opposite to the Earth's, and is inclined at the angle $\approx 10°$ with respect to the rotation axis of the planet. Irregularity of the Jovian magnetic field appears as its complex distribution over the planet surface with considerable variation of the field strength in the range 3–14 G.

Radio measurements made near the moments of spacecraft eclipse by and its subsequent appearance out of the planet disk yilded the data on the electron density and temperature distributions in Jovian ionosphere. Maximal values of N are of the order of $10^5\,\mathrm{cm^{-3}}$. There exists a pronounced equatorial anomaly — the decrease of N by the order of magnitude with respect to the values at high latitudes. The maximal electron temperature is in excess of 10^3 K.

As for the Jovian magnetosphere extending up to 50–100 planet radii, the maximal values $N \simeq 3{\cdot}10^3\,\mathrm{cm^{-3}}$ are observed within the Io plasma torus located along the orbit of this satellite. Generally, Io is the main source of plasma and energetic particles in the magnetosphere. The inner part of the torus facing the planet is formed by the relatively cold plasma with mean particle energy of several eV, whereas the outer, more extended and tenuous, part is comprised of hot plasma (~ 80 eV per particle). There is also the high-energy component formed by the particles trapped by the planetary magnetic field and constituting the radiation belts of the planet.

Note that the preliminary information on the magnetic field and radiation belts of Jupiter has been initially derived from radio astronomical observations of its decametric radio emission, long before direct studies by spacecrafts [3,23,24].

The magnetic field of Saturn is more than an order of magnitude weaker compared to the Jovian one: the surface field strength ranges from 0.2 to 0.8 G. There is no pronounced inclination of the magnetic dipole with respect to the rotation axis. The maximal electron density in the Saturn ionosphere reaches the value of $2{\cdot}10^4\,\mathrm{cm}^{-3}$. The density distribution probably exhibits the equatorial anomaly similar to the Jovian one. The temperature reaches the value of 750 K. The magnetosphere of Saturn is considerably smaller than that of Jupiter: its size is about 16–22 planet radii. Plasma in the region of Dione-Tethys torus (4 to 8 Saturn radii from the planet center) has the density of several electrons per cm^3 and the temperature of several dozen eV. There also exists the energetic particle population in Saturn's magnetosphere. As well as in the Jovian one, particle drift in the inhomogeneous magnetic field gives rise to azimuthal electric current (ring current) which substantially distorts the field structure in the magnetosphere. For further details on magnetic fields and physical conditions in the plasma envelopes of outer planets, see [25–28].

Turning to the solar corona, we note that the Baumbach-Allen formula

$$N(R) = 10^8 \left[1.55 \left(\frac{R}{R_\odot} \right)^{-6} + 2.99 \left(\frac{R}{R_\odot} \right)^{-16} \right] \ \mathrm{cm}^{-3}, \tag{1.16}$$

where $R_\odot$ is the radius of the photosphere, R — the distance from the solar center, is frequently used for estimates of the electron density and related calculations. Fig. 1.1 shows the plot of electron density $N(R/R_\odot)$ and plasma frequency distributions corresponding to the formula (1.16). Note that the values yielded with this formula fall in between the equatorial zone densities $N(R)$ during solar maximum and solar minimum periods. The figure also shows the so-called Newkirk distributions of N and f_L along a coronal ray resting upon a photospheric active region [29] and the corresponding distributions obtained with radio data [30]. Note that both Baumbach-Allen and Newkirk distributions are quite well justified by the measurements of the intensity of sunlight scattering by coronal electrons. The large size of the corona, intensive radio emission and the observed broad emission lines of highly ionized atoms are evidence for the high kinetic temperature $T \sim 10^6$ K of the coronal plasma. Generally, the temperature of the corona is nearly independent on the height above the photosphere, but increases several times above active regions. The high coronal temperature ensures the coronal matter (primarily, hydrogen) to be nearly completely ionized, the fraction of the neutrals not exceeding 10^{-7}.

Note that observations of other stars in X-rays have revealed higher coronal temperatures $\sim 10^7$ K. Hot coronas have been discovered around main sequence stars of practically all spectral classes [31].

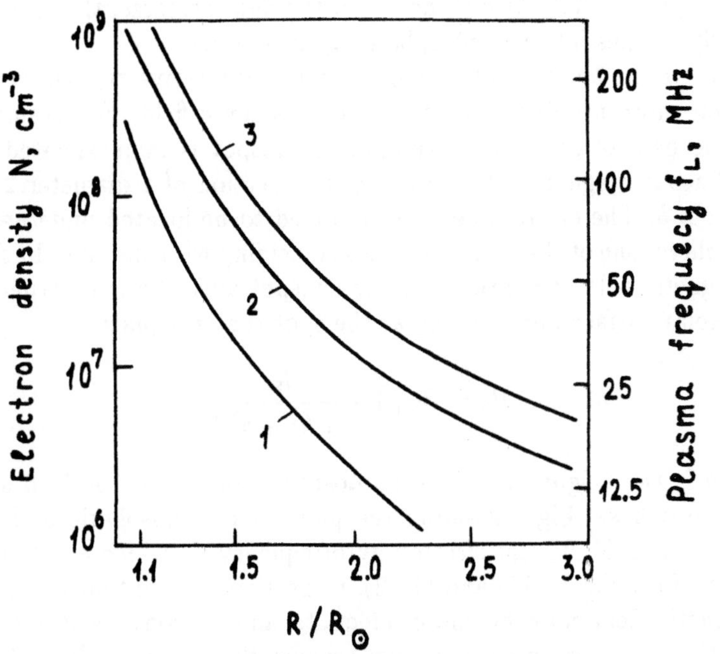

Figure 1.1. Distribution of electron density N and plasma frequency f_L in a quiet corona and along a coronal ray: *1* — corona (Baumbach-Allen); *2* — coronal ray (Newkirk); *3* — distribution above an active region obtained using radio data.

General structure of solar magnetic field at high heliographic latitudes, far from active regions, is probably close to the dipole one, the field strength at the photospheric level not exceeding 1 G. Stronger sunspot fields (up to $3 \cdot 10^3$ G) mask this component at lower latitudes.

Observations of flaring stars of the UV Ceti type give evidence for higher activity. These stars are main-sequence red dwarfs of the dMe class. There are spots of strong magnetic field at the surface of these stars. Contrary to sunspots, such stellar spots cover up to 10–20 % of the star surface and, probably, have stronger magnetic fields up to 10^4 G. UV Ceti type stars have been detected in X-rays during *Einstein* survey. It indicates the presence of hot plasmas (coronas) around these stars.

To determine the magnetic field structure in the solar corona is quite a complicated problem. Optical and/or X-ray images of the corona yield an indirect information about this. Coronal magnetic fields can also be calculated using the surface (photospheric) field distribution known from the optical studies and neglecting electric currents in coronal plasma [32]. Surely, such an assumption will go wrong at distances $R > 2.5$–$3\,R_\odot$ where the energy density of the magnetic field $B_0^2/8\pi$ becomes smaller than the kinetic energy density $\rho v^2/2$ of the solar wind. Thus, the magnetic field becomes quasi-radial. As distance from the Sun increases, the solar corona gradually merges into the interplanetary medium.

Preliminary analysis of emission from solar active regions is usually based on more primitive models of the magnetic field. For example, the magnetic field of an isolated (unipolar) sunspot is approximated by the field of a bar magnet with a circular cross-section of a diameter $2b$ and a length $l \gg b$. The entire magnet is assumed to be located underneath the photosphere, one of the magnetic poles coinciding with the spot. In this case the magnetic field strength along the sunspot axis, the line perpendicular to the solar surface and starting at the spot center, equals

$$B_0 = B_{\rm b}\left(1 - \frac{h}{\sqrt{h^2 + b^2}}\right), \tag{1.17}$$

where h is the height above the photosphere and $B_{\rm b}$ is the field strength at the spot base. Fig. 1.2 shows the plots of chromospheric and coronal values of the field B_0 and electron gyrofrequency f_B which are determined using the formulas (1.17) and (1.12), respectively. Note that other models of magnetic field may be more adequate and/or convenient for specific calculations. For example, one may assume the magnetic field above the photosphere to be formed by elementary dipoles underneath.

Configuration of magnetic field lines above bipolar groups of sunspots is illustrated by the Fig. 1.3. It is taken into account here that high in the corona, even above an active region, the magnetic field becomes quasi-radial. This may lead to the formation of the so-called neutral current sheets which contain surfaces of zero magnetic field. Such surfaces at which $B_0 = 0$ may also separate neighboring bipolar groups or have more complicated shapes stipulated by the structure of magnetic fields in active regions. If the hydrostatic equilibrium of the coronal plasma near current sheets is provided due to the increase of its density, such layers can be seen on the optical images of the corona. Few observations of this kind have been reported in [33]. The possibility to observe current sheets by radio astronomical methods is discussed in the sections 3.2 and 6.2.

The global magnetic field of the Sun (~ 1 G) is much weaker than the fields of sunspots or active regions. Peculiar Ap stars exhibit stronger fields up to 10^4 G giving rise to observable Zeeman splitting of narrow spectral lines in optics. Line profiles imply that the magnetic field of these stars has a global structure rather than the local one similar to sunspots. The most adequate model of the magnetic field of an Ap star is a centered dipole inclined with respect to the rotation axis, though more sophisticated models are also discussed in the literature [34].

It is not clear yet whether Ap stars have coronas. According to X-ray observations by *Einstein*, normal A stars do. X-ray emission has also been

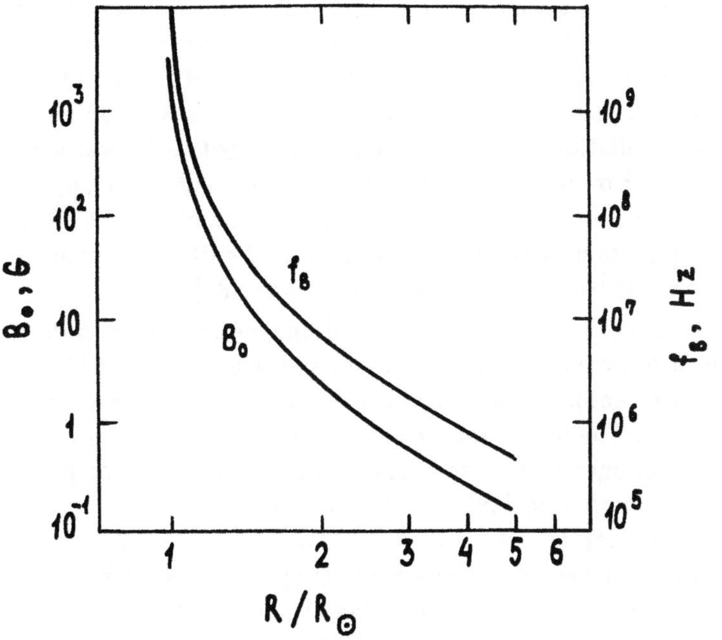

Figure 1.2. Distribution of the magnetic field B_0 and gyrofrequency f_B over the axis of a unipolar sunspot with $B_b = 2.5 \cdot 10^3$ G and $b = 10^{19}$ cm.

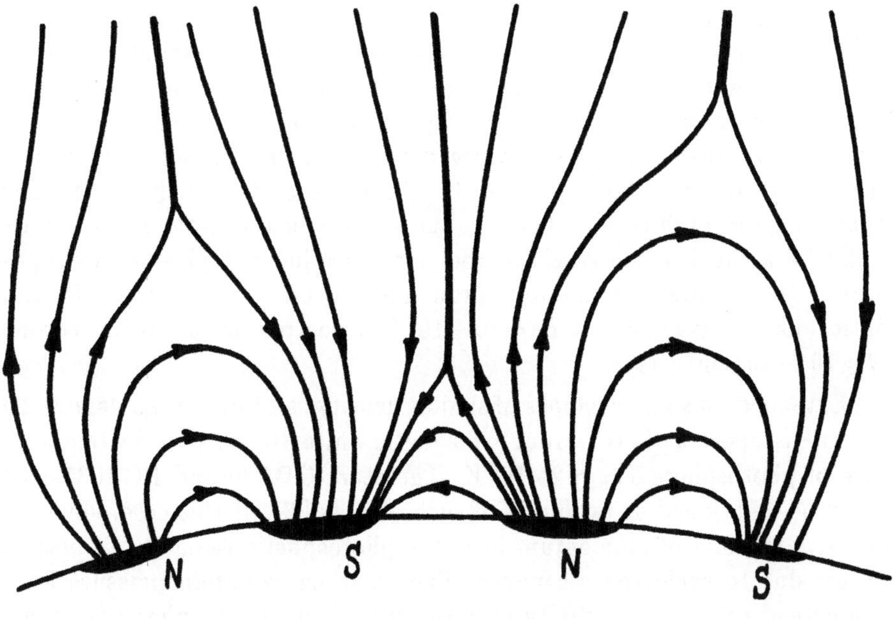

Figure 1.3. Schematic view of magnetic field lines over bipolar groups of sunspots. Thick lines mark intersections of neutral current sheets with the plane of figure.

observed from binary systems α CVn and 46 Dra. These binaries comprise magnetic Ap stars as one of the companions. Unfortunately, binary components of these systems are not resolved in X-rays, so it is not evident if the observed emission is indeed related to the hot corona of the Ap star. Note that hot plasma in strong magnetic field of a star can emit cyclotron microwave radiation which may be a more sensitive indicator of a corona than X-rays. However, a few attempts to detect such an emission have not succeeded until now [35,36]. The absence of the observable thermal radio emission constrains kinetic temperature and density of coronal electrons by the values $5 \cdot 10^6$ K and $10^9\,\mathrm{cm}^{-3}$, respectively [37].

As far as field strength is concerned, magnetic white dwarfs occupy the interval in between magnetic Ap stars and neutron stars. The majority of two dozen of known magnetic white dwarfs have magnetic fields of about 10^7–$5 \cdot 10^8$ G. The field strength is determined using the observed Zeeman-split spectral lines and/or the degree of linear and circular polarization of the optical continuum [38]. The latter values can be as large as 10% for the dwarfs having the strongest field. Additional information on magnetic fields can be obtained for the stars exhibiting depression bands in the spectrum originated due to cyclotron scattering. For example, according to [39], the magnetic white dwarf GD 229 with a broad absorption band in the ultraviolet domain should have the magnetic field $B_0 \sim 10^9$ G.

The existence of hot coronas around magnetic white dwarfs is an open question now. The scale height of a hydrogen plasma with temperature $T \sim 10^7$ K in strong gravitational field of a white dwarf is only about 1 % of the star radius, so the corona would be localized near the star surface. Pointed X-ray observations made by *ROSAT* posed an upper limit to the soft X-ray flux by isolated magnetic white dwarf GR 290 and thus constrained the density of the corona (if any) by the value of $4 \cdot 10^{11}\mathrm{cm}^{-3}$ [40]. The appearance of cyclotron features in the spectra of magnetic white dwarfs could be a much more sensitive tool for detecting hot plasma envelopes around these stars. For example, pronounced cyclotron features in IR and optical bands may occur at $B_0 \sim 10^7$–10^8 G if electron density in the corona $N \gtrsim 10^7 - 10^8\,\mathrm{cm}^{-3}$ [41].

Extended plasma envelopes of similar density can form broad depression features in spectra of hot, strongly magnetic white dwarfs with the temperature of photosphere $T_{\mathrm{ph}} \gtrsim 2 \cdot 10^4$ K. The stars GD 229 and PG 1031+234 could be mentioned as examples of such objects [39]. It this case, however, effective transfer of momentum from the photospheric radiation to plasma occurs due to cyclotron resonance. The resulting radiation pressure force can exceed the force of gravity near the star. Thus, photospheric radiation at cyclotron frequencies can both drive the ejection of plasma from the photosphere and support a dense plasma envelope of a few stellar radii in diameter which is accumulated in the closed part of the star magnetosphere. The density profile in the envelope may have a maximum near the magnetic equator plane. Such an object as a whole is called radiation-driven diskon [39]. Cyclotron scattering of the star radiation in the diskon envelope may form the UV depressions observed in the spectra of the stars GD 229 and PG 1031+234.

As for parameters, structure and dynamics of plasmas on neutron stars, a number of problems are to be solved. These issues are clearly quite important not only to explain such phenomena as radio and X-ray pulsars or cosmic gamma-ray bursts but also from the more general point of view: to understand the behavior of relativistic astrophysical plasmas in superstrong magnetic fields. Possible conditions in neutron star magnetospheres are discussed, for example, in the monographs [9, 42]; therefore, in what follows only short remarks are given.

Note, first of all, that the expected values of the surface magnetic fields of neutron stars can reach the values 10^{12}–10^{13} G. Such an estimate is based upon a consideration on magnetic flux conservation inside a star with high conductivity during its collapse to a neutron star state [43]. Similar values result from the observed spin down of a neutron star if one attributes such an effect to the energy loss due to low-frequency radiation by a rotating magnetic dipole in vacuum. The observed polarization of radio and optical radiation by pulsars evidences the existence of a strong magnetic field of these neutron stars, as well. Direct measurement of the magnetic field strength can be made using the frequencies of cyclotron lines observed in the spectra of X-ray pulsars, i.e., neutron stars in binary systems. Such measurements yield the values $4 \cdot 10^{12}$ G and 10^{12} G for the sources Her X-1 and 4U 0115+69, respectively [44–46].

Beginning with the *KONUS* experiment onboard Soviet *Venera* probes [47], cyclotron absorption features were seen in the spectra of some cosmic gamma-ray bursts. The field strengths deduced with the line frequencies are similar to X-ray pulsar ones. Later on, gamma-burst detectors onboard *Ginga* satellite and *Phobos-2* probe have registered absorption features at multiple frequencies [48,49]. This has become an additional strong argument in favor of the interpretation of gamma-burst absorption features as first and second (and even, probably, the third) harmonics of electron gyrofrequency. All these data strengthen the hypothesis that cosmic gamma-ray bursts are related to Galactic neutron stars. Recently, however, *BATSE* gamma-ray burst detector onboard Compton Gamma Ray Observatory (*GRO*) has accumulated copious data on burst positions and spectra which contradict earlier results. First, no lines have yet been detected in the burst spectra. Second, spatial distribution of the bursts seemed to be surprisingly isotropic but, at the same time, apparently inhomogeneous. Since no Galactic source population can fit such a distribution and no evidence for the cyclotron harmonic features has been obtained, an alternative hypothesis on the cosmological origin of the burst sources has been reanimated [50]. Comparative discussion of both classes of gamma-ray burst models (Galactic and cosmological) can be found in [51].

Not much is known about plasma density in the vicinity of neutron stars. Since a neutron star radius is only about 10 km while its mass is of the order of the solar one, the corresponding free-fall acceleration is as large as $g \sim 10^{11}\,\mathrm{cm\,s^{-2}}$. In such a strong gravitational field the scale height of a hot plasma atmosphere with $T \sim 10^6$ K is only about 1 cm.

Therefore, an extended plasma envelope formed solely by ram pressure and gravity cannot exist on a neutron star. However, spatial distribution of particles in the vicinity of such an object can be maintained by electromagnetic forces rather than the above mentioned factors. For example, a neutron star with strong magnetic field $B_0 \sim 10^{12}$ G rotating around the axis $\mathbf{\Omega}$ aligned with the magnetic moment $\boldsymbol{\mu}$ generate the unipolar induction field $E_0 \sim 10^{10}$ CGSE [14]. It is this strong electric field that mainly affects the formation of the magnetosphere in such a neutron star. If the density of charged particles in the magnetosphere increases up to $N \sim 10^{12}\,\mathrm{cm^{-3}}$, the induction field E_0 could be effectively compensated by the space charge. Charged particles produced due to either accretion of an interstellar plasma or electron-positron pair creation near the star surface [52, 53] constitute the neutron star magnetosphere.

Hot plasma may also form a relatively thin, dense layer on a neutron star surface. For example, if *Ginga* gamma-ray bursts GB 870303 and GB 880205 with multiple absorption features really originate from magnetized neutron stars, one could deduce plasma parameters in the source using the observed positions, widths and depths of the lines. The observed features can be formed as a result of cyclotron scattering in a plasma layer with longitudinal (with respect to the magnetic field) electron temperature 20–140 keV and column density $(3-9)\cdot 10^{20}\,\mathrm{cm^{-2}}$ [54]. If the height of the layer is about the scale height ($\sim 10^3$ cm for the adopted plasma temperature), then plasma density is estimated as $(3-9)\cdot 10^{17}\,\mathrm{cm^{-3}}$.

1.1.3. PLASMA UNDER EXTREME ASTROPHYSICAL CONDITIONS

Physical conditions in astrophysical plasmas can be regarded as extreme in respect of particle density, gravitational field, or magnetic field. High densities of matter (about the nuclear one) are reached in neutron star interiors. Even higher values occurred at early stages of the cosmological expansion. Strong gravitational fields with large gradients appear near the Schwarzschild sphere of black holes. Finally, strong magnetic fields arise at neutron stars and white dwarfs — the latest stages of stellar evolution. Analysis of the role of first two factors is beyond the scope of this book. In what follows attention will be paid to the plasma properties gained in the presence of a super-strong magnetic field, since such fields proved to change radically the interaction between plasma and radiation.

Magnetic fields of astrophysical objects span over an enormous range, from 10^{-6} G in an interstellar medium (and even less in intergalactic one) up to $\gtrsim 10^{12}$ G in neutron stars. The very width of this interval implies possible qualitative differences in plasma properties at largely different magnetic fields. Exploring various parts of this interval which covers more than 18 orders of magnitude, one can encounter both classical plasmas, similar to ones found on the Earth under laboratory or natural conditions, and (quantized) plasmas with unusual properties caused by an extremely strong magnetic field which can alter electromagnetic wave propagation and interaction between plasma and radiation.

Interstellar medium, chromospheres and coronas of the Sun and non-degenerate stars, planetary ionospheres and magnetospheres are comprised of classical (non-quantized) plasmas with $\kappa T \gg \hbar\omega_B$ which can be very well described by the physics developed for laboratory plasmas and Earth ionosphere. These plasmas are usually non-relativistic, $\kappa T \ll m_e c^2$. On the contrary, plasmas on neutron stars in magnetic fields $B_0 \sim 10^{12}$–10^{13} G are, in many cases, relativistic (particle energies $\mathcal{E} \gg m_e c^2$), quantized ($\kappa T_\perp \ll \hbar\omega_B$, where $T_\perp$ is the transverse temperature with respect to the magnetic field) and submerged into magnetized, birefringent vacuum.

If the above mentioned condition $\kappa T \ll \hbar\omega_B$ is satisfied, the majority of electrons occupy the lowest Landau level. Then electron motion in the plane perpendicular to the magnetic field becomes quantized. In the opposite case, $\kappa T \gg \hbar\omega_B$, quantum effects still can greatly influence particle motion in rather dense plasma in which the de-Broglie wavelength of an electron $\bar{\lambda} = \hbar/m_e v_T$ is large compared to the mean distance between particles, $r_0 \sim N^{-1/3}$. Such a criterion of quantum plasma can be presented in the form

$$\kappa T \ll \frac{\hbar^2 N^{2/3}}{2m_e} \simeq \mathcal{E}_F, \tag{1.18}$$

where $\mathcal{E}_F$ is Fermi energy. Electron gas satisfying the condition (1.18) becomes degenerate, and its equilibrium distribution essentially differs from the Maxwell one. For astrophysical plasmas this criterion holds only for the interiors of white dwarfs, neutron stars, and massive stars at their last evolution stages. These regions hidden from an observer are of practically no interest for the theory of the formation of the observed radiation by these objects. Therefore, this book deals only with plasmas in which the inequality opposite to (1.18) is satisfied.

The influence of vacuum birefringence on propagation and absorption of radiation becomes important if

$$\frac{\omega_L^2 c}{\omega^2 v_T} \ll \frac{1}{45\pi}\frac{e^2}{\hbar c}\frac{B_0^2}{B_c^2}, \tag{1.19}$$

where $B_c \simeq 4\cdot 10^{13}$ G is the critical field at which $\hbar\omega_B = m_e c^2$ (see the details in section 2.2). For example, this condition is met at frequencies $\omega \simeq \omega_B$ falling into X-ray band for plasma with $\kappa T \simeq 10$ keV and density $N \lesssim 10^{23}$ cm^{-3} which is situated near the surface of the rotating neutron star with $B_0 \simeq 4\cdot 10^{12}$ G in the X-ray pulsar Her X-1.

The above qualitative differences from the case of laboratory plasmas allows one to treat the strong magnetic fields $B_0 \sim 10^{10}$–10^{13} G of neutron stars as an extreme condition. However, the interaction of magnetized plasma with radiation at cyclotron frequencies $\omega \simeq \omega_B$ is altered at much lower values $B_0 \sim 10^7$–10^9 G relevant to magnetic white dwarfs. In fact, the character of this interaction depends on the ratio of t_c, the time of radiative transition of an electron between adjacent Landau levels, and $t_{\rm eff} = 1/\nu_{\rm eff}$, the time between collisions. Since magnetic fields of the Sun and non-degenerate stars are relatively weak, $B_0 \lesssim 10^4$ G, plasmas on these objects are "collisional":

$$t_c \nu_{\rm eff} \gg 1. \tag{1.20}$$

Strong magnetic fields of degenerate stars (white dwarfs and neutron stars) significantly lower the time $t_c \propto B_0^{-2}$, so that the "collisionless" limit:[2]

$$t_c \nu_{\rm eff} \ll 1 \tag{1.21}$$

is realized.

In the first limiting case of collisional plasma (1.20) the effect of the cyclotron radiation on the electron distribution is negligible, and this distribution is determined mainly by particle collisions, acceleration mechanism etc. In the second case (1.21) of collisionless plasma the effective influence of radiation on plasma electrons occurs due to the relatively small value of the time t_c. Here electron distribution (first and foremost, over transverse velocities) is governed by the intensity of radiation at cyclotron frequencies, whereas the role of collisions becomes minor (see section 7.1).

The pointed out differences between the above limiting cases is reflected by the form of the relevant equation of radiation transfer. In the case (1.20) the transfer of cyclotron radiation occurs via resonance absorption and emission by plasma with a given electron distribution. On the other hand,

[2]One should bear in mind that the definitions of collisional and collisionless limits characterize interaction of plasma with radiation. In plasma physics these terms are usually used to separate the cases $\omega \ll \nu_{\rm eff}$ and $\omega \gg \nu_{\rm eff}$.

due to the leading influence of radiation on particle distribution in the second case (1.21), the corresponding transfer equation has the form typical for scattering media. Thus, interaction of radiation with plasma is radically changed in the latter case as compared to the former one, and one can claim plasmas at white dwarfs with $B_0 \sim 10^7 - 10^9$ G being under extreme conditions. The additional rationale for this conclusion can be the fact that such strong fields cannot be produced in laboratories on the Earth.

In addition to pure astrophysical interest, investigation of cosmic plasmas under extreme conditions is a powerful boost for the promotion of new branches of plasma physics as a whole.

1.2. Electromagnetic radiation and its main characteristics

Consider now general characteristics of electromagnetic radiation without any respect to the conditions at which it is generated. Such characteristics are the frequency and angular spectrum of radiation and its polarization parameters. These values contain the main information brought by extraterrestrial radiation. The analysis of the above characteristics is the principal source of our knowledge about the nature of astrophysical objects.

1.2.1. ENERGY DENSITY AND SPECIFIC INTENSITY

Cosmic radiation sources produce electromagnetic fields $\boldsymbol{E}(\boldsymbol{r},t)$, $\boldsymbol{H}(\boldsymbol{r},t)$. The following expansion of this field over plane waves can be made in a homogeneous medium:

$$\begin{aligned} \boldsymbol{E}(\boldsymbol{r},t) &= \int \boldsymbol{E}_{\mathbf{k}} e^{i\mathbf{kr}-i\omega t} d^3k, \\ \boldsymbol{H}(\boldsymbol{r},t) &= \int \boldsymbol{H}_{\mathbf{k}} e^{i\mathbf{kr}-i\omega t} d^3k. \end{aligned} \tag{1.22}$$

Here $\boldsymbol{E}$ and $\boldsymbol{H}$ are electric and magnetic field intensities, respectively; $\boldsymbol{r}$ and t are radius-vector and time; ω and $\boldsymbol{k}$ are frequency and wave vector of a wave with the complex amplitudes $\boldsymbol{E}_{\mathbf{k}}$, $\boldsymbol{H}_{\mathbf{k}}$; and d^3k is the volume element in the wave vector space. The frequency ω is related to the wave vector $\boldsymbol{k}$ by the function $\omega = \omega(\boldsymbol{k})$. This relation is the solution of the dispersion equation introduced further in sections 2.1, 2.2.[3]

[3]Dispersion equation of a medium can have more than one solution $\omega = \omega(\mathbf{k})$ which corresponds to the modes with different polarizations. Therefore, all the values introduced in this section are related to waves with certain polarization and one-to-one correspondence between ω and k. Besides, absorption of waves in a media is neglected or assumed to be sufficiently small.

If the phases of the amplitudes $\boldsymbol{E}_{\mathbf{k}}$ and $\boldsymbol{H}_{\mathbf{k}}$ are random, the energy of electromagnetic field in a volume V can be written as

$$W = \int W_{\mathbf{k}}\, d^3k dV = \int W_k dk d\Omega dV. \tag{1.23}$$

Here the spectral energy density $W_{\mathbf{k}}$ equals the energy of electromagnetic waves per unit volume of coordinate and wavevector spaces. Respectively, W_k is the spectral energy density found per unit solid angle Ω in wavevector space and unit interval of wave numbers k. Since

$$d^3k = k^2 dk d\Omega, \tag{1.24}$$

where dk is the wave number interval and $d\Omega$ is the solid angle element enclosing the wave vectors from d^3k, the relation (1.23) yields

$$W_k = k^2 W_{\mathbf{k}}. \tag{1.25}$$

In addition to the energy density W_k per unit wave number interval, the spectral energy density per unit frequency band, W_ω, may also be introduced. To relate W_k and W_ω, one should demand $W_k|dk| = W_\omega|d\omega|$. This leads to the following relation:

$$W_\omega = W_k \left|\frac{\partial k}{\partial \omega}\right|. \tag{1.26}$$

As well as W_k, the value W_ω characterizes the energy density per unit solid angle along $\boldsymbol{k}$.

The specific intensity of radiation $I_{\mathbf{k}}$ is also widely used in the theory of radiation. This value is defined as the energy of radiation with wave vectors $\boldsymbol{k}$ from the unit volume in the $\boldsymbol{k}$-space which passes through a surface of unit area oriented perpendicular to the group velocity vector $\boldsymbol{v}_{\mathrm{gr}} = d\omega/d\boldsymbol{k}$.

The radiation energy flux F through a surface S, i.e., the amount of energy passing through this surface per unit time, can be conveniently written using the value $I_{\mathbf{k}}$:

$$F = \int I_{\mathbf{k}} \cos\delta\, d^3k dS. \tag{1.27}$$

Here δ is the angle between $\boldsymbol{v}_{\mathrm{gr}}$ and the normal to the surface element dS. The integral (1.27) is taken over those parts of the surface S and those wave vectors $\boldsymbol{k}$ for which $\cos\delta > 0$. If $\boldsymbol{k}$ and $\boldsymbol{v}_{\mathrm{gr}}$ are collinear (for example, in vacuum or isotropic medium; see section 2.2), the above wave vectors span over the solid angle $\Omega = 2\pi$ at one side of the surface S. In astrophysics the radiation flux is sometimes defined using the solid angle intergal over

4π. This quantity equals the difference of two integrals (1.27) over the solid angle 2π in which $\cos\delta > 0$ and $\cos\delta < 0$.

Taking into account the relation (1.24), one can present the formula (1.27) for the energy flux in the form

$$F = \int I_k \cos\delta \, dk dS. \tag{1.28}$$

The intensity I_k used in this formula is related to $I_{\mathbf{k}}$ by the expression

$$I_k = k^2 I_{\mathbf{k}}, \tag{1.29}$$

as seen by the comparison of (1.27) and (1.28). Both $I_{\mathbf{k}}$ and I_k describe the energy flux through the same surface, though the latter is related to the unit wave number interval k and unit solid angle along $\boldsymbol{k}$. However, the most commonly used value is the specific intensity I_ω related to unit frequency range. It is introduced by the formula

$$I_\omega = I_k \left| \frac{\partial k}{\partial \omega} \right| \tag{1.30}$$

Similar to I_k, the intensity I_ω corresponds to the unit solid angle along $\boldsymbol{k}$ or, equivalently, along the phase velocity vector $\boldsymbol{v}_{\rm ph} = (\omega/k)\boldsymbol{s}$, where $\boldsymbol{s} = \boldsymbol{k}/k$.

The radiation flux density per unit frequency range, F_ω, is defined in the following way:

$$F_\omega = \int I_\omega \cos\delta d\omega dS \tag{1.31}$$

is also frequently used. It is related to F by the evident formula: $F = \int F_\omega d\omega dS$.

The above definition for radiation intensity I_ω is not quite traditional, though it has been in use since the publishing of the monograph [3]. Usually another definition is adopted which relates the specific intensity to the unit solid angles along the group velocity vector $\boldsymbol{v}_{\rm gr}$. This latter value differs from the above defined I_ω by the factor $d\Omega/d\Omega_{\rm gr}$, where $d\Omega$ and $d\Omega_{\rm gr}$ are solid angle elements along phase and group velocities, respectively. Being the same in vacuum[4] or isotropic medium where $\boldsymbol{v}_{\rm ph} \parallel \boldsymbol{v}_{\rm gr}$ and $d\Omega/d\Omega_{\rm gr} = 1$, these definitions become different in magnetoactive plasma (see section 2.2). The use of radiation intensity related to the unit solid angle along $\boldsymbol{k}$ is justified by the resulting considerable simplification of the formulas comprising I_ω (see, for example, the relations (1.37)-(1.39)). Otherwise, the expressions containing the radiation intensity would also

[4]The term "vacuum" is the synonym of "empty space". In cases when the influence of strong magnetic field on vacuum properties is taken into account, the terms "magnetized vacuum" or "birefringent vacuum" will be used.

include the factor $d\Omega/d\Omega_{gr}$ which depends in a complex way on parameters of magnetoactive plasma and the type of the waves considered.

Energy density and radiation intensity are related by the expressions

$$I_{\mathbf{k}} = W_{\mathbf{k}} \left|\frac{d\omega}{d\mathbf{k}}\right|, \qquad I_k = W_k \left|\frac{d\omega}{d\mathbf{k}}\right|, \qquad I_\omega = W_\omega \left|\frac{d\omega}{d\mathbf{k}}\right|, \tag{1.32}$$

which point out the fact that in a transparent or weakly absorbing medium the radiation energy is "transferred" with the group velocity $\boldsymbol{v}_{gr} = d\omega/d\mathbf{k}$. Then, evidently, the derivative $\partial\omega/\partial k$ entering the relations (1.26), (1.30) is the projection of the group velocity onto the vector $\boldsymbol{k}$:

$$\frac{\partial\omega}{\partial k} = v_{gr}\cos\vartheta. \tag{1.33}$$

Here ϑ is the angle between $\boldsymbol{k}$ and $\boldsymbol{v}_{gr}$, and $v_{gr} = |d\omega/d\mathbf{k}|$. It is interesting to note that, according to (1.30), (1.32) and (1.33),

$$I_\omega = W_k|\cos\vartheta|^{-1} = W_{\mathbf{k}}k^2|\cos\vartheta|^{-1}. \tag{1.34}$$

Therefore, $W_k = I_\omega$ in an isotropic medium where $\vartheta = 0$.

The spectrum of electromagnetic radiation is characterized by the functional dependence of the energy density $W_{\mathbf{k}}$ on the wave vector $\boldsymbol{k}$. One can also judge the spectrum using the dependence of I_ω on the frequency ω and $\boldsymbol{s} = \boldsymbol{k}/k$, the direction of the wave vector (i.e., on phase velocity $\boldsymbol{v}_{ph}$). The dependence of the function $I_\omega(\omega, \boldsymbol{s})$ on ω determines the frequency spectrum, while that on the direction $\boldsymbol{s}$, the angular pattern of radiation.

To determine $W_{\mathbf{k}}(\boldsymbol{k})$ or $I_\omega(\omega, \boldsymbol{s})$ at any point of a medium with known type and distribution of radiation sources is quite a complex problem. Some aspects of this problem are considered in this book. However, in a specific case of total thermodynamic equilibrium the functions $W_{\mathbf{k}}(\boldsymbol{k})$ and $I_\omega(\omega, \boldsymbol{s})$ become uniquely determined and may be found in quite a simple way.

Under the condition $\hbar\omega \ll \kappa T$ the influence of quantum effects on equilibrium radiation is negligible. Here the energy per mode equals κT (see, e.g., [55]). It is not difficult to see that the number of modes with definite polarization (see section 2.1 for more detail about this) which satisfy the boundary conditions on the surface enclosing a large volume V equals

$$dN = \frac{V}{(2\pi)^3}d^3k = \frac{Vk^2}{(2\pi)^3}dkd\Omega. \tag{1.35}$$

The above value is the number of modes with wave vectors in the wave number interval dk and in the solid angle element $d\Omega$. It follows from the above that the equilibrium energy density is given by

$$W_{\mathbf{k}}^{(0)} = \frac{\kappa T}{(2\pi)^3}, \qquad W_k^{(0)} = \frac{k^2\kappa T}{(2\pi)^3}, \tag{1.36}$$

while the corresponding intensity equals

$$I_\omega^{(0)} = \frac{n^2\omega^2\kappa T}{(2\pi)^3c^2|\cos\vartheta|}. \tag{1.37}$$

To obtain the last expression, one should use the formula (1.34) and the relation $k = \omega n/c$ which defines the refractive index $n = n(\omega, s)$ of a plane wave in a transparent (non-absorbing) medium.

In radio astronomy the specific intensity of radiation $I_\omega(\omega, s)$ is usually characterized by the effective, or brightness, temperature T_b. By definition, it is the temperature of equilibrium radiation at which the intensity (1.37) equals $I_\omega(\omega, s)$. According to this definition, T_b is the function of ω and s. The brightness temperature relates to $I_\omega(\omega, s)$ by the following expression:

$$I_\omega(\omega, s) = \frac{n^2\omega^2\kappa T_\mathrm{b}(\omega, s)}{(2\pi)^3c^2|\cos\vartheta|}. \tag{1.38}$$

This relation acquires the most simple form in vacuum ($n = 1$, $\vartheta = 0$) in terms of the wavelength $\lambda = 2\pi c/\omega$ and the radiation intensity $I_f = 2\pi I_\omega$ related to the unit interval of the frequency $f = \omega/2\pi$. The desired intensity in this case is given by

$$I_f(f, s) = \frac{\kappa T_\mathrm{b}(f, s)}{\lambda^2}. \tag{1.39}$$

Of course, brightness temperature may also be defined using the expression (1.36) for radiation energy density; here

$$W_\mathbf{k} = \frac{\kappa T_\mathrm{b}}{(2\pi)^3}. \tag{1.40}$$

The inequality $\hbar\omega \ll \kappa T$ determines the validity range of the so-called Rayleigh-Jeans law in the spectrum of equilibrium radiation. It is very well satisfied at radio frequencies. For example, this condition is valid for the wave with $\omega \lesssim 2\cdot 10^{11}\,\mathrm{s}^{-1}$ (or, equivalently, $\lambda \gtrsim 1\,\mathrm{cm}$) at temperatures $T \gg 1\,\mathrm{K}$. In optics, however, the relation $\hbar\omega \sim \kappa T$ is more typical at temperatures $T \sim 10^3 - 10^5$ K relevant to stellar atmospheres. The Rayleigh-Jeans limit still holds for stellar coronae in which temperatures rise up to $10^6 - 10^7$ K. In X-ray range the Wien limit $\hbar\omega \gg \kappa T$ is usually realized.

In case of arbitrary ratio $\hbar\omega/\kappa T$ the number of photons per one mode equals

$$N_\mathbf{k} = \left(e^{\hbar\omega/\kappa T} - 1\right)^{-1}, \tag{1.41}$$

while energy density and specific intensity of equilibrium radiation are given by

$$W_{\mathbf{k}}^{(0)} = \frac{\hbar\omega}{(2\pi)^3}\frac{1}{e^{\hbar\omega/\kappa T}-1}, \qquad W_k^{(0)} = \frac{\hbar\omega^3}{(2\pi)^3c^2}\frac{1}{e^{\hbar\omega/\kappa T}-1} \tag{1.42}$$

and

$$I_\omega^{(0)} = \frac{n^2\hbar\omega^3}{(2\pi)^3c^2|\cos\theta|}\frac{1}{e^{\hbar\omega/\kappa T}-1}, \tag{1.43}$$

respectively. In the limit $\hbar\omega \ll \kappa T$ the formulas (1.42), (1.43) reduce to (1.36), (1.37), respectively. In the frequency range where the refraction index n and $\cos\theta$ for electromagnetic waves in astrophysical plasmas are close to unity, the formula (1.43) looks like

$$I_\omega^{(0)} = \frac{\hbar\omega^3}{(2\pi)^3c^2}\frac{1}{e^{\hbar\omega/\kappa T}-1}. \tag{1.44}$$

Of course, an effective, or brightness, temperature T_b can also be defined for the region $\hbar\omega \gtrsim \kappa T$. It is related to the specific intensity of radiation by the formula (cf. (1.38):

$$I_\omega(\omega, s) = \frac{\hbar\omega^3}{(2\pi)^3c^2}\frac{1}{e^{\hbar\omega/\kappa T_b}-1}. \tag{1.45}$$

The equation (1.45) yields the following expression:

$$T_b(\omega, s) = \frac{\hbar\omega/\kappa}{\ln\left\{1+\dfrac{\hbar\omega^3}{(2\pi)^3c^2}\dfrac{1}{I_\omega(\omega,s)}\right\}}. \tag{1.46}$$

It should be especially stressed that the specific intensity and energy density of radiation with a given polarization (of a certain normal wave in a medium; see sections 2.1, 2.2) are used in the above formulas. Thus, to get the total intensity of unpolarized radiation in vacuum one should double the value (1.39).

The rationale for the use of brightness temperature can be seen in the example of thermal radiation when T_b relates in a quite simple way to the kinetic temperature of a source (see section 4.1). If the size of the source is sufficiently large, both temperatures become equal. The latter also explains the reason for using the value T_b to characterize the radiation intensity of a non-thermal source: this value shows the temperature the source should be heated to for its thermal radiation at the frequency considered to reach the observed level.

1.2.2. POLARIZATION OF RADIATION. POLARIZATION TENSOR AND STOKES PARAMETERS

To judge the polarization characteristics of radiation, the electric field (see (1.22)) should be presented as the integral over frequency ω and solid angle Ω:

$$\boldsymbol{E}(\boldsymbol{r},t) = \int \boldsymbol{E}_\omega e^{-i\omega t + i\mathbf{kr}}\, d\omega d\Omega. \tag{1.47}$$

Since $d^3k = k^2 dk d\Omega = k^2|\partial k/\partial\omega| d\omega d\Omega$, the amplitude $\boldsymbol{E}_\omega$ is determined by the following formula:

$$\boldsymbol{E}_\omega = \boldsymbol{E}_{\mathbf{k}} k^2 \left|\frac{\partial \boldsymbol{k}}{\partial \omega}\right|.$$

Next, consider the component of the field (1.47) comprised of the plane waves with frequencies within the narrow band $\Delta\omega \ll \omega$ which propagate the same direction $\boldsymbol{s} = \boldsymbol{k}/k$. Such waves constitute the electric field

$$\boldsymbol{E}(t) = \int_{\Delta\omega} \boldsymbol{E}_\omega e^{-i\omega t + i\mathbf{kr}}\, d\omega$$

at a point $\boldsymbol{r}$. This field can be presented in the form

$$\boldsymbol{E}(t) = \boldsymbol{E}_0(t) e^{-i\omega_0 t}, \tag{1.48}$$

where $\boldsymbol{E}_0(t)$ is the complex amplitude of the electric field and ω_0 is some mean value from the interval $\Delta\omega$. According to (1.48), the orthogonal x- and y-components of the projection of the electric field onto the wavefront plane (the plane perpendicular to the wave vector $\boldsymbol{k}$) oscillate as follows:

$$\begin{aligned} E_x(t) &= E_{0x}(t) e^{-i\omega_0 t + i\psi_x(t)}, \\ E_y(t) &= E_{0y}(t) e^{-i\omega_0 t + i\psi_y(t)}. \end{aligned} \tag{1.49}$$

The characteristic time scale of the phases ψ_x, ψ_y and the amplitudes E_{0x}, E_{0y}, as well as the complex amplitude $\boldsymbol{E}_0(t)$, is about $1/\Delta\omega$. Due to the condition $\Delta\omega \ll \omega_0$ all these values are slowly varying compared to the oscillation period $2\pi/\omega_0$. Thus, the projection of the vector $\boldsymbol{E}(t)$ onto the xy-plane circumscribes an ellipse in a time $2\pi/\omega_0$. The ellipse size and shape fluctuate with characteristic time $\Delta t \sim 1/\Delta\omega$.

Generally, an extra-terrestrial radiation comprises both polarized and unpolarized components. These components have continuous frequency spectra, but the oscillations $E_x(t)$ and $E_y(t)$ in the unpolarized component (which is also called naturally, or chaotically polarized) are non-coherent: the condition

$$\overline{E_x(t) E_y^*(t)} = 0 \tag{1.50}$$

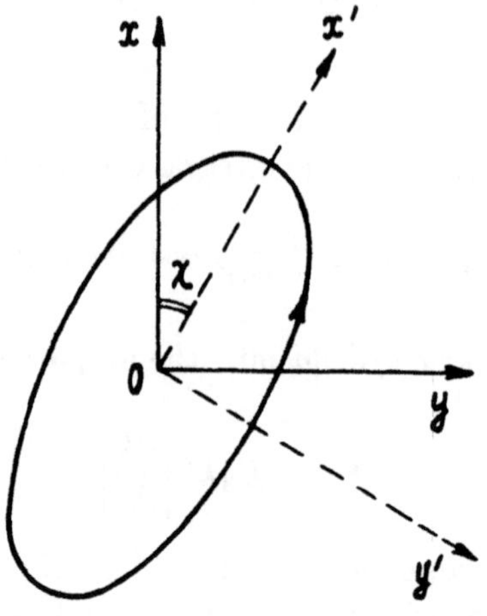

Figure 1.4. Polarization ellipse for radiation with the wave vector directed away from the reader.

holds for arbitrary frame (x, y). The bar here denotes averaging over the time interval $\Delta t \gg 1/\Delta\omega$, and the asterisk marks the complex conjugate. Alternatively, the components $E_x(t)$ and $E_y(t)$ are coherent in the other, polarized, component: the value (1.50) is non-zero in arbitrary frame.

The condition $\overline{E_x(t)E_y^*(t)} \neq 0$ for the polarized radiation is explained by the fact that the corresponding ratio E_y/E_x in such a radiation is time-independent, i.e., the amplitude ratio E_{0y}/E_{0x} and phase difference $\psi_{xy} = \psi_x - \psi_y$ in (1.49) remain constant. So, the ellipse circumscribed by the electric filed vector of a polarized radiation in the x, y-plane fluctuates in such a way that its orientation and eccentricity are hold fixed.

It is thus clear that polarized radiation can be characterized by three parameters: intensity I_{pol}, orientation of the polarization ellipse χ, and its eccentricity p (Fig. 1.4). The quantity χ is defined as the angle between the x-axis and the major axis of the ellipse. The direction of electric field rotation is described by the sign of the parameter p: $p > 0$ for the rotation shown at the figure.

Unpolarized radiation is characterized by the chaotic variation of both orientation and eccentricity of polarization ellipse. Thus the unpolarized component can be described by the single parameter, the intensity I_{nat}. In general, radiation in a frequency band $\Delta\omega$ which propagates in the direction $\boldsymbol{s}$ is completely described by the four values: the total intensity $I = I_{\text{pol}} + I_{\text{nat}}$, the degree of polarization $\zeta = I_{\text{pol}}/I$, the eccentricity p of the polarization ellipse, and its orientation χ. The case $\zeta = 0$ corresponds to unpolarized radiation, whereas $\zeta = 1$ for completely polarized radiation. Any other value of ζ in between 0 and 1 corresponds to partially polarized radiation.

The polarization of radiation is often described with the so-called Stokes parameters I, Q, U, V which are related to the above defined I, ζ,p and χ by the formulas

$$\begin{aligned} I &= I, \\ Q &= \zeta I \cos(2\sigma)\cos(2\chi), \\ U &= \zeta I \cos(2\sigma)\sin(2\chi), \\ V &= \zeta I \sin(2\sigma). \end{aligned} \tag{1.51}$$

These formulas in which $\sigma = \arctan p$ can be considered as definitions for the Stokes parameters. According to (1.51),

$$\begin{aligned} \zeta &= (Q^2+U^2+V^2)^{1/2}I^{-1}, \\ \sin(2\sigma) &= V(Q^2+U^2+V^2)^{-1/2}, \\ \tan(2\chi) &= UQ^{-1}. \end{aligned} \tag{1.52}$$

The Stokes parameters, at the same time, relate in quite a simple way to the polarization tensor [56]:

$$I_{\alpha\beta} = \overline{E_\alpha E_\beta^*}, \tag{1.53}$$

in which $\alpha,\beta = x, y$ in the frame with z-axis directed along the wave vector $\boldsymbol{k}$, and E_α, E_β are the corresponding components of the electric field (1.49). As before, the averaging in (1.53) is made over the interval $\Delta t \gg 1/\Delta\omega$. The following expressions relate the components $I_{\alpha\beta}$ to the Stokes parameters:[5]

$$\begin{aligned} I &= I_{xx} + I_{yy}, \\ Q &= I_{xx} - I_{yy}, \\ U &= I_{xy} + I_{yx}, \\ V &= i(I_{yx} - I_{xy}). \end{aligned} \tag{1.54}$$

The definition (1.53) for the polarization tensor is usually introduced for transverse waves, but electric field in a magnetoactive plasma has also the longitudinal component E_z. Therefore, the components I_{xz}, I_{yz} and I_{zz} should also be used to describe completely the polarization characteristics of radiation in a magnetoactive plasma. Alternatively, one can alter the

[5]The definition (1.53) and relations (1.54) reveal that the total intensity I equals the mean square of the projection of the electric field intensity $\mathbf{E}$ onto xy-plane. On the contrary, the specific intensity used earlier (see, e.g., (1.27),(1.28),(1.30)) has been defined another way, as the energy flux through the unit area into the unit solid angle and unit frequency range. The latter definition will also be used in section 4.1 and thereafter when dealing with radiation transfer problems. These two values are different in a magnetoactive plasma, though they are identical (to be correct, proportional) in some cases, for example, in vacuum.

definition of polarization tensor in magnetoactive plasma in the following way [57]:

$$I_{\alpha\beta} = \overline{D_\alpha D^*_\beta}. \tag{1.55}$$

Here D_α and D_β are the components of the electric induction vector $\boldsymbol{D}$ of electromagnetic radiation. This vector remains transverse, $(\boldsymbol{kD})=0$, even in the magnetoactive plasma (see section 2.1). With such a definition of the polarization tensor $I_{\alpha\beta}$ all the formulas (1.51), (1.52), (1.54) for the Stokes parameters remain valid, while the values I, ζ, p, χ should be related to the vector $\boldsymbol{D}$, not $\boldsymbol{E}$.

Let us now prove the relations (1.54) using the definitions (1.51), (1.53) of the Stokes parameters and polarization tensor.

The first formula in (1.54) means that two orthogonal linear component have orthogonal polarizations, their intensities being additive. [6]

To prove the remaining three relations, introduce the new orthogonal frame with the coordinate axes oriented along the axes of polarization ellipse of the polarized radiation component. Thus, the angles between the new axes x', y' and the corresponding old ones x, y equal χ (see Fig. 1.4). The electric field components in the primed frame are evidently written as

$$E_{x'} = E_0(t)e^{-i\omega_0 t}\cos\sigma, \qquad E_{y'} = E_0(t)e^{-i\omega_0 t - i\pi/2}\sin\sigma, \tag{1.56}$$

where $\sigma = \arctan 2p$ and the value $E_0(t)$ fluctuates with a characteristic time $1/\Delta\omega$. Then the oscillations of x- an y-component of the field $\boldsymbol{E}(t)$ are determined by the expressions

$$\begin{aligned} E_x &= E_{x'}\cos\chi - E_{y'}\sin\chi = \\ &\qquad E_0(\cos\sigma\cos\chi e^{-i\omega_0 t} - \sin\sigma\sin\chi e^{-i\omega_0 t - i\pi/2}), \\ E_y &= E_{x'}\sin\chi + E_{y'}\cos\chi = \\ &\qquad E_0(\cos\sigma\sin\chi e^{-i\omega_0 t} - \sin\sigma\cos\chi e^{-i\omega_0 t - i\pi/2}). \end{aligned} \tag{1.57}$$

The following relations ensure the identity of (1.57) and (1.49):

$$\begin{aligned} E_{0x} &= E_0(\cos^2\sigma\cos^2\chi + \sin^2\sigma\sin^2\chi)^{1/2} \\ E_{0y} &= E_0(\cos^2\sigma\sin^2\chi + \sin^2\sigma\cos^2\chi)^{1/2}, \\ \tan\psi_x &= \tan\sigma\tan\chi, \qquad \tan\psi_y = -\tan\sigma\cot\chi. \end{aligned} \tag{1.58}$$

[6] Polarizations of two waves are orthogonal if the corresponding electric vectors circumscribe similar ellipses with mutually orthogonal axes. In particular, two circular polarizations with opposite signs and two linear polarizations oriented along perpendicular axes are orthogonal. Regardless of the phase relations between two waves with orthogonal polarizations, their intensities are additive.

It follows from (1.58) that

$$\begin{aligned} \overline{E_{0x}^2} + \overline{E_{0y}^2} &= \overline{E_0^2}, \\ \overline{E_{0x}^2} - \overline{E_{0y}^2} &= \overline{E_0^2}\cos(2\sigma)\cos(2\chi), \\ 2\overline{E_{0x}E_{0y}}\cos\psi_{xy} &= \overline{E_0^2}\cos(2\sigma)\sin(2\chi), \\ 2\overline{E_{0x}E_{0y}}\sin\psi_{xy} &= \overline{E_0^2}\sin(2\sigma). \end{aligned} \tag{1.59}$$

Since $\overline{E_{0x}^2} + \overline{E_{0y}^2} = I_{\rm pol} = \zeta I$, the value $\overline{E_0^2} = \zeta I$ and, therefore, the right-hand sides of the last three relations in (1.59) are equivalent to the Stokes parameters Q, U and V. On the other hand, the corresponding left-hand sides are equal to $I_{xx} - I_{yy}$, $I_{xy} + I_{yx}$ and $i(I_{yx} - I_{xy})$, respectively. For completely polarized radiation the last statement immediately follows from (1.49) and (1.53). The above arguments prove the validity of the relations (1.54) for completely polarized radiation. But the presence of unpolarized components does not contribute to the values $I_{xx} - I_{yy}$, $I_{xy} + I_{yx}$ and $i(I_{yx} - I_{xy})$. Thus, the relations (1.54) are proved for arbitrary polarized radiation.

Note that the use of the Stokes parameters to describe the polarization of radiation is a kind of tradition. The clear physical sense is attributed to the parameters I, ζ, p and χ, while the measurable quantities are the components of the polarization tensor (1.53). In the radiowave domain the corresponding measurements can be made with the system of two orthogonal dipoles.

Antennas used in radio astronomy are frequently built to detect the left- or right-handed circular polarizations.[7] The intensities of these components equal

$$I_r = \overline{E_{0r}^2} + \frac{1}{2}I_{\rm nat}, \qquad I_l = \overline{E_{0l}^2} + \frac{1}{2}I_{\rm nat}. \tag{1.60}$$

Here E_{0r} and E_{0l} are the amplitudes of left- and right-hand polarizations fluctuating with a characteristic time of about $\Delta t \sim 1/\Delta\omega$, and the constant phase shift between these components is denoted by $\psi_{\rm rl}$. In this case the Stokes parameters are related to the above defined values through the formulas

$$\begin{aligned} I &= I_r + I_l \\ Q &= 2\overline{E_{0r}E_{0l}}\cos\psi_{rl}, \\ U &= 2\overline{E_{0r}E_{0l}}\sin\psi_{rl}, \\ V &= I_r - I_l. \end{aligned} \tag{1.61}$$

[7] Such an antenna can be made using two orthogonal dipoles with the outputs connected via a quarter-wave line element. The latter is necessary to introduce the phase shift $\pi/2$ between the signals received by individual dipoles.

These formulas can be proved in a similar way as (1.54); see [58].

If only some of the Stokes parameters are determined by polarization measurements, it is impossible to yield the complete information on the polarization of the observed radiation. For example, only the intensities I_r and I_l of two circularly polarized components are usually measured in observations of solar radio emission. As a result, the "degree of circular polarization"

$$\zeta_c = \frac{I_l - I_r}{I_l + I_r} = V/I \tag{1.62}$$

is found. This quantity is related to the true degree of polarization $\zeta = I_{pol}/I$ as follows:

$$\zeta_c = \zeta \sin(2\sigma) \tag{1.63}$$

(see (2.23)). The absolute value of ζ_c equals ζ only at $\sigma = \pm\pi/4$, i.e. if $Q = U = 0$. Since $\sigma = \arctan p$, the equality $|\zeta_c| = \zeta$ is only possible if the observed radiation is circularly polarized, partially or completely ($p = \pm 1$). In a general case of elliptically-polarized radiation $|\zeta_c| < \zeta$. This seems quite natural since it is impossible to account for the linear polarization in the observed radiation using the difference $I_l - I_r$ which just characterize the intensity of the circularly polarized component.

When measuring the intensities of linearly-polarized components,[8] the degree of linear polarization is defined by the following formula:

$$\zeta_l = \frac{I_{xx} - I_{yy}}{I_{xx} + I_{yy}} \frac{1}{\cos(2\chi)} = \frac{Q}{I\cos(2\chi)}. \tag{1.64}$$

According to (1.51),

$$\zeta_l = \zeta \cos(2\sigma) \tag{1.65}$$

and, generally, $\zeta_l \leq \zeta$. The equality is realized in the case $2\sigma = \pi$, i.e. if the observed radiation is linearly polarized, partially or completely ($p = 0, \infty$). Note that in optical astronomy, as a rule, both linear and circular polarization are measured. As for X- or gamma-ray astronomy, polarization measurements are not yet performed.

1.3. Radiation sources and radiation mechanisms. Definitions and general properties

Basic terms and definitions are introduced in this section which characterize sources of electromagnetic radiation both from the point of view of total energy radiated and according to the means and processes of emission of radiation (so-called radiation mechanisms). It is shown that the relation

[8]This is the case when observing synchrotron radiation of the Galaxy, supernova remnants and extragalactic discrete radio sources.

between the radiation losses of a source and the radiative energy flux through a fixed surface depends essentially on the source motion relative to this surface. As for radiation mechanisms, the definitions given below allow one to establish their general classification according to the qualitative differences of the conditions under which extra-terrestrial radiation is emitted.

1.3.1. SPECTRAL POWER AND FLUX DENSITY OF RADIATION FROM STATIONARY AND MOVING SOURCES

The energy of radiation emitted by a source per unit time is called the total radiation power $\mathcal{P}$. Of course, it is equal to the radiative energy loss of the source per the same time. In addition, the spectral power, $\mathcal{P}_{\omega\Omega}$, is introduced for more detailed description of the source radiation. By definition, it equals the energy radiated per unit of frequency ω, of time t, and of solid angle Ω. The spectral power is related to the total radiation power by the formula

$$\mathcal{P} = \int \mathcal{P}_{\omega\Omega}\, d\omega d\Omega. \tag{1.66}$$

Further, the radiation power per unit frequency range,

$$\mathcal{P}_{\omega} = \int \mathcal{P}_{\omega\Omega} d\Omega, \tag{1.67}$$

will be used together with $\mathcal{P}$ and $\mathcal{P}_{\omega\Omega}$. The formulas (1.66) and (1.67) are related to the source as a whole. As for the radiation by individual particles, the notation P (and, similarly, P_{ω}, $P_{\omega\Omega}$ etc.) will stand for $\mathcal{P}$. Note that the direction $\boldsymbol{s}$, which the solid angle element $d\Omega$ corresponds to, is determined here by the wave vector $\boldsymbol{k}$, the same way as in section 1.2: $\boldsymbol{s} = \boldsymbol{k}/k$.

The spectral power of radiation depends on the motion of a source. It changes as transformation is made from one inertial frame to another. Let

$$d\mathcal{E}' = \mathcal{P}_{\omega'\Omega'}(\omega', \mathrm{s}')d\Omega' d\omega' dt'. \tag{1.68}$$

be the energy radiated by the source into a solid angle $d\Omega'$, in a frequency range $d\omega'$, and in a time dt'. The primed quantities correspond to the co-moving frame A' in which the source is motionless at a given time t'. Let the frame A' and the source move with the velocity $\boldsymbol{v}_0$ with respect to another inertial frame A. Then an observer in the frame A will measure the radiated energy [9]

$$d\mathcal{E} = \mathcal{P}_{\omega\Omega}(\omega, \mathrm{s})d\Omega d\omega dt. \tag{1.69}$$

[9] All the quantities in this expression are related to the frame A.

The relativistic rules of transformation from A' to A for the quantities under consideration have the following form [56]:

$$d\mathcal{E} = d\mathcal{E}' \frac{\sqrt{1-\beta^2}}{1-\beta\cos\theta}, \tag{1.70}$$

$$d\omega = d\omega' \frac{\sqrt{1-\beta^2}}{1-\beta\cos\theta}, \tag{1.71}$$

$$dt = \frac{dt'}{\sqrt{1-\beta^2}}, \tag{1.72}$$

$$\sin\theta = \frac{\sqrt{1-\beta^2}\sin\theta'}{1+\beta\cos\theta'} \qquad \phi = \phi'. \tag{1.73}$$

Here $\beta = v_0/c$, and the angles θ, ϕ and θ', ϕ' set the vectors $\boldsymbol{s}$ $\boldsymbol{s}'$ in A and A', respectively: θ is the angle between $\boldsymbol{s}$ and $\boldsymbol{v}_0$ while ϕ determines the position of the projection of $\boldsymbol{s}$ onto the plane perpendicular to $\boldsymbol{v}_0$. The same definitions hold for θ' and ϕ' which determine the direction of $\boldsymbol{s}'$ in the frame A'.

The relation (1.70) reflects the invariance of the photon number in different frames: $dN' = dN$. Indeed, this transformation law can be obtained by putting $d\mathcal{E} = \hbar\omega dN$, $d\mathcal{E}' = \hbar\omega' dN'$, and taking into account that radiation frequency (and, therefore, photon energy) changes due to the Doppler effect in the following way:

$$\omega = \frac{\omega'\sqrt{1-\beta^2}}{1-\beta\cos\theta}. \tag{1.74}$$

Then, the relation (1.71) is the straightforward consequence of the Doppler formula (1.74). The next equality, (1.72), reflects the time-dilation effect: a time interval measured in a frame A in which the source is moving is larger than the corresponding one in the co-moving frame A'. Finally, (1.73) represents the aberration law which characterizes the change of the phase velocity direction upon transformation from one inertial frame to another.

With (1.70)-(1.74), the relation between radiation spectral powers in frames A' and A follows from (1.68) and (1.69) [59]:

$$\mathcal{P}_{\omega\Omega}(\omega,\theta,\phi) = \frac{(1-\beta^2)^{3/2}}{(1-\beta\cos\theta)^2} \mathcal{P}_{\omega'\Omega'}(\omega',\theta',\phi'). \tag{1.75}$$

This formula is valid if radiation emitted in the direction θ', ϕ' has the continuous frequency spectrum, and so $\mathcal{P}_{\omega'\Omega'}$ describes the energy radiated in unit frequency range. But if the source is monochromatic (i.e., only the single frequency ω' is radiated in the direction θ', ϕ'), then the quantity $\mathcal{P}_{\Omega'}$ should be used instead of $\mathcal{P}_{\omega'\Omega'}$. It is defined as the power of radiation with frequency ω' in unit solid angle Ω'. The transformation law from $\mathcal{P}_{\Omega'}$ to the corresponding value, $\mathcal{P}_{\Omega}$, in the frame A can be easily obtained by replacing $\mathcal{P}_{\omega'\Omega'}, \mathcal{P}_{\omega\Omega}$ with $\mathcal{P}_{\Omega'}, \mathcal{P}_{\Omega}$ in (1.68),(1.69) and dropping the intervals $d\omega'$ and $d\omega$ there. As a result [56], [10]

$$\mathcal{P}_{\Omega}(\theta, \phi) = \frac{(1-\beta^2)^2}{(1-\beta\cos\theta)^3}\mathcal{P}_{\Omega'}(\theta', \phi'). \tag{1.76}$$

If an isotropic source of radiation has a power-law spectrum

$$\mathcal{P}_{\omega'\Omega'} = \tilde{\mathcal{K}}(\omega')^{-\alpha} \tag{1.77}$$

in the frame co-moving frame A', the corresponding spectrum in the rest frame A will remain a power-law with the same spectral index α:

$$\mathcal{P}_{\omega\Omega} = \frac{(1-\beta^2)^{(3+\alpha)/2}}{(1-\beta\cos\theta)^{(2+\alpha)}}\tilde{\mathcal{K}}\omega^{-\alpha} \tag{1.78}$$

Therefore, a power-law spectrum is invariant of the transformation (1.75).

According to the formula (1.78), the width of the radiation pattern of a radiation source which moves with respect to the frame A is determined by the factor $(1-\beta\cos\theta)^{2+\alpha}$. The corresponding full width at half-maximum (FWHM) value, $\Delta\theta = 2\theta_{1/2}$, can be found from the relation

$$(1-\beta\cos\theta_{1/2})^{2+\alpha} = 2(1-\beta)^{2+\alpha},$$

which is reduced to

$$\Delta\theta = 2\theta_{1/2} \simeq 2a\sqrt{1-\beta^2}; \qquad a^2 = \sqrt[2+\alpha]{2} - 1 \tag{1.79}$$

for $1-\beta \ll 1$.

Transformation of spectral power from one reference frame to another is important when dealing with sources moving at relativistic velocities in the observer frame. In astrophysical conditions such sources are observed, for example, as plasma envelopes and jets ejected from quasars and galactic nuclei. In some cases the velocity of relative motion of radio emitting components of quasars and active galactic nuclei in the plane perpendicular to the line of sight exceeds c, the speed of light. According to VLBI observations, distinct ejections in quasar nuclei can move apart with a relative

[10]The formulas (1.75),(1.76) are valid for a source moving in vacuum. The effect of a medium on the transformation laws of the radiation power is taken into account in [60].

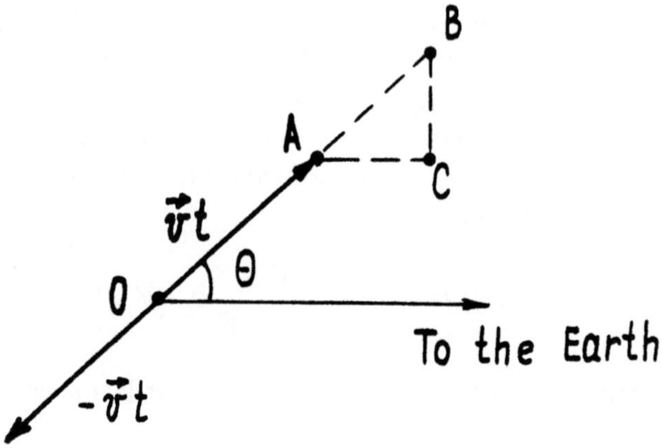

Figure 1.5. A sketch of counter-streaming jets ejecting from a central object at the point O.

velocity of about 4–20 c. For example, two components of the compact circumnuclear radio source in the quasar 3C 345 are observed to move apart at a rate 0.17 marcsec/yr. With the distance to the object $\simeq 1800$ Mpc taken into account, this corresponds to the relative velocity $\sim 6\,c$. Recently, the superluminal motion with the velocity $\simeq 1.25\,c$ has been discovered in the jets of the Galactic X-ray source GRS 1915+105 [61].

The following considerations clarify the origin of superluminal velocities in astrophysical sources [62]. Assume that a source at the point O ejects a pair of counter-streaming jets with the same velocity v_0 (see Fig. 1.5). The matter of a jet travels the distance $AB = v\Delta t$ along $\boldsymbol{v}$ during a time Δt and thus shifts toward the Earth by the interval $AC = v\Delta t \cos\theta$. Light passes the distance AC in a time $\Delta t' = (v/c)\Delta t \cos\theta$. As a result, the observed motion of the jet from the point A to B occurs in a time $\Delta t - \Delta t'$. Therefore, the observed jet velocity over the selectial sphere equals

$$v_{\rm obs} = \frac{BC}{(\Delta t - \Delta t')} = \frac{v \sin\theta}{1 - (v/c)\cos\theta}. \tag{1.80}$$

The corresponding value for the opposite jet is given by

$$v_{\rm obs} = \frac{v \sin\theta}{1 + (v/c)\cos\theta}. \tag{1.81}$$

The sum of (1.80) and (1.81) equals the relative expansion velocity of the matter in counter-streaming jets:

$$v_{\rm exp} = \frac{2v \sin\theta}{1 - (v/c)^2 \cos^2\theta}. \tag{1.82}$$

The expansion velocity of relativistic jets, $1-(v/c)^2 \ll \sin^2\theta$, may become superluminal:

$$v_{\mathrm{exp}} = \frac{2c}{\sin\theta}. \tag{1.83}$$

The account of relativistic motion of a radiation source is also very important in the theory of pulsar radiation (first and foremost, the famous Crab pulsar). It seems to be quite probable [59,63,64] that optical and X-ray radiation of pulsars may originate in a region which follows the rotation of a neutron star, i.e. moves around it with velocity comparable to c. Due to this fact the source radiation, even if emitted isotropically in the co-moving frame, is seen as strongly beamed in the frame of an observer on the Earth (see (1.75)). This effect of the "relativistic beaming" of radiation pattern can explain the pulsed nature of the observed radiation from a pulsar. It directly relates the narrowness of its angular pattern to the rotation of a neutron star and the location of emitting regions quite far from the star surface, somewhere near the light cylinder. [11] The attractive aspect of this hypothesis is that the narrow angular pattern and pulsed nature of pulsar radiation are explained by the cause independent of the particular emission mechanism.

Consider now the radiation flux through the fixed surface S produced by a radiating source. It will be shown to depend strongly on the source motion.

Note, first of all, that the spectral power of radiation of a steady-state source at rest, $\mathcal{P}_{\omega\Omega}$, can be easily found using the specific intensity, $I_{\omega\Omega}$, on the surface S enclosing the source. To do this one should take into account that the total power of radiation of such a source, $\mathcal{P}$, equals the radiation flux, F, through the surface mentioned:

$$F = \int I_{\omega\Omega} \cos\delta \, d\Omega d\omega dS \tag{1.84}$$

(see in this respect (1.28),(1.31)); here δ is the angle between the group velocity, $\boldsymbol{v}_{\mathrm{gr}}$, and the outward normal to the surface S. Comparison of (1.84) and (1.66) leads to the relation

$$\mathcal{P}_{\omega\Omega} = \int I_{\omega\Omega} \cos\delta \, dS. \tag{1.85}$$

The integral in (1.85) is taken over the part of the surface S where $\cos\delta > 0$.

[11] The light cylinder is the cylindrical surface with the axis coincident with the rotation axis of a star. Any point of this surface rotates around the axis with the linear velocity equal to c.

At first sight, it seems that the source need not be steady-state to ensure the equality $\mathcal{P}=F$. However, this is not true. In general,

$$\mathcal{P} = F + \frac{dW}{dt}, \tag{1.86}$$

where W is the energy of electromagnetic radiation inside the surface S. For a steady-state source at rest the spatial distribution of the intensity $I_{\omega\Omega}$ is time-independent, hence $W=$ const. Therefore, indeed, $\mathcal{P}=F$ which, in its turn, makes valid the formula (1.85). On the contrary, for a non-stationary (in particular, moving) source the value $W=W(t)$. As a result, $\mathcal{P}\neq F$, so the relation between $\mathcal{P}_{\omega\Omega}$ and $I_{\omega\Omega}$ becomes more complicated.

To acquaint oneself with this relation, consider a source, steady-state in the co-moving frame A', which moves as a whole with the velocity $\boldsymbol{v}_0$ relative to another frame, A. Let us find the radiation flux through a fixed surface S at rest in A which is situated far enough from the source. Suppose, for the sake of simplicity, that the source radiates in vacuum. Then the radiation emitted at an instant $t_{\rm rad}$ will reach some point B on S at the instant $t=t_{\rm rad}+r(t_{\rm rad})/c$, where $r(t)\gg L$ is the distance from the source to B, and L is the linear size of the source. So, the radiation emitted during an interval $dt_{\rm rad}$ will pass through the surface S during

$$dt = dt_{\rm rad} + \frac{1}{c}\frac{dr}{dt_{\rm rad}}dt_{\rm rad} = (1-\beta\cos\theta)dt_{\rm rad}. \tag{1.87}$$

To come to the last expression, one should account for the variation of the distance r due to the source motion; this yields $dr/dt_{\rm rad} = -v_0\cos\theta$, where $v_0=\beta c$ is the velocity of the source, θ is the angle between $\boldsymbol{v}_0$ and $\boldsymbol{r}$, the radius-vector from the source to the point B on the surface S. The energy radiated during the interval $dt_{\rm rad}$ into the solid angle $d\Omega$ is equal to

$$d\mathcal{E} = \mathcal{P}_{\omega\Omega}d\Omega d\omega dt_{\rm rad}, \tag{1.88}$$

where $\mathcal{P}_{\omega\Omega}$ is the spectral power of the radiation in the frame A. Upon reaching the surface S, this energy provides the radiation flux density F_ω, (1.31), such that

$$d\mathcal{E} = F_\omega dS d\omega dt. \tag{1.89}$$

Here dt is related to $dt_{\rm rad}$ according to (1.87), and $dS=r^2d\Omega$ is the surface element perpendicular to $\boldsymbol{r}$ which solid angle $d\Omega$ rests upon. Now it is not difficult to find from (1.88),(1.89) that

$$F_\omega = \frac{\mathcal{P}_{\omega\Omega}}{r^2(1-\beta\cos\theta)}. \tag{1.90}$$

The relation obtained means that if the velocity $\boldsymbol{v}_0$ of the source motion is comparable to c, the value of the radiation flux density at a stationary point B will be altered radically as compared to the corresponding value $F_\omega = P_{\omega\Omega}/r^2$ for a stationary source at the same distance r. Namely, F_ω increases if the source is moving towards an observer ($\cos\theta > 0$) at a moment of emission $dt_{\rm rad}$, and, inversely, F_ω decreases if the source is moving away from an observer ($\cos\theta < 0$). Recalling the above example of an object with counter-streaming jets (see Fig. 1.5), we should note that this effect leads to the significant difference between the intensities of radiation emitted by jets moving apart from each other. Such a difference is indeed observed in jets from quasars and Galactic sources.

It should be emphasized that the effect considered, which has been pointed out in [65–67], is related neither to the relativistic transformation of time intervals similar to (1.72) nor, generally, to the transition from one reference frame to another. Such a transition has already been made when finding the spectral power of radiation by a moving source, $P_{\omega\Omega}$, from the known co-moving frame radiation power of this source, $P_{\omega'\Omega'}$ (see (1.75)). The effect described by the formula (1.90) appears only in the frame A in which the source is moving. In fact, this formula is the consequence of time delay caused by the finite speed of electromagnetic wave propagation.[12] In fact, emission time of a moving source, $dt_{\rm rad}$, differs from the corresponding detection time (1.87) of the emitted radiation. This leads to the formula (1.90).

Consider now the role of this effect in the case of the circular (orbital) motion of a source. This case can be relevant to the above mentioned pulsar models with relativistic beaming of the source radiation pattern.

The angle between the velocity of an orbiting source and the direction to an observer is given by $\theta = (2\pi/\tau)t_{\rm rad}$, where τ is the orbital (pulsar) period. Since this angle enters the expression (1.90) for the flux density F_ω of the observed radiation, the latter will vary periodically in time. This holds even for a source with isotropic radiation pattern $\mathcal{P}_{\omega\Omega}(\theta) = \text{const}$ in the observer frame A in which the source is moving. But if the radiation pattern is isotropic in the co-moving frame A': $\mathcal{P}_{\omega'\Omega'}(\theta') = \text{const}$, the corresponding angular pattern $\mathcal{P}_{\omega\Omega}(\theta)$ in A will be stretched along the velocity $\boldsymbol{v}_0$ due to the effect of relativistic beaming. The width of the radiation pattern by a power-law source with the spectrum (1.77) is given by the formula (1.79), while the dependence of $\mathcal{P}_{\omega\Omega}$ on the angle θ, by (1.78). The combined influence of the beaming effect and the shortening of the detection time for radiation emitted by a source approaching an observer leads to the following behavior of the observed radiation flux density (1.90):

[12]This is also true for the formulas (1.81), (1.82) which describe the effect of "superluminal" motion of radio sources.

$$F_\omega(t) = \frac{(1-\beta^2)^{(3+\alpha)/2}}{r^2(1-\beta\cos\theta)^{(3+\alpha)}}\tilde{K}\omega^{-\alpha}, \tag{1.91}$$

where $\theta = \frac{2\pi}{\tau}t_{\rm rad}$, and $t_{\rm rad}$ is related to t according to the following:

$$t = t_{\rm rad} + \frac{r(t_{\rm rad})}{c}.$$

The pulse duration Δt of the observed radiation (i.e., the FWHM value for $F_\omega(t)$) can be easily found from the equation

$$(1-\beta\cos\theta_{1/2})^{3+\alpha} = 2(1-\beta)^{3+\alpha}.$$

Its solution for $1-\beta \ll 1$ has the form:

$$\theta_{1/2} \simeq b\sqrt{1-\beta^2}, \qquad b^2 = \sqrt[3+\alpha]{2} - 1. \tag{1.92}$$

In its turn,

$$\Delta t_{\rm rad} = \frac{\tau}{2\pi}\cdot 2\theta_{1/2}. \tag{1.93}$$

The value of Δt can be finally found by integrating (1.87):

$$\Delta t = \int_{-\Delta t_{\rm rad}/2}^{+\Delta t_{\rm rad}/2}(1-\beta\cos\theta)dt_{\rm rad}.$$

The ansatz $t_{\rm rad} = (\tau/2\pi)\theta$ leads to [59]:

$$\Delta t = \frac{\tau}{\pi}(\theta_{1/2} - \beta\sin\theta_{1/2}). \tag{1.94}$$

In the relativistic limit $1-\beta \ll 1$ this formula looks like

$$\Delta t = \frac{\tau}{2\pi}(1-\beta^2)^{3/2}b(1+\frac{b^2}{3}). \tag{1.95}$$

Thus, it determines the pulse duration Δt from a remote source with a power-law spectrum which moves along a circular orbit at a relativistic linear velocity.

For such high velocities the relative pulse duration $\Delta t/\tau \sim (1-\beta^2)^{3/2}$ is small, even if the radiation pattern of a source is rather wide: $\Delta\theta \sim (1-\beta^2)^{1/2}$ (cf. (1.95) and (1.79)). This is caused by the effect of compression of pulses emitted at the moments when a source is approaching an observer. According to the pulsar model with relativistic shaping of the radiation pattern, this makes it possible to detect the radiation by a large fraction (up to one third) of existing pulsars despite their stretched (pencil-beam) radiation patterns [64].

Note, finally, that the pulse compression effect does not change the mean radiation flux density from a source moving along a closed orbit. For example, averaging of the flux density (1.90) from a circularly moving source over the orbital period yields

$$\bar{F}_\omega = \frac{1}{\tau}\int_0^\tau F_\omega(t)dt = \frac{1}{\tau}\int_0^\tau \frac{P_{\omega\Omega}(\omega,\theta)}{r^2(1-\beta\cos\theta)}dt. \tag{1.96}$$

Since dt relates to dt_{rad} by the formula (1.87), the ansatz $t_{\text{rad}} = (\tau/2\pi)\theta$ leads to [68]

$$\bar{F}_\omega = \frac{1}{2\pi}\int_0^{2\pi} \frac{P_{\omega\Omega}(\omega,\theta)}{r^2}d\theta. \tag{1.97}$$

1.3.2. RADIATION MECHANISMS

Consider now terminology and basic concepts which are used in astrophysics (and, especially, in theoretical radio astronomy) to describe radiation mechanisms which function in astrophysical plasmas as well as characteristics of radiation escaping the source. The term "radiation mechanism" usually means a physical process (or a group of processes) resulting in the emission of electromagnetic radiation. Note that the elementary process of emission by a single particle may widely vary from cyclotron or synchrotron to Čerenkov radiation, bremsstrahlung etc.

Radiation is called thermal if it is generated by a particle system with the equilibrium velocity distribution (the Maxwell one for non-relativistic particles). But if particle distribution is non-equilibrium, the emitted radiation will be non-thermal. Radiation is classified as thermal or non-thermal according to the type of distribution function of radiating particles, not the type of elementary emission process.

The equilibrium radiation with the intensity (1.37), (1.43) and the energy density (1.36), (1.42) is the particular case of thermal radiation. The brightness, or effective, temperature of the equilibrium radiation T_{b}, equals T, the kinetic temperature of radiating particles; the latter, in its turn, is estimated as $T \simeq \bar{\mathcal{E}}/\kappa$, where $\bar{\mathcal{E}}$ is the mean energy of thermal motion of radiating particles. The brightness temperature of thermal radiation can not exceed T (see section 4.1):

$$T_{\text{b}} \leq T \simeq \frac{\bar{\mathcal{E}}}{\kappa}. \tag{1.98}$$

Astrophysical radiation mechanisms can be divided into coherent and incoherent [3,69]. A mechanism is coherent if the spectral power of a source radiation, $\mathcal{P}_{\omega\Omega}$, exceeds the sum of spontaneous[13] radiation powers by individual particles constituting the source:

$$\mathcal{P}_{\omega\Omega} > \sum_i P_{\omega\Omega}^{(i)}. \tag{1.99}$$

Here $P_{\omega\Omega}^{(i)}$ is the spectral power of radiation by the i-th particle at frequency ω; the sum is taken over all the particles in the source which radiate at a given frequency and direction.

It is the inequality (1.99) that, by definition, discriminates between coherent and incoherent mechanisms. For the latter ones, the spectral power of radiation of a source does not exceed the sum of the spectral powers of spontaneous radiation by individual particles:

$$\mathcal{P}_{\omega\Omega} \leq \sum_i P_{\omega\Omega}^{(i)}. \tag{1.100}$$

An object which contains radiating particles with the equilibrium velocity distribution is an example of a source in which an incoherent radiation mechanism is realized. Reabsorption in such a source is always positive: radiation is partially absorbed before escaping it (see section 4.1). This ensures the validity of the criterion of (1.100). In fact, the same reason leads to the relation (1.98) as well. Note that the condition $T_{\mathrm{b}} \lesssim \bar{\mathcal{E}}/\kappa$ can be used for any incoherent mechanism as a rough estimate of the expected value of T_{b} if the mean particle energy $\bar{\mathcal{E}}$ is known (the reverse is also true). Of course, this may be done only if particle distribution in a source does not differ much from the equilibrium one (or, in other words, the system is far from the instability threshold; see the chapter 7).

It is clear from (1.98) that an incoherent mechanism can only yield large brightness temperatures if energies of radiating particles are high. For example, radio emission from pulsars is characterized by $T_{\mathrm{b}} \sim 10^{30}\,\mathrm{K}$. In case of an incoherent mechanism this demands the mean energy of radiating particles to be estimated as $\bar{\mathcal{E}} \gtrsim 10^{26}\,\mathrm{eV}$. Therefore, coherent mechanisms seem to be preferable for the interpretation of radiation with high brightness temperatures from some objects (in particular, pulsars radio emission). These mechanisms do not require extremely high energies of radiating particles to provide a radiation flux as large as the observed one.

Of course, it is impossible to offer the universal rule to distinguish between coherent and incoherent mechanisms when explaining the radiation from cosmic sources. For example, the above arguments in favor of the coherent mechanism of pulsar radio emission sound quite reasonable taking into account extremely high values of T_{b}. However, the same arguments

[13]Spontaneous radiation is considered in section 7.1.

appear to be much less convincing when used for the solar radio emission. The reason is the much lower values of T_b in the observed radiation. Here the coherent mechanisms are favored by the complicated frequency spectra, extremely non-stationary events and so forth. The choice of the specific mechanism, coherent or incoherent, to interpret the radiation characteristics of any source (including the Sun and pulsars) can finally be approved solely by the successful explanation of the entire complex of the main features of the observed radiation, not some distinct parameters (for example, T_b), however important they are.

As far as coherent mechanisms are concerned, note that, generally, two kinds of such mechanisms, antenna and maser ones, are possible [63, 70]. In an antenna mechanism the coherence is achieved by the initial phasing (bunching) of radiating particles. The phases of spontaneous radiation of individual particles in this case are no longer random, so the resulting total radiation power can exceed the sum of their spontaneous radiation powers. On the contrary, maser mechanisms work even if there is no initial phasing of particles in the source. Here the specific particle distribution over energy or momentum is created initially and maintained thereon which provides the inverted population of energy states. Reabsorption in such a system is negative. As a result, spontaneous radiation emitted inside the source is amplified when propagating to its boundary. So, the inequality (1.99) is satisfied and, in this sense, the coherent mechanism is realized. This means that the radiation of an object with maser generation mechanism is, to some extent, phased too. However, contrary to antenna mechanisms, the phasing here is established "self-consistently" under the action of the amplified radiation itself.

Of course, at some stage a maser mechanism can be reduced to an antenna one. Indeed, a steady-state motion of electrons inside a source is equivalent to some complicated system of external currents with amplitudes and phases varying significantly over a distance of about the wavelength of emitted radiation. If such a current system is created and maintained with some external e.m.f., the resulting radiation would be identical to that produced by a maser mechanism. However, it is practically impossible to create such a complex current distribution necessary to generate powerful radiation. On the contrary, it inevitably arises in a maser system under the action of its radiation. The above arguments indicate that antenna mechanisms are quite improbable in cosmic plasma, though they cannot exclude this possibility.

That is why maser mechanisms are favored in this book. They are considered in detail in chapter 7. Here we only note that coherent maser mechanisms can be divided into two groups. The first group is constituted by mechanisms in which the waves capable of escaping freely from the radiating source are amplified and then observed from the Earth. Radiation mechanisms comprising the second group are the ones in which the amplified waves (in particular, plasma ones) cannot escape from the dense plasma of the source to the rarefied interstellar or interplanetary medium; in this case the observed radiation occurs as a result of wave conversion, or transformation (for example, plasma waves into electromagnetic ones). It is thus quite important to investigate in detail such conversion processes in cosmic plasmas. The corresponding results are given in section 3.2 and in chapter 8.

Electromagnetic waves in homogeneous plasma

This chapter gives the basics of the linear theory of electromagnetic waves in homogeneous plasma. The issues necessary to understand the further material are favoured. Complexity and the wide variety of wave processes in plasma, especially the magnetized one, forced us to restrict ourselves to only the theory of high-frequency waves. Their features are completely determined by the electrons, whereas the influence of ions is negligible. Such a restriction is somehow justified by the great importance of the high-frequency waves for the generation of extra-terrestrial radiation. Of course, this does not mean that the lower-frequency waves are of minor importance for astrophysical and space plasmas. As an example, we mention the magnetosphere of the Earth where the processes of lower-frequency wave excitation determine the characteristics of the observed emissions and significantly affect the dynamics of the radiation belts. The problem of generation and propagation of lower-frequency waves in astrophysical and space plasmas is worth a separate book. We refer the reader to monographs [71–73] which give general theory of low-frequency waves in plasma and also to a number of original papers.

The presence of strong magnetic fields of about 10^{12}–10^{13} G on neutron stars leads to polarization and magnetization of vacuum and can drastically alter electromagnetic wave propagation. In certain cases mode polarizations and refractive indices are dominated by the effect of vacuum birefringence. Thus, in a special section of this chapter the electromagnetic wave propagation in a system comprising plasma and birefringent vacuum is analyzed.

2.1. Dielectric permittivity of plasma. Dispersion equations

2.1.1. ELECTROMAGNETIC FIELD EQUATIONS IN A MEDIUM

Analysis of electromagnetic wave propagation in plasma is based upon Maxwell equations. These equations for harmonic processes ($\propto e^{-i\omega t}$) can be written in the following form:

$$\operatorname{curl} \boldsymbol{H} = \frac{4\pi}{c}\boldsymbol{j} - \frac{i\omega}{c}\boldsymbol{D}, \tag{2.1}$$

$$\operatorname{curl} \boldsymbol{E} = \frac{i\omega}{c}\boldsymbol{B}, \tag{2.2}$$

$$\operatorname{div} \boldsymbol{D} = 4\pi\rho_e, \tag{2.3}$$

$$\operatorname{div} \boldsymbol{B} = 0, \tag{2.4}$$

where

$$\boldsymbol{D} = \boldsymbol{E} + 4\pi\boldsymbol{P}, \tag{2.5}$$

$$\boldsymbol{B} = \boldsymbol{H} + 4\pi\boldsymbol{M}, \tag{2.6}$$

$\boldsymbol{E}$ and $\boldsymbol{H}$ are electric and magnetic field intensities, $\boldsymbol{D}$ and $\boldsymbol{B}$ are the corresponding inductions, $\boldsymbol{P}$ and $\boldsymbol{M}$ are polarization and magnetization vectors, $\boldsymbol{j}$ and ρ_e are current and charge densities, respectively. The continuity equation for currents and charges,

$$-i\omega\rho_e + \operatorname{div} \boldsymbol{j} = 0, \tag{2.7}$$

follows from the equations (2.1) and (2.3). To obtain (2.7), one should apply the divergence operator to the equation (2.1) and substitute div$\boldsymbol{D}$ from the equation (2.3).

For processes varying in space and time as $\exp(i\boldsymbol{kr} - i\omega t)$ the system (2.1)-(2.4) is reduced to:

$$[\boldsymbol{k} \times \boldsymbol{H}] = -\frac{\omega}{c}\left(\boldsymbol{D} + i\frac{4\pi}{\omega}\boldsymbol{j}\right), \tag{2.8}$$

$$[\boldsymbol{k} \times \boldsymbol{E}] = \frac{\omega}{c}\boldsymbol{B}, \tag{2.9}$$

$$\boldsymbol{k}\left(\boldsymbol{D} + i\frac{4\pi}{\omega}\boldsymbol{j}\right) = 0, \tag{2.10}$$

$$\boldsymbol{kB} = 0. \tag{2.11}$$

Vectors $\boldsymbol{D}$, $\boldsymbol{B}$ and $\boldsymbol{j}$ are related to $\boldsymbol{E}$ and $\boldsymbol{H}$ via the constitutive relations. The form of these relations is determined by the properties of a medium. For isotropic motionless plasma in which, due to the absence of a static magnetic field $\boldsymbol{H}_0$, no specific direction can be pointed out, the constitutive relations look like:

$$\boldsymbol{D} = \epsilon(\omega, \boldsymbol{k})\boldsymbol{E}, \quad \boldsymbol{B} = \mu(\omega, \boldsymbol{k})\boldsymbol{H}, \quad \boldsymbol{j} = \sigma(\omega, \boldsymbol{k})\boldsymbol{E}, \tag{2.12}$$

Here ϵ, μ and σ are dielectric permittivity, magnetic permeability and electrical conductivity, respectively.

Dependence of ϵ, μ and σ on frequency ω and/or wave vector $\boldsymbol{k}$ characterizes time and/or spatial dispersion of a plasma. Time (or frequency) dispersion is the consequence of the fact that the values of $\boldsymbol{D}$, $\boldsymbol{B}$ and $\boldsymbol{j}$ in a medium depend not only on the field intensities $\boldsymbol{E}$, $\boldsymbol{H}$ at a given moment but also on their temporal histories. Respectively, spatial dispersion reveals that local values of $\boldsymbol{D}$, $\boldsymbol{B}$ and $\boldsymbol{j}$ depend on the spatial distribution of the fields $\boldsymbol{E}$ and $\boldsymbol{H}$.

With the constitutive relations (2.12) the system (2.8)-(2.11) can be rewritten in the form

$$[\boldsymbol{k} \times \boldsymbol{H}] = -\mu\frac{\omega}{c}\epsilon'\boldsymbol{E}, \tag{2.13}$$

$$[\boldsymbol{k} \times \boldsymbol{E}] = \frac{\omega}{c}\boldsymbol{B}, \tag{2.14}$$

$$\epsilon'\boldsymbol{kE} = 0, \tag{2.15}$$

$$\boldsymbol{kB} = 0, \tag{2.16}$$

where ϵ' denotes complex dielectric constant:

$$\epsilon' = \epsilon + i\frac{4\pi\sigma}{\omega}. \tag{2.17}$$

The Lorenz force

$$\boldsymbol{f} = e\boldsymbol{E} + \frac{e}{c}[\boldsymbol{v} \times \boldsymbol{B}] \tag{2.18}$$

acts upon a point charge moving with velocity $\boldsymbol{v}$ in plasma. The formula (2.18) takes into account that the electric field acting on a particle in plasma is equal to the macroscopic mean field (i.e., to the vector $\boldsymbol{E}$[1]), whereas the macroscopic mean value of the magnetic field in a medium equals $\boldsymbol{B}$, not $\boldsymbol{H}$. Thus, only vectors $\boldsymbol{E}$ and $\boldsymbol{B}$ in plasma have genuine physical meaning, while $\boldsymbol{D}$ and $\boldsymbol{H}$ are only of subordinate importance. This gives a choice to define the vector of electric induction in different ways to arrive at the most convenient form of the field equations. For example, the analysis of electromagnetic processes in plasma, especially in an anisotropic, magnetized one where $\boldsymbol{B}_0 \neq 0$, is frequently laid upon the system

$$\operatorname{curl}\boldsymbol{B} = -i\frac{\omega}{c}\boldsymbol{D}', \tag{2.19}$$

$$\operatorname{curl}\boldsymbol{E} = i\frac{\omega}{c}\boldsymbol{B}, \tag{2.20}$$

$$\operatorname{div}\boldsymbol{D}' = 0, \tag{2.21}$$

$$\operatorname{div}\boldsymbol{B}' = 0. \tag{2.22}$$

which differs from (2.1)–(2.4) by the absence of the field $\boldsymbol{H}$ and the permeability μ. This is possible because of the redefinition of the electric induction vector. The vector $\boldsymbol{D}'$ which will also be called electric induction is related to the previously used vector $\boldsymbol{D}$ as

$$-i\frac{\omega}{4\pi}\boldsymbol{D}' = -i\frac{\omega}{4\pi}\boldsymbol{D} + \boldsymbol{j} + c\operatorname{curl}\boldsymbol{M}. \tag{2.23}$$

[1]This has been clarified in [74–76].

Together with (2.5), this relation clarifies the meaning of the newly defined vector $\boldsymbol{D}'$. The value of $-(i\omega/4\pi)\boldsymbol{D}'$ is the sum of the displacement current, $-(i\omega/4\pi)\boldsymbol{E}$, and the total current, $\boldsymbol{j}_{\text{tot}}$, induced by the electromagnetic field in plasma. Thus,

$$\boldsymbol{D}' = \boldsymbol{E} + i\frac{4\pi}{\omega}\boldsymbol{j}_{\text{tot}} \tag{2.24}$$

where $\boldsymbol{j}_{\text{tot}}$ comprises polarization current $-i\omega\boldsymbol{P}$, conduction current $\boldsymbol{j}$ and magnetization current $c\,\text{curl}\,\boldsymbol{M}$.

Substituting $\boldsymbol{H}$ from (2.6) and taking into account the definition of the vector $\boldsymbol{D}'$ (equation (2.23)), it is easy to see that the equation (2.19) follows from (2.1). Then, the equation (2.20) and (2.22) are identical to (2.2) and (2.4). As for the equations (2.21) and (2.22), they follow from (2.19) and (2.20)), respectively. Thus, the equations in these pairs are not independent in the system (2.19)-(2.22).

For vectors $\boldsymbol{B}$, $\boldsymbol{D}' \propto \exp(i\boldsymbol{kr} - i\omega t)$ the system (2.19)-(2.22) takes the form:

$$[\boldsymbol{k} \times \boldsymbol{B}] = -\frac{\omega}{c}\boldsymbol{D}' \tag{2.25}$$

$$[\boldsymbol{k} \times \boldsymbol{E}] = \frac{\omega}{c}\boldsymbol{B}, \tag{2.26}$$

$$\boldsymbol{kD}' = 0, \tag{2.27}$$

$$\boldsymbol{kB} = 0. \tag{2.28}$$

These equations should be supplemented with the constitutive relations which are taken in the form[2]

$$D'_\alpha = \epsilon_{\alpha\beta}(\omega, \boldsymbol{k})E_\beta. \tag{2.29}$$

Here D'_α and E_β are the orthogonal components of the corresponding vectors, $\epsilon_{\alpha\beta}(\omega, \boldsymbol{k})$ is the dielectric permittivity tensor. This dielectric tensor is the most important characteristic which determines the features of an electromagnetic process in a medium.

Tensor origin of the relation (2.29) between $\boldsymbol{D}'$ and $\boldsymbol{E}$ seems quite natural when dealing with magnetoactive plasma (i.e., in plasma with constant magnetic field $\boldsymbol{H}_0$ which makes it anisotropic). It is not surprising that in such a medium the total current density $(4\pi)^{-1}\partial\boldsymbol{D}'/\partial t = -i\omega\boldsymbol{D}'/4\pi$ induced by electric field $\boldsymbol{E}$ does not coincide with the vector $\boldsymbol{E}$. However, the same tensor relation between $\boldsymbol{D}'$ and $\boldsymbol{E}$ holds also in isotropic plasma.

[2]Here, as usual, the summation over the identical scripts is supposed, i.e., the relation (2.29) is the reduced form of the following equality:

$$D'_\alpha = \epsilon_{\alpha\alpha}E_\alpha + \epsilon_{\alpha\beta}E_\beta + \epsilon_{\alpha\gamma}E_\gamma.$$

Here it results from the definition of the vector $\boldsymbol{D}'$ which includes the magnetization current $c\,\mathrm{curl}\,\boldsymbol{M}$ (see (2.23)). To find the tensor $\epsilon_{\alpha\beta}(\omega,\boldsymbol{k})$ in isotropic plasma, we first rewrite it in the form [77]

$$\epsilon_{\alpha\beta}(\omega,\boldsymbol{k}) = \left(\delta_{\alpha\beta} - \frac{k_\alpha k_\beta}{k^2}\right)\epsilon_\perp(\omega,\boldsymbol{k}) + \frac{k_\alpha k_\beta}{k^2}\epsilon_\parallel(\omega,\boldsymbol{k}) \qquad (2.30)$$

This expression takes into account that only two tensors, Kronecker delta $\delta_{\alpha\beta}$ and $k_\alpha k_\beta$ can be relevant to such a medium. [3]. As follows from (2.29), (2.30), the relation between $\boldsymbol{D}'$ and $\boldsymbol{E}$ can be represented in the following form:

$$\boldsymbol{D}' = \epsilon_\perp \boldsymbol{E} - (\epsilon_\perp - \epsilon_\parallel)\frac{\boldsymbol{k}(\boldsymbol{k}\boldsymbol{E})}{k^2}. \qquad (2.31)$$

The value $\epsilon_\perp$ is called the transverse, while $\epsilon_\parallel$ — the longitudinal dielectric permittivity. These terms reveal that $\epsilon_\perp(\omega,\boldsymbol{k})$ and $\epsilon_\parallel(\omega,\boldsymbol{k})$ determine the properties of transverse and longitudinal waves in plasma, respectively (see the following chapter). The relation (2.31) allows one to very easily obtain the explicit relation between the pairs ϵ', μ and $\epsilon_\perp$, $\epsilon_\parallel$. Such relations must exist, since the systems (2.13)-(2.16) and (2.25)-(2.28) are identical in an isotropic plasma.

With the expression (2.31) taken into account, the equations (2.25)-(2.28) take the following form:

$$[\boldsymbol{k}\times\boldsymbol{B}] = -\frac{\omega}{c}\left(\epsilon_\perp \boldsymbol{E} - (\epsilon_\perp - \epsilon_\parallel)\frac{\boldsymbol{k}(\boldsymbol{k}\boldsymbol{E})}{k^2}\right), \qquad (2.32)$$

$$[\boldsymbol{k}\times\boldsymbol{E}] = \frac{\omega}{c}\boldsymbol{B}, \qquad (2.33)$$

$$\epsilon_\parallel \boldsymbol{k}\boldsymbol{E} = 0, \qquad (2.34)$$

$$\boldsymbol{k}\boldsymbol{B} = 0 \qquad (2.35)$$

The identity of the equations (2.15) and (2.34) yields

$$\epsilon' = \epsilon_\parallel. \qquad (2.36)$$

Substituting the magnetic induction $\boldsymbol{B}$ from the equation $[\boldsymbol{k}\times\boldsymbol{E}]=(\omega/c)\boldsymbol{B}$ into (2.13) and (2.32), subtracting (2.32) from (2.13), and taking into account the equality $\epsilon'=\epsilon_\parallel$, we find

$$\left(1-\frac{1}{\mu}\right)[\boldsymbol{k}\times[\boldsymbol{k}\times\boldsymbol{E}]] = \frac{\omega^2}{c^2k^2}(\epsilon_\parallel - \epsilon_\perp)\left(k^2\boldsymbol{E} - \boldsymbol{k}(\boldsymbol{k}\boldsymbol{E})\right) \qquad (2.37)$$

Since $[\boldsymbol{k}\times[\boldsymbol{k}\times\boldsymbol{E}]]=-k^2\boldsymbol{E}+\boldsymbol{k}(\boldsymbol{k}\boldsymbol{E})$, the above equality is unconditionally valid if [78]

$$1-\frac{1}{\mu} = \frac{\omega^2}{c^2k^2}(\epsilon_\perp - \epsilon_\parallel) \qquad (2.38)$$

[3]The tensor $\delta_{\alpha\beta}$ has the following properties: $\delta_{\alpha\beta}=0$ if $\alpha\neq\beta$ and $\delta_{\alpha\beta}=1$ if $\alpha=\beta$.

This formula relates the magnetic permeability of an isotropic plasma to longitudinal and transverse dielectric permittivities.

It is clear from (2.38) that the tensor relation between $\boldsymbol{D}'$ and $\boldsymbol{E}$ arises since the magnetic permeability μ differs from unity. If $\mu=1$, then $\epsilon_\perp=\epsilon_\parallel$ and, according to (2.30), (2.36), $\epsilon_{\alpha\beta}=\delta_{\alpha\beta}\epsilon_\perp=\delta_{\alpha\beta}\epsilon'$. Hence, the constitutive relations are reduced to $D_{\alpha\beta}=\delta_{\alpha\beta}\epsilon' E_\beta$ or, in vector form, $\boldsymbol{D}'=\epsilon'\boldsymbol{E}$.

Note that in an ionized gas with spatial dispersion $\epsilon_\perp \neq \epsilon_\parallel$ (see the formulas (2.88), (2.89) further on). Therefore, the magnetic permeability μ of an isotropic plasma generally differs from unity. However, if the parameter $Z=\omega/\sqrt{2}kv_T \to \infty$ (i.e. for relatively long waves in plasma with $v_T\neq 0$), the spatial dispersion vanishes: $\epsilon_\perp$ and $\epsilon_\parallel$ do not depend on k. Here $\epsilon_\perp = \epsilon_\parallel = 1-\omega_L^2/\omega^2$ and, thus, $\mu=1$. For static fields $\epsilon_\perp\neq\epsilon_\parallel$; nevertheless, as $\omega\to 0$, the right-hand side of (2.38) is nullified, so we arrive again at the equality $\mu=1$. The last case corresponds to the well-known Bohr–van Leven theorem on the absence of magnetic susceptibility of a classical equilibrium particle system in a static magnetic field.

2.1.2. DISPERSION RELATIONS FOR NORMAL WAVES

To determine the characteristics of waves propagating in plasma we start from the system of equations (2.25)–(2.28) supplemented with the constitutive relation (2.29). These equations hold for $\boldsymbol{B}$, $\boldsymbol{E}$, $\boldsymbol{D}' \propto \exp(i\boldsymbol{kr}-i\omega t)$. Equation (2.25) allows to exclude the magnetic induction $\boldsymbol{B}$ from (2.26). As a result,

$$[\boldsymbol{k}\times[\boldsymbol{k}\times\boldsymbol{E}]] = -\frac{\omega^2}{c^2}\boldsymbol{D}'$$

or, equivalently,

$$k^2\boldsymbol{E}-\boldsymbol{k}(\boldsymbol{kE})-\frac{\omega^2}{c^2}\boldsymbol{D}'=0. \qquad (2.39)$$

With (2.29) this vector equation can be written as the system of three linear algebraic equations for the components of the electric field E_β:

$$\left(k^2\delta_{\alpha\beta}-k_\alpha k_\beta-\frac{\omega^2}{c^2}\epsilon_{\alpha\beta}(\omega,\boldsymbol{k})\right)E_\beta=0. \qquad (2.40)$$

The homogeneous system (2.40) has non-trivial (non-zero) solution only if its determinant is equal to zero:

$$\det\left(k^2\delta_{\alpha\beta}-k_\alpha k_\beta-\frac{\omega^2}{c^2}\epsilon_{\alpha\beta}(\omega,\boldsymbol{k})\right)=0. \qquad (2.41)$$

Here det denotes the 3×3 determinant in which an element in α-th row and β-th column is equal to the expression in curly brackets.

The equation (2.41) is called dispersion equation. It relates the frequency ω of a plane monochromatic wave to its wave vector $\boldsymbol{k}$:

$$\omega = \omega_j(\boldsymbol{k}). \tag{2.42}$$

A dispersion equation may have more than one solution of the form (2.42). Each solution specified with the number $j = 1, 2, \ldots$ determines a normal wave, or mode, which can propagate in plasma. A normal wave with a given wave vector $\boldsymbol{k}$ differs from the other ones by the frequency ω and polarization, the ratio of the components of the electric field $\boldsymbol{E}$. Mode polarization can be found from the system (2.40) using ω and $\boldsymbol{k}$ which satisfy the dispersion relation (2.41).

If the dielectric tensor does not depend on the electric field, the system (2.40) is linear with respect to $\boldsymbol{E}$.[4] In this case each normal wave propagates independently of the others. General solution for electromagnetic field in plasma is given by a superposition (linear) of all normal waves.

In isotropic plasma with the tensor $\epsilon_{\alpha\beta}(\omega, \boldsymbol{k})$ in the form (2.30) the system (2.40) which determines normal waves is written as:

$$(k^2 - \frac{\omega^2}{c^2}\epsilon_{\perp})E_x = 0, \qquad (k^2 - \frac{\omega^2}{c^2}\epsilon_{\perp})E_y = 0, \tag{2.43}$$

$$\frac{\omega^2}{c^2}\epsilon_{\parallel}E_z = 0. \tag{2.44}$$

The electric field components in this equations correspond to the frame with the z-axis directed along the wave vector $\boldsymbol{k}$. With such a choice the resulting dispersion equation

$$(k^2 - \frac{\omega^2}{c^2}\epsilon_{\perp})^2\epsilon_{\parallel} = 0, \tag{2.45}$$

splits into a pair of the following ones:

$$k^2 - \frac{\omega^2}{c^2}\epsilon_{\perp}(\omega, \boldsymbol{k}) = 0, \tag{2.46}$$

$$\epsilon_{\parallel}(\omega, \boldsymbol{k}) = 0. \tag{2.47}$$

The formulas (2.43) and (2.44) clearly reveal that the equation (2.46) determines a transverse electromagnetic wave in which $\boldsymbol{E} \perp \boldsymbol{k}$. Indeed, the component $E_z = 0$ whereas $E_x, E_y \neq 0$. It is also seen from (2.43) that the ratio E_x/E_y can be arbitrary. In other words, polarization of transverse electromagnetic wave with $\boldsymbol{E} \perp \boldsymbol{k}$ in an isotropic plasma can be arbitrary: linear, circular or elliptical.

[4]This is valid for relatively weak fields (see the following chapter).

In a constant magnetic field $\boldsymbol{B}_0$ plasma becomes magnetoactive, and the tensor $\epsilon_{\alpha\beta}$ in the form (2.30) is no longer valid. Then the modes are determined by the equations (2.40), (2.41) in their general form. Choosing again the Cartesian frame x, y, z with z-axis directed along $\boldsymbol{k}$, we write the equations (2.40) in the following way:

$$\begin{aligned} (\frac{c^2k^2}{\omega^2} - \epsilon_{xx})E_x - \epsilon_{xy}E_y - \epsilon_{xz}E_z = 0 \\ -\epsilon_{yx}E_x + (\frac{c^2k^2}{\omega^2} - \epsilon_{yy})E_y - \epsilon_{yz}E_z = 0 \\ -\epsilon_{zx}E_x - \epsilon_{zy}E_y - \epsilon_{zz}E_z = 0 \end{aligned} \tag{2.48}$$

Evidently, the modes determined by this system are neither pure longitudinal nor strictly transverse. With few exceptions discussed in section 3.1, all the components, E_x, E_y, and E_z, are generally non-zero. Here we discuss the conditions at which normal waves in magnetoactive plasma are close to longitudinal.

At large c^2k^2/ω^2 and finite $\epsilon_{\alpha\beta}$ the first two equations in (2.48) become:

$$\frac{c^2k^2}{\omega^2}E_x = 0, \qquad \frac{c^2k^2}{\omega^2}E_y = 0.$$

It is seen from the above relations that $E_x \to 0$ and $E_y \to 0$ at $c^2k^2/\omega^2 \to \infty$, i.e. the mode becomes longitudinal. The third equation of the system (2.48) is then reduced to

$$\epsilon_{zz}E_z = 0. \tag{2.49}$$

The corresponding dispersion relation,

$$\epsilon_{zz}(\omega, \boldsymbol{k}) = 0, \tag{2.50}$$

ensures the existence of longitudinal waves with $E_z \neq 0$. It is clear that the sufficient condition which makes possible the transition from the dispersion relation (2.41), (2.48) to the equation (2.47) for longitudinal waves looks like $c^2k^2/\omega^2 \gg |\epsilon_{ij}|$ for any i, j.[5] Note that the equation (2.49) is similar to the equation (2.44) for isotropic plasma. The role of $\epsilon_{\parallel}$ is played by the component $\epsilon_{zz}(\omega, \boldsymbol{k})$.

Let us now find the equation for the electric field in such waves for arbitrary orientation of the coordinate axes with respect to the wave vector $\boldsymbol{k}$ assuming the sufficient condition for the existence of longitudinal waves to be satisfied. Multiplying the equation (2.40) by k_α and summing over α, we get

$$k_\alpha \epsilon_{\alpha\beta}(\omega, \boldsymbol{k})E_\beta = 0. \tag{2.51}$$

[5] Here and thereafter ϵ_{ij} denotes the components of the tensor $\epsilon_{\alpha\beta}$ in the frame i, j, k with k-axis along the wave vector $\mathbf{k}$.

Taking into account that the electric field is longitudinal, we substitute $\boldsymbol{E}= E_k\boldsymbol{k}/k$ into (2.51). As a result, this equation is transformed to

$$\frac{k_\alpha \epsilon_{\alpha\beta}(\omega, \boldsymbol{k}) k_\beta}{k^2} E_k = 0. \tag{2.52}$$

Note that (2.52) is formally similar to (2.44), the coefficient of E_k playing the role of $\epsilon_{\|}$. Therefore, we use the same notation for this coefficient:

$$\epsilon_{\|} = \frac{k_\alpha \epsilon_{\alpha\beta}(\omega, \boldsymbol{k}) k_\beta}{k^2} \tag{2.53}$$

Then, as before, the dispersion equation for longitudinal waves in magnetoactive plasma is written as

$$\epsilon_{\|}(\omega, \boldsymbol{k}) = 0. \tag{2.54}$$

It is easy to see that in the frame with z-axis directed along the vector $\boldsymbol{k}$ the expression (2.53) for $\epsilon_{\|}$ is reduced to ϵ_{zz}.

Dispersion equations considered in this section relate complex values of the frequency ω to complex wave vectors $\boldsymbol{k}$. If $\boldsymbol{k}$ is real for real ω, the medium is called transparent for the given mode at a given frequency. But if the wave vector $\boldsymbol{k}$ is complex for real ω, then $\boldsymbol{k}=\boldsymbol{k}'+i\boldsymbol{k}''$, where the vectors $\boldsymbol{k}'$ and $\boldsymbol{k}''$ are real. A wave $\exp(i\boldsymbol{k}\boldsymbol{r}-i\omega t)=\exp(-\boldsymbol{k}''\boldsymbol{r}+i\boldsymbol{k}'\boldsymbol{r}-i\omega t)$ with $\boldsymbol{k}' \parallel \boldsymbol{k}''$ is called homogeneous. In such a wave a plane of constant phase, $\boldsymbol{k}'\boldsymbol{r}=\mathrm{const}$, coincides with that of constant amplitude, $\boldsymbol{k}''\boldsymbol{r}=\mathrm{const}$.[6] Waves for which $\boldsymbol{k}'$ and $\boldsymbol{k}''$ are not parallel are called inhomogeneous. Inhomogeneous waves appear in a smoothly inhomogeneous medium or if homogeneous media are separated by sharp boundaries.

The wave vector $\boldsymbol{k}$ of a homogeneous wave with $\boldsymbol{k}' \parallel \boldsymbol{k}''$ can be written in the form $\boldsymbol{k}=k\boldsymbol{s}$ where $\boldsymbol{s}$ is the unit vector along $\boldsymbol{k}$ and

$$k = k' + ik'' = \frac{\omega}{c}(n + i\eta). \tag{2.55}$$

The values n and η introduced by this relation are called refraction and absorption indices, respectively. The value of k (and, therefore, n and η) at a given direction $\boldsymbol{s}$ can be found from the dispersion relation which is treated as a function of real frequency ω.

If, on the contrary, the frequency ω of a normal wave is complex for a real wave vector $\boldsymbol{k}$, we may write

$$\omega = \Omega - i\gamma, \tag{2.56}$$

[6] Evidently, waves with $\mathbf{k}''=0$ are always homogeneous.

where γ is called decrement. If $\gamma > 0$, the wave amplitude decreases in time as $e^{-\gamma t}$.

In the case of weak absorption, when either $\gamma \ll \Omega$ for real $\boldsymbol{k}$ or $\eta \ll n$ for real ω, the decrement γ and absorption index η are related by the simple formula [79]

$$\gamma = \frac{\Omega}{c}\eta v_{\mathrm{gr}\mathbf{k}}. \tag{2.57}$$

To derive it, one should start from the dispersion relation

$$\omega = \omega_j(\boldsymbol{ks}).$$

for real $k = n\Omega/c$. The corresponding frequency is then equal to $\omega = \Omega - i\gamma$. An infinitesimal increase dk yields the corresponding variation of the frequency:

$$\omega + d\omega = \omega_j((k + dk)\boldsymbol{s}).$$

Let us now choose the value of dk which makes $\omega + d\omega$ real. Putting $\omega + d\omega = \Omega$, or $d\omega = i\gamma$, leads to the required $dk = i\eta\Omega/c$. Finally, the relation between $d\omega$ and dk is provided by the linear term of power series expansion of the function $\omega_j(\boldsymbol{k})$:[7]

$$d\omega = \frac{d\omega_j}{d\boldsymbol{k}} \simeq \frac{d\Omega}{d\boldsymbol{k}}\boldsymbol{s}dk.$$

Taking into account the expressions for $d\omega$ and dk and recognizing the projection of the group velocity onto the direction of the wave vector $\boldsymbol{k}$: $(d\Omega/d\boldsymbol{k})\boldsymbol{s} = v_{\mathrm{gr}\mathbf{k}}$, we immediately arrive at the relation (2.57).

2.1.3. DIELECTRIC TENSOR OF PLASMA

Electromagnetic characteristics of plasma are determined by the specific form of constitutional relations $D'_\alpha = \epsilon_{\alpha\beta}(\omega, \boldsymbol{k})E_\beta$, i.e. dielectric tensor $\epsilon_{\alpha\beta}(\omega, \boldsymbol{k})$. More rigorous calculation of the components of this tensor yields more accurate and complete description of electromagnetic processes in plasma.

When calculating the tensor $\epsilon_{\alpha\beta}(\omega, \boldsymbol{k})$ one should take into account that, according to (2.24),

$$\boldsymbol{D}^0 = \boldsymbol{E} + i\frac{4\pi}{\omega}\boldsymbol{j}_{\mathrm{tot}}. \tag{2.58}$$

This expression for a process $\propto e^{i\mathbf{kr} - i\omega t}$ can be rewritten in the following way (see 2.29):

$$\epsilon_{\alpha\beta}(\omega, \boldsymbol{k})E_\beta = E_\alpha + i\frac{4\pi}{\omega}j_{\alpha\,\mathrm{tot}}. \tag{2.59}$$

---

[7]Note that the resulting formula (2.57) is applicable as long as this approximation at $dk = i(\Omega/c)n$ is valid.

The above expressions reveal that the form of the dielectric tensor $\epsilon_{\alpha\beta}(\omega, \boldsymbol{k})$ can be found using the expression for the total current $\boldsymbol{j}_{\text{tot}}$ induced in plasma by wave fields $\boldsymbol{E} \propto \exp(i\boldsymbol{kr} - i\omega t)$ and $\boldsymbol{B} = (c/\omega)[\boldsymbol{k} \times \boldsymbol{E}]$.

The total current $\boldsymbol{j}_{\text{tot}}$, in its turn, may be found using the equations of charged particle motion in the electromagnetic field. Depending on the required accuracy, the three following methods are typically applied to calculate the values $\boldsymbol{j}_{\text{tot}}$ and $\epsilon_{\alpha\beta}(\omega, \boldsymbol{k})$ in plasma: (i) elementary, (ii) quasi-hydrodynamic, and (iii) kinetic. The first, elementary, approach, completely ignores thermal motion of plasma particles. The quasi-hydrodynamic one somehow takes it into account by the use of electron pressure. The consistent description of plasma particle motion is only possible in the framework of the kinetic equation which explicitly accounts for the existence of a widespread spectrum of particle velocities at any elementary volume.

Prior to description of the above methods of derivation the dielectric tensor $\epsilon_{\alpha\beta}$ in plasma, we stress that the following consideration is restricted by the analysis of high-frequency electromagnetic processes in the range

$$\omega \gg \omega_{B\text{i}}, \tag{2.60}$$

where $\omega_{B\text{i}} = eB/m_{\text{i}}c$ is ion gyrofrequency. This criterion typically allows to neglect the motion of heavy ions ($m_{\text{i}} \gg m_{\text{e}}$) under the action of high-frequency fields $\boldsymbol{E}, \boldsymbol{B}$.[8] The values $\boldsymbol{j}_{\text{tot}}$ and $\epsilon_{\alpha\beta}$ are calculated in this case assuming that ion motion does not depend on the fields $\boldsymbol{E}$ and $\boldsymbol{B}$. This leads to remarkable simplification of the analysis, since the equation of motion for ions is not needed any more.

The calculation of the dielectric tensor is started here by the description of the elementary approach. According to this method, the total current density in a motionless plasma,

$$\boldsymbol{j}_{\text{tot}} = eN_0\boldsymbol{v}, \tag{2.61}$$

where N_0 is unperturbed electron density, is found by the solution of the non-relativistic equation of motion of an electron:[9]

$$m_{\text{e}}\dot{\boldsymbol{v}} = -e(\boldsymbol{E} + \frac{1}{c}[\boldsymbol{v} \times \boldsymbol{B}]) - \frac{e}{c}[\boldsymbol{v} \times \boldsymbol{B}_0]. \tag{2.62}$$

The right-hand side of this equation comprises the Lorenz force (2.18) on electron originating due to high-frequency fields $\boldsymbol{E}$ and $\boldsymbol{B}$ and the force

[8]The condition (2.60) is not always sufficient to neglect the contribution of ions to the dielectric tensor $\epsilon_{\alpha\beta}$. For example, in plasma with $\omega_{\text{L}}^2 \gg \omega_B^2$ and for waves propagating across the magnetic field $\mathbf{B}_0$ this is only possible under the more severe condition: $\omega \gg \sqrt{\omega_B\omega_{B\text{i}}} = \omega_B\sqrt{m_{\text{e}}/m_{\text{i}}}$ [71].

[9]The influence of collisions on $\epsilon_{\alpha\beta}$ can be taken into account by adding the term $-\nu_{\text{eff}}m_{\text{e}}\mathbf{v}$ to the right-hand side of the equation (2.62). This term describes the loss of electron momentum due to its collisions with other particles (see section 6.1).

$-(e/c)[\boldsymbol{v}\times\boldsymbol{B}_0]$ taking into account the influence of the static magnetic field $\boldsymbol{B}_0$ on the motion of plasma electrons.

Assuming the high-frequency field to be weak enough, we neglect the term $(e/c)[\boldsymbol{v}\times\boldsymbol{B}]$ in the equation (2.62). Such an approximation is justified by the resulting proportionality of $\boldsymbol{v}$ and $\boldsymbol{E}$. Since, according to the Maxwell equation (2.26), the field $\boldsymbol{B}$ is proportional to $\boldsymbol{E}$, the term $(e/c)[\boldsymbol{v}\times\boldsymbol{B}]\propto E^2$. Thus, as $\boldsymbol{E}\to 0$ the term $(e/c)[\boldsymbol{v}\times\boldsymbol{B}]$ in the equation (2.62) is negligible compared to the other ones.

Substituting now $\boldsymbol{E}\propto\exp(i\boldsymbol{kr}-i\omega t)$ into (2.62), one may solve the resulting equation

$$i\omega\boldsymbol{v}=\frac{e\boldsymbol{E}}{m_{\rm e}}+[\boldsymbol{v}\times\boldsymbol{\omega}_B], \tag{2.63}$$

in which $\boldsymbol{\omega}_B=e\boldsymbol{B}_0/m_{\rm e}c$. The desired solution is readily found by taking scalar and vector products of equation (2.63) and the vector $\boldsymbol{\omega}_B$ [80]:

$$\begin{aligned} i\omega[\boldsymbol{v}\times\boldsymbol{\omega}_B] &= \frac{e}{m_{\rm e}}[\boldsymbol{E}\times\boldsymbol{\omega}_B]-\boldsymbol{v}\omega_B^2+\boldsymbol{\omega}_B(\boldsymbol{v}\,\boldsymbol{\omega}_B)\\ i\omega(\boldsymbol{v}\,\boldsymbol{\omega}_B) &= \frac{e}{m_{\rm e}}(\boldsymbol{E}\,\boldsymbol{\omega}_B) \end{aligned} \tag{2.64}$$

Excluding the products $[\boldsymbol{v}\times\boldsymbol{\omega}_B]$ and $(\boldsymbol{v}\,\boldsymbol{\omega}_B)$ from (2.63), (2.64) we then find:

$$\boldsymbol{v}=\frac{ie\omega}{m_{\rm e}(\omega^2-\omega_B^2)}\left(\frac{\boldsymbol{\omega}_B(\boldsymbol{E}\,\boldsymbol{\omega}_B)}{\omega^2}-\boldsymbol{E}\right)-\frac{e}{m_{\rm e}}\frac{[\boldsymbol{E}\times\boldsymbol{\omega}_B]}{\omega^2-\omega_B^2}. \tag{2.65}$$

Substituting the obtained velocity $\boldsymbol{v}$ into the expression (2.61) and then the resulting total current into the formula (2.58) for the electric induction vector $\boldsymbol{D}'$ we finally obtain:

$$\boldsymbol{D}'=\left(1-\frac{\omega_{\rm L}^2}{\omega^2-\omega_B^2}\right)\boldsymbol{E}+\frac{\omega_{\rm L}^2}{\omega^2-\omega_B^2}\frac{\boldsymbol{\omega}_B(\boldsymbol{E}\,\boldsymbol{\omega}_B)}{\omega^2}+\frac{i\omega_{\rm L}^2}{\omega^2-\omega_B^2}\frac{[\boldsymbol{E}\times\boldsymbol{\omega}_B]}{\omega}. \tag{2.66}$$

Here $\omega_{\rm L}=\sqrt{4\pi e^2N/m_{\rm e}}$ is the plasma frequency.

With (2.29) and (2.66), it is possible to derive the plasma dielectric tensor $\epsilon_{\alpha\beta}(\omega,\boldsymbol{k})$. Its components in the Cartesian frame I (Fig. 2.1, a) with z-axis orientated along the magnetic field $\boldsymbol{B}_0$ look like:[10]

$$\epsilon_{\alpha\beta}=\begin{pmatrix}\epsilon_1 & ig & 0\\ -ig & \epsilon_1 & 0\\ 0 & 0 & \epsilon_2\end{pmatrix}, \tag{2.67}$$

[10]Since the tensor $\epsilon_{\alpha\beta}$ in the form (2.67) does not depend on $\mathbf{k}$, the rotation of the frame I around z-axis does not alter its components. Note also that the gyrotropy of electron plasma is determined by the factor g in the tensor (2.67). This factor depends on the sign of particle charge. For example, in electron-positron plasma with equal densities of the species the factor $g=0$ and gyrotropy vanishes.

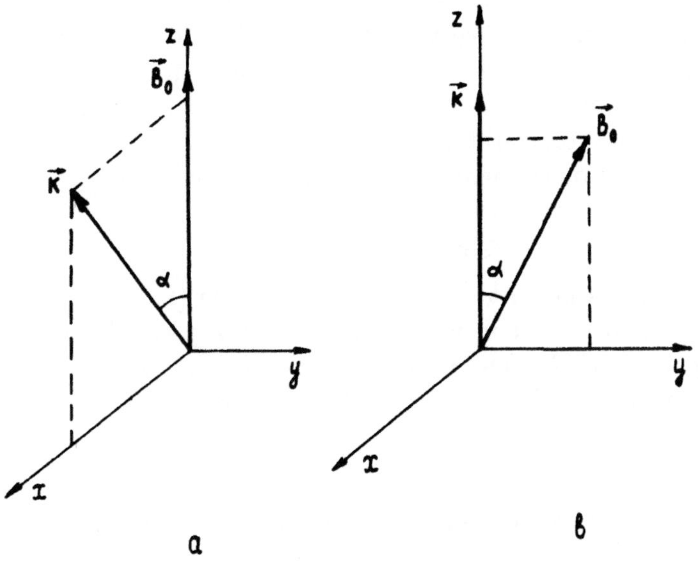

Figure 2.1. Orientation of the magnetic field vector $\mathbf{B}_0$ and the wave vector $\mathbf{k}$ with respect to the coordinate axes: *a*, the frame I; *b*, the frame II

where

$$\epsilon_1 = 1 - \frac{v}{1-u}, \qquad \epsilon_2 = 1 - v, \qquad g = \frac{v\sqrt{u}}{1-u}. \tag{2.68}$$

The notations

$$v = \frac{\omega_L^2}{\omega^2}, \qquad u = \frac{\omega_B^2}{\omega^2}, \tag{2.69}$$

which will be extensively used thereafter, are introduced in the above formulas. In what follows the components of the dielectric tensor in the coordinate system II with the magnetic field $\boldsymbol{B}_0$ lying in the yz-plane (see Fig. 2.1, b) will also be used. These components may be found by either writing the relation (2.66) in coordinate form or transforming the components of the tensor (2.67) according to the rotation of the coordinate system. Passing over the details of this procedure (see [71]), we just give the resulting components of the desired tensor:

$$\begin{aligned} \epsilon_{xx} &= 1 - \frac{v}{1-u}, & \epsilon_{yy} &= 1 - \frac{v(1-u\sin^2\alpha)}{1-u}, \\ \epsilon_{zz} &= 1 - \frac{v(1-u\cos^2\alpha)}{1-u}, & \epsilon_{xy} &= -\epsilon_{yx} = i\frac{v\sqrt{u}\cos\alpha}{1-u}, \\ \epsilon_{xz} &= -\epsilon_{zx} = -i\frac{v\sqrt{u}\sin\alpha}{1-u}, & \epsilon_{yz} &= \epsilon_{zy} = \frac{uv\cos\alpha\sin\alpha}{1-u}, \end{aligned} \tag{2.70}$$

where α is the angle between $\boldsymbol{k}$ and $\boldsymbol{B}_0$. It is clear from (2.67)-(2.70) that the components of the dielectric tensor found by the elementary approach do not depend on the wave vector $\boldsymbol{k}$. Therefore, no spatial dispersion appears in the framework of this theory. Rather, the dependence of $\epsilon_{\alpha\beta}$ on the frequency ω indicates for the time dispersion. Therefore, the relation $D'_\alpha = \epsilon_{\alpha\beta}E_\beta$ is valid not only for $E_\beta \propto \exp(i\boldsymbol{kr})$ but for a monochromatic field E_β of arbitrary spatial configuration.

In the absence of the magnetic field $\boldsymbol{B}_0$ the tensor $\epsilon_{\alpha\beta}(\omega)$ becomes diagonal:

$$\epsilon_{\alpha\beta}(\omega) = (1 - v)\delta_{\alpha\beta}. \tag{2.71}$$

This expression can be obtained by putting $u = 0$ in (2.67), (2.68), (2.70). Note that the above components are identical in any coordinate frame.[11] Next, comparison of (2.71) with the general expression (2.30) for $\epsilon_{\alpha\beta}(\omega, \boldsymbol{k})$ shows that in the considered case of isotropic plasma

$$\epsilon_{\parallel} = \epsilon_{\perp} = 1 - v. \tag{2.72}$$

Taking into account the relations (2.36) and (2.38) we find that

$$\epsilon' = 1 - v = 1 - \frac{\omega_L^2}{\omega^2}, \qquad \mu = 1. \tag{2.73}$$

The formulas (2.67) for the components of the tensor $\epsilon_{\alpha\beta}$ and the one following from these have been derived in the framework of the elementary theory which only takes into account the forced motion of plasma particles under the action of a high-frequency field. This means that the theory developed is valid for a static "cold" plasma in which thermal motion is neglected.

Thermal motion can be approximately taken into account by the quasi-hydrodynamic description of a non-relativistic plasma. It models electron and ion components of plasma as "fluids" freely penetrating each other and interacting via electromagnetic fields. Each "fluid" obeys the corresponding hydrodynamic equations: Euler and continuity ones. Such equations for "electron fluid" can be written the following way:

$$\begin{aligned} \frac{\partial \boldsymbol{v}}{\partial t} + (\boldsymbol{v}\nabla)\boldsymbol{v} &= -\frac{\nabla p}{m_e N} - \frac{e}{m_e}(\boldsymbol{E} + \frac{1}{c}[\boldsymbol{v} \times \boldsymbol{B}]), \\ \frac{\partial N}{\partial t} + \nabla(N\boldsymbol{v}) &= 0. \end{aligned} \tag{2.74}$$

Here $\boldsymbol{v}$ is the velocity of the electron component, N and p are electron density and pressure. The total current density $\boldsymbol{j}_{tot}$ is expressed via electron

[11] It is clear from (2.69), (2.70) that in the framework of the elementary theory the influence of the magnetic field $\mathbf{B}_0$ on the tensor $\epsilon_{\alpha\beta}$ (i.e., on electromagnetc wave propagation in plasma) can be neglected if $\omega_B \ll \omega$. There are, however, some effects, for example, the Faraday effect in magnetoactive plasma, for which this condition is insufficient (see section 4.2).

and ion densities N, $N_{\rm i}$ and velosities $\boldsymbol{v}$, $\boldsymbol{v}_{\rm i}$ as

$$\boldsymbol{j}_{\rm tot} = -eN\boldsymbol{v} + eN_{\rm i}\boldsymbol{v}_{\rm i} \tag{2.75}$$

The equations similar to (2.74) hold also for ion "fluid". Since the contribution of ions to the current density induced by a high-frequency field is negligible (see the remark following the condition (2.60)), the equations of ion motion are not used thereon. We also omit the details of the quasi-hydrodynamic procedure and refer the reader to [3, 71]. Note only that the method discussed accounts for thermal motion of plasma particles in a quite simplified way, with electron pressure only. The quasi-hydrodynamic description thus yields a rather approximate description of the effects of thermal motion which is only valid in some specific cases.

The consistent account of thermal motion can only be provided with the kinetic method based on the Boltzmann–Vlasov equation for particle distribution function $f(\boldsymbol{p}, \boldsymbol{r}, t)$, where $\boldsymbol{p}$ and $\boldsymbol{r}$ are the momentum and radius-vector of a particle. This function determines particle density at a point $(\boldsymbol{p}, \boldsymbol{r})$ of the phase space at a time t. The number of particles in a phase volume element near this point equals

$$dN = f(\boldsymbol{p}, \boldsymbol{r}, t) d^3r d^3p.$$

Here $d^3r = dV$ and d^3p are volume elements in coordinate and momentum spaces, respectively. The density of particles (electrons) is equal to $N = \int f(\boldsymbol{p}, \boldsymbol{r}, t) d^3p$.

The kinetic equation valid even for relativistic plasma electrons is written in the form

$$\frac{\partial f}{\partial t} + \boldsymbol{v}\frac{\partial f}{\partial \boldsymbol{r}} = e\left(\boldsymbol{E} + \frac{1}{c}[\boldsymbol{v} \times \boldsymbol{B}]\right)\frac{\partial f}{\partial \boldsymbol{p}} - I_{\rm st}, \tag{2.76}$$

where the last term $I_{\rm st}$ denotes the so called integral of collisions. This term describes the variation of particle distribution function due to collisions of neighboring particles. The corresponding equation for the ion component is not considered in this book due to the above mentioned reasons. The influence of high-frequency fields on ions is neglected, so ion motion and the corresponding distribution function $f_{\rm i}$ are assumed to be given.

The current density $\boldsymbol{j}_{\rm tot}$ which enters the expression (2.24) for the electric induction $\boldsymbol{D}'$ is related to the distribution function $f(\boldsymbol{p}, \boldsymbol{r}, t)$ as

$$\boldsymbol{j}_{\rm tot} = -e\int \boldsymbol{v}\,(f - f_{\rm i})\, d^3p. \tag{2.77}$$

The equation (2.76) is non-linear, and it is quite difficult to obtain its solution. However, this equation for weak electromagnetic perturbations may be linearized. This is done by assuming $f = f_0 + f'$, where $f_0(\boldsymbol{p}, \boldsymbol{r})$ is steady-state unperturbed distribution function obeying the equation

$$\boldsymbol{v}\frac{\partial f_0}{\partial \boldsymbol{r}} - e\left(\boldsymbol{E}_0 + \frac{1}{c}[\boldsymbol{v}\times\boldsymbol{B}_0]\right)\frac{\partial f_0}{\partial \boldsymbol{p}} + I_0 = 0, \qquad (2.78)$$

and f' is the small perturbation caused by the high-frequency fields $\boldsymbol{E}'$, $\boldsymbol{B}'$. Taking into account (2.78) and keeping in (2.76) the linear terms with respect to $\boldsymbol{E}'$, $\boldsymbol{B}'$, and f', we get the following linearized equation:

$$\frac{\partial f'}{\partial t} + \boldsymbol{v}\frac{\partial f'}{\partial \boldsymbol{r}} - e\left(\boldsymbol{E}_0 + \frac{1}{c}[\boldsymbol{v}\times\boldsymbol{B}_0]\right)\frac{\partial f'}{\partial \boldsymbol{p}} = e\left(\boldsymbol{E}' + \frac{1}{c}[\boldsymbol{v}\times\boldsymbol{B}']\right)\frac{\partial f_0}{\partial \boldsymbol{p}} - I', \qquad (2.79)$$

in which $\boldsymbol{E}_0$ and $\boldsymbol{B}_0$ are stationary fields obeying the system of Maxwell equations (2.19)-(2.22) with the current $\boldsymbol{j}_{0\,\mathrm{tot}} = -e\int \boldsymbol{v}(f_0 - f_{\mathrm{i}})d^3p$ (see (2.77)).

In many cases the effect of collisions on electromagnetic disturbances in equilibrium plasma with Maxwell distribution function f_0 can be described rather accurately using the perturbed collision integral in the form $I' = \nu_{\mathrm{eff}} f'$ with relevant value of effective collisional frequency ν_{eff}.[12] With this approximation the analysis of dissipative processes in plasma with the kinetic equation (2.79) is greatly simplified (see [71,78,81]).

The following procedure [82] provides a convenient way to derive the dielectric tensor $\epsilon_{\alpha\beta}(\omega, \boldsymbol{k})$ of a homogeneous, collisionless ($\nu_{\mathrm{eff}} = 0$) plasma using the kinetic equation (2.79). Consider the temporal variation of the function $f'(\boldsymbol{p}, \boldsymbol{r}, t)$ along the unperturbed trajectory of a particle. Such a trajectory is described by the equations

$$\frac{d\boldsymbol{r}}{dt} = \boldsymbol{v}(t), \qquad \frac{d\boldsymbol{p}}{dt} = -e\boldsymbol{E} - \frac{e}{c}[\boldsymbol{v}(t)\times\boldsymbol{B}_0] \qquad (2.80)$$

with initial conditions $\boldsymbol{r}(t_0) = \boldsymbol{r}_0$, $\boldsymbol{v}(t_0) = \boldsymbol{v}_0$.

The full derivative of the function $f'(\boldsymbol{p}(t), \boldsymbol{r}(t), t)$ is given by

$$\frac{df'}{dt} = \frac{\partial f'}{\partial t} + \frac{d\boldsymbol{r}}{dt}\frac{\partial f'}{\partial \boldsymbol{r}} + \frac{d\boldsymbol{p}}{dt}\frac{\partial f'}{\partial \boldsymbol{p}}. \qquad (2.81)$$

It follows from (2.79) and (2.80) that df'/dt can be presented in the form

$$\frac{df'}{dt} = e\left\{\boldsymbol{E}'(\boldsymbol{r}(t), t) + \frac{1}{c}[\boldsymbol{v}(t)\times\boldsymbol{B}'(\boldsymbol{r}(t), t)]\right\}\frac{\partial f_0}{\partial \boldsymbol{p}}.$$

[12]This expression describes the variation of the function f' per unit time due to particle loss from unit phase space volume caused by collisions with effective frequency ν_{eff}.

If $f' \to 0$ at $t \to -\infty$, the solution of this equation looks like:[13]

$$f'(\boldsymbol{p}(t), \boldsymbol{r}(t), t) =$$
$$e\int_{-\infty}^{t}\left\{\boldsymbol{E}'(\boldsymbol{r}'(\tilde{t}),\tilde{t}) + \frac{1}{c}[\boldsymbol{v}(\tilde{t})\times\boldsymbol{B}'(\boldsymbol{r}'(\tilde{t}),\tilde{t})]\right\}\frac{\partial f_0}{\partial \boldsymbol{p}(\tilde{t})}\,d\tilde{t}. \quad (2.82)$$

The desired tensor $\epsilon_{\alpha\beta}(\omega, \boldsymbol{k})$ is found using the expression for the total current $\boldsymbol{j}'_{\rm tot} = -e\int \boldsymbol{v} f' d^3p$ which is induced in plasma by high-frequency fields $\boldsymbol{E}', \boldsymbol{B}' \propto \exp[i\boldsymbol{k}\boldsymbol{r} - i\omega t]$. Substituting $\boldsymbol{E}' = \boldsymbol{E}_{\rm a}\exp(i\boldsymbol{k}\boldsymbol{r}(t) - i\omega t)$ and taking into account the relation $\boldsymbol{B}' = (c/\omega)[\boldsymbol{k} \times \boldsymbol{E}']$ (see (2.26)), we get

$$j'_{\alpha\,\rm tot}(\boldsymbol{r}(t), t) = -e^2 E_{\rm a\beta}\exp[i\boldsymbol{k}\boldsymbol{r}(t) - i\omega t]\times$$
$$\int d^3p(t)\int_{-\infty}^{t} d\tilde{t}\, v_\alpha(t)\exp\{i\boldsymbol{k}[\boldsymbol{r}(\tilde{t}) - \boldsymbol{r}(t)] - i\omega(\tilde{t} - t)\}\times$$
$$\left\{(1 - \frac{k_\gamma v_\gamma(\tilde{t})}{\omega})\frac{\partial f_0}{\partial p_\beta(\tilde{t})} + \frac{k_\gamma v_\beta(\tilde{t})}{\omega}\frac{\partial f_0}{\partial p_\gamma(\tilde{t})}\right\}. \quad (2.83)$$

Introducing the new variable in (2.83): $\tilde{t} \to \xi = t - \tilde{t}$, we arrive at the integral over ξ from 0 to infinity. Changing then the order of integration, we notice that the integral over ξ should not explicitly depend on t for arbitrary stationary distribution f_0. Indeed, any possible temporal variation of the current $\boldsymbol{j}'_{\rm tot}$ similar to the one of the field $\boldsymbol{E}'(t)$ is always taken into account by the factor before the integral. Since the limits of the integral over ξ do not depend on t, the integrand should also not depend on t. Therefore, the value of t in this expression can be chosen arbitrarily. It is convenient to put $\xi = t$. Then, taking into account (2.80), we yield from (2.83), (2.59) the tensor $\epsilon_{\alpha\beta}$ in a relativistic plasma:

$$\epsilon_{\alpha\beta}(\omega, \boldsymbol{k}) = \delta_{\alpha\beta} - i\frac{4\pi e^2}{\omega}\int d^3p_0\int_0^{\infty} dt\, v_\alpha\times$$
$$\exp\left(i\omega t - i\int_0^t \boldsymbol{k}\boldsymbol{v}(t')dt'\right)\left\{\left(1 - \frac{\boldsymbol{k}\boldsymbol{v}_0}{\omega}\right)\frac{\partial f_0}{\partial p_{0\beta}} + \frac{\boldsymbol{k}v_{0\beta}}{\omega}\frac{\partial f_0}{\partial \boldsymbol{p}_0}\right\}. \quad (2.84)$$

Here $\boldsymbol{p}_0$ and $\boldsymbol{v}_0$ are particle velocity and momentum at $t = 0$, and $\boldsymbol{v}(t)$ is the unperturbed velocity given by the equations (2.80).

Since the field dependence on coordinates and time has been chosen in the form $\exp(i\boldsymbol{k}\boldsymbol{r}(t) - i\omega t)$, the relation (2.84) has been proved only along a given trajectory of the unperturbed motion which originates at a point $\boldsymbol{v}(t_0) = \boldsymbol{v}_0$, $\boldsymbol{r}(t_0) = \boldsymbol{r}_0$. However, by varying the initial values $\boldsymbol{r}_0$ and $\boldsymbol{v}_0$ one can prove the validity of (2.84) at any $\boldsymbol{r}$ and t.

[13] We have used tilde to differentiate the integration variable $\tilde{t}$ from its value t at the upper boundary of integration range.

If plasma is comprised of several particle species (electrons, ions, etc.) or if there is single species which can be conveniently separated into several components with different distributions f_0, the tensor $\epsilon_{\alpha\beta}$ should be derived by summing the second term in (2.84) over all species (or components).

Note that (2.84) has been obtained under the initial condition $f' \to 0$ at $t \to -\infty$. This means that $\mathrm{Im}\,\omega > 0$, so that the wave is growing in time, and the decrement $\gamma = -\mathrm{Im}\,\omega$ is negative. For damping waves with $\mathrm{Im}\,\omega < 0$ the corresponding dielectric tensor should be considered as analytical continuation of the function $\epsilon_{\alpha\beta}(\omega, k)$ in the form (2.84) onto the lower part of the plane of complex ω. Such a rule for derivation of $\epsilon_{\alpha\beta}(\omega, k)$ in the region $\mathrm{Im}\,\omega < 0$ is justified by the rigorous solution of the problem of generation of a current $\boldsymbol{j}_{\mathrm{tot}}(\boldsymbol{r}, t)$ in plasma with instantaneous appearance of the fields at $t = t_0$ (see in this respect [83]).

The tensor (2.84) is simplified for isotropic plasma where $\boldsymbol{E}_0 = 0$, $\boldsymbol{B}_0 = 0$, and the distribution function $f_0(\boldsymbol{p})$ does not depend on the direction of $\boldsymbol{p}$. In such a plasma $\boldsymbol{v}(t) = \mathrm{const} = \boldsymbol{v}_0$ and

$$\epsilon_{\alpha\beta}(\omega, \boldsymbol{k}) = \delta_{\alpha\beta} - i\frac{4\pi e^2}{\omega} \int d^3p_0 \int_0^{\infty} v_\alpha^0 \frac{\partial f_0}{\partial p_{0\beta}} \exp\left[i(\omega - \boldsymbol{k}\boldsymbol{v}_0)t\right] dt.$$

Here the relations

$$\frac{\partial f_0}{\partial \boldsymbol{p}} = \frac{\partial f_0}{\partial p}\frac{\boldsymbol{p}}{p}, \qquad \frac{\partial f_0}{\partial p_\alpha} = \frac{\partial f_0}{\partial p}\frac{p_\alpha}{p}$$

have been taken into account. Then, at $\mathrm{Im}\,\omega > 0$ the integral

$$\int_0^{\infty} \exp\left[i(\omega - \boldsymbol{k}\boldsymbol{v}_0)t\right] dt = \frac{i}{\omega - \boldsymbol{k}\boldsymbol{v}_0},$$

so [14]

$$\epsilon_{\alpha\beta}(\omega, \boldsymbol{k}) = \delta_{\alpha\beta} + \frac{4\pi e^2}{\omega} \int \frac{v_\alpha \dfrac{\partial f_0}{\partial p_\beta}}{\omega - \boldsymbol{k}\boldsymbol{v}} d^3p_0 \tag{2.85}$$

Comparison of (2.85) and (2.30) allows one to write down the longitudinal and transverse dielectric permittivities of isotropic plasma [78]:

$$\begin{aligned} \epsilon_{\|}(\omega, \boldsymbol{k}) &= 1 + \frac{4\pi e^2}{\omega k^2} \int \frac{(\boldsymbol{k}\boldsymbol{v})^2}{\omega - \boldsymbol{k}\boldsymbol{v}} \frac{df_0}{d\mathcal{E}} d^3p, \\ \epsilon_{\perp}(\omega, \boldsymbol{k}) &= 1 + \frac{2\pi e^2}{\omega k^2} \int \frac{[\boldsymbol{k}\times\boldsymbol{v}]^2}{\omega - \boldsymbol{k}\boldsymbol{v}} \frac{df_0}{d\mathcal{E}} d^3p. \end{aligned} \tag{2.86}$$

[14] The subscript "0" has dropped for velocities and momenta to simplify the resulting formula.

Here $df_0/d\mathcal{E}$ is the derivative over particle energy; evidently, $\partial f_0/\partial \boldsymbol{p} = \boldsymbol{v} df_0/d\mathcal{E}$.

The formulas (2.86) have been obtained for the fields growing in time: $\mathrm{Im}\,\omega > 0$, so the corresponding integrals have been taken over the real momenta $\boldsymbol{p}$. But if $\mathrm{Im}\,\omega < 0$, the pole $v_{\mathbf{k}} = \omega/k$ shifts from the upper to the lower half-plane of complex $p_{\mathbf{k}}$ (the values $v_{\mathbf{k}}$ and $p_{\mathbf{k}}$ are the projections of $\boldsymbol{v}$ and $\boldsymbol{p}$ onto $\boldsymbol{k}$). To obtain the analytical continuation of $\epsilon_{\|}$, $\epsilon_{\perp}$ from the region $\mathrm{Im}\,\omega > 0$ to $\mathrm{Im}\,\omega < 0$, one should demand that the pole shifting into the lower half-plane ω does not cross G, the contour of intergation. To assure it, the contour G at $k > 0$ should always envelope the pole from below.

For non-relativistic plasma with Maxwell velocity distribution

$$F_0(\boldsymbol{v}) = \frac{N_0}{(2\pi\kappa T/m_{\mathrm{e}})^{3/2}} \exp\left(-\frac{m_{\mathrm{e}} v^2}{2\kappa T}\right) \tag{2.87}$$

the functions $\epsilon_{\|}$ and $\epsilon_{\perp}$ take the form [15]

$$\epsilon_{\|}(\omega, \boldsymbol{k}) = 1 + \frac{\omega_{\mathrm{L}}^2}{k^2 v_T^2}\left(1 - \frac{Z}{\sqrt{\pi}} \int_G \frac{e^{-y^2}}{Z - y}\, dy,\right) \tag{2.88}$$

$$\epsilon_{\perp}(\omega, \boldsymbol{k}) = 1 - \frac{\omega_{\mathrm{L}}^2}{k^2 v_T^2} \frac{1}{2\sqrt{\pi} Z} \int_G \frac{e^{-y^2}}{Z - y}\, dy, \tag{2.89}$$

where $v_T = \sqrt{\kappa T/m_{\mathrm{e}}}$, $y = v_{\mathbf{k}}/\sqrt{2} v_T$, $Z = \omega/\sqrt{2} k v_T$. To transform (2.86) to the last formulas, the relation $f_0(\boldsymbol{p}) d^3 p = F_0(\boldsymbol{v}) d^3 v$ has been taken into account. Since $p = m_{\mathrm{e}} \boldsymbol{v}$, the function $f_0(\boldsymbol{p}) = m_{\mathrm{e}}^{-3} F_0(\boldsymbol{v})$. The integral entering (2.88), (2.89) equals

$$\int_G \frac{e^{-y^2}}{Z - y}\, dy = -i\pi w(Z), \tag{2.90}$$

where

$$w(Z) = e^{-Z^2}\left(1 + \frac{2i}{\sqrt{\pi}} \int_0^Z e^{x^2}\, dx\right). \tag{2.91}$$

is the Kramp function, the probability integral of complex argument. The following series represents this function at $|Z| \ll 1$:

$$w(Z) = 1 + \frac{2i}{\sqrt{\pi}} Z + \ldots, \tag{2.92}$$

[15] It is not difficult to check that $\epsilon_{\|}$ and $\epsilon_{\perp}$ become identical and $\mu \to 1$ in the limit $|Z| \gg 1$, $|\mathrm{Im} Z| \ll 1$. At $|Z| \lesssim 1$, however, $\epsilon_{\|}$ and $\epsilon_{\perp}$ are different and, therefore, the corresponding magnetic susceptibility $\mu \neq 1$.

while the asymptotic formula

$$w(Z) = e^{-Z^2} + \frac{i}{\sqrt{\pi}}\left(\frac{1}{Z} + \frac{1}{2Z^3} + \frac{3}{4Z^5} + \ldots\right) \tag{2.93}$$

holds at $|Z| \gg 1$, $|\text{Im}\, Z| \ll 1$.

If isotropic plasma with Maxwellian velocity distribution moves as a whole with the bulk velocity $\boldsymbol{v}_0$, the corresponding distribution in a rest "laboratory" frame is obtained from (2.87) by using $(\boldsymbol{v} - \boldsymbol{v}_0)^2$ instead of v^2 in the exponent. For waves with $\boldsymbol{k} \parallel \boldsymbol{v}_0$ the form of $\epsilon_\perp$ is not changed (see (2.86), (2.89)), while $\epsilon_\parallel$ can be found from (2.88) by the ansatz $Z = (\omega - kv_0)/\sqrt{2}kv_T$.

In a constant magnetic field $\boldsymbol{B}_0$ the unperturbed particle trajectories are helical lines with axes parallel to the magnetic field. Bearing this in mind and transforming appropriately the expression (2.84) for the tensor $\epsilon_{\alpha\beta}$, one can derive the dielectric tensor of a non-equilibrium magnetoactive plasma in the coordinate system I (Fig. 2.1, a):

$$\epsilon_{\alpha\beta}(\omega, \boldsymbol{k}) = \delta_{\alpha\beta} - i^{\alpha+\beta} L_{\alpha\beta}(\omega, \boldsymbol{k}) \tag{2.94}$$

(see, e.g., [84]). The following notations are adoped in this expression:

$$L_{\alpha\beta} = \sum_{l=-\infty}^{\infty} \int_0^{\infty} dp_\perp \int_{-\infty}^{-\infty} dp_\parallel \frac{D^l_{\alpha\beta}(\omega, \boldsymbol{k}, p_\parallel, p_\perp)}{\omega - \omega_l(\boldsymbol{k}, p_\parallel, p_\perp)},$$

$$D^l_{\alpha\beta} = \frac{8\pi^2 e^2 p_\perp}{m_e^2 \omega^2} \times \tag{2.95}$$

$$\left\{ \begin{array}{ccc} Ap_\perp J_l^2 l^2 \xi^{-2} & -Ap_\perp J_l J_l' l \xi^{-1} & Cp_\perp J_l^2 l^2 \xi^{-2} - Bp_\perp J_l^2 l \xi^{-1} \\ Ap_\perp J_l J_l' l \xi^{-1} & -Ap_\perp J_l'^2 & Cp_\perp J_l J_l' l \xi^{-1} - Bp_\perp J_l J_l' \\ -Ap_\parallel J_l^2 l \xi^{-1} & Ap_\parallel J_l J_l' & -Cp_\parallel J_l^2 l \xi^{-1} + Bp_\parallel J_l^2 \end{array} \right\},$$

$$A = (m\omega - k_\parallel p_\parallel)\frac{\partial f_0}{\partial p_\perp} + k_\parallel p_\perp \frac{\partial f_0}{\partial p_\parallel},$$

$$B = m\omega \frac{\partial f_0}{\partial p_\parallel}, \qquad C = k_\perp \left(p_\perp \frac{\partial f_0}{\partial p_\parallel} - p_\parallel \frac{\partial f_0}{\partial p_\perp}\right),$$

$$\omega_l = l\Omega_B + k_\parallel \frac{p_\parallel}{m}, \qquad \xi = \frac{k_\perp v_\perp}{m\Omega_B}.$$

In the above expressions J_l and J_l' are Bessel function of the argument ξ and its derivative; $p_\parallel$ and $p_\perp$ are projections of the momentum $\boldsymbol{p}$ onto $\boldsymbol{B}_0$ and onto the plane perpendicular to this vector, respectively; the same meanings hold for the projections $k_\parallel$ and $k_\perp$ of the wave vector $\boldsymbol{k}$; finally, $\Omega_B = eB_0/mc$ is the relativistic gyrofrequency, $m = m_e/\sqrt{1-\beta^2}$ is electron relativistic mass and $\beta = v/c$. The unperturbed distribution function over electron momenta $f_0(\boldsymbol{p})$ is normalized in the following way:

$$\int f_0(\boldsymbol{p})d^3p = 2\pi \int_{-\infty}^{-\infty} dp_{\|} \int_0^{\infty} p_{\perp} f(p_{\|}, p_{\perp}) dp_{\perp} = N_0.$$

The longitudinal dielectric permeability in the adopted frame I can be written as

$$\epsilon_{\|} = \epsilon_{xx} \sin^2 \alpha + \epsilon_{zz} \cos^2 \alpha + (\epsilon_{xz} + \epsilon_{zx}) \sin \alpha \cos \alpha, \qquad (2.96)$$

where α is the angle between $\boldsymbol{k}$ and $\boldsymbol{B}_0$. Together with (2.94), (2.95), this formula yields:

$$\epsilon_{\|} = 1 + \frac{8\pi^2 e^2 m_e}{k^2} \sum_{l=-\infty}^{\infty} \int_0^{\infty} dp_{\perp} \int_{-\infty}^{+\infty} dp_{\|} J_l^2 \left(\frac{k_{\perp} v_{\perp}}{m\Omega_B}\right) \times \frac{k_{\|} p_{\perp} \dfrac{\partial f_0}{\partial p_{\|}} + lm\Omega_B \dfrac{\partial f_0}{\partial p_{\perp}}}{m\omega - lm\Omega_B - k_{\|} p_{\|}} \qquad (2.97)$$

(see, e.g., [85]).

If $\operatorname{Im}\omega > 0$, the integrals in (2.94) and (2.97) should be taken over the real values of $p_{\|}$ and $p_{\perp}$. The analytical continuation of $\epsilon_{\alpha\beta}(\omega, \boldsymbol{k})$ and $\epsilon_{\|}(\omega, \boldsymbol{k})$ onto the region $\operatorname{Im}\omega < 0$ is performed by the deformation of the integration contour in the complex $p_{\|}$- or $p_{\perp}$-plane in such a way that the poles of the corresponding integrand, while shifting to the lower half-plane, do not cross this contour.

For equilibrium, non-relativistic plasma with Maxwell distribution function (2.87) the integrals over $p_{\|}$ and $p_{\perp}$ in the tensor $\epsilon_{\alpha\beta}$ can be evaluated, and its components get the following form [86]:

$$\begin{aligned}
\epsilon_{xx} &= 1 + \sum_{l=-\infty}^{\infty} \frac{i\omega_{\mathrm{L}}^2}{\omega k_{\|} v_T} \sqrt{\frac{\pi}{2}} w_l e^{-\chi} \frac{l^2}{\chi} I_l, \\
\epsilon_{xy} = -\epsilon_{yx} &= -\sum_{l=-\infty}^{\infty} \frac{\omega_{\mathrm{L}}^2}{\omega k_{\|} v_T} \sqrt{\frac{\pi}{2}} w_l e^{-\chi} l (I_l - I_l'), \\
\epsilon_{yy} &= \epsilon_{xx} + \sum_{l=-\infty}^{\infty} \frac{i\omega_{\mathrm{L}}^2}{\omega k_{\|} v_T} \sqrt{\frac{\pi}{2}} w_l e^{-\chi} 2\chi (I_l - I_l'), \\
\epsilon_{xz} = \epsilon_{zx} &= \sum_{l=-\infty}^{\infty} \frac{\omega_{\mathrm{L}}^2}{\omega\omega_B} \frac{k_{\perp}}{k_{\|}} (1 + i\sqrt{\pi} w_l Z_l) \frac{l}{\chi} e^{-\chi} I_l,
\end{aligned} \qquad (2.98)$$

$$\epsilon_{yz} = -\epsilon_{zy} = -\sum_{l=-\infty}^{\infty} \frac{i\omega_{\rm L}^2}{\omega\omega_B}\frac{k_\perp}{k_\parallel}(1+i\sqrt{\pi}w_l Z_l)e^{-\chi}(I_l - I_l'),$$

$$\epsilon_{zz} = 1 + \sum_{l=-\infty}^{\infty} \frac{\sqrt{2}\omega_{\rm L}^2}{\omega k_\parallel v_T} e^{-\chi} I_l Z_l (1 + i\sqrt{\pi} w_l Z_l).$$

Here $w_l(Z_l)$ is the probability integral (2.91); $I_l(\chi)$ and $I_l'(\chi)$ are modified Bessel function of the order l and its derivative;

$$Z_l = \frac{\omega - l\omega_B}{\sqrt{2}k_\parallel v_T}, \qquad \chi = \frac{k_\perp^2 v_T^2}{\omega_B^2}, \qquad v_T^2 = \frac{\kappa T}{m_e}. \tag{2.99}$$

At $v_T \to 0$ the above tensor is reduced to the one (2.67)-(2.69) of a "cold" magnetoactive plasma.

The longitudinal dielectric permittivity (2.97) in a Maxwellian plasma is determined by the formula

$$\epsilon_\parallel = 1 + \frac{\omega_{\rm L}^2}{k^2 v_T^2}\left\{1 + \sum_{l=-\infty}^{\infty} i\sqrt{\pi} Z_0 I_l(\chi) e^{-\chi} w(Z_l)\right\}. \tag{2.100}$$

This formula for magnetoactive plasma can be reduced to the formula (2.88) for the longitudinal dielectric permittivity of isotropic plasma provided one of the following conditions is met [87]:

$$\frac{\operatorname{Im}\omega}{\omega_B} \gg 1, \qquad \frac{k_\parallel v_T}{\omega_B} \gg 1. \tag{2.101}$$

The first inequality means that a particle cannot perform a full rotation in the magnetic field during the time of variation of the field amplitude. The approximation $B_0 = 0$ then holds for arbitrary angle α between $\boldsymbol{B}_0$ and $\boldsymbol{k}$. If, however, this inequality is violated, the effect of the magnetic field on the characteristics of longitudinal waves in negligible only for small α for which the second criterion (2.101) is met. Provided this condition is valid, thermal motion of plasma electrons contribute to the formation of longitudinal waves more than the magnetic field.

On the contrary, at angles α close to $\pi/2$ the magnetic field should necessarily be taken into account if $\operatorname{Im}\omega \lesssim \omega_B$. For strictly transverse propagation [88]

$$\epsilon_\parallel = 1 - \sum_{l=-\infty}^{\infty} \frac{2\omega_{\rm L}^2 l^2 I_l(\chi) e^{-\chi}}{(\omega^2 - l^2\omega_B^2)\chi}. \tag{2.102}$$

This expression can be obtained by letting $k_\parallel \to 0$ (i.e., $|Z_l| \to \infty$) in (2.100) and using the identity $\sum_{l=-\infty}^{\infty} I_l(\chi)e^{-\chi} = 1$.

The expressions for plasma dielectric permittivity will be extensively used further to consider the properties of the modes, their absorption or amplification, the conditions of wave propagaion in plasmas etc.

2.2. Electromagnetic waves in homogeneous equilibrium plasma

2.2.1. WAVES IN ISOTROPIC PLASMA

It has been clarified in section 2.1.2 that only two distinct modes can exist in isotropic plasma: transverse electromagnetic waves and longitudinal plasma ones. The vectors $\boldsymbol{E}$ and $\boldsymbol{B}$ in transverse waves are perpendicular to the wave vector $\boldsymbol{k}$, and their polarization can be arbitrary. Any transverse electromagnetic field can therefore be presented as a linear superposition of two waves $\boldsymbol{E} \propto \exp(i\boldsymbol{kr} - i\omega t)$ with arbitrary orthogonal polarization – the transverse modes of isotropic plasma (concerning waves with orthogonal polarizations, refer the section 1.2). If, however, waves in isotropic plasma are considered as the limiting case of the ones in magnetoactive plasma at $\boldsymbol{B}_0 \to 0$, the mode polarization becomes quite definite (see the following paragraph). Finally, $\boldsymbol{E} \parallel \boldsymbol{k}$ and $\boldsymbol{B} = 0$ in longitudinal waves.

The frequency and the wave vector of an electromagnetic wave are related via the dispersion equation (2.46). The transverse dielectric permittivity for a "cold" plasma which enters this equation is given by (2.72). Thus, the following dispersion relation describes transverse waves in isotropic plasma:[16]

$$\omega^2 = \omega_{\mathrm{L}}^2 + k^2 c^2. \tag{2.103}$$

This equation yields real k for real $\omega \geq \omega_{\mathrm{L}}$. One can define the refraction index of electromagnetic waves as $n_{\mathrm{em}} = ck/\omega$ (see (2.55)). According to (2.103) and (2.73),

$$n_{\mathrm{em}}^2 = \epsilon_{\perp} = \epsilon = 1 - \omega_{\mathrm{L}}^2/\omega^2, \tag{2.104}$$

where ϵ is the dielectric constant of isotropic plasma with thermal motion neglected. The plot of n_{em}^2 as a function of the parameter $v = \omega_{\mathrm{L}}^2/\omega^2$ is shown in Fig. 2.2.

The wavelength and phase velocity of an electromagnetic wave equal, respectively, $\lambda_{\mathrm{em}} = 2\pi/k = 2\pi c/\omega\sqrt{\epsilon}$ and

$$\boldsymbol{v}_{\mathrm{ph}} = \frac{\omega}{c}\boldsymbol{s} = \frac{c}{\sqrt{\epsilon}}\boldsymbol{s}, \tag{2.105}$$

[16] Taking account of thermal motion with the kinetic equation leads to the appearance of the additional term $v_T^2 k^2$ at the right-hand side of (2.103). In a non-relativiastic plasma, however, this term can be neglected compared to $c^2 k^2$ [78].

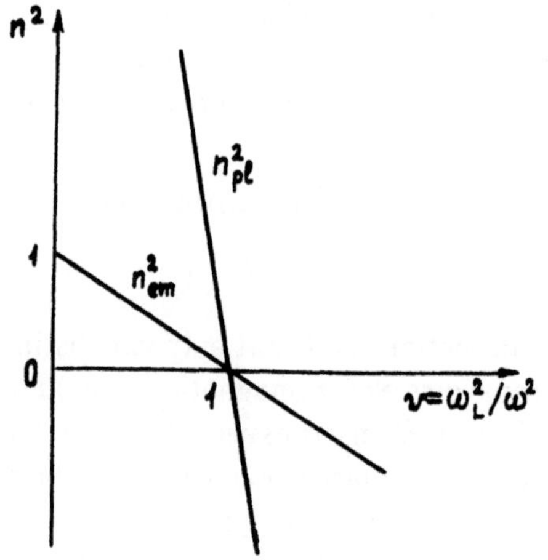

Figure 2.2. Dependence of n^2 on the parameter v in isotropic plasma.

where $\boldsymbol{s}$ is the unit vector along $\boldsymbol{k}$. It follows from (2.105) that $v_{\text{ph}} \geq c$. In its turn, the group velocity

$$\boldsymbol{v}_{\text{gr}} = \frac{d\omega}{dk} = c\frac{d\omega}{d(\omega n_{\text{em}})}\boldsymbol{s} \tag{2.106}$$

is directed along $\boldsymbol{v}_{\text{ph}}$. With (2.104), the group velocity equals $v_{\text{gr}} = c\sqrt{\epsilon} \leq c$, and the product $v_{\text{ph}} v_{\text{gr}} = c^2$.

Note that $n^2_{\text{em}} < 0$ at $\omega < \omega_L$, i.e., the value n_{em} becomes pure imaginary. According to the definition (2.55) of refraction and absorption indices, in the frequency range considered $n = 0$, $\eta^2 = 1 - \omega_L^2/\omega^2$ and $\text{Re}\, k = 0$. The spatial structure of the electromagnetic field in this case is of a non-oscillatory, exponentially decaying type, and the phase shift between the temporal oscillations of $\boldsymbol{E}$ and $\boldsymbol{B}$ at a given spatial point equals $\pi/2$).[17] Such a phase shift nullifies the time-average energy flux in a transverse wave at $\omega < \omega_L$.

Energy density of random-phased transverse waves equals [71,89]

$$W_\perp = \frac{1}{8\pi}\int (\boldsymbol{E}^2)_{\mathbf{k}} \left(\frac{1}{\omega}\frac{\partial}{\partial\omega}(\omega^2\epsilon_\perp)\right)_{\omega=\omega(\mathbf{k})} d^3k. \tag{2.107}$$

Here $(E^2)_{\mathbf{k}}$ is determined by the relation $\langle E_{\mathbf{k}} E^*_{\mathbf{k}'}\rangle = (E^2)_{\mathbf{k}}\delta(\boldsymbol{k}-\boldsymbol{k}')$ in which $(\boldsymbol{E}^2)_{\mathbf{k}}$ is the complex amplitude of the expansion (1.22), the brackets $\langle\,\rangle$ denote averaging over phases of plane waves, and the function $\omega = \omega(\boldsymbol{k})$

[17] This follows from the equation (2.26) at imaginary $\mathbf{k}$.

is determined by the dispersion equaiton (2.103).[18] It follows from (2.107) that spectral energy density of transverse waves

$$W_{\mathbf{k}} = \frac{1}{8\pi}(E^2)_{\mathbf{k}} \left(\frac{1}{\omega} \frac{\partial}{\partial \omega}(\omega^2 \epsilon_{\perp}) \right)_{\omega=\omega(\mathbf{k})} \tag{2.108}$$

With (2.104) this yields

$$W_{\mathbf{k}} = \frac{1}{4\pi}(E^2)_{\mathbf{k}} \tag{2.109}$$

Since $v_{\text{gr}} = c\sqrt{\epsilon}$, the spectral intensity of the waves considered is equal to

$$I_{\mathbf{k}} = W_{\mathbf{k}} v_{\text{gr}} = \frac{c\sqrt{\epsilon}}{4\pi}(E^2)_{\mathbf{k}} \tag{2.110}$$

(see (1.32)).

Let us turn now to plasma waves for which $E_{\mathbf{k}} \parallel k$. The dispersion equation (2.47) for these waves in the quasi-hydrodynamic approximation,

$$\omega^2 = \omega_{\text{L}}^2 + v_T^2 k^2, \tag{2.111}$$

is valid for arbitrary ω and k. The approximate nature of the quasi-hydrodynamic approximation influences the values of the coefficients in (2.111) and restricts the validity range of this equation. Indeed, the longitudinal dielectric permeability $\epsilon_{\parallel}$ in equilibrium plasma with thermal motion taken into account in the framework of the kinetic approach is determined by (2.88). With the use of (2.90) the dispersion equation for longitudinal waves in isotropic plasma is written in the form

$$\epsilon_{\parallel} = 1 + \frac{\omega_{\text{L}}^2}{k^2 v_T^2}[1 + i\sqrt{\pi} Z w(Z)] = 0, \tag{2.112}$$

where $Z = \omega/\sqrt{2}kv_T$. In the region of weak damping and large phase velocities $v_{\text{ph}} \gg v_T$,

$$Z \gg 1, \qquad |\text{Im}\, Z| \ll 1, \tag{2.113}$$

the equation (2.112) can be presented in the following form:

$$\epsilon_{\parallel} = 1 - \frac{\omega_{\text{L}}^2}{\omega^2} - \frac{3\omega_{\text{L}}^2 k^2 v_T^2}{\omega^4} + \frac{\omega_{\text{L}}^2}{k^2 v_T^2} i\sqrt{\pi} Z e^{-Z^2} = 0. \tag{2.114}$$

The asymptotic expansion (2.92) of the Kramp function $w(Z)$ has been used to transform (2.112) to (2.114).

[18] In order to assure that the expansion (1.22) comprises transverse waves ($\mathbf{E}_{\mathbf{k}} \perp \mathbf{k}$), one should use only the component div $\mathbf{E} = 0$ of the electric field $\mathbf{E}(\mathbf{r}, t)$. The expansion of another component for which curl $\mathbf{E} = 0$ comprises longitudinal waves ($\mathbf{E}_{\mathbf{k}} \parallel \mathbf{k}$).

The imaginary term with the small exponent in (2.112) gives rise to weak damping of plasma waves (the appearance of $\mathrm{Im}\,\omega < 0$ at real k). This is so-called Landau damping. The discussion of its role and physical meaning is postponed to chapter 6 in which the absorption of waves in cosmic plasmas is considered. At the same time, taking into account that the frequency ω is close to ω_{L} (see (2.113)), we present the dispersion equation (2.114) in the following approximate form:

$$\omega^2 = \omega_{\mathrm{L}}^2 + 3v_T^2 k^2. \tag{2.115}$$

This equation determines the refractive index of plasma waves:

$$n_{\mathrm{pl}}^2 = \frac{1 - \omega_{\mathrm{L}}^2/\omega^2}{3\beta_T^2} = \frac{n_{\mathrm{em}}^2}{3\beta_T^2}, \tag{2.116}$$

where $\beta_T = v_T/c$. The plot of n_{pl}^2 versus $v = \omega_{\mathrm{L}}^2/\omega^2$ is shown in Fig. 2.2. The wavelength $\lambda_{\mathrm{pl}} = 2\pi\sqrt{3}v_T/\omega\sqrt{\epsilon}$ of a plasma wave with frequency ω is $(\sqrt{3}\beta_T)^{-1}$ times less than that of an electromagnetic wave of the same frequency. Respectively, the refractive index of plasma waves is $(\sqrt{3}\beta_T)^{-1}$ times greater than the one for electromagnetic waves. In solar corona conditions where $T \sim 10^6$ K, $v_T \sim 4 \cdot 10^8\,\mathrm{cm\,s^{-1}}$ and $\beta_T \sim 10^{-2}$, the value n_{pl} is two orders of magnitude greater than n_{em}. In cooler objects (for example in solar chromosphere or interstellar H II regions) this difference is even larger.

Similar to electromagnetic transverse waves, $\boldsymbol{v}_{\mathrm{ph}} \parallel \boldsymbol{v}_{\mathrm{gr}} \parallel \boldsymbol{k}$ for plasma waves. The phase and group velocities of a plasma wave are equal to

$$\begin{aligned} v_{\mathrm{ph}} &= \frac{\omega}{k} = \sqrt{3}\,\frac{v_T}{\sqrt{\epsilon}}, \\ \boldsymbol{v}_{\mathrm{gr}} &= \frac{d\omega}{dk} = \frac{c\,d\omega}{d(\omega n_{\mathrm{pl}})} = \sqrt{3\epsilon}\,v_T, \end{aligned} \tag{2.117}$$

respectively, their product being equal to $v_{\mathrm{ph}} v_{\mathrm{gr}} = 3v_T^2$.

Finally, we write the formulas for energy density and energy flux for plasma waves. The value of spectral energy density is determined by the expression similar to (2.108) [71,89]:

$$W_{\mathbf{k}} = \frac{1}{8\pi}(\boldsymbol{E}^2)_{\mathbf{k}} \left(\frac{1}{\omega}\frac{\partial}{\partial\omega}(\omega^2\epsilon_{\|})\right)_{\omega=\omega(\mathbf{k})}.$$

However, contrary to (2.108), the value $(\boldsymbol{E}^2)_{\mathbf{k}}$ characterizes the longitudinal electric field $\mathrm{curl}\,\boldsymbol{E} = 0$, and the function $\omega(\boldsymbol{k})$ is given by the solution of the dispersion equation $\epsilon_{\|}=0$ for plasma waves (see (2.112)). Provided the conditions (2.113) are met, the value $\epsilon_{\|}$ is determined by (2.114), so

$$W_{\mathbf{k}} = \frac{1}{4\pi}(\boldsymbol{E}^2)_{\mathbf{k}}(1+\epsilon) \approx \frac{1}{4\pi}(\boldsymbol{E}^2)_{\mathbf{k}}. \tag{2.118}$$

The above relations have been obtained by neglecting the small exponential term in (2.114). Since, according to (1.32), $I_{\mathbf{k}} = W_{\mathbf{k}} v_{\mathrm{gr}}$, and $v_{\mathrm{gr}} = \sqrt{3\epsilon} v_T$, the corresponding specific intensity of plasma waves equals

$$I_{\mathbf{k}} = \frac{\sqrt{3\epsilon} v_T}{4\pi}(\boldsymbol{E}^2)_{\mathbf{k}}(1+\epsilon) \approx \frac{\sqrt{3\epsilon} v_T}{4\pi}(\boldsymbol{E}^2)_{\mathbf{k}}. \tag{2.119}$$

To get the last equalities in (2.118), (2.119), one should take into account that the frequency ω is close to ω_{L} and $\epsilon \ll 1$ in the region $\omega \gg \sqrt{2} k v_T$ which satisfies the conditions (2.113).

The formula (2.115) for plasma wave frequency fails in the region $\omega \sim \sqrt{2} k v_T$ in which the phase velocity of a wave $v_{\mathrm{ph}} = \omega/k$ becomes comparable to the electron thermal velocity v_T. At the same time, the Landau damping becomes very strong in this region. This leads to the absorption of a wave in a time of about $1/\omega \sim 1/\omega_{\mathrm{L}}$. This fact can be easily explained by the following simple considerations. Let a longitudinal wave with a wavelength $1/k \sim v_T/\omega_{\mathrm{L}} \sim v_T/\omega$ (i.e., with phase velocity $v_{\mathrm{ph}} \sim v_T$) exist in plasma at some initial moment. Thermal electrons travel a distance $\sim 1/k$ in a time $1/\omega$. By this motion, the regular velocity gained due to the action of the wave field of the phase $\omega t - \boldsymbol{k}\boldsymbol{r}$ is transferred to the region where the phase of a wave is essentially different. This leads to the strong damping of the field $\boldsymbol{E}$ over a time $t \sim 1/\omega$.

Note that, due to the reasons discussed in the section 6.1, the Landau damping exists only for waves with phase velocities less than the speed of light. Thus, the Landau damping is absent for plasma waves with $n_{\mathrm{pl}} = c/v_{\mathrm{ph}} < 1$. The Landau damping is impossible for electromagnetic waves since n_{em} is always less than unity.

2.2.2. ORDINARY AND EXTRAORDINARY WAVES IN MAGNETOACTIVE PLASMA

In the presence of a static magnetic field the electromagnetic properties of plasma are radically changed. The magnetic field makes plasma magnetoactive, i.e., gyrotropic and anisotropic medium. Plasma gyrotropy appears as elliptic polarization of the normal waves, while anisotropy leads to the dependence of the mode characteristics on the direction of propagation.

In a "cold" magnetoactive plasma with dielectric tensor $\epsilon_{\alpha\beta}(\omega, \boldsymbol{k})$ in the form (2.67) the dispersion equation (2.41) yields the refractive index n_j of

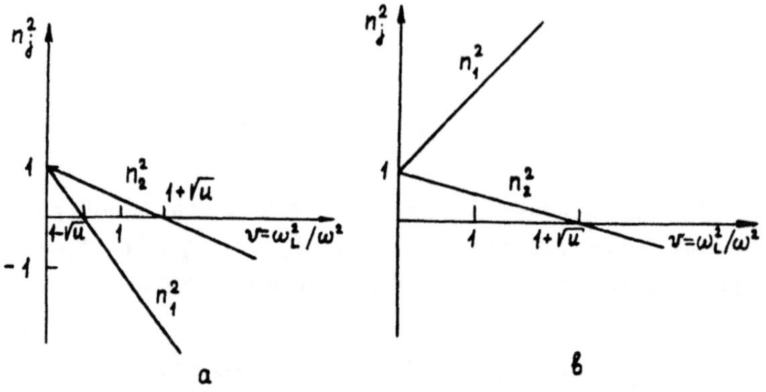

Figure 2.3. The dependence $n_j^2(v)$ for longitudinal propagation in magnetized plasma: (a), $u<1$; (b), $u>1$.

each normal wave j. With particle collisions neglected, the absorption in such a plasma is absent: the absorption index $\eta_j = 0$ and $n_j = ck_j/\omega$. The dispersion equation (2.41) in this case is reduced to [71]

$$(1-u-v+uv\cos^2\alpha)n^4 - [2(1-v)^2 \\ -u(2-v-v\cos^2\alpha)]n^2 - (1-v)[u-(1-v)^2] = 0, \quad (2.120)$$

where α is the angle between the wave vector $\boldsymbol{k}$ and the magnetic field $\boldsymbol{B}_0$. This equation determines the following two values of n_j^2 ($J = 1,2$) which correspond to a couple of "magneto-ionic modes" propagating in homogeneous "cold" magnetoactive plasma:

$$n_j^2 = 1 - \frac{2v(1-v)}{2(1-v) - u\sin^2\alpha \mp \sqrt{u^2\sin^4\alpha + 4u(1-v)^2\cos^2\alpha}}. \quad (2.121)$$

Here minus sign and $j = 1$ correspond to extraordinary mode, while the ordinary one is denoted by $j=2$. The above expression for n_j^2 is reduced to the one (2.104) for transverse waves in isotropic plasma if $u = \omega_B^2/\omega^2 \ll 1$. The plots of the functions $n_j^2(v)$ for longitudinal, $\alpha = 0$, and transverse, $\alpha = \pi/2$, propagation with respect to the magnetic field $\boldsymbol{B}_0$ are shown in Figs. 2.3, 2.4. The next Fig. 2.5 illustrates the behavior of the functions $n_j^2(v)$ at arbitrary but small α.

The mode refractive indices are nullified at $v = v_0$, where

$$v_0 = 1, \qquad v_0 = 1 \mp \sqrt{u}, \quad (2.122)$$

and go to infinity at u, v satisfying the equation

$$1 - u - v + uv\cos^2\alpha = 0, \quad (2.123)$$

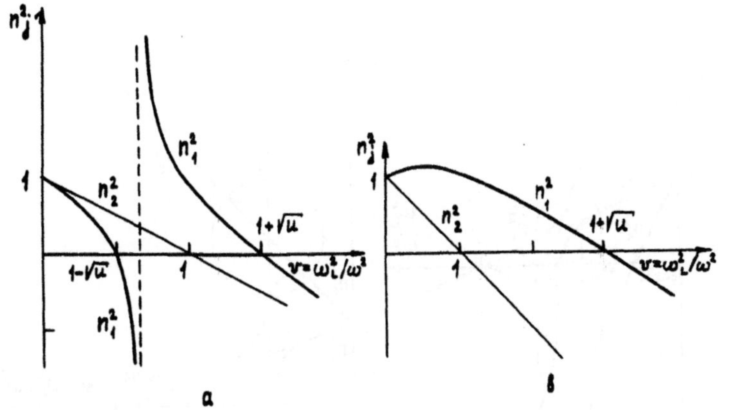

Figure 2.4. The dependence $n_j^2(v)$ for transverse propagation in magnetized plasma: (*a*), $u < 1$; (*b*), $u > 1$.

i.e. at

$$v = v_\infty = \frac{1-u}{1-u\cos^2\alpha}. \tag{2.124}$$

From the figures mentioned it is clear that the forms of dispersion curves $n_j^2(v)$ in the regions $u<1$ and $u>1$ are drastically different. In the last case one should distinguish two possible variants: $u\cos^2\alpha<1$ and $u\cos^2\alpha>1$. Since the parameter v cannot be negative, the functions $n_j^2(v)$ do not have poles in the region $u>1$, $u\cos^2\alpha<1$ (see (2.124)). At small α this region is insignificant, since it corresponds to the quite narrow interval of the parameter u.

To prevent an ambiguity concerning the above plots of the dispersion curves, we should note the following. Regarding the plots in Fig. 2.3 as the limiting transition $\alpha\to 0$ of the ones in Fig. 2.5, one should denote by $n_2^2(v)$ the part $v>1$ of the dispersion line $n_1^2(v)$ to ensure the consistency between the notations of the modes at $\alpha=0$ and $\alpha\neq 0$. Then the relevant waves should be called ordinary, despite the waves corresponding to the part $v<1$ of the same curve being called extraordinary. In the same way, the part $v>1$ of the dispersion line $n_2^2(v)$ of ordinary wave in Fig. 2.3 should be denoted as $n_1^2(v)$ and the corresponding wave should be called extraordinary. However, the designations of Figs. 2.3–2.5 are traditionally accepted in the theory of electromagnetic wave propagation in plasmas and will be adopted thereafter. Nevertheless, we will bear in mind that it is necessary to redesignate the dispersion curves in the region $u>1$ to match the case of longitudinal propagation and the one $\alpha\neq 0$.

In fact, the Figs. 2.3-2.5 illustrate the variation of the mode refractive indices with plasma frequency ω_L at ω, ω_B = const. It is also interesting to see the functions $n_j^2(\omega)$ at fixed ω_L and ω_B. The corresponding plots for

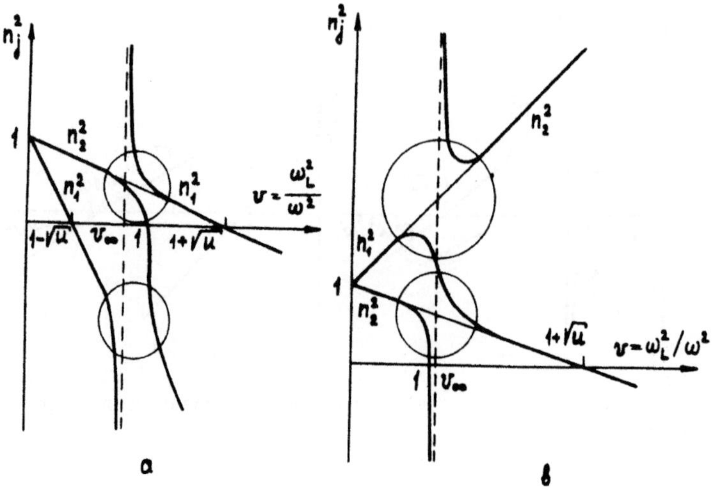

Figure 2.5. The plots of $n_j^2(v)$ for $\alpha \ll 1$: (a), $u < 1$; (b), $u \cos^2 \alpha > 1$. The circles mark the regions of linear coupling (see the section 3.2).

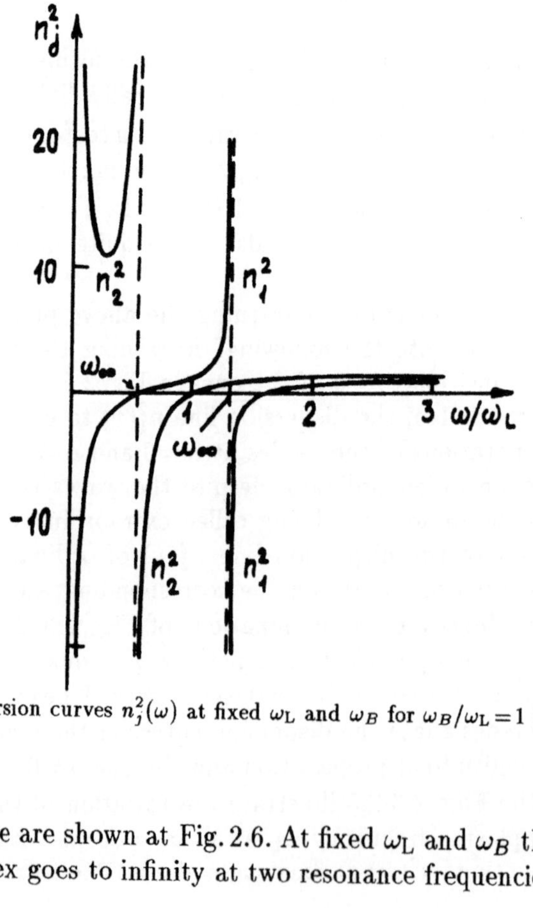

Figure 2.6. Dispersion curves $n_j^2(\omega)$ at fixed ω_L and ω_B for $\omega_B/\omega_L = 1$ and $\alpha = \pi/4$.

one particular case are shown at Fig. 2.6. At fixed ω_L and ω_B the square of the refractive index goes to infinity at two resonance frequencies,

$$\omega_\infty^2 = \frac{1}{2}(\omega_B^2 + \omega_\mathrm{L}^2) \pm \left[\frac{1}{4}(\omega_B^2 + \omega_\mathrm{L}^2)^2 - \omega_B^2\omega_\mathrm{L}^2 \cos^2\alpha\right]^{1/2}, \qquad (2.125)$$

denoted by the equation (2.123). The above value of ω_∞^2 is always non-negative, i.e., the functions $n_j^2(\omega)$ have two singular points at any ω_L, ω_B and α. At $\alpha = 0$ one of the resonance frequencies, $\omega_\infty = \omega_L$, corresponds to plasma resonance, while another one, $\omega_\infty = \omega_B$, to cyclotron resonance. If $\alpha = \pi/2$, one of the resonance values $\omega_\infty = \sqrt{\omega_\mathrm{L}^2 + \omega_B^2}$ corresponds to upper hybrid resonance. If ion thermal motion is taken into account, the lower hybrid resonance $\omega_\infty = \sqrt{\omega_B \omega_{B\mathrm{i}}} = \omega_B\sqrt{m_\mathrm{e}/m_\mathrm{i}}$ appears in the low-frequency range.

The expression (2.121) for the refractive index is quite complicated. However, it gets simplified in two limiting cases called quasi-longitudinal and quasi-transverse propagation [71]. Quasi-longitudinal propagation is realized in plasma under the following conditions: [19]

$$\begin{aligned} \frac{u\sin^4\alpha}{4\cos^2\alpha} &\ll (1-v)^2, \\ |1 \mp |\sqrt{u}\cos\alpha|| &\gg \frac{u\sin^2\alpha}{|1-v|}, \end{aligned} \qquad (2.126)$$

while the quasi-transverse propagation is realized if

$$\begin{gathered} \frac{u\sin^4\alpha}{4\cos^2\alpha} \gg (1-v)^2, \qquad \tan^2\alpha \gg |1-v|, \\ |1-v-u\sin^2\alpha| \gg (1-v)^2\cot^2\alpha. \end{gathered} \qquad (2.127)$$

The corresponding refractive indices are given by

$$n_{1,2}^2 = 1 - \frac{v}{1 \mp |\sqrt{u}\cos\alpha|} \qquad (2.128)$$

in case of quasi-longitudinal and by

$$n_1^2 = 1 - \frac{v(1-v)}{1-v-u\sin^2\alpha}, \qquad n_2^2 = 1 - v \qquad (2.129)$$

in case of quasi-transverse propagation.

[19] Note that the first inequalities in (2.126), (2.127) are related to both modes. As for the other ones, the second inequality in (2.126) with upper sign corresponds to extraordinary, and with lower sign to ordinary wave. In (2.127) the second criterion stands only for ordinary while the third only to extraordinary waves.

The expression (2.121) is simplified greatly in the region of sufficiently low frequencies

$$v\cos^2\alpha \gg u\cos^2\alpha \gg 1. \tag{2.130}$$

In this case

$$n_1^2 \approx -\frac{v}{|\sqrt{u}\cos\alpha|}, \qquad n_2^2 \approx \frac{v}{|\sqrt{u}\cos\alpha|}. \tag{2.131}$$

Thus, the extraordinary wave is non-propagating ($n_1^2 < 0$ and $\mathrm{Re}\, k = 0$), contrary to the ordinary one for which $n_1^2 > 0$. The last equation becomes evident if one takes into account that, according to (2.130), $v \gg u \gg 1$ and, moreover, $v \gg \sqrt{u}$. The conditions (2.130) are typical for "whistlers" propagating in the Earth magnetosphere. That is why this name is also used for the ordinary wave with the refractive index (2.131). In Fig. 2.5, b whistler waves correspond to the part $v \gg 1$ of the dispersion curve n_2^2.

Under the additional condition $v \gg |\sqrt{u}\cos\alpha| - 1$ the square of the whistler refractive index looks like [20]

$$n_1^2 \approx \frac{v}{|\sqrt{u}\cos\alpha| - 1}. \tag{2.132}$$

Contrary to (2.131), the condition $u\cos^2\alpha \gg 1$ is not required for the validity of the expression (2.132).

Note also another case in which the expressions for n_2^2 have the most simple form [90,91]. Similar to whistlers, it is also realized at sufficiently low frequencies but at different relations between the parameters ω_B and ω_L. Namely, at the conditions

$$u\sin^4\alpha \gg 4v^2\cos^2\alpha, \qquad v\cos^2\alpha \gg 1 \tag{2.133}$$

the formula (2.121) for ordinary wave transforms to

$$n_2^2 \approx \cos^{-2}\alpha. \tag{2.134}$$

The refractive index is independent on frequency in this case, and the phase velocity $v_{\mathrm{ph}} = c\cos\alpha$.

Generally, ordinary and extraordinary waves in magnetoactive plasma are elliptically polarized. In the frame with z-axis along the wave vector $\boldsymbol{k}$ (the frame II shown in Fig. 2.1 b in the section 2.1) the relation between the electric field components is given by [3]:

$$\frac{E_y}{E_x} = iK_j, \qquad \frac{E_z}{E_x} = i\Gamma_j, \tag{2.135}$$

[20]The comparison of (2.132) with (2.131) should be made bearing in mind that the notations for the modes in the region $v > 1$ are changed as $\alpha \to 0$.

where the polarization coefficients

$$K_j = \frac{2\sqrt{u}(1-v)\cos\alpha}{u\sin^2\alpha \pm \sqrt{u^2\sin^4\alpha + 4u(1-v)^2\cos^2\alpha}},$$

$$\Gamma_j = -\frac{v\sqrt{u}\sin\alpha + K_j uv\sin\alpha\cos\alpha}{1-u-v+uv\cos^2\alpha}. \tag{2.136}$$

The upper sign in the first formula is related to extraordinary wave $j=1$, the lower, to the ordinary one $j=2$.

The definitions (2.135), (2.136) of polarization coefficients clearly show that K_j characterizes the eccentricity of the ellipse circumscribed by the end of the electric field vector in a plane perpendicular to the wave vector $\boldsymbol{k}$, i.e., the ellipticity $p=\tan\sigma=K_j$ of normal waves in plasma (see section 1.2). The coefficient Γ_j, in its turn, determines the relative value of the longitudinal (with respect to the vector $\boldsymbol{k}$) component of the electric field in the mode. Since $K_1K_2=-1$, the polarization ellipses of ordinary and extraordinary waves are similar, their larger axes are orthogonal, and electric field vectors $\boldsymbol{E}$ rotate in opposite directions. Such a polarization of two waves is called orthogonal.

At $\alpha=\pi/2$ the modes are linearly polarized, the electric field in the ordinary wave being directed along the magnetic field $\boldsymbol{B}_0$, and the one in the extraordinary wave lying in the plane perpendicular to $\boldsymbol{B}_0$ and circumscribing the ellipse in this plane. At $\alpha=0$ the normal waves become transverse and circularly polarized, the electric field vectors of ordinary and extraordinary waves rotating counterclockwise and clockwise, respectively, as seen along the direction of propagation. These directions of electric field rotation are the same for any acute angle between $\boldsymbol{k}$ and $\boldsymbol{B}_0$, the case of propagation along $\boldsymbol{B}_0$ included. But if the electromagnetic wave propagates opposite to the direction of the field $\boldsymbol{B}_0$ or at an obtuse angle to this vector, the signs of electric field rotation in extraordinary and ordinary mode change with respect to the former case. Anyway, at $v<1$ the projection of the electric field of extraordinary mode onto the plane perpendicular to $\boldsymbol{k}$ rotates in the same direction as the electron in the magnetic field $\boldsymbol{B}_0$ while the rotation of the same projection in the ordinary mode is opposite to electron gyration.

At arbitrary $\alpha\neq 0,\pi/2$ the mode polarization becomes purely linear only for $v=1$ and remains elliptical at $v\neq 1$. However, the polarization of the modes can be well approximated by the linear one over the interval (2.127) of the angles α which correspond to quasi-perpendicular propagation. In a similar way, the mode polarizations are close to circular in the case of quasi-longitudinal propagation.

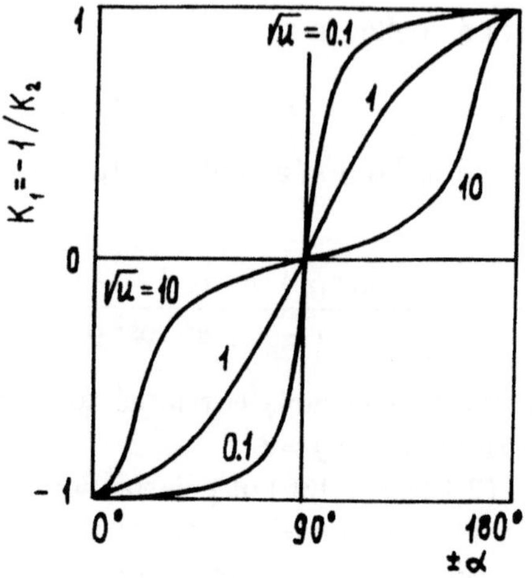

Figure 2.7. Polarization coefficients $K_{1,2}$ as functions of the angle α.

The above ellipticity of modes at $v \neq 1$ holds also in the vacuum limit $v \to 0$ where $\Gamma_j \to 0$ and

$$K_j \to \frac{2\sqrt{u}\cos\alpha}{u\sin^2\alpha \pm \sqrt{u^2\sin^4\alpha + 4u\cos^2\alpha}}. \qquad (2.137)$$

The ratio of polarization ellipse axes as a function of the angle α at various values of the parameter $u = \omega_B^2/\omega^2$ is shown in Fig. 2.7. We mention here that the validity of the formula (2.137) is restricted by the criterion of "dense plasma",

$$\omega_L^2/\omega^2\beta_T \gg 1, \qquad (2.138)$$

which can be met even for $v \ll 1$ provided β_T is small enough. This criterion is typically satisfied for radio waves in solar corona and chromosphere. In the opposite limit of "rarefied plasma" (see the criterion (2.144) below) mode polarizations and the refractive index of extraordinary mode are different.

Consider now the mode polarization at frequencies $\omega \to \omega_\infty$, where the value ω_∞ is determined by the equation (2.123) and $n_{1,2}^2 \to \infty$. The formula (2.136) for the coefficient Γ_j then reveals that the ratio E_z/E_x for the normal waves in this region is large, i.e., the modes become nearly longitudinal.[21]

The case of longitudinal propagation ($\alpha = 0$) at frequencies close to ω_B is an exception in a sense that here the coefficients $\Gamma_j = 0$, $K_j = 1$.

[21] As we see below, thermal motion should necessarily be taken into account at frequencies $\omega \approx \omega_\infty$. As a result, the dispersion curves $n_j^2(\omega, \alpha)$ are radically changed in this region.

This means that the circularly polarized extraordinary wave is transverse, despite $n_1^2 \to \infty$ at $\omega \to \omega_B$. The ordinary waves corresponding to the whistler band are not longitudinal, though $n_2^2 \gg 1$. Since $E_{x,y} \gg E_z$, $n_j^2 \gg 1$ in both of these peculiar cases, the magnetic field of a wave $B_{x,y} \propto n_j E_{x,y}$ exceeds the electric one (see the equation (2.26).

For some applications it is necessary to know the ratio of electric field components of the modes in another frame, the one with z-axis along the magnetic field $\boldsymbol{B}_0$ and the wave vector $\boldsymbol{k}$ in the yz-plane. In this frame the ratio of electric field components is given by

$$\frac{E_y}{E_x} = ig_j, \qquad \frac{E_z}{E_x} = ih_j, \tag{2.139}$$

where

$$g_j = K_j \cos\alpha - \Gamma_j \sin\alpha, \qquad -h_j = \Gamma_j \cos\alpha + K_j \sin\alpha, \tag{2.140}$$

and the coefficients K_j and Γ_j are determined by the formulas (2.136).

Note the interesting feature of mode polarization in a "cold" magnetoactive plasma: $g_j \to -1$ for both modes as $u \to 1$. This means that at $\omega \to \omega_B$ the vector $\boldsymbol{E}$ circumscribes the circle in the plane perpendicular to the field $\boldsymbol{B}_0$ and rotates in the same direction for both modes. It is very important that this direction is opposite to electron gyration in a magnetic field. As a result, the intensity of cyclotron radiation of an electron at frequencies $\omega \approx \omega_B$ is strongly depressed (see the details in the section 10).

Due to anisotropy of magnetoactive plasma the vectors of group and phase velocity, $\boldsymbol{v}_{\text{gr}} = d\omega/d\boldsymbol{k}$ and $\boldsymbol{v}_{\text{ph}} = (\omega/k)\boldsymbol{s}$, where $\boldsymbol{s} = \boldsymbol{k}/k$, are not collinear. The projection of the group velocity onto the direction of the wave vector $\boldsymbol{k}$ is given by

$$v_{\text{gr}\,\mathbf{k}} = \frac{d\omega}{d\boldsymbol{k}}\boldsymbol{s} = \frac{d\omega}{d\boldsymbol{k}}\frac{\partial \boldsymbol{k}}{\partial k} = c\,\frac{\partial \omega}{\partial(\omega n_j)}. \tag{2.141}$$

The angle ϑ_j between $\boldsymbol{v}_{\text{ph}}$ and $\boldsymbol{v}_{\text{gr}}$ is introduced by the relation

$$\tan\vartheta_j = -\frac{1}{2n_j^2}\frac{\partial n_j^2}{\partial \alpha} \tag{2.142}$$

which is derived in [71]. It is clear that $\vartheta_j = 0$ at the angles $\alpha = 0, \pi/2$ (longitudinal and transverse directions with respect to the magnetic field $\boldsymbol{B}_0$) at which the function $n_j^2(\alpha)$ has an extremum.

The formula (2.142) yields $\tan\vartheta_j = -(1/2)\tan\alpha$ for the whistler refractive index n_2^2 determined by (2.131). Since the vectors $\boldsymbol{B}_0$, $\boldsymbol{k}$ and $d\omega/d\boldsymbol{k}$ are coplanar, the angle between $\boldsymbol{v}_{\text{gr}}$ and $\boldsymbol{B}_0$ equals $\vartheta_j + \alpha$. The maximal value

$\vartheta_j + \alpha \approx 19.5°$ can be found from the relation $\tan\vartheta_j = -(1/2)\tan\alpha$. The extremum appears at $\cos\alpha = 1/\sqrt{3}$. Thus, the angle between group velocity of whistlers and magnetic field $\boldsymbol{B}_0$ cannot exceed 19.5°. This statement constitutes the so-called Story theorem [92].

Finally, the group velocity of the ordinary wave with the refractive index (2.134) is directed exactly along the field $\boldsymbol{B}_0$. Indeed, the expression $n_2^2 = k^2c^2/\omega^2$ yields $\omega = \pm(\boldsymbol{kB}_0)B_0^{-1}c$ from which we immediately obtain [22]

$$\frac{d\omega}{d\boldsymbol{k}} = \pm c\frac{\boldsymbol{B}_0}{B_0}. \tag{2.143}$$

The behavior of $\boldsymbol{v}_{gr}$ in this case is similar to the one for the Alfven (magnetohydrodynamic) waves. Dispersion equation of such waves in an ideally conducting medium with mass density ρ looks like [93] $\omega = \pm\boldsymbol{kB}_0/\sqrt{4\pi\rho}$. The corresponding group velocity vector is given by the expression $d\omega/d\boldsymbol{k} = \pm\boldsymbol{B}_0/\sqrt{4\pi\rho}$. The orientation of this vector is the same as of (2.143). Such an analogy, however, is not extended further since the values of group velocity are greatly different.

Note that thermal motion has been totally neglected in the above analysis. However, with $\beta_T = 0$ it is impossible to get the true expressions for refractive indices, polarization coefficients, as well as cyclotron emissivity, absorption or scattering coefficients for electromagnetic waves in plasma at frequencies $\omega \approx \omega_B$; see sections 5.1,6.2,8.1. It is especially true for rarefied plasma for which the following criterion is satisfied: [23]

$$\omega_L^2/\omega^2\beta_T \simeq 0.8\, N B_0^{-2} T^{-1/2} \ll 1 \tag{2.144}$$

This criterion at frequencies $\omega \approx \omega_B$ can be easily realized for plasmas in strong magnetic fields $B_0 \sim 10^7 - 10^{13}$ G of white dwarfs and neutron stars. For, example, (2.144) is valid for the hot ($T \sim 10^7$ K) plasma envelope of a magnetic white dwarf with $B_0 \sim 10^7 - 10^8$ G at densities $N \ll 10^{18} - 10^{20}\,\text{cm}^{-3}$.

Since the refractive indices of both modes in rarefied plasma are close to unity, the wave number $k = \omega n/c \simeq \omega/c$. Therefore, the parameter χ entering the expression (2.99) for the dielectric tensor $\epsilon_{\alpha\beta}$ becomes small:

$$\chi = (k^2 v_T^2/\omega_B^2)\sin^2\alpha \simeq \beta_T^2 \sin^2\alpha \ll 1.$$

Expanding the modified Bessel functions and their derivatives in (2.99) in power series over a small parameter χ and keeping the leading terms, we

[22] The signs in these formulas are fixed by the requirement that the phase velocity $\mathbf{v}_{ph} = (\omega/c)\mathbf{s}$ is directed along $\mathbf{s} = \mathbf{k}/k$. To satisfy this requirement, the sign in the resulting formulas should be the same as the one of the scalar product $(\mathbf{kB}_0)$.

[23] The formulas (5.60) and (5.61) in section 5.1 give more rigorous criteria for the limits of rarefied and dense plasma.

find the following expressions for the components of the dielectric tensor of a non-relativistic, rarefied plasma in the frame I (see Fig. 2.1, *a*):

$$\epsilon_{xx} = \epsilon_{yy} = 1 + ib, \qquad \epsilon_{xy} = -\epsilon_{yx} = b,$$
$$\epsilon_{zz} = 1, \qquad \epsilon_{xz} = \epsilon_{zx} = \epsilon_{yz} = \epsilon_{zy} = 0, \tag{2.145}$$

where

$$b = \sqrt{\frac{\pi}{8}} \frac{\omega_{\mathrm{L}}^2 w(Z)}{\omega^2 \beta_T |\cos\alpha|}, \tag{2.146}$$

and $w(Z)$ is the Kramp function (2.91) of the argument

$$Z = \frac{\omega - \omega_B}{\sqrt{2}\beta_T \omega |\cos\alpha|}. \tag{2.147}$$

The solution of the dispersion equation (2.41) with the tensor (2.145) yields the refractive index of the extraordinary wave:

$$n_1 = 1 - \frac{\omega_{\mathrm{L}}^2 (1 + \cos^2\alpha)}{\sqrt{8}\omega^2 \beta_T |\cos\alpha|} e^{-Z^2} \int_0^Z e^{\xi^2}\, d\xi. \tag{2.148}$$

The corresponding solution for the ordinary mode in the adopted approximation is simply $n_2^2 = 1$. However, keeping the terms of the order of $b\beta_T^2$ in the tensor $\epsilon_{\alpha\beta}$, which have been neglected while deriving (2.146) from (2.99), leads to the following expression valid in the region of cyclotron resonance $\omega \approx \omega_B$:

$$n_2^2 = 1 - \frac{\omega_{\mathrm{L}}^2}{\omega^2 (1 + \cos^2\alpha)}.$$

This formula coincides with the corresponding one for cold plasma.

The mode polarization at gyrofrequency is determined by the system of equations (2.40) with the tensor (2.145). For the extraordinary mode it is given by the relations

$$\frac{E_x}{E_y} = -i\cos^2\alpha, \qquad \frac{E_z}{E_y} = i\sin\alpha\cos\alpha, \tag{2.149}$$

and for the ordinary one, by

$$\frac{E_x}{E_y} = i, \qquad \frac{E_z}{E_y} = -i\tan\alpha. \tag{2.150}$$

Both formulas correspond to the frame I (Fig. 2.1 *a*).

2.2.3. PLASMA WAVES

The dispersion equation (2.120) for waves in magnetoactive plasma has been written in the framework of the elementary theory which neglects completely thermal motion of plasma particles. With the use of the quasi-hydrodynamic approximation (see section 2.1), which accounts for the effect of thermal motion by the electron pressure term in the relevant equations, we obtain the following dispersion relation instead of (2.120) [71]:

$$\begin{aligned}\beta_T^2(1-u\cos^2\alpha)n^6 - [1-u-v+uv\cos^2\alpha+\\ 2\beta_T^2(1-v-u\cos^2\alpha)]n^4 + [2(1-v)^2 - u(2-v-v\cos^2\alpha)+\\ \beta_T^2(1-2v+v^2-u\cos^2\alpha)]n^2 + (1-v)[u-(1-v)^2] = 0, \quad (2.151)\end{aligned}$$

where $\beta_T^2 = \kappa T/m_e c^2$. The equation (2.151) is reduced to (2.120) as $\beta_T^2 \to 0$.

The equation (2.151) is cubic with respect to n^2. Thus, generally, it yields three solutions (versus two in elementary theory) for the refraction index $n_j = k_j\omega/c$ which correspond to normal waves with a given frequency ω. As a result, three dispersion curves which thereafter are marked by the Roman I, II, III, appear at the graphs $n_j^2(v)$ ($j = 1,2,3$). Fig. 2.8 shows these curves for the case of propagation at small angle α to the field $\boldsymbol{B}_0$. The waves corresponding to the distinct branches of the dispersion curves and separated by the encircled regions are usually called extraordinary (n_1^2), ordinary (n_2^2) and plasma ones (n_3^2). So, the subscripts 1,2,3 denote the name of the mode, while the numbers I, II, III indicate the branch of the dispersion curve which this or that mode is related to.

The encircled regions are not distinguished in a homogeneous plasma, thus the separation of ordinary, extraordinary and plasma modes is rather artificial. This separation, however, is better justified in an inhomogeneous plasma: the process of linear coupling of different modes with their "transition" from one curve to another is possible in the regions where the dispersion curves of different numbers come close to each other (refer to chapter 3 for more detail on linear mode coupling).

By comparison of the dispersion curves at Figs. 2.5, 2.8 it is clear that thermal motion considerably influences the behavior of $n_j^2(v)$ in the region $n_j^2(v) \gg 1$ and gives rise to the longitudinal plasma waves. The regions of large $n_j^2(v)$ near the poles (2.124) of the function (2.121) at $\beta_T^2 \to 0$ correspond to the plasma branch of the dispersion curves. In other regions the influence of thermal motion on refractive indices, and polarization coefficients, is usually negligible.

At $\beta_T^2 \ll 1$ all the terms except the one proportional to n^6 can be dropped in the dispersion equation (2.151). The remaining term becomes important at large n^2 when only the higher-order terms comprising n^4 and n^6 can be kept in the equation (2.151). Then one may obtain

$$n_3^2 = \frac{1-u-v+uv\cos^2\alpha}{\beta_T^2(1-u\cos^2\alpha)}. \quad (2.152)$$

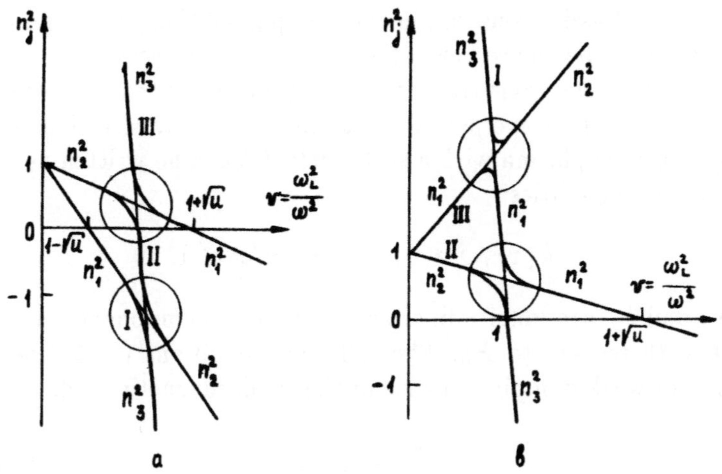

Figure 2.8. Dispersion curves $n_j^2(v)$ ($j = I, II, III$) for $\alpha \ll 1$ in plasma with thermal motion taken into account: (a), $u < 1$; (b), $u\cos^2\alpha > 1$. The circles indicate the regions of mode coupling in inhomogeneous plasma, the thin lines inside these circles show the functions $n_j^2(v)$ in the case of longitudinal propagation.

This formula describes the square of refractive index for plasma wave branch of the dispersion curves in the region $n_3^2(v) \gg 1$. The formula (2.152) is obtained in quasi-hydrodynamic approximation. More rigorous kinetic calculation made with the use of the dielectric tensor $\epsilon_{\alpha\beta}(\omega, \boldsymbol{k})$ in the form (2.99) shows that the dispersion equation of plasma waves, $\epsilon_{\parallel} = 0$, where $\epsilon_{\parallel}$ is determined by the formula (2.100), can be reduced to the form

$$n_3^2 \approx \frac{(1 - u - v + uv\cos^2\alpha)/(v\beta_T^2)}{\dfrac{3\sin^4\alpha}{1-4u} + \left[1 + \dfrac{5-u}{(1-u)^2}\right]\cos^2\alpha\sin^2\alpha + 3(1-u)\cos^4\alpha} \tag{2.153}$$

in the region $n_3^2(v) \gg 1$ [94].

The expression (2.153) is essentially different from the previous one. This illustrates the "conventional" nature of quasi-hydrodynamic approximation in a magnetoactive plasma. The use of this approach to account for thermal motion can only be justified by its relative simplicity and clarity when dealing with the complicated problems of wave propagation in an inhomogeneous plasma etc. The most important drawback of the quasi-hydrodynamic approximation is that it misses the collisionless Landau damping which, under certain conditions, can radically change the wave propagation leading to their damping at a time of about $1/\omega$.

The problem of absorption of waves in equilibrium plasma is the subject of special section 6.2. Here we only point out the criterion ensuring that the

Landau damping is not too strong and does not make the formula (2.153) invalid. It has already been said in the first part of this paragraph that the Landau damping becomes weak (the corresponding decrement $\gamma \ll \Omega$, see (2.56)) if thermal electrons travel a distance much less than the wavelength $\lambda = 2\pi/k$ during the wave period $2\pi/\omega$, i.e., at $kv_T/\omega \ll 1$. This criterion for magnetoactive plasma with a static field $\boldsymbol{B}_0$ can be written as a pair of the following inequalities:

$$k_\perp v_T/\omega \ll 1, \qquad k_\| v_T/\omega \ll 1, \tag{2.154}$$

where $k_\|$ and $k_\perp$ are longitudinal and transverse components of the wave vector $\boldsymbol{k}$ with respect to $\boldsymbol{B}_0$. The sufficient conditions (2.154) determine the region of weak damping in case of the small magnetic field:

$$\frac{k_\perp v_T}{\omega_B} \gg 1. \tag{2.155}$$

If, on the contrary, the opposite inequality is satisfied:

$$\frac{k_\perp v_T}{\omega_B} \ll 1, \tag{2.156}$$

the radius of electron gyration in the magnetic field, $r_B = v_T/\omega_B$ becomes smaller than the characteristic spatial scale $1/k_\perp = \lambda_\perp/2\pi$ of the high-frequency field in the direction perpendicular to the magnetic field ($\lambda_\perp$ is the "transverse" wavelength). Therefore, an electron cannot shift a distance $2\pi/k_\perp$ across the field $\boldsymbol{B}_0$. This evidently makes the first of the conditions (2.154) unnecessary. As a result, the criteria of weak damping in a strong magnetic field are (2.156) and the second inequality in (2.154).

2.2.4. BERNSTEIN MODES

In the case of the transverse propagation the mode structure gets more complicated than described in previous paragraphs. First and foremost, this is true for the longitudinal waves with the dispersion relation $\epsilon_\|(\omega, k) = 0$, where $\epsilon_\|$ is given by (2.100). As $|Z_l| = |(\omega - l\omega_B)/k_\| v_T| \to \infty$, i.e., $k_\| \to 0$, the longitudinal dielectric constant is reduced to (2.102). Thus, in case of transverse propagation ($\alpha = \pi/2$) the dispersion equation for longitudinal waves looks like

$$\epsilon_\| = 1 - \sum_{l=1}^{\infty} \frac{2\omega_L^2 l^2 I_l(\chi) e^{-\chi}}{(\omega^2 - l^2\omega_B^2)\chi} = 0, \tag{2.157}$$

where l is the harmonic number, $\chi = k_\perp^2 v_T^2/\omega_B^2$, and $I_l(\chi)$ is the modified Bessel function. The equation (2.157) describes the dispersion properties of the so-called electron cyclotron waves, or Bernstein modes [88]. The interest to these waves in cosmic plasma physics has arisen due to their possible role in generation of certain types of fine-structure sporadic radio emissions observed from the Sun (see section 7.3).

We repeat that the equation (2.157) is only valid under the condition $c^2k_\perp^2/\omega^2 \gg 1$, i.e. in the region

$$\chi \gg \beta_T^2 \frac{\omega^2}{\omega_B^2}. \tag{2.158}$$

Besides, this equation has been derived by the limiting transition to strictly transverse propagation: $k_\parallel \to 0$ ($|Z_l| \to \infty$), though it is certainly valid in the small range of α close to $\pi/2$ where $|Z_l| \gg 1$ [24] i.e., at

$$|\omega - l\omega_B| \gg k_\parallel v_T. \tag{2.159}$$

Violation of this criterion leads to the strong cyclotron absorption of longitudinal waves (see section 6.2). If the condition $\omega_B \ll k_\parallel v_T$ (see (2.101)) becomes valid at larger $k_\parallel$, the magnetic field $\boldsymbol{B}_0$ does not influence the characteristics of longitudinal waves any longer, and these waves become similar to the ones in isotropic plasma.

Let us now turn to the analysis of the dispersion equation (2.157). This equation can be written the following way

$$\epsilon_\parallel = 1 - \frac{\omega_{\rm L}^2}{\omega^2 - \omega_B^2} - \frac{3\omega_{\rm L}^2\omega_B^2\chi}{(\omega^2 - 4\omega_B^2)(\omega^2 - \omega_B^2)} - \frac{\omega_{\rm L}^2}{\omega^2 - s^2\omega_B^2}\frac{s}{(s-1)!}\left(\frac{\chi}{2}\right)^{s-1} = 0. \tag{2.160}$$

for the waves with wavelength much larger than the gyration radius $r_B = v_T/\omega_B$ of thermal electrons, i.e., for $\chi = k_\perp^2 v_T^2/\omega_B^2 \ll 1$. The number $s \leq 3$ in the above equation is the number of one of the harmonic l. The equation (2.160) has the solution both at frequencies close to cyclotron harmonic frequencies $s\omega_B$:

$$\omega^2 - s^2\omega_B^2 = \frac{\omega_{\rm L}^2\omega_B^2}{(s^2-1)\omega_B^2 - \omega_{\rm L}^2}\frac{s(s+1)}{(s-2)!}\left(\frac{\chi}{2}\right)^{s-1}, \tag{2.161}$$

and at the frequency

$$\omega^2 = \omega_{\rm h}^2 + 3\chi\omega_B^2(1 - 3\omega_B^2/\omega_{\rm L}^2)^{-1} = \omega_{\rm h}^2 + 3k_\perp^2 v_T^2(1 - 3\omega_B^2/\omega_{\rm L}^2)^{-1} \tag{2.162}$$

[24] The condition constraining the range of the angles α in which propagation of weakly-damped, longitudinal waves is possible can be derived by combining the criterion (2.158) with the solutions of the dispersion equation (2.161), (2.164) (see below). In the narrow angular interval $\cos^2\alpha \ll 1$ the condition (2.158) is satisfied for any χ.

close to the upper hybrid frequency $\omega_{\rm h} = \sqrt{\omega_{\rm L}^2 + \omega_B^2}$.[25]

The solution simplifies also in the limit of large χ when the wavelength of a longitudinal wave is small compared to the gyroradius, and the frequency ω is again close to one of the harmonic $s\omega_B$. In this case the dispersion equation (2.157) is reduced to

$$\epsilon_{\|} = 1 - \frac{\omega_{\rm L}^2}{\omega^2 - s^2\omega_B^2} \frac{2s^2}{\chi\sqrt{2\pi\chi}} = 0, \tag{2.163}$$

while the corresponding solution is written in the form

$$\omega^2 - s^2\omega_B^2 \simeq \frac{2s^2\omega_L^2}{\chi\sqrt{2\pi\chi}}. \tag{2.164}$$

The dispersion curves of Bernstein modes for arbitrary χ from the interval (2.158) and various definite values of the ratio $\omega_{\rm L}^2/\omega_B^2$ are plotted in Fig. 2.9 [95]. It is clearly seen at this figure that the behavior of the dispersion curves in the band $(s-1)\omega_B < \omega < s\omega_B$ are qualitatively different for the cases when the upper hybrid frequency $\omega_{\rm h} = \sqrt{\omega_{\rm L}^2 + \omega_B^2}$ lies above, below or within this band. In the last this band is called the hybrid band.

If the frequency interval between two adjacent harmonic $(s-1)\omega_B$ and $s\omega_B$ lies below the hybrid band, the dispersion curve at small and large χ approaches the harmonic $(s-1)\omega_B$ and $s\omega_B$, or the frequencies of electron-cyclotron resonance. The frequency of a Bernstein mode varies according to the equations (2.164) and (2.161). However, at $\chi \sim s^2$ the dispersion curves come from one harmonic to another. In particular, at $\chi \approx s^2$ the dispersion equations yield the frequencies close to the middle of the interval between the adjacent harmonic. The frequency monotonously decreases with increasing $k_\perp$ in this case, i.e. Bernstein modes exhibit anomalous dispersion.

The behavior of the dispersion curves in the bands above the hybrid one is different. The frequency of Bernstein modes is close to the lower boundary of the band $(s-1)\omega_B$ both at small and at large χ, and deviates from this value at $\chi \sim s^2$. As the harmonic number s increases, the dispersion curves approach closer and closer to the lower boundary of the band, even in the above region of the parameter χ.

Finally, in the hybrid band the dispersion curves at $\chi > s$ are nearly the same as the ones for Bernstein modes in the bands below the hybrid one.

[25] If the frequency ω is close to one of the cyclotron harmonic $s\omega_B$ or to upper hybrid frequency $\omega_{\rm h}$, the equation (2.157) containing the infinite sum over harmonic can be simplified. Only the terms $l = 1$ and $l = 2$ determining the solution of the dispersion equation hear hybrid frequency and at the second harmonic, respectively, and, for $s \geq 3$, the resonance term $l = s \simeq \omega/\omega_B$ have been kept while transforming (2.157) to (2.160)

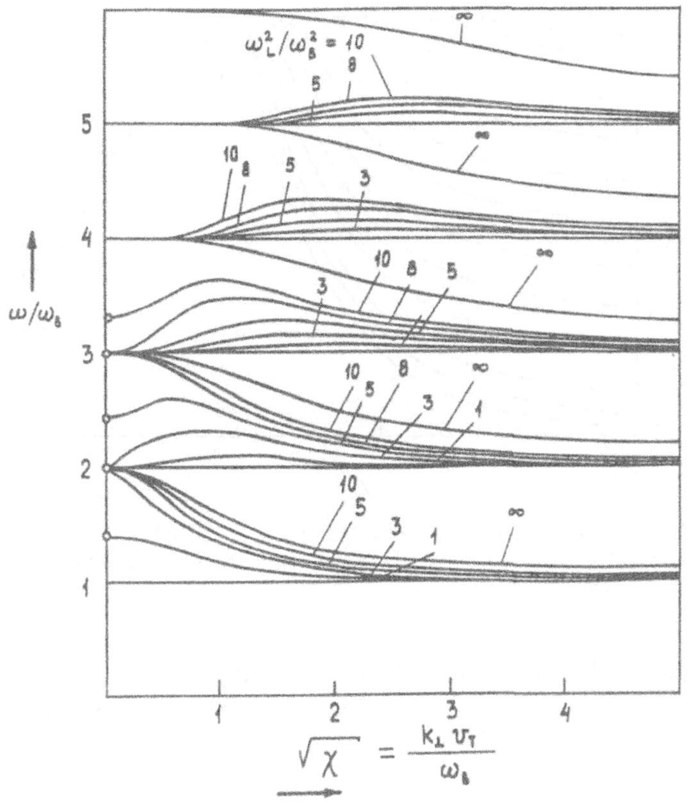

Figure 2.9. Dispersion curves of the Bernstein modes in non-relativistic equilibrium plasma. The light dots on the axis ω/ω_B mark the points where $\omega = \omega_h$.

The qualitative difference appears at $\chi < s$ where the derivative $d\omega/dk_\perp$ changes the sign. At small χ, according to the equation (2.162), the curves approach the value ω_h. The evolution of the dispersion curves in the hybrid band with varying ratio ω_h/ω_B is seen in Fig. 2.10 [96–98].

Note that the coincidence of an electron-cyclotron resonant frequency $s\omega_B$ with the resonance frequency ω_∞, (2.125), is usually called the double plasma resonance. In the case considered, when $\alpha \approx \pi/2$, the frequency ω_∞ corresponds to the upper hybrid frequency ω_h, so the double plasma resonance occurs if $\omega_h = s\omega_B$. As frequencies ω_L or ω_B change, this condition may be realized twice, at the upper and lower boundary of the hybrid band.

The above consideration has been restricted to longitudinal electron-cyclotron waves. The detailed analysis of the dispersion equation (2.41) with the tensor $\epsilon_{\alpha\beta}$ determined by (2.99) shows that, in addition to the

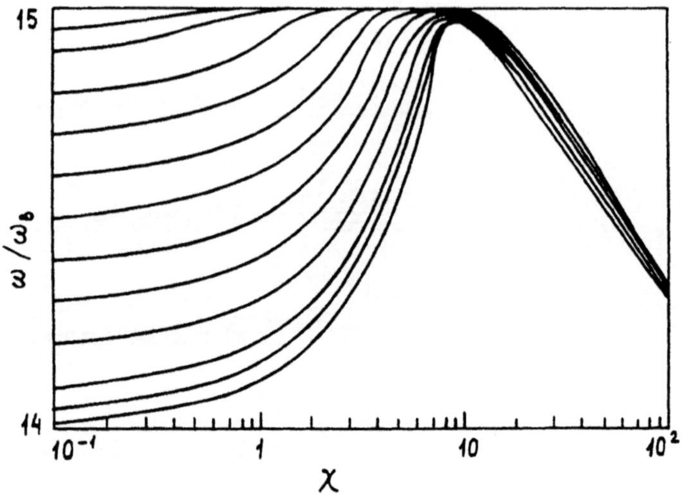

Figure 2.10. Dispersion curves of longitudinal waves in the hybrid band at fixed ω_h/ω_B within the interval $14 < \omega_h/\omega_B < 15$ (the angle $\alpha = \pi/2$). At small χ the value of ω_h/ω_B for each curve is close to ω/ω_B.

above considered modes, the ordinary and extraordinary cyclotron waves with frequency $\omega \approx s\omega_B$ exist in the case of transverse propagation. These waves are not considered here, and for more detailed information we refer the reader to the monograph [72,99].

2.2.5. WAVES IN RELATIVISTIC PLASMA

The properties of electromagnetic waves in a non-relativistic or weakly relativistic plasmas ($\beta_T^2 = v_T^2/c^2 \ll 1$) have been considered in previous paragraphs of this section. This approximation is quite sufficient for the majority of astrophysical objects. Due to the discovery of pulsars, interest has grown considerably in the investigations of relativistic plasmas. The latter is explained by the possible existence of intense electric fields in the pulsar magnetospheres capable to accelerate particles up to relativistic energies (section 1.1). It is quite possible that the magnetospheric plasma as a whole is relativistic. Of course, when analyzing the mode properties in pulsar magnetosphere one can not ignore the influence of the super-strong magnetic field of the neutron star. However, for the sake of simplicity, we firstly consider the waves in relativistic isotropic plasma.

Similar to the non-relativistic case, transverse electromagnetic and longitudinal plasma waves can propagate in a relativistic plasma. These waves are described by the dispersion equations (2.46) and (2.47) with $\epsilon_{\parallel}$ and $\epsilon_{\perp}$ in the form (2.86). The latter formulas for isotropic distribution $f_0(\boldsymbol{p}) = f_0(p)$

of relativistic particles with $v \approx c$ and $\mathcal{E} \approx pc$ can be transformed to

$$\epsilon_{\perp} \approx 1 - \frac{16\pi^2 e^2}{k^2\omega^2} \times \int_0^{\infty} f_0(p) p \left[\frac{\omega^2}{k^2c^2} - \left(1 - \frac{\omega^2}{k^2c^2}\right) \frac{\omega}{kc} \ln \left| \frac{1 - kc/\omega}{1 + kc/\omega} \right| \right] dp, \tag{2.165}$$

$$\epsilon_{\parallel} \approx 1 + \frac{16\pi^2 e^2}{\omega^2} \times \int_0^{\infty} f_0(p) pv \left[2\frac{\omega^2}{k^2v^2} + \frac{\omega^3}{k^3v^3} \ln \left| \frac{1 - kv/\omega}{1 + kv/\omega} \right| \right] dp. \tag{2.166}$$

For monoenergetic particle spectrum

$$f_0(p) = \frac{N}{4\pi p^2} \delta(p - p_0) \tag{2.167}$$

we find from (2.165) the refractive index

$$n_{\rm em}^2 = \epsilon_{\perp} \approx 1 - \frac{4\pi e^2 N c}{\omega^2 p_0} = 1 - \frac{\omega_{\rm L}^2}{\omega^2} \frac{m_e c^2}{\mathcal{E}_0} \tag{2.168}$$

valid at $|1 - k^2c^2/\omega^2| \ll 1$ ($\omega_{\rm L}$ is the non-relativistic plasma frequency). The refractive index of plasma waves for the same particle spectrum (2.166) equals

$$n_{\rm pl}^2 \approx 1 - 4\exp\left(-2 - \frac{\omega^2}{\omega_{\rm L}^2} \frac{\mathcal{E}_0}{m_e c^2}\right). \tag{2.169}$$

This formula is valid in the interval $1 \gg |1 - n_{\rm pl}^2| \gg (m_e c^2/\mathcal{E}_0)^2$, where $\mathcal{E}_0$ is the energy of relativistic electrons.

The distribution function of the equilibrium relativistic plasma ($\kappa T \gg m_e c^2$) differs from the Maxwell one and has the form [55]

$$f_0(\mathbf{p}) = \frac{N}{8\pi} \left(\frac{c}{\kappa T}\right)^3 e^{-cp/\kappa T}. \tag{2.170}$$

Substitution of (2.170) into (2.165) leads to the following expressions for refraction index of electromagnetic waves [78]:

$$\begin{aligned} n_{\rm em}^2 &\approx 1 - \frac{\omega_{\rm L}^2}{\omega^2} \frac{m_e c^2}{2\kappa T}, && (1 - n \ll 1), \\ n_{\rm em}^2 &\approx \tfrac{5}{6}\left(1 - \frac{\omega_{\rm L}^2}{\omega^2} \frac{m_e c^2}{3\kappa T}\right), && (n \ll 1). \end{aligned} \tag{2.171}$$

Respectively, the formulas (2.166), (2.170) for plasma waves yield [78]

$$n_{\rm pl}^2 \approx 1 - 4\exp\left(-2 - \frac{\omega^2}{\omega_{\rm L}^2}\frac{2\kappa T}{m_{\rm e}c^2}\right), \quad (1-n \ll 1),$$
$$n_{\rm pl}^2 \approx \tfrac{5}{3}\left(1 - \frac{\omega_{\rm L}^2}{\omega^2}\frac{m_{\rm e}c^2}{3\kappa T}\right) \qquad (n \ll 1). \tag{2.172}$$

The above formulas indicate that in a relativistic plasma $n_{\rm em}^2 < 1$, so that, similar to the non-relativistic case, the phase velocity of transverse waves exceeds c. The formulas for $n_{\rm pl}^2$ also correspond to the case $n_{\rm pl} < 1$. But if the phase velocity of plasma waves, $v_{\rm ph}$, becomes less than c, strong Landau damping should occur at particles with velocities $v \approx c$. However, it is not clear whether the solutions of the dispersion equation $\epsilon_{||}(\omega, k) = 0$ of equilibrium plasma exists in this region.

The characteristics of electromagnetic waves in a relativistic plasma can be studied using the particle distribution functions (2.167) or (2.170). It is however quite unreal that such distributions are realized in some specific astrophysical objects with strong magnetic field, for example, in neutron star magnetospheres. However, among all the non-equilibrium distributions imaginable the following one can be distinguished :

$$f_0(\boldsymbol{p}) = \mathcal{F}(p_{||})\frac{\delta(p_\perp)}{2\pi p_\perp}, \qquad \mathcal{F}(p_{||}) = \frac{N}{2m_{\rm e}c^2}\left[1 + \frac{p_{||}^2}{m_{\rm e}^2c^2}\right]^{-3/2}. \tag{2.173}$$

Any isotropic relativistic distribution should relax to (2.173) because of the magnetic bremsstrahlung energy losses in a strong magnetic field $\boldsymbol{B}_0$ (see section 5.1 and the formula (5.38) in particular). In a plasma with such one-dimensional electron distribution the longitudinal waves with $\boldsymbol{k}||\boldsymbol{B}_0$ are described by the following dispersion equation [100]:

$$\omega^2 = \frac{c^2k^2}{4}\left(\frac{ck}{\tilde{\omega}_{\rm L}} + \frac{\tilde{\omega}_{\rm L}}{ck}\right)^2, \qquad \tilde{\omega}_{\rm L}^2 = \frac{3\pi}{4}\omega_{\rm L}^2. \tag{2.174}$$

As follows from this equation, the frequencies of plasma waves span over the wide range from $\tilde{\omega}_{\rm L}/2$ at $k=0$ to infinity as $k \to \infty$. The phase velocity of these waves exceeds c at any k, so the Landau damping of these waves is impossible. It should be emphasized in this respect that the existence of longitudinal waves with $v_{\rm ph} < c$ in a relativistic plasma depends on the specific form of the distribution function $f_0(\boldsymbol{p})$. The examples of the distribution $f_0(\boldsymbol{p})$ can be pointed out for which plasma waves with $v_{\rm ph} < c$ certainly exist and these are subject to the Landau damping (see [9,10]).

2.2.6. ELECTROMAGNETIC WAVES IN A SYSTEM COMPRISING BIREFRINGENT VACUUM AND RAREFIED PLASMA

Strong magnetic fields close to the critical value

$$B_{\rm cr} = m_e^2 c^3/\hbar e \simeq 4.44 \cdot 10^{13}\,{\rm G} \tag{2.175}$$

are known to alter physical properties of vacuum causing its polarization and magnetization. The value $B_{\rm cr}$ corresponds to a magnetic field strength at which the gyrofrequency quantum energy $\hbar\omega_B$ becomes equal to the electron rest energy $m_e c^2$. Such strong fields typical for neutron stars may affect the propagation of electromagnetic waves. The refractive indices and polarization of the modes turn out to be dependent on the direction and strength of the magnetic field $\boldsymbol{B}_0$ which now determine the components of the tensors $\epsilon_{\alpha\beta}$ and $\mu_{\alpha\beta}$ of the anisotropic, birefringent "medium" — magnetized vacuum. The components of the tensors $\epsilon_{\alpha\beta}$ and $\mu^{-1}_{\alpha\beta}$ in the frame II (Fig. 2.1, *b*) have the following form:[26]

$$\epsilon_{\alpha\beta} = \delta_{\alpha\beta}(1-2a) + 7a\frac{B_\alpha B_\beta}{B_0^2}, \tag{2.176}$$

$$\mu^{-1}_{\alpha\beta} = \delta_{\alpha\beta}(1-2a) + 4a\frac{B_\alpha B_\beta}{B_0^2}, \tag{2.177}$$

where $\delta_{\alpha\beta}$ is the Kronecker delta and the parameter

$$a = \frac{1}{45\pi}\frac{e^2}{\hbar c}\frac{B_0^2}{B_{\rm cr}^2} \simeq 5.16 \cdot 10^{-5} B_0^2 B_{\rm cr}^{-2}. \tag{2.178}$$

These formulas are valid in the limit $B_0 \ll B_{\rm cr}$ when the electron-positron pair creation and the corresponding absorption of electromagnetic waves can be neglected. Note that the form of the tensors (2.176) and (2.177) indicate that the birefringent vacuum is equivalent to a uniaxial anisotropic, non-gyrotropic medium. The non-zero components of dielectric and inverse magnetic tensors in the frame I (Fig. 2.1, *a*) look like

$$\begin{aligned} \epsilon_{xx} = \epsilon_{yy} = 1-2a, &\quad \epsilon_{zz} = 1+5a, \\ \mu^{-1}_{xx} = \mu^{-1}_{yy} = 1-2a, &\quad \mu^{-1}_{zz} = 1-6a. \end{aligned} \tag{2.179}$$

The influence of a strong magnetic field on electromagnetic wave propagation in vacuum has been considered in [101]. This paper was recognized later on in [102], and the attention was drawn to the importance of the

[26] The inverse tensor $\mu^{-1}_{\alpha\beta}$ is defined by the relation $H_\alpha = \mu^{-1}_{\alpha\beta} B_\beta$ between the components H_α and B_β of the vectors **H** and **B**.

vacuum birefringence effect to the theory of X-ray emission from neutron stars.

Consider now the modes in a homogemeous system comprising a rarefied plasma and birefringent vacuum [103, 104]. These waves are described by the system of Maxwell equations for the fields $\boldsymbol{E}, \boldsymbol{B} \propto \exp(-i\omega t)$:

$$\begin{aligned} \operatorname{curl}(\hat{\mu}^{-1}\boldsymbol{B}) &= -\frac{i\omega}{c}\left(\hat{\epsilon} + \hat{\epsilon}^{\mathrm{pl}} - \hat{\delta}\right)\boldsymbol{E}, \\ \operatorname{curl}\boldsymbol{E} &= \frac{i\omega}{c}\boldsymbol{B}, \end{aligned} \tag{2.180}$$

where the operators are introduced according to the following rules:

$$\begin{aligned} \hat{\epsilon} \to \epsilon_{\alpha\beta}E_\beta, \quad & \hat{\epsilon}^{\mathrm{pl}} \to \epsilon^{\mathrm{pl}}_{\alpha\beta}E_\beta, \\ \hat{\mu}^{-1} \to \mu^{-1}_{\alpha\beta}B_\beta, \quad & \hat{\delta} \to \delta_{\alpha\beta}E_\beta, \end{aligned} \tag{2.181}$$

the summation over repeating subscripts being assumed. The tensor $\epsilon_{\alpha\beta}$ characterizes the influence of birefringent vacuum, while $\epsilon^{\mathrm{pl}}_{\alpha\beta}$ stands for the effect of the rarefied plasma (see (2.145)). The system (2.180) for a wave electric field $\boldsymbol{E} \propto \exp(i\boldsymbol{kr} - i\omega t)$ is reduced to the following equation

$$[\boldsymbol{k} \times \hat{\mu}^{-1}[\boldsymbol{k} \times \boldsymbol{E}]] = -\frac{\omega^2}{c^2}\left(\hat{\epsilon} + \hat{\epsilon}^{\mathrm{pl}} - \hat{\delta}\right)\boldsymbol{E}. \tag{2.182}$$

This vector equation is equivalent to the homogeneous linear system of three algebraic equations, the corresponding projections onto the coordniate axes of the frame I. The non-trivial solution of this equation can only exist if its determinant is equal to zero:

$$\begin{aligned} 4\eta^2 - 2\eta\frac{\omega}{c}\left[ib(1+\cos^2\alpha) + 11a\sin^2\alpha\right] + \\ \frac{\omega^2}{c^2}\sin^2\alpha\left[iba(7+4\cos^2\alpha) + 28a^2\sin^2\alpha\right] = 0, \end{aligned} \tag{2.183}$$

where $\eta = k - \omega/c$. The obtained dispersion relation yields the following refractive indices of two electromagnetic modes, ordinary and extraordinary, in a system comprising birefringent vacuum and rarefied plasma:

$$n_{1,2} = 1 + \frac{c}{\omega}\operatorname{Re}\eta, \tag{2.184}$$

as well as the corresponding absorption coefficients

$$\mu_{1,2} = 2\operatorname{Im}\eta. \tag{2.185}$$

In a pure vacuum $b = 0$, so the equation (2.183) yields $\mathrm{Im}\,\eta = 0$. Thus, there is no absorption of electromagnetic waves in this case. The corresponding vacuum refractive indices $n = kc/\omega$ are given by [101]:

$$n_1 = 1 + \frac{7}{2} a \sin^2 \alpha, \qquad n_1 = 1 + 2a \sin^2 \alpha. \tag{2.186}$$

Both modes are linearly polarized. The vector $\boldsymbol{E}_1$ of the extraordinary mode lying in the xz-plane at a right angle to the wave vector $\boldsymbol{k}$, the radio $E_x/E_z = -\cot\alpha$. The electric field vector $\boldsymbol{E}_2$ of the ordinary wave is directed along the y-axis.

In a system of birefringent vacuum and extremely rarefied plasma (cf. the criterion (2.144)) which satisfy the condition

$$\frac{\omega_L^2}{\omega^2 \beta_T} \ll \frac{1}{45\pi} \frac{e^2}{\hbar c} \frac{B_0^2}{B_{\mathrm{cr}}^2} \tag{2.187}$$

the parameter $|b| \ll a$, so that the solution of the dispresion equation (2.183) takes the following form:

$$\begin{aligned} \eta_1 &= \frac{\omega}{2c}\left(7a \sin^2 \alpha + ib \cos^2 \alpha\right), \\ \eta_2 &= \frac{\omega}{2c}\left(4a \sin^2 \alpha + ib\right). \end{aligned} \tag{2.188}$$

According to these formulas, the refractive indices, as before, are determined by the expressions (2.186) with the accuracy of the order b/a. In addition, however, cyclotron absorption appears at frequencies $\omega \approx \omega_B$. Such an absorption of collisionless nature originates due to the influence of plasma. This will be considered in detail in section 6.2.

The linear polarization of the modes under the condition (2.187) is caused by birefringent vacuum, whereas rarefied plasma gives rise to the absorption in the system. The condition (2.187) can be realized in the strong magnetic fields of neutron stars. For example, at a star with $B_0 \sim 0.1\, B_{\mathrm{cr}}$ this inequality is valid for $\beta_T \simeq 0.15$ and plasma densities $N \ll 5 \cdot 10^{19}\,\mathrm{cm}^{-3}$ at frequencies $\omega \sim 1.5 \cdot 10^{18}\,\mathrm{s}^{-1}$ which correspond to hard X-rays.

On the contrary, under the condition opposite to (2.187) the effect of birefrinent vacuum can be neglected. Here the parameter $|b| \gg a$, so polarization, refractive indices, and absorption coefficients of the modes are entirely determined by the rarefied plasma. This case has already been analysed above (see the formulas (2.148)–(2.150)). It is realized in white dwarf conditions at frequencies $\omega \approx \omega_B$ which, typically, fall into the optical band. Note that the criterion (2.187) is applicable to radiation in the spectral band $(\omega - \omega_B)^2 \lesssim \omega^2 \beta_T^2$ near the gyrofrequency. At frequencies outside this interval electromagnetic wave propagation is dominated by birefringent vacuum if [105]

$$\frac{\omega_{\mathrm{L}}^2}{\omega^2}\frac{\omega_B^2}{|\omega^2-\omega_B^2|} \ll \frac{1}{15\pi}\frac{e^2}{\hbar c}\frac{B_0^2}{B_{\mathrm{cr}}^2}. \tag{2.189}$$

Electromagnetic waves in inhomogeneous astrophysical plasma

Inhomogeneities of astrophysical plasmas and magnetic fields significantly affect electromagnetic wave propagation. Waves in regularly inhomogeneous astrophysical plasmas are considered in this chapter. At the same time, the characteristics of the observed astrophysical radiation can be altered because of the scattering by chaotic macroscopic inhomogeneities. We mean here the increase of pulse duration of pulsars at low frequencies due to scattering of radiation by an interstellar medium, the increase of angular size of discrete radio sources due to their eclipse by solar "extra-corona", scintillations of discrete sources due to scattering by the interplanetary medium etc. Unfortunately, the limited size of this book prevents consideration of these effects here, so we refer the reader to the reviews [106, 107], original papers [108–113] and bibliography therein. Note also that scattering by plasma particles (electrons) in strong magnetic fields may significantly affect optical and X-ray radiation of magnetic degenerate stars. These effects are considered in detail in chapter 8.

3.1. Geometrical optics approximation in an inhomogeneous medium

3.1.1. THE FORM OF THE GEOMETRICAL OPTICS SOLUTION

Electromagnetic wave propagation in a homogeneous, infinite medium has been considered in the previous chapter. Such waves are described by the system of Maxwell equations (2.19)-(2.22). The corresponding solutions are "normal waves" (or modes) $\boldsymbol{B}$, $\boldsymbol{E} \propto \exp(i\boldsymbol{k}_j\boldsymbol{r} - i\omega t)$ or their superposition. However, such solutions do not satisfy the same Maxwell equations in an inhomogeneous medium. Nevertheless, these equation can be solved in the so-called geometrical optics approximation if plasma parameters vary relatively weakly over the size of about the wavelength. Geometrical optics approximation explores an intuitively clear assumption that the propagation of electromagnetic waves in the relatively small region of such a smoothly inhomogeneous medium does not differ much from that in the homogeneous medium with similar parameters. Thus, the corresponding geometrical optics solution for the fields $\boldsymbol{E}$ and $\boldsymbol{B}$ should generally resemble the one for the homogeneous medium. One should only bear in mind that the wave vector $\boldsymbol{k}_j$ varies with $\boldsymbol{r}$, so the phase variation due to propagation over a distance r equals $\int \boldsymbol{k}_j d\boldsymbol{r}$, not simply $\boldsymbol{k}_j \boldsymbol{r}$. Therefore,

$$\boldsymbol{E} = \boldsymbol{E}_a(\boldsymbol{r}) \exp(i \int \boldsymbol{k}_j d\boldsymbol{r} - i\omega t), \qquad \boldsymbol{B} = \boldsymbol{B}_a(\boldsymbol{r}) \exp(i \int \boldsymbol{k}_j d\boldsymbol{r} - i\omega t) \quad (3.1)$$

The subscript j in (3.1) indicates the mode: ordinary, extraordinary or plasma. In addition, the above formulas account for the possible variation of the wave amplitude in the considered approximation.

Note that the vector $\boldsymbol{k}_j$ and the ratio of the components of the vector $\boldsymbol{E}_a$ (the wave polarization) are found from the dispersion equation (2.41) and the system of linear equations (2.40) which describe the homogeneous plasma with the same parameters as in the inhomogeneous medium at a given point $\boldsymbol{r}$. [1] It is thus clear that geometrical optics approximation does not alter refractive indices n_j, the number and polarization of modes in an inhomogeneous medium, as compared to the homogeneous one.

If there are no regions of violation of the geometrical optics in an inhomogeneous medium or such regions are not reached by waves, the solutions (3.1) are linearly independent. Then arbitrary electromagnetic field can be presented as a superposition of the solutions (3.1) with different values of $\boldsymbol{k}_j$ and amplitudes $\boldsymbol{E}_a$, $\boldsymbol{B}_a$. That is why the geometrical optics solutions are called "normal waves" (modes) in a smoothly inhomogeneous medium.

On the contrary, if waves propagate through a region where geometrical optics fails and the expressions for the fields $\boldsymbol{E}, \boldsymbol{B} \propto e^{-i\omega t}$ are radically different from (3.1) or from a linear superposition of these functions, the geometrical optics solutions (3.1) are not independent outside such a region. The mode amplitudes are linked by a certain relation, the form of which depends on the solution inside the region where the geometrical optics approximation is violated. Such an effect, called "linear mode coupling" in plasma, is considered in section 3.2. The geometrical optics has a rather wide validity range if applied to smoothly inhomogeneous media. It is violated only in relatively small regions. So, in the present section the features of wave propagation in the framework of the geometric optics are discussed.

3.1.2. GEOMETRICAL OPTICS APPROXIMATION IN ISOTROPIC AND MAGNETOACTIVE PLASMA

The dependence of refractive indices $n_j = k_j c/\omega$ on the parameter $v = \omega_L^2/\omega^2$ in isotropic plasma is shown in Fig 2.2 in section 2.2. If electron density N

[1]The ratio of the components of the magnetic field $\mathbf{B}_a$ can be determined with equation (2.33) using the known amplitude $\mathbf{E}_a$.

of inhomogeneous plasma varies linearly with a Cartesian coordinate z,[2] the plots at this figure show, in some scale, the spatial dependence $n_j(z)$ as well.

For waves propagating along the z-axis in a smoothly inhomogeneous, plane-parallel medium the geometrical optics approximation is valid anywhere except the point at which $v=1$, i.e., $\epsilon=0$ and $n_j^2=0$; see (2.104), (2.116). The latter is not surprising since the wavelength $\lambda_j = 2\pi c/\omega n_j$ of an electromagnetic or Langmuir wave in plasma grows monotonically, approaching the level $n_j^2=0$. Therefore, the relative variation of the medium properties over a distance $\lambda_j/2\pi$ gets larger, and finally the main validity criterion of geometrical optics [71],

$$\frac{d\epsilon}{dz}\frac{\lambda}{2\pi} \ll \epsilon \tag{3.2}$$

fails.

Approximating the function $\epsilon(z)=1-\omega_L^2/\omega^2$ in the layer $\epsilon\approx 0$ by the linear one, $\epsilon=-za$, where the constant $a=|d\epsilon/dz|$, reduces the inequality (3.2) to the form

$$|z| \gg (c^2/\omega^2 a)^{1/3} \tag{3.3}$$

for transverse waves with $\lambda_{\rm em}=2\pi c/(\omega\sqrt{\epsilon})$ and to

$$|z| \gg (3v_T^2/\omega^2 a)^{1/3} \tag{3.4}$$

for longitudinal waves with $\lambda_{\rm pl}=2\pi\sqrt{3}v_T/(\omega\sqrt{\epsilon})$. For example, in the solar corona where $|d\epsilon/dz|\sim 10^{-10}\,{\rm cm}^{-1}$, the geometrical optics at frequencies $\omega\sim 2\pi\cdot 10^8\,{\rm s}^{-1}$ is violated only near the level $\epsilon(\omega)=0$ in the narrow layer of the size $|z|\lesssim 2\cdot 10^4\,{\rm cm}$ for electromagnetic and $|z|\lesssim 10^3\,{\rm cm}$ for plasma waves.[3]

The geometrical optics approximation for electromagnetic waves propagating along z-axis in the regions (3.3) has the following form [71]:

$$\boldsymbol{E} = c_1\frac{\boldsymbol{e}_1}{\sqrt{n_1}}e^{i\frac{\omega}{c}\int n_1 dz - i\omega t} + c_2\frac{\boldsymbol{e}_2}{\sqrt{n_2}}e^{-i\frac{\omega}{c}\int n_2 dz - i\omega t} \tag{3.5}$$

The factor $1/\sqrt{n_j}$ determines the dependence of electric field amplitude on the coordinate z, while the constants c_1 and c_2, the corresponding absolute values of the normal wave amplitudes. The vector $\boldsymbol{e}_j$ characterizes the polarization of the j-th mode with the refractive index $n_j(\omega)$. Any two waves with the same $n_j(\omega)=n_{\rm em}$ determined by (2.104) and orthogonal

[2] Such a medium is called plane-parallel.

[3] According to the rigorous solution of the problem of wave propagation in a linear plane-parallel medium given, for example, in [71], the right-hand sides of (3.3), (3.4) are of about the maximal possible wavelength $\lambda_{\max}$ in plasma.

polarizations e_j can be considered as the "normal" ones ($j = 1, 2$) in isotropic plasma. The electric field of a plasma wave ($j = 3$) in a region (3.4) is determined by the same expression (3.5) with $e_j \parallel k_j$ and the factor $\omega_L/\omega\sqrt{n_{pl}}$ instead of $1/\sqrt{n_{em}}$ [114, 115].

In a region $z < 0$, where $n_j^2 > 0$ and the refractive index $n_j(z)$ is real, the first term in (3.5) describes a wave propagating into the plasma in the direction of positive z, while the second one corresponds to a wave in the opposite direction. Alternatively, at $z > 0$ the value $n_j^2 < 0$ and the refractive index $n_j(z)$ is pure imaginary.[4] Here the first term in (3.5) corresponds to a wave exponentially decaying with increasing z while the second term to an exponentially growing one. The violation of geometrical optics approximation (3.5) in the region $n_j^2 \approx 0$ causes the "linear coupling" of counter waves and gives rise to the reflection of waves. The relation between the coefficients c_1 and c_2 in the regions $z < 0$ and $z > 0$ is set owing to this effect. For example, a wave with the amplitude c_1 infalling a linear layer gives rise to the reflected one with the amplitude $c_2 = c_1 e^{-i\pi/2}$ escaping from the region $n_j^2 \approx 0$ into $z < 0$ and the exponentially decaying (non-propagating) field with the same amplitude $c_1 e^{-i\pi/2}$ in the region $z > 0$.

In case of an oblique propagation in a plane-parallel medium with respect to $\nabla\epsilon$ the electric field of a wave can be written in the form $\boldsymbol{E} = \boldsymbol{F}(z)e^{ik_x x + ik_y y}$ [71]. The coordinate system can be rotated around z-axis to align x-axis and the projection of $\boldsymbol{k}_j$ onto the plane perpendicular to z. Such a procedure eliminates the k_y projection, so without loss of generosity, the above expression for the electric field can be reduced to

$$\boldsymbol{E} = \boldsymbol{F}(z)e^{ik_x x}, \qquad k_x = \text{const}. \tag{3.6}$$

Since $k_x = \text{const}$, the function $\boldsymbol{E}(z)$ is entirely characterized by the function $\boldsymbol{F}(z)$.

The electromagnetic field with the vector $\boldsymbol{E}$ lying, together with $\boldsymbol{k}_j$, in the plane xz and the field with $\boldsymbol{E}$ perpendicular to this plane are described by mutually independent equations. This means that the waves with the above indicated orientations of electric field propagate independently in a plane-parallel medium. The function $\boldsymbol{F}(z)$ for an electromagnetic wave with $\boldsymbol{E}$ perpendicular to the xz-plane is generally similar to the case of propagation along z-axis which is described by the function (3.5). However, the position of the reflection point at which the propagating wave with oscillatory spatial structure turns out to the exponential non-propagating

[4] See section 2.2 on the properties of waves in the region of pure imaginary values of n_j.

one is now determined by the condition

$$n_j(z) = \frac{ck_x}{\omega}, \tag{3.7}$$

contrary to $n_j(z)=0$ in case of normal incidence.

The function $\boldsymbol{F}(z)$ is more peculiar for waves with the vector $\boldsymbol{E}$ in the propagation plane. Of course, the corresponding electromagnetic wave is again reflected from the point $n_{\rm em}=ck_x/\omega$. But, in addition, the geometrical optics also fails for this wave at the point of plasma wave reflection: $n_{\rm pl}(z)=ck_x/\omega$. Therefore, the coupling between electromagnetic and plasma wave occurs near this point. It causes the transformation from one mode to another: electromagnetic to plasma wave or vice versa. Such an effect is discussed in section 3.2.

Consider now the case of magnetoactive plasma. Note, first of all, that, since $\omega_{\rm L}^2=4\pi e^2N/m_{\rm e}$ and so the parameter $v=\omega_{\rm L}^2/\omega^2 \propto N$, the dispersion curves $n_j^2(v)$ plotted in Figs. 2.3–2.5 in section 2.2 show at the same time the dependence of n_j^2 on plasma density N in a constant magnetic field $\boldsymbol{B}_0$ and at fixed frequency ω and the angle α between $\boldsymbol{k}_j$ and $\boldsymbol{B}_0$. The same curves, with the appropriate abscissa scale, also show the function $n_j^2(z)$ in an inhomogeneous, plane-parallel, "linear" plasma region with the parameters $u=\text{const}$ and $v=az$, where a is a constant. The general behavior of the dispersion curves: the relative position of the null points and the poles of n_j^2, the regions of coupling between different modes, remains similar for any monotonous function $v(z)$.

Geometrical optics approximation in the form (3.1) is usually valid in magnetoactive plasma under the condition

$$\frac{\omega}{c}n_j\Lambda \gg 1, \tag{3.8}$$

where Λ is the characteristic scale for those parameters of a medium which determine the wave propagation.

Investigation of electromagnetic wave propagation in astrophysical plasma gets more complicated if, in addition to the electron density N (or the parameter v), the magnetic field $\boldsymbol{B}_0$ exhibits considerable spatial variation thus causing the corresponding variation of u and α. The examples of such conditions are brought by pulsar or planetary magnetospheres, active regions above sunspots etc. The case of pulsars is even more complicated since plasma is, quite probably, relativistic and non-equilibrium there. Theory of wave propagation in such plasmas is far from being completed (see section 2.2), so, despite the apparent importance of this problem to the theory of pulsar radiation, it is not considered in this book.

The case of active regions in the solar corona and chromosphere is simpler since the bulk of the plasma is non-relativistic and equilibrium.

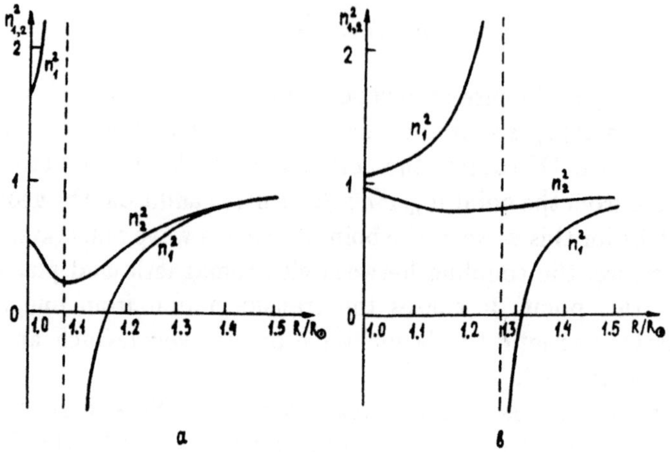

Figure 3.1. The functions $n^2_{1,2}(R/R_\odot)$ in coronal plasma above a sunspot ($\alpha = 0$): *a*, $B_b = 2.5 \cdot 10^2$ G, *b*, $B_b = 2.5 \cdot 10^3$ G.

Nevertheless, the distribution of n_j^2 is quite complicated even in this case. To get an impression on the character of this function, consider, as the simplest example, the profile of n_j^2 as a function of the distance from the solar center, $R > R_\odot$, along the axis of a unipolar sunspot ($R_\odot$ is the radius of the photosphere) [116]. The magnetic field of the spot at the line considered is determined by the formula (1.17), and electron density in the corona is assumed to be given by the Baumbach-Allen formula (1.16).

Figs. 3.1–3.3 show the refractive indices n_j^2 for extraordinary ($j=1$) and ordinary ($j=2$) mode at frequency $\omega = 2\pi \cdot 10^8\,\mathrm{s}^{-1}$ and various angles α between $\boldsymbol{B}_0$ and $\boldsymbol{k}_j$ versus $R/R_\odot$ in the corona above a very large sunspot with the magnetic pole area $b^2 = 3 \cdot 10^{19}\,\mathrm{cm}^2$ and different values of the magnetic field strength B_b in the sunspot center.

In a sunspot model with a strong magnetic field, Fig. 3.2, *b*, the level $\omega_B = \omega$ in the solar atmosphere lies above the level $\omega_L = \omega$, whereas the opposite configuration is realized for the weak field model, Fig. 3.2, *a*. This explains the qualitative difference between the plots $n^2_{1,2}(R/R_\odot)$ in these figures.

The refractive indices $n^2_{1,2}(R/R_\odot)$ in the "cold plasma" approximation $\beta_T = v_T/c \to 0$ become infinite at the coronal levels where the condition (2.123) is satisfied. This condition, in particular, reveals that the functions $n^2_{1,2}(R/R_\odot)$ have two poles $u \approx 1$ and $v \approx 1$ at small angles α (see Fig. 3.2). As soon as this angle increases, the ordinary wave pole situated closer to the photosphere shifts to the inner corona and then disappears beneath the photosphere. On the contrary, the more distant extraordinary wave pole

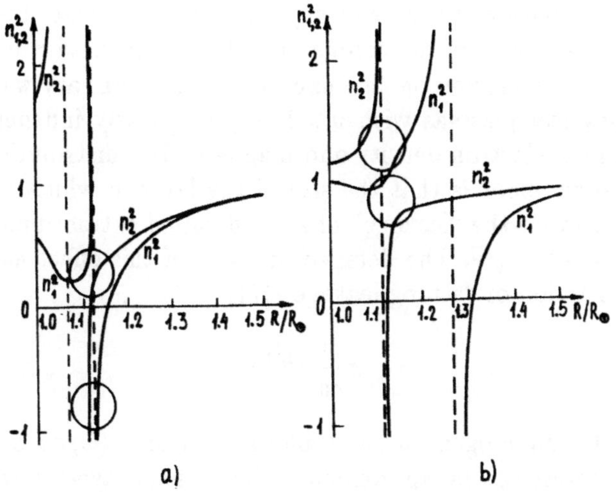

Figure 3.2. The same as in Fig. 3.1 but for $\alpha = 15°$. The circles indicate the regions of coupling of different modes.

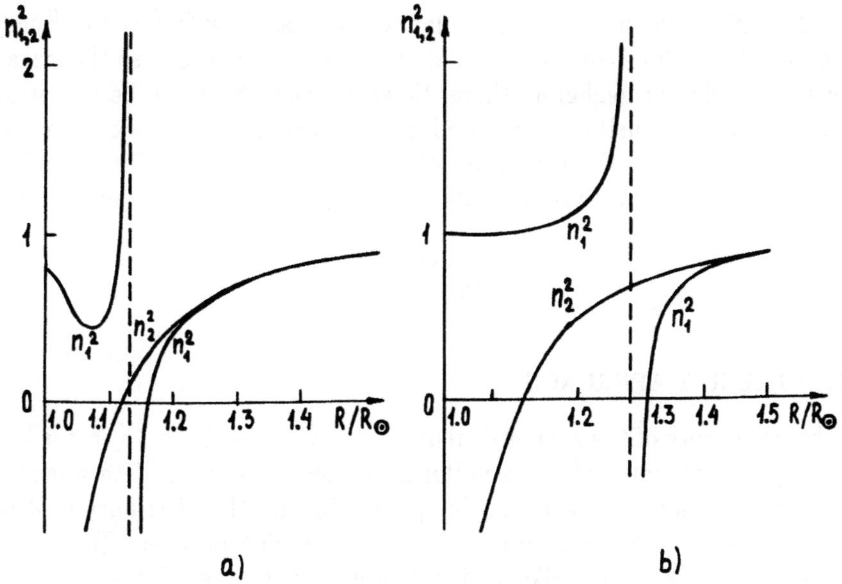

Figure 3.3. The same as in Fig. 3.1 but for $\alpha = 90°$.

exists at any α and shifts to the outer corona as this angle increases. The position of this pole at $\alpha = \pi/2$ is determined by the equation $1 - u - v = 0$, so it is located in the region $u < 1$, $v < 1$.

The function $n_1^2(R/R_\odot)$ for the extraordinary wave is negative in the region between the above mentioned pole and the null of this function which is situated higher in the corona at $v = 1 - \sqrt{u}$. Therefore, the extraordinary wave is non-propagating (exponentially damped) in this region. Thus, extraordinary radiation at the frequency ω generated beneath the level $v = 1 - \sqrt{u}$ can not escape from the corona into the interplanetary medium.

The above conclusion on the inability of extraordinary waves to escape the astrophysical plasmas with $v > 1 - \sqrt{u}$ is nearly independent on the specific form of electron density and magnetic field distributions. The only essential requirement is that the size of the layer in which $n_1^2 < 0$ is large enough to prevent the "leakage" of an undamped extraordinary wave from the region $v > 1 - \sqrt{u}$. The detailed analysis of this effect has yielded the following criterion of "strong leakage" [71]

$$\pi \frac{\omega}{ca} u^{3/4} \sim 1 \tag{3.9}$$

for smoothly inhomogeneous astrophysical plasma ($\omega/ca \gg 1$) when the region of interest can be approximated by a linear layer: $v = az$, $u =$ const. The above condition can only be realized in very weak magnetic fields in which $u \ll 1$. For example, (3.9) can only be met for radiation at frequency $\omega \sim 2\pi \cdot 10^8\,\mathrm{s}^{-1}$ in the stationary solar corona plasma with $a \sim 10^{-10}\,\mathrm{cm}^{-1}$ if $B_0 \sim 5{\cdot}10^{-5}$ G. The coronal magnetic fields are much larger, so the leakage effect is absolutely negligible there. However, this effect can be important at large electron density gradients which can arise in astrophysical plasmas, for example, due to the formation of shocks. Finally, the influence of the magnetic field on wave propagation in the region $v \approx 1$ can be neglected if

$$\pi \frac{\omega}{ca} u^{3/4} \ll 1. \tag{3.10}$$

3.1.3. THE RAY APPROACH

In the framework of the geometrical optics the value of the wave vector $\boldsymbol{k}_j = (\omega/c) n_j \boldsymbol{s}$ at any point of a non-uniform isotropic medium is determined by the refractive index $n_j(\omega, \boldsymbol{r})$ at this point. To find the distribution of the unit vectors $\boldsymbol{s}$ in such a medium is a more complex problem. To solve it, the function $\psi(\boldsymbol{r})$ called eikonal is introduced with the relation $\int \boldsymbol{k}_j d\boldsymbol{r} = (\omega/c)\psi(\boldsymbol{r})$. It is clear from this definition that

$$\nabla\psi \equiv \frac{d\psi}{d\boldsymbol{r}} = n_j \boldsymbol{s}. \tag{3.11}$$

As a result, we arrive at the eikonal equation

$$(\nabla\psi)^2 = n_j^2. \tag{3.12}$$

Supplemented with appropriate boundary conditions, this equation determines the function $\psi(\boldsymbol{r})$ in an isotropic medium with the known distribution of the refractive index $n_j(\omega, \boldsymbol{r})$. The distribution $\boldsymbol{s}(\boldsymbol{r})$ can then be found using the relation (3.11).

Geometrical optics approximation in an inhomogeneous medium allows also for the so-called "ray approach" to wave propagation which is especially important when dealing with the analysis of radiation transfer (see chapter 4). The term "ray" here means the trajectory of the "center of mass" of a wave packet, or signal, bounded both spatially and temporally. In the framework of the geometrical optics, as well as in a homogeneous medium, the direction of the packet motion is determined by the vector of group velocity $\boldsymbol{v}_{\rm gr} = d\omega/d\boldsymbol{k}$. The latter, in its turn, is collinear to the time-average energy flux in the wave considered. In an isotropic medium the vector $\boldsymbol{v}_{\rm gr}$ is directed along $\boldsymbol{k}$, i.e., the vector $\boldsymbol{s}$. In other words, $\boldsymbol{s} = d\boldsymbol{r}/dl$, where $\boldsymbol{r}(l)$ is the trajectory of a ray in isotropic plasma and l is the distance along a ray.

The equation

$$n_j \frac{d\boldsymbol{s}}{dl} = \nabla n_j - \boldsymbol{s}(\boldsymbol{s}\nabla n_j) \tag{3.13}$$

for the vector $\boldsymbol{s}(l)$ along a ray can be derived from the eikonal equation (3.12). The derivation procedure has been described in [117]. However, the equation (3.13) can be derived in a simpler way using the Fermat principle [118].

It follows from the equation (3.13) that in a plane-parallel medium in which $n_j = n_j(z)$ a ray lies in a plane parallel to z-axis, its shape being described by the equation

$$\frac{1}{n_j}\frac{dn_j}{dz} + \cot\varphi \frac{d\varphi}{dz} = 0, \tag{3.14}$$

where φ is the angle between $\boldsymbol{s}$ and z-axis. The solution of this equation,

$$n_j \sin\varphi = \text{const}, \tag{3.15}$$

where the constant is independent on z, is called the Cartesius–Snellius law of refraction. Of course, the relation (3.15) is an immediate consequence of the form of the geometrical optics solution in a plane-parallel medium, (3.6), according to which $k_x = (\omega/c)n_j \sin\varphi$ =const.[5] The constant in (3.15) is fixed by setting the angle $\varphi = \varphi_0$ between $\boldsymbol{k}_j$ and z-axis at some level $z = z_0$:

$$n_j(z)\sin\varphi(z) = n_j(z_0)\sin\varphi_0. \tag{3.16}$$

---

[5]This relation also holds for a plane-parallel magnetoactive plasma. However, the ray pattern becomes quite complex in such a medium because of the difference in the directions of the vectors $\mathbf{v}_{\rm ph}$ and $\mathbf{v}_{\rm gr}$. Some results on ray tracing in magnetoactive plasma are given in [71].

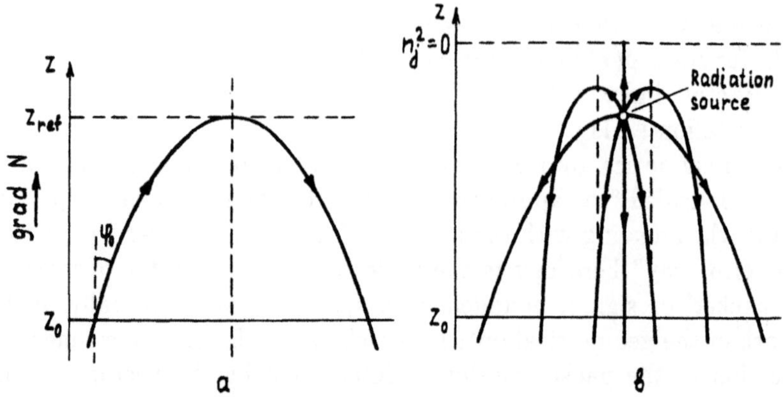

Figure 3.4. Refraction in a plane-parallel plasma with ∇N directed along the z-axis: (a) the shape of a ray entering the plasma at an angle φ_0; (b), ray pattern by a point source in plasma.

The boundary of plasma layer, where $\epsilon = 1$ and, thus, the transition from vacuum to plasma appears, is frequently used as the level z_0. In this case the angle φ is called the angle of incidence of a wave onto the layer. The refractive index $n_j(z_0)$ at this level equals unity for electromagnetic and $1/\sqrt{3}\beta_T$ for plasma waves(see (2.104) and (2.116)).[6]

Fig. 3.4, a illustrates the shape of a ray in case of an oblique incidence of a wave onto a plane-parallel medium. The top of the ray corresponds to the reflection point $z_{\rm ref}$. Its position is found from (3.16) at $\varphi = \pi/2$:

$$n_j(z_{\rm ref}) = n_j(z_0)\sin\varphi_0 \tag{3.17}$$

or, equivalently,

$$\epsilon(z_{\rm ref}) = 1 - \frac{\omega_{\rm L}^2(z_{\rm ref})}{\omega^2} = \sin^2\varphi_0. \tag{3.18}$$

The refraction guides radiation of an isotropically emitting source in plasma. The maximum escape angle $\varphi \le \varphi_{\rm max}$ of the radiation by such a source situated at a point $z = z_{\rm src}$ can be determined from the condition that the top of the corresponding ray coincides with the source position (Fig. 3.4 b). Then, according to (3.18),

$$\varphi_{\rm max} = \arcsin\sqrt{\epsilon(z_{\rm src})} = \mathrm{arcsec}\frac{\omega}{\omega_{\rm L}{}^{\rm src}}, \tag{3.19}$$

where $\omega_{\rm L}{}^{\rm src} = \omega_{\rm L}(z_{\rm src})$. Hence, the beaming of the escaped radiation is increased as the source approaches the level $\epsilon(z,\omega) = 0$, either due to the

[6] Of course, plasma waves cannot exist in vacuum. Moreover, these waves cannot propagate in the regions far enough from the level $\epsilon = 0$, where $v_{\rm ph} \sim v_T$, because of their strong Landau damping at a distance of about the wavelength. Thus, the incidence angle for these waves can be introduced only by convention in the framework of the quasi-hydrodynamic approximation.

shift of the source inside the plasma at fixed frequency or due to the decrease of the frequency generated by the fixed source. The formula (3.19) provides an estimate for the electron density near a source using the width of the observed angular pattern at a given frequency.

Now let the electron density distribution in plane-parallel plasma be non-monotonous with the maximum at $z = z_{\max}$. This case is obviously interesting for planetary ionospheres, in particular, the Jovian one which is probably the source of the observed decametric radio emission.

A wave incident in the ionospheric layer at the angle φ_0 is reflected from the layer $z = z_{\text{ref}}$ where the plasma frequency is found according to (3.18):

$$\omega_{\text{L}}(z_{\text{ref}}) = \omega \cos \varphi_0. \tag{3.20}$$

As radiation frequency increases, the reflection point shifts deeper inside the plasma to the layer with larger ω_{L}. The layer with maximal electron density $N_{\max}$ is reached at the critical frequency

$$\omega_{\text{cr}} = \omega_{\text{L}}^{\max} \sec \varphi_0, \tag{3.21}$$

where the plasma frequency $\omega_{\text{L}}^{\max}$ corresponds to the density $N_{\max}$. At higher frequencies $\omega > \omega_{\text{cr}}$ the reflection vanishes, and radiation penetrates the ionospheric layer freely. This inequality written at fixed frequency ω sets the interval of the angles φ_0 for which the ionosphere is transparent, the maximum angle being given by the following formula:

$$\varphi_{\max} = \operatorname{arcsec} \frac{\omega}{\omega_{\text{L}}^{\max}}. \tag{3.22}$$

This value decreases with the ratio $\omega/\omega_{\text{L}}^{\max}$ and nullifies at $\omega/\omega_{\text{L}}^{\max} = 1$.

Owing to its simplicity, the plane-parallel model for an inhomogeneous plasma is convenient for the analysis of the principal aspects of wave propagation in planetary ionospheres, solar corona etc. In some cases, however, the investigation of the ray pattern, absorption of radiation along a ray etc. requires more rigorous treatment of ray tracing in the ionosphere or corona with the account of the effects of sphericity.

Ray pattern in a spherically-symmetric refractive medium is determined by the refraction law which follows from the equation (3.13). Assuming $n_j = n_j(R)$, where R is the distance from the center of symmetry (for instance, the center of the Sun or planet) to the point considered, one may write a ray length element as $dl = dR/\cos\varphi$, where φ is the angle between the ray direction and the radius R. With this formula and Fig. 3.5 taken into account, the equation (3.13) can be transformed to

$$\frac{1}{n_j}\frac{dn_j}{dR} + \cot\varphi \frac{d\varphi}{dR} = -\frac{1}{R}. \tag{3.23}$$

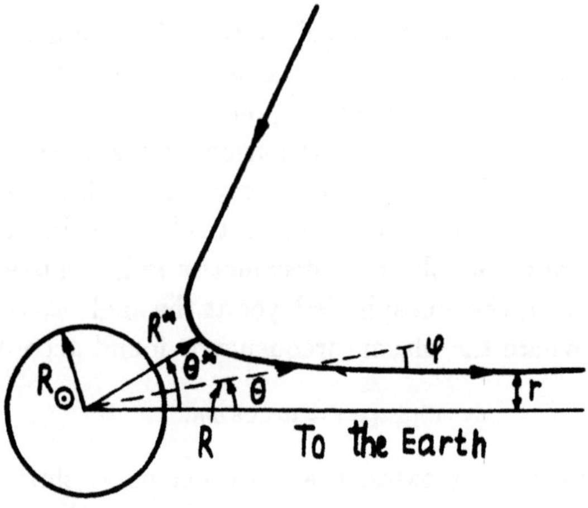

Figure 3.5. Toward the calculation of ray shape in a spherically-symmetric solar corona.

The solution of this equation,

$$n_j(R)R\sin\varphi = \text{const}, \tag{3.24}$$

is the desired refraction law for the spherically-symmetric medium. If R_∞ corresponds to a point on a ray outside the medium, the relation (3.24) can be written in the form

$$n_j(R)R\sin\varphi = n_j(R_\infty)R_\infty \sin\varphi(R_\infty) = n_j(R_\infty)r, \tag{3.25}$$

where r is the impact parameter which characterizes the distance from a ray entering the medium to the parallel radius.[7] Since $n_j(R_\infty) = 1$ for electromagnetic waves outside the corona, the formal account of the difference of this value from unity is needed for the case of plasma wave refraction. The typical ray path in the solar corona with electron density monotonously decreasing with increasing R is shown in Fig.,3.5.

Using the formula (3.25) and this figure, the following expression for the ray length element is obtained:

$$dl = \frac{dR}{\cos\varphi} = \frac{dR}{\sqrt{1 - r^2/n_j^2R^2}}. \tag{3.26}$$

The ray trajectory in polar coordinates R, θ is given by

$$\theta = \theta_\infty + \int_R^{R_\infty} \frac{r\,dR}{R\sqrt{n_j^2R^2 - r^2}}. \tag{3.27}$$

[7]The impact parameter r in solar conditions is usually called the distance from the solar disk center.

Evidently, $\theta_\infty = \varphi(R_\infty)$. Then, the reflection ("turnover") point at which $\varphi = \pi/2$, $R = R^*$ and $\theta = \theta^*$ is determined by the equation

$$R^* n_j(R^*) = r. \tag{3.28}$$

Ray trajectory is symmetrical with respect to the radius R^* passing through the turnover point and lies in the same plane as the symmetry center. Fig. 3.6 shows the ray pattern calculated in [119] using the formula (3.27) as well as the Baumbach-Allen formula (1.16) for the electron density distribution.

The graphs clearly demonstrate that the smaller the impact parameter r at a given frequency, the deeper this ray penetrates into the corona. In other words, the envelope of the reflection points, though tangential to the layer $n_j(\omega) = 0$, or $\omega = \omega_L$, at $r = 0$, rises above this level with the increase of the parameter r. Since this envelope is the lower boundary of the coronal region from which radiation at a given frequency is able to escape at a given direction, the sources of the radiation observed from the disk center are deeper than the ones providing the radiation from the limb. The turnover point envelope lowers deeper into the corona as radiation frequency increases. This enlarges the coronal region which can contribute to the observed radiation. It is refraction that makes the apparent position of a radio source in the solar corona to be closer to the disk than the actual one.

3.1.4. EFFECT OF THE GROUP DELAY OF SIGNALS

Consider now the role of the group velocity of waves in propagation of signals along a ray. If a wide-band radiation with a pulse-shape envelope is emitted by a source in plasma, the arrival times of the signal at different frequencies are different due to frequency dispersion of the group velocity: $v_{gr} = v_{gr}(\omega)$. The group delay of radiation at frequency ω in a medium compared to vacuum equals

$$\Delta t_{gr} = \int \left(\frac{1}{v_{gr}} - \frac{1}{c} \right) dl. \tag{3.29}$$

The integral is taken over the ray connecting the source and the receiver.

The group velocity of electromagnetic waves in isotropic plasma equals $v_{gr} = c\sqrt{\epsilon}$; see section 2.2. Thus, the maximal delay in a linear layer occurs at the frequency ω at which the refractive index at source position is nullified: $n_{em} = \sqrt{\epsilon} \approx 0.$[8] In this case

$$(\Delta t_{gr})_{max} = \frac{1}{c|\nabla\epsilon|}. \tag{3.30}$$

[8]The condition $v_{gr} = 0$ at the source position does not prevent the escape of radiation from the level $n_{em} = 0$, since the geometrical optics fails near this point thus making the

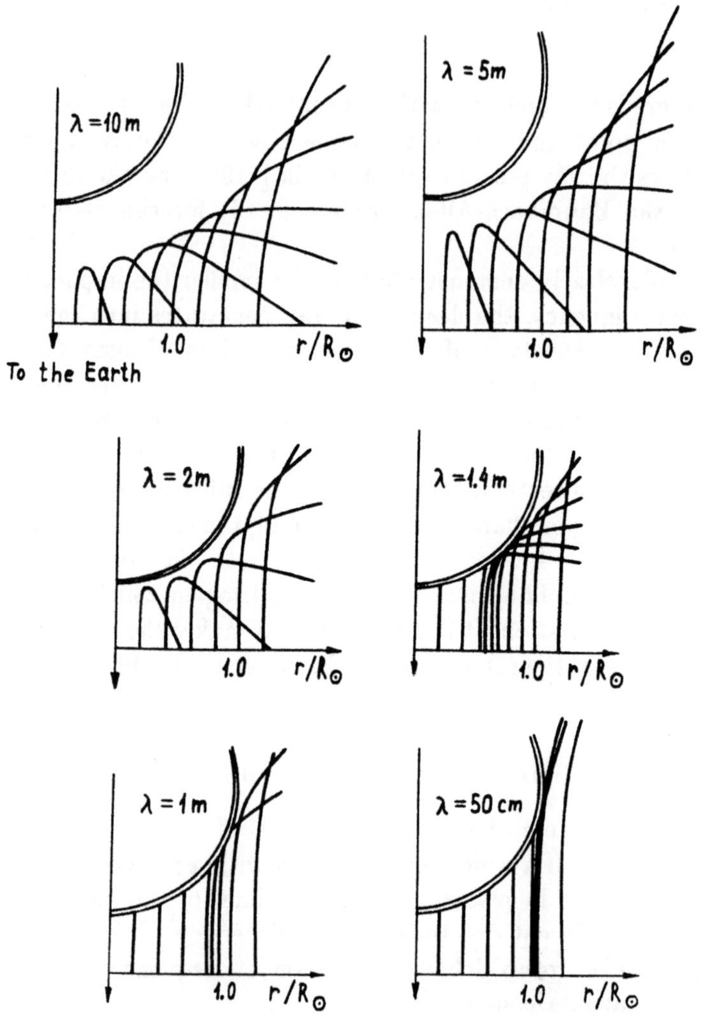

Figure 3.6. Ray pattern in solar corona at different wavelengths from 50 cm to 10 m.

This effect is usually of minor importance for observations of solar radio emission. Since $|\nabla\epsilon| \sim 10^{10}$ cm for the coronal conditions, the corresponding $(\Delta t_{gr})_{max}$ is not larger than a second or a fraction of a second. However, the above estimate increases if one accounts for the great extent of the corona far from the level $\epsilon \approx 0$, i.e., the deviation from the linear layer

above formula for v_{gr} invalid. Therefore, a more rigorous formula than (3.30) should be used to describe the value Δt_{gr}. However, the contribution to $(\Delta t_{gr})_{max}$ by the region of the size of a few wavelengths near the point $n_{em} = 0$ is negligible compared to the one by the extended region in which the geometrical optics approximation is valid.

approximation in the outer corona.

At the same time, it is hard to overestimate the importance of the effect of group delay in the studies of pulsar radio emission. This effect allows one to estimate the distance to these sources and thus pose the crucial argument in favor of their Galactic origin [120]. In case of rarefied interstellar medium where $\omega_L^2 \ll \omega^2$ the group velocity of radio waves equals $v_{gr} = c\sqrt{\epsilon} \approx c(1 - \omega_L^2/2\omega^2)$. Therefore, the formula (3.29) turns into

$$\Delta t_{gr} \approx \frac{1}{2c\omega^2} \int \omega_L^2 dl. \tag{3.31}$$

A pulse emitted by a pulsar at a time $t_{rad} = 0$ is registered by an observer on the Earth at a time $t_{obs} = L/c + \Delta t_{gr}$, the lower-frequency signature of the pulse appearing later than the high-frequency one: $t_{obs} - L/c \propto \omega^{-2}$. As a result, the derivative

$$\frac{dt_{obs}}{d\omega} \approx -\frac{1}{c\omega^3} \int \omega_L^2 dl \tag{3.32}$$

is inversely proportional to the frequency cubed. This scaling can be checked using the data on the pulse arrival times at two nearby frequencies to yield $dt_{obs}/d\omega$. The investigations [120, 121] show that, indeed, $dt_{obs}/d\omega \propto \omega^{-3}$ over a wide range of frequencies. Hence, the arrival time of pulses by a pulsar is really determined by the group delay in a rarefied plasma with $\omega_L^2 \ll \omega^2$.

The measured value $dt_{obs}/d\omega$ can be used to determine the "dispersion measure" $\mathrm{DM} = \int N dl$ between a source and a receiver point from the equation (3.32). The measurements yielded $\mathrm{DM} \approx 12\,\mathrm{pc\,cm^{-3}}$ for PSR 1919, the first pulsar discovered [121]. Assuming that the obtained value of DM is contributed mainly by the interstellar medium with average electron density $N \sim 0.1\,\mathrm{cm^{-3}}$, the distance to the pulsar $L \sim 1.2 \cdot 10^3\,\mathrm{pc}$ can be estimated. Of course, such estimates strongly depend on the value of N in the interstellar medium. Alternatively, if an independent estimate of the distance to some pulsar is known, the dispersion measure is useful for the conclusions on the average electron density $\bar{N}$ of the interstellar medium.

For example, the distances to PSR 0532 in the Crab nebula and PSR 0833 in Vela, the pulsars in supernova remnants, are estimated as $1.7 \cdot 10^3\,\mathrm{pc}$ and $5 \cdot 10^2\,\mathrm{pc}$, respectively. Hence, the known dispersion measures for these objects [122], $57\,\mathrm{pc\,cm^{-3}}$ and $69\,\mathrm{pc\,cm^{-3}}$, yield the values $\bar{N} \simeq 3.3 \cdot 10^{-2}\,\mathrm{cm^{-3}}$ and $\bar{N} \simeq 1.4 \cdot 10^{-1}\,\mathrm{cm^{-3}}$ in the direction of PSR 0532 and PSR 0833, respectively.

The role of the group delay effect turns out to be most important if $v_{gr} \ll c$ along the entire ray. Such a case can be realized for the whistler branch of the dispersion curves of magnetoactive plasma. The corresponding value

n_2^2 in this case is determined by the formula (2.131), while (2.142) gives the angle ϑ between the group and phase velocities. The latter formula yields the relation $\tan\vartheta = -(1/2)\tan\alpha$, where α is the angle between $\boldsymbol{k}$ and $\boldsymbol{B}_0$. Using (2.131) and (2.142) it is easy to derive the projection of $\boldsymbol{v}_{gr}$ onto the vector $\boldsymbol{k}$:

$$v_{gr\mathbf{k}} = \frac{2c}{\omega_L}\sqrt{\omega\omega_B|\cos\alpha|}. \tag{3.33}$$

Obviously, $v_{gr} = v_{gr\mathbf{k}}|\cos\alpha|^{-1}$, so, with the relation $\tan\vartheta = -(1/2)\tan\alpha$, it leads to the following expression for the time of arrival of a whistler radiated at a point A at $t=0$ to the observation point B [92]:

$$t_{obs} = \int\frac{dl}{v_{gr}} = \frac{1}{c}\int n_2\sqrt{4+\tan^2\alpha}\,dl, \tag{3.34}$$

where the refractive index n_2 is determined by (2.131), and the integral is taken over the ray connecting the points A and B. To derive t_{obs} at lower frequencies where the condition $u\cos^2\alpha \gg 1$ fails, one should use the formulas (2.132) for the refractive indices n_j^2. Then the formula

$$t_{obs} = \frac{1}{c}\int n_1\sqrt{4\left(\frac{|\sqrt{u}\cos\alpha| - 1}{\sqrt{u}\cos\alpha}\right)^2 + \tan^2\alpha} \tag{3.35}$$

follows instead of (3.34) [123].

If a ray shape does not vary with frequency, the formula (3.34) means that $t_{obs} \propto \omega^{-1/2}$. The latter, strictly speaking, is valid only in a homogeneous medium. However, since $\boldsymbol{v}_{gr}$ is close to the direction of the magnetic field (see section 2.2), the use of the formula (3.34) is well justified in a majority of cases when the magnetic field in an inhomogeneous medium is spatially inhomogeneous, too.

The dependence $t_{obs} \propto \omega^{-1/2}$ is typical for lower-frequency signals in the band 1 – 10 kHz generated by the atmospheric discharge on the Earth. In "sound receiving" these signals are heard as whistles of gradually lowering tone. Such a feature gave rise to the term "whistlers", or "whistling atmospherics". The above proportionality between t_{obs} and $\omega^{-1/2}$ is explained by the fact that the signal propagating through the Earth magnetosphere along the arc which constitutes a small angle with the geomagnetic field lines corresponds to the waves with n_2^2 from the dispersion branch (2.131).

It is clear from the more general formula (3.35) that the scaling $t_{obs} \propto \omega^{-1/2}$ holds only for low frequencies at which $|\sqrt{u}\cos\alpha| \gg 1$. According to (3.35), the function $t_{obs}(\omega)$ in the case of longitudinal propagation has the minimum at the frequency

$$\omega_{min} = \frac{\omega_B}{4} \tag{3.36}$$

and increases at higher frequencies [124]. This fact can be used to determine the value of the magnetic field in plasma by the whistler arrival time at different frequencies.

3.2. Linear mode coupling in magnetoactive plasma

As was shown in the previous section, refractive indices and polarizations of ordinary and extraordinary waves in an inhomogeneous plasma vary according to the formulas (2.121), (2.135), (2.136) in which the parameters $v = \omega_L^2/\omega^2$, $u = \omega_B^2/\omega^2$ and the angle α are the functions of spatial coordinates. This is true, however, as long as the geometrical optics approximation is valid and a mode propagates independently on the other ones. In a smoothly inhomogeneous medium in which variation of any parameter over a scale of about a wavelength is negligible, this approximation may fail (and the field structure could become more complicated) only within a few localized regions. In this case arbitrary electromagnetic field can only be composed of exact solutions of the system of equations which describes electromagnetic waves in an entire plasma, including the regions at which geometrical optics is invalid. Such solutions outside these regions are asymptotically close to linear superposition of geometrical optics waves (3.1) with certain relation between the amplitudes. Therefore, if waves propagate through a region of violation of the geometrical optics, a "coupling" between them occurs, so that certain relations between the amplitudes of the geometrical optics solution are established. This amplitude ratio depends on the conditions of wave propagation in the region of geometrical optics violation. In particular, a radiation which enters the coupling region in a certain mode gives rise to a pair of coherent normal waves of both types leaving this region. In other words, partial transformation of waves from one mode to another is possible here. Such a mode coupling is called "linear", since it does not break the superposition principle for electromagnetic fields in plasma, contrary to non-linear interaction considered in chapter 8. [9]

Analysis shows that linear coupling of various modes occurs in the regions of plasma at which the dispersion curves $n_j^2(\omega)$ are close to each other. It seems quite natural, since the "transition" of a wave from one dispersion curve to another is favored in the regions with similar mode properties. For example, it is clear from Figs. 2.5, 2.8 that such a transition for waves propagating at small angles α to a constant magnetic field $\boldsymbol{B}_0$ is possible in regions where $v \approx 1$.

The studies on linear mode coupling in magnetoactive plasma have been

[9] It is clear from the above that the effect of linear coupling occurs in a region where geometrical optics is violated for both modes. However, this condition is generally not sufficient, and a special analysis is necessary in any specific case to judge the existence of mode coupling.

initiated in [125, 126] in connection with the effect of "tripling" of signals in the ionosphere [71]. Linear mode coupling at $v \approx 1$ is important for the problem of escape of radiation from dense plasmas, for example, from the solar corona. The effect of the transformation of one mode to another in a dense inhomogeneous plasma has drawn the attention of controlled fusion researchers in connection with the problem of laboratory plasma heating. One of possible schemes of the heating can be the injection of electromagnetic radiation (optical or microwave) into plasma followed by its transformation to plasma waves. Absorption of the latter provides the required heating of the plasma. [10]

The linear mode coupling can also occur in an inhomogeneous magnetic field if the plasma is rarefied enough ($v \ll 1$) or the magnetic field is sufficiently weak ($u \ll 1$). Mutual transformation of extraordinary and ordinary modes under such "weakly anisotropic" conditions can influence the polarization of radiation which propagates in plasma. It is with this effect that we begin the analysis of linear mode coupling in plasma.

3.2.1. EQUATIONS OF QUASI-ISOTROPIC APPROXIMATION AND BASIC PROBLEMS OF LINEAR MODE COUPLING THEORY

As we noted earlier, the geometrical optics approximation in the form (3.1) is valid in the major part of a smoothly inhomogeneous plasma in which the criterion (3.8) is satisfied. This approximation may nevertheless fail in the regions in which the condition

$$\frac{\omega}{c}|n_1 - n_2|\Lambda \gg 1 \tag{3.37}$$

is not satisfied. In such regions the dispersion curves $n_1(\boldsymbol{r})$ and $n_2(\boldsymbol{r})$ come close to each other, so the properties of the medium change significantly over the "wavelength" $(c/\omega)|n_1 - n_2|^{-1}$ of the beatwave between the modes.

Investigation of the problem of the linear mode coupling in three-dimensional, inhomogeneous, magnetoactive plasma is simplified if the criterion of weak anisotropy

$$|\epsilon_{\alpha\beta} - \epsilon\delta_{\alpha\beta}| \ll \epsilon \simeq \frac{1}{3}\,\mathrm{Sp}\,\epsilon_{\alpha\beta} \tag{3.38}$$

is satisfied in addition to the criterion (3.8). Here $\epsilon_{\alpha\beta}$ is dielectric tensor with the trace $\mathrm{Sp}\,\epsilon_{\alpha\beta}$, $\delta_{\alpha\beta}$ is Kronecker delta, and

$$\epsilon = \frac{1}{2}\,\frac{2(1-v)(u+v-1) + uv\sin^2\alpha}{u - 1 + v - uv\cos^2\alpha}. \tag{3.39}$$

[10] Linear mode coupling in laboratory plasmas is reviewed in [127, 128] and studied in a number of original papers.

Following [129] (see also [130]), we use Maxwell equations (2.19), (2.20) and constitutive relation (2.29) with tensor $\epsilon_{\alpha\beta}(\omega)$ in the form (2.70) to find asymptotic expansions for the electric and magnetic inductions, $\boldsymbol{D}$ and $\boldsymbol{B}$, in powers of $1/k_0$, where $k_0 = \omega/c$. Taking into account that $(k_0 n\Lambda)^{-1}$ and $\epsilon/|\epsilon_{\alpha\beta} - \epsilon\delta_{\alpha\beta}|$ are small parameters for a smoothly inhomogeneous, weakly anisotropic medium, we obtain the following system of equations

$$\begin{aligned} \frac{df_1}{d\zeta} + in_1 f_1 &= -\Psi f_2, \\ \frac{df_2}{d\zeta} + in_2 f_2 &= \Psi f_1 \end{aligned} \tag{3.40}$$

for complex "amplitudes" $f_{1,2}$. Here

$$\Psi = -\frac{1}{2(1+q^2)}\frac{dq}{d\zeta}, \qquad q = \frac{\sqrt{u}\sin^2\alpha}{2(1-v)\cos\alpha}, \tag{3.41}$$

$\zeta = k_0 l$, and l is the distance along a ray in a medium with the refraction index $n = (n_1 + n_2)/2 \simeq \epsilon^{1/2}$. The shape of such a "quasi-isotropic" ray is determined by the eikonal equation (3.12) with the refraction index n. The electric field components E_x, E_y in a monochromatic wave $\propto e^{-i\omega t}$ are determined by the amplitudes $f_{1,2}$ via the relations

$$\begin{aligned} E_x &= \Phi_0 \left\{ \frac{f_1}{(1+K_1^2)^{1/2}} + \frac{f_2}{(1+K_2^2)^{1/2}} \right\}, \\ E_y &= \Phi_0 \left\{ \frac{iK_1 f_1}{(1+K_1^2)^{1/2}} + \frac{iK_2 f_2}{(1+K_2^2)^{1/2}} \right\}. \end{aligned} \tag{3.42}$$

Here $K_{1,2}$ are the polarization coefficients of extraordinary and ordinary modes. If the wavevector $\boldsymbol{k}$ is directed along a quasi-isotropic ray, these coefficients are equal to

$$K_{1,2} = -i\frac{E_y^{(1),(2)}}{E_x^{(1),(2)}} = q \mp \sqrt{1+q^2} \tag{3.43}$$

in the frame shown in Fig. 2.1, b. [11] The upper and lower sign in this formula correspond to the extraordinary and ordinary mode, respectively. The function $\Phi_0(l)$ obeys the equation

$$\operatorname{div}\left(n\Phi_0^2 \boldsymbol{l}\right) = 0 \tag{3.44}$$

[11] The use of this frame in which the magnetic field $\mathbf{B}_0$ lies in a fixed xy-plane is approved by the achieved simplification in the form and analysis of resulting equations. However, with this frame it is impossible to account for the effects of sheared magnetic field, i.e., variation of the orientation of the projection of $\mathbf{B}_0$ onto the xy-plane. Linear mode coupling in plasma with a sheared magnetic field is considered in [131–133].

in which $\boldsymbol{l}$ is unit vector along a ray. Since (3.40) yields the relation

$$|f_1|^2 + |f_2|^2 = \text{const}, \tag{3.45}$$

the equation (3.44) is equivalent to the energy conservation law for a monochromatic wave: $\text{div}\boldsymbol{S} = 0$, where $\boldsymbol{S}$ is Poynting vector. Therefore, the formulas (3.42) merely introduce the transformation of variables when deriving the approximate system (3.40) from the Maxwell equations.

The criterion (3.38) of weak anisotropy which ensures the validity of the system (3.40) is satisfied for either a weak magnetic field:

$$\sqrt{u} \ll 1 - v, \qquad v < 1, \tag{3.46}$$

or rarefied plasma:

$$v \ll 1, \qquad |1 - \sqrt{u}| \gg v. \tag{3.47}$$

In both cases $\epsilon \simeq 1 - v$ at $u \ll 1$, while in the second case of rarefied plasma $\epsilon \simeq 1$ at $u \gg 1$. The condition (3.46) corresponds to quasi-isotropic plasma, the second inequality in it ensures that the medium is transparent for electromagnetic waves. The criteria (3.47) mean that dispersion properties of cold plasma are close to vacuum, and the latter inequality eliminates the gyroresonance region $u \simeq 1$ ($\omega \simeq \omega_B$) in which, as long as thermal motion is neglected, strong anisotropy remains even at $v \ll 1$ (see the condition (2.138)).

The system (3.40) has been derived in [134] for $u \ll 1$, $v \ll 1$. Previously, in [135, 136], this system in a one-dimensional form describing the case of wave propagation along the gradient of n_j^2 in a smoothly inhomogeneous, plane-parallel medium has also been derived under weaker constraints, namely: (i) reflected waves are neglected, and (ii) the values of refractive indices of different modes are relatively close: $|n_1 - n_2| \ll n_1 + n_2$. This condition is equivalent to the following pair of inequalities:

$$uv \sin^2 \alpha \ll |1 - v| \cdot |1 - v - u|, \qquad v\sqrt{u}|\cos\alpha| \ll |1 - v - u|, \tag{3.48}$$

which can be met even in a strongly anisotropic plasma in which the criterion (3.38) is violated.

In regions where geometrical optics is valid the functions f_1 and f_2 describe the amplitudes of extraordinary and ordinary wave, respectively. On the contrary, in the linear coupling regions the correct separation of electromagnetic field into two modes is impossible because of the rapid variation of f_1 and f_2. By solving the system of equations (3.40) one can determine the amplitudes $f_{1,2}^{\text{out}}$ of the waves escaping the linear coupling region for given amplitudes of incoming radiation $f_{1,2}^{\text{inc}}$. It allows one to calculate the so-called transformation coefficient Q. This is defined as the relative intensity of a mode escaping the coupling region provided that another mode enters it. According to this definition and the equality (3.45),

$$Q = \frac{|f_1^{\rm out}|^2}{|f_1|^2 + |f_2|^2} = \frac{|f_1^{\rm out}|^2}{|f_2^{\rm inc}|^2} \tag{3.49}$$

if $f_1^{\rm inc} = 0$ and

$$Q = \frac{|f_1^{\rm out}|^2}{|f_1|^2 + |f_2|^2} = \frac{|f_2^{\rm out}|^2}{|f_1^{\rm inc}|^2} \tag{3.50}$$

if $f_2^{\rm inc} = 0$. Evidently, $0 \leq Q \leq 1$ and $1 - Q$ is the coefficient of "transformation" into the same mode: for example, the ratio $|f_1^{\rm out}|^2/|f_1^{\rm inc}|^2$ at $f_2^{\rm in} = 0$. Due to the symmetry of the system (3.40) the values (3.49) and (3.50) are identical.

Note that the linear mode coupling vanishes and the corresponding transformation coefficient $Q = 0$ in case of transverse ($\alpha = \pi/2$) and longitudinal ($\alpha = 0$) propagation. The function (3.41) nullifies in these cases, and the system (3.40) splits into independent equations which describe propagation of two decoupled modes. Moreover, the form of the equations (3.40) implies that, as long as this system is valid, linear coupling occurs between the waves propagating into the same direction, while counter waves remain independent.

It is more convenient to perform further analysis of linear mode coupling using the Riccatti equation for the ratio $P = -if_1/f_2$,

$$\frac{dP}{d\eta} = i(P^2 - 1) - 2iG(\eta)P, \tag{3.51}$$

which follows from the system (3.40). In this equation $\eta = -\int \Psi \, d\zeta$ is the new variable,

$$G(\eta) = \frac{n_2 - n_1}{2\Psi} = \frac{n_1 - n_2}{2\eta'}, \tag{3.52}$$

and the prime denotes the derivative with respect to the dimensionless coordinate $\zeta = k_0 l$. According to (3.41) and (3.43),

$$\eta = \frac{\pi}{4} + \frac{1}{2}\arctan q, \qquad K_2 = -K_1^{-1} = \tan\eta. \tag{3.53}$$

If the ratio $P(\eta)$ is found from the equation (3.51), polarization coefficient for radiation at any point of the ray can be determined the following way:

$$K = -iE_x/E_y = K_1(K_2 - iP)(K_1 - iP)^{-1}. \tag{3.54}$$

Note that the second-order equation

$$\frac{d^2 f}{d\eta^2} + I(\eta)f = 0, \tag{3.55}$$

where

$$I(\eta) = 1 - i\frac{dG(\eta)}{d\eta} + G^2(\eta),$$

can also be used along with (3.51). It can be obtained from (3.40) by eliminating the function f_1 and introducing the new unknown

$$f = f_2 \exp\left(i \int n\, d\zeta\right)$$

and new independent variable (3.53). This equation clearly shows that the variation of polarization of radiation along the entire ray, the linear coupling region included, is determined by the behavior of the "characteristic" function $G(\eta)$ over the interval $\Delta\eta$ which corresponds to the range of the variable $\eta(l)$ in the plasma layer considered. Thus, a solution of this equation with a given function $G(\eta)$ describes not only the linear coupling in the specific plasma region with definite functions $n_{1,2}(\eta)$ and $q(\eta)$, but holds also for any other layer with the same characteristic function $G(\eta)$, whatever $n_{1,2}(\eta)$ and $q(\eta)$ are. This favors the use of the standard problem method in which an exact analytical solutions of the linear coupling equations for a wide class of inhomogeneous plasmas are found. Indeed, settling of the specific profile of inhomogeneity leads, in most cases, to a differential equation with some complex function $G(\eta)$ for which no analytical solution is known.[12] On the contrary, if one starts by choosing such a standard function $G(\eta)$ that reduces the resulting equation (3.51) to one of the well-studied cases for which the solution is known, analysis of linear coupling is greatly simplified. Fortunately, the choice of such standard problems is rather wide. According to (3.52), a given $G(\eta)$ imposes one differential relation between the functions $n_{1,2}(\zeta)$ and $q(\zeta)$ which describe the profiles of the parameters of an inhomogeneous medium. When integrated, this relation yields a class of functions corresponding to various specific cases of an inhomogeneous layer. Thus, by starting the analysis of linear coupling with the choice of the standard function $G(\eta)$ it is possible to reduce the initial complex problem to a pair of simpler ones: first, the known analytical solution of the equation (3.55) is obtained, and, second, the group of inhomogeneous layers in which linear coupling is described by the selected function $G(\eta)$ is determined using (3.52). The efficiency of this approach has been demonstrated in the review [137] and in papers cited therein.

Let us now turn to the qualitative analysis of linear coupling using equation (3.51). Note, first of all, that value of the variable η is close to $\pi/4$ over the interval $q^2 \ll 1$ of quasi-longitudinal propagation with circular mode polarization. On the other hand, η is close to 0 or $\pi/2$ in the region

[12] A few exceptions are discussed later.

$q^2 \gg 1$ of quasi-transverse propagation with linear mode polarization (see section 2.2). The value η varies mainly in the region where $q \sim 1$ and the coefficients $K_{1,2}$ correspond to elliptical polarization of extraordinary and ordinary waves.

If the inequality $|G(\eta)| \gg 1$ is satisfied in the plasma layer considered (this is possible when $n_1 \neq n_2$ and the variable η varies slowly along the ray; see (3.52)), the term $i(P^2 - 1)$ in the Riccati equation can be omitted. Then the solution of the resulting equation takes the form

$$P(\eta) = P(\eta_0) \exp\left[-2i \int_{\eta_0}^{\eta} G(\eta)\, d\eta\right]$$
$$= P(\zeta_0) \exp\left[-i \int_{\zeta_0}^{\zeta} (n_1 - n_2)\, d\zeta\right], \tag{3.56}$$

where ζ_0 is the coordinate of the ray origin. If $f_1(\zeta_0) = 0$ and $f_2(\zeta_0) \neq 0$ (i.e., $P(\zeta_0) = 0$), then $P(\zeta) = 0$ and $f_1(\zeta) = 0$ over the entire ray. Thus, if the radiation enters such a plasma layer in a mode of certain type (the ordinary one in the case considered), another mode does not emerge on the other side of the layer. Therefore, the transformation coefficient $Q = 0$ at $|G(\eta)| \gg 1$. This case corresponds to the geometrical optics approximation.

In the opposite limiting case $|G(\eta)| \ll 1$, when $n_1 = n_2$ but η significantly varies over the ray, one can neglect the last term in the right-hand side of the equation (3.51). The resulting equation yields

$$P(\eta) = i \tan\{\eta - \eta_0 + \arctan[iP(\eta_0)]\}. \tag{3.57}$$

In this case polarization ellipse of radiation does not change along the ray: K = const. Therefore, "isotropic approximation" is realized: wave propagation is similar to the case of vacuum or isotropic plasma. Since $K_{1,2}$ vary along the ray due to variation of the parameters v, u, and α, the relative amount of the modes changes, too. If the electromagnetic field at the ray origin, $\eta = \eta_0$, corresponds to a certain mode (for example, $K = K_1$), then, due to variation of $K_{1,2}$, the fixed polarization ellipse at some other point of the ray should comprise both modes. Therefore, the transformation coefficient $Q \neq 0$. This coefficient,

$$Q = \sin^2 \Delta\eta, \tag{3.58}$$

where $\Delta\eta = \eta - \eta_0$, can be easily found by expanding the fixed polarization ellipse over the mode polarizations at the initial and final points of the ray which penetrates the plasma layer. Initially, the polarization ellipse corresponds to a certain mode, while the one of the escaping radiation is comprised of two orthogonal polarization ellipses corresponding to both modes.

If mode polarizations $K_{1,2}$ are essentially changed over the layer: $\Delta\eta \sim 1$, the case $|G(\eta)| \ll 1$ corresponds to strong coupling. Here the formula (3.58) determines the maximal possible coefficient $Q_{\max}$. If the transition from the quasi-transverse propagation ($q^2 \gg 1$) to quasi-longitudinal one ($q^2 \ll 1$), or vice versa, occurs in a single region along a ray, then $\Delta\eta = \pm\pi/4$ and $Q_{\max} = 1/2$. The value $Q_{\max} = 1$ corresponds to the case $\Delta\eta = \pm\pi/2$ when the character of wave propagation is altered twice (for example, transition from the quasi-longitudinal to quasi-transverse propagation and then again to the quasi-longitudinal one).

Note that a small variation of η along a ray leads only to weak coupling $Q \ll 1$, even at the interval $|G(\eta)| \ll 1$ (see (3.58)).

To sum up, linear coupling is practically absent if $|G(\eta)| \gg 1$ along a ray. The coupling remains weak ($Q \ll 1$) also at $|G(\eta)| \ll 1$ if the range of the variable η is small: $|\Delta\eta| \ll 1$. Effective coupling which increases the transformation coefficient up to the values $Q \sim 1$ can occur if $|G(\eta)| \sim 1$ over the interval $|\Delta\eta| \sim 1$. Finally, the coupling becomes strong if $|G(\eta)| \ll 1$ over this interval. Transformation coefficient Q reaches its maximal values 1/2 and 1 at $\Delta\eta = \pm\pi/4$ and $\Delta\eta = \pm\pi/2$, respectively.

It is thus clear that if variation of the function $G(\eta)$ over the interval $|\Delta\eta| \sim 1$ is similar to or less than the value of this function, the efficiency of linear coupling can be characterized by the parameter $G = |G(\eta^*)|$, where η^* is some value from the interval $\Delta\eta \sim 1$. Since η varies slowly at $q^2 \gg 1$ or $q^2 \ll 1$, this interval corresponds to $q^2 \sim 1$. Thus, the coupling parameter can be determined if one puts $\eta^* = 3\pi/8$ at $q > 0$ and $\eta^* = \pi/2$ otherwise. Note that both of these values correspond to $q^2 = 1$. As a result,

$$G = |G(\eta)|_{q^2=1} = \left|\frac{n_2 - n_1}{2\Psi}\right|_{q^2=1} . \tag{3.59}$$

The above definition can be used in the following way. Let a region of considerable variation of mode polarizations $K_{1,2}$ along a ray, $|\Delta\eta| \sim 1$, exist in plasma. Let also the relative variation of the characteristic function $G(\eta)$ over this region be comparable to or less than unity. Then the coupling parameter (3.59) allows one to judge the efficiency of linear mode coupling according to the following rule:

$$\begin{aligned} Q &\ll 1 \quad && \text{if } G \gg 1 \\ Q &\sim 1 && \text{if } G \sim 1 \\ Q &\simeq Q_{\max} && \text{if } G \ll 1. \end{aligned} \tag{3.60}$$

The meaning of the introduced coupling parameter G is clarified if one takes into account that

$$2\int_{\eta_1}^{\eta_2} G(\eta)\, d\eta = \frac{\omega}{c}\int_{l_2}^{l_1} (n_1 - n_2)\, dl \tag{3.61}$$

and introduces the characteristic size of the coupling region. The latter is determined by the scale of q in the region $q^2 \sim 1$ along a ray:

$$\Lambda_q = \left|\frac{1}{q}\frac{dq}{dl}\right|^{-1}_{q^2=1} . \tag{3.62}$$

If the size $l_2 - l_1$ of the coupling region ($q^2 \sim 1$) is similar to Λ_q, the variation $\Delta\eta = \eta_2 - \eta_1$ is about unity. Therefore, the equation (3.63) can be approximated as

$$2G \simeq \frac{\omega}{c}|n_1 - n_2|\Lambda_q . \tag{3.63}$$

Since the relation

$$\frac{\omega}{2c}|n_1 - n_2|l \sim 1$$

determines the spatial period of the beatwave between extraordinary and ordinary modes, the equation (3.63) means that the value of the parameter G equals the number of spatial beats in the coupling region. On the other hand, the size Λ_q characterizes the scale of considerable variation of the mode polarization. Therefore, linear mode coupling is weak if $G \gg 1$, i.e., the variation of mode polarization is small over the spatial period of the beatwave. On the contrary, the coupling becomes strong ($G \ll 1$) if mode polarization varies considerably over this scale. In the latter case, as we have seen, $K = \text{const}$ inside the coupling region, i.e., the resulting polarization cannot follow the variations of the polarizations K_1 and K_2 which correspond to geometrical-optical waves.

3.2.2. LINEAR COUPLING IN AN INHOMOGENEOUS MAGNETIC FIELD

Linear coupling in inhomogeneous magnetic fields plays an important role in the formation of polarization characteristics of extra-terrestrial radio emission. This is especially true for solar radio emission and, probably, for the Jovian one. Therefore, this process is considered in detail in this paragraph. It is assumed that electron density is constant ($v = \text{const}$), and plasma inhomogeneity is entirely related to variation of the magnetic field in strength and/or direction. Coupling parameter in a magnetoactive plasma is determined by the expression

$$\begin{aligned} G^2 &= [2(n_1 - n_2)\frac{\omega}{c}\Lambda_q]^2_{q^2=1} \\ &= \frac{8uv^2(1-v)\cos^2\alpha}{(1-v-u)(1-v-u+uv\cos^2\alpha)} \\ &\quad \times \left\{\frac{d}{d\eta}\left[\frac{\sqrt{u}\sin^2\alpha}{2(1-v)\cos\alpha}\right]\right\}^{-2}_{q^2=1} , \end{aligned} \tag{3.64}$$

where the scale Λ_q is given by (3.62). If $v \ll 1$ and $u \sim |1-u| \sim \alpha \sim 1$, then, according to (3.64),

$$G \sim \frac{\omega}{c} v \Lambda_q. \tag{3.65}$$

Analysis of the general expression (3.64) shows that the parameter G is minimized under one of the following conditions:

$$\sqrt{u} \ll 1 - v, \qquad \cos^2\alpha \ll 1, \qquad v < 1 \tag{3.66}$$

or

$$u \gg 1, \qquad \sin^2\alpha \ll 1, \qquad v \ll \sqrt{u}. \tag{3.67}$$

The inequality $v < 1$ in the former one ensures that plasma is transparent, while $v \ll \sqrt{u}$ in the latter means that the dispersion curves $n_{1,2}(l)$ are close to each other.

If the scale Λ_q is determined mainly by variation of the magnetic field strength along a ray,[13] one can put $u = u(\zeta)$, $v = \text{const}$, $\alpha = \text{const}$ in the expression (3.64). Then, under the condition (3.66) the interaction parameter

$$G_0 = 4\sqrt{2}\frac{\omega}{c} v(1-v)^{1/2}\cos^2\alpha \left|\frac{\sqrt{u}}{d\sqrt{u}/dl}\right|_{q^2=1}, \tag{3.68}$$

while in the region (3.67) this parameter

$$G_\infty = \sqrt{2}\frac{\omega}{c}\frac{v}{|1-v|}\sin^2\alpha \left|\frac{\sqrt{u}}{d\sqrt{u}/dl}\right|_{q^2=1}. \tag{3.69}$$

The parameter G_0 describes, in particular, linear mode coupling in a neutral current sheet in rarefied plasma ($v \ll 1$) (see [138] and the further section). The case described by the parameter G_∞ can be realized under certain conditions in a transition layer where the quasi-longitudinal propagation changes to the quasi-transverse one [130] (see also the further consideration of this case).

Let now the scale Λ_q be determined mainly by variation of the magnetic field direction. Putting $\alpha = \alpha(\zeta)$, $u = \text{const}$, $v = \text{const}$ in (3.64), one obtains the following interaction parameter valid under the criterion (3.66):

$$G_\perp = \sqrt{2}\frac{\omega}{c}\frac{uv}{(1-v)^{3/2}}\left|\frac{\cos\alpha}{d\cos\alpha/dl}\right|_{q^2=1} \tag{3.70}$$

---

[13]The corresponding criterion can be easily found using the definition (3.62) which relates Λ_q to the derivative of q along a ray.

This parameter describes the linear coupling in a transverse magnetic field [3,131,134,139,140]. Then, in the region (3.88) the parameter

$$G_{\|} = \sqrt{2}\frac{\omega}{c}\frac{v}{\sqrt{u}}\left|\frac{\sin\alpha}{d\sin\alpha/dl}\right|_{q^2=1}. \tag{3.71}$$

This parameter is introduced in [12]. Similar to G_∞, it relates to transition layers.

If plasma is sufficiently rarefied ($v \ll 1$), the small factors appear in all the above coupling parameters. Thus, strong coupling ($G \lesssim 1$) is realized if either the scale Λ_q or the parameter v is larger than in the general case (3.65) when no small parameter enters the coupling coefficient.

When applying the formulas (3.68)-(3.71) to solve a specific problem, one should bear in mind that the values α, u and v in the expression for the coupling parameter are not independent. These values are related through the condition $q^2 = 1$, where q is given by the second of the formulas (3.41).

Taking into account the above, we present the coupling parameters in the form convenient for practical use [3,130,138–141]:

$$G_0 = 8\sqrt{2}\frac{\omega}{c}\frac{v|\cos^3\alpha|}{|d\sqrt{u}/dl|}(1-v)^{3/2}, \tag{3.72}$$

$$G_\infty = 2\sqrt{2}\frac{\omega}{c}\frac{v}{|d\sqrt{u}/dl|}, \tag{3.73}$$

$$G_\perp = \frac{1}{\sqrt{2}}\frac{\omega}{c}\frac{vu^{3/2}}{(1-v)^{5/2}|d\alpha/dl|}, \tag{3.74}$$

$$G_{\|} = 2\frac{\omega}{c}\frac{vu^{-3/4}(1-v)^{1/2}}{|d\alpha/dl|}. \tag{3.75}$$

In some of these formulas the corrected numerical factors are given.

3.2.3. TRANSITION LAYERS AND "LIMITING" POLARIZATION OF RADIATION

The problem on limiting polarization of radiation escaping a magnetoactive plasma is related to the following. It is clear that the influence of the magnetic field on mode polarization (the form of the polarization ellipse) becomes negligible in a weakly anisotropic medium, at sufficiently small v or u. If $v \to 0$, wave propagation is similar to the vacuum case, while at $u \to 0$, to isotropic plasma. Therefore, the polarization coefficient K (see the formula (3.54)) is unchanged in the region of small v or u, so that radiation escapes in vacuum or isotropic plasma having nearly the same polarization ellipse as the one of the radiation entering the region of small anisotropy. On the contrary, according to the geometrical optics approximation, the form of the polarization ellipse should vary even in this region in accord with variation of the parameters v, u and α (see the expressions (3.43) for the coefficients $K_{1,2}$). Therefore, geometrical optics solutions for ordinary and extraordinary modes become invalid at small u or v. So, the problem on the limiting polarization of a wave proparating from a region with large v, u into the region where one of these parameters is small should be treated more rigorously using the equations which describe electromagnetic wave propagation in inhomogeneous magnetoactive plasma.

Let an electromagnetic wave propagate from a region of geometrical optics $|G(\eta)| \gg 1$ to the one $|G(\eta)| \ll 1$ in which "isotropic" approximation is realized. Initially, a mode of a certain type (for example, the extraordinary one) propagates toward isotropic plasma ($u \to 0$) or vacuum ($v \to 0$) in full accord with geometrical optics approximation. But as soon as it penetrates the layer $|G(\eta)| \sim 1$, variation of polarization stops, and the resulting polarization is thereafter conserved. Thus, the limiting polarization K of radiation escaping to the region $u \to 0$ or $v \to 0$ will generally differ from the values (3.43) inferred by the geometrical optics solutions. Such a phenomenon constitutes the effect of limiting polarization. The polarization ellipse in this region can be composed of those for ordinary and extraordinary waves. Therefore, a wave of a certain type infalling onto the layer $|G(\eta)| \sim 1$ gives rise to coherent waves of both types which escape to isotropic plasma or vacuum. This can be treated as a result of linear mode coupling in the region where geometrical optics is invalid.

What are the conditions at which linear coupling considerably changes polarization compared to the geometrical optics values? First of all, it is clear that the limiting polarization corresponds to geometrical optics approximation if mode polarizations $K_{1,2}$ are only weakly changed: $\Delta\eta \ll 1$ (see (3.53)). The effect of limiting polarization is absent in this case. But if transition from an anisotropic medium to an isotropic one or vacuum is connected with considerable variation of the mode polarizations $|\Delta\eta| \sim 1$, the linear mode coupling may become strong. This is possible when a wave propagates from a region of the quasi-transverse ($q^2 \ll 1$) to quasi-longitudinal propagation ($q^2 \gg 1$), and, moreover, such a transition occurs in the region where the geometrical optics approximation is violated. In other words, the strong effect of linear mode coupling takes place if there exists a layer $q^2 \sim 1$ in the region of "isotropic" propagation $|G(\eta)| \ll 1$. On the contrary, this effect becomes weak if geometrical optics approximation holds for the region $q^2 \sim 1$ as well. For the latter it is necessary that the condition $|G(\eta)| \gg 1$ is realized in this region. Finally, at $|G(\eta)| \sim 1$ the critical combination of parameters is realized which separate the cases in which the effect of limiting polarization is strong or weak.

If the magnetic field $\boldsymbol{B}_0$ is constant in the transition region from plasma to vacuum, the appearance of a considerable effect of limiting polarization due to variation of the parameter v (i.e., electron density N) is impossible. Here the value q^2 and geometrical-optical polarization ellipse of a mode $K_{1,2}$, as well, do not vary much ($|\Delta\eta| \ll 1$) in the region $v \ll 1$. Of course, polarization ellipse is also constant in the "isotropic" approximation if the latter is realized at the region where the radiation escapes from the plasma. On the contrary, inhomogeneity of the magnetic field in the region $|G(\eta)| \ll 1$ may lead to considerable variation of q^2 at which $|\Delta\eta| \sim 1$ and, thus, to the strong effect of limiting polarization at $v \ll 1$.

The above qualitative description of the limiting polarization effect is confirmed by the solution of standard problems on linear coupling in a transition layer (i.e., plasma layer in which smooth, monotonous transition from some $q = q_{\rm inc}$ to $q = q_{\rm out}$ takes place). As an example, consider a layer with the function (3.52) in the form

$$G(\eta) = -\frac{p}{2\sqrt{2}} \frac{(q_{\rm inc} - q_{\rm out})[1 + \cot^2 2\eta]^{3/2}}{[q_{\rm inc} + \cot 2\eta][q_{\rm out} + \cot 2\eta]}, \tag{3.76}$$

where $q_{\rm inc} > q_{\rm out} > 0$ [130]. If there is a transition from the quasi-transverse ($q_{\rm inc} \ll 1$) to quasi-longitudinal ($q_{\rm out} \gg 1$) propagation, the role of the coupling parameter G is played by $p = G(3\pi/8) = |G(\eta)|_{q^2=1}$. The function $G(\eta) \gg 1$ at the layer boundaries where

$$\eta_{\rm inc} = \frac{\pi}{4} + \frac{1}{2}\arctan q_{\rm inc}, \qquad \eta_{\rm out} = \frac{\pi}{4} + \frac{1}{2}\arctan q_{\rm out},$$

while it reaches the minimum $G_{\rm min} > 0$ inside. According to the above analysis, such a profile of $G(\eta)$ ensures the validity of the geometrical optics approximation at the layer boundaries. If $G_{\rm min} \gg 1$, the effect of linear coupling is negligible. At $G_{\rm min} \sim 1$ the coupling becomes significant, and $G_{\rm min} \ll 1$ corresponds to the strong coupling. According to (3.52), the distribution of q in a layer $G(\eta)$ is given by

$$q(\sigma) = \frac{q_{\rm inc} - q_{\rm out}}{1 + \exp(2^{5/2}\sigma/p)}, \qquad \sigma = \int \frac{n_1 - n_2}{2(q^2 + 1)^{1/2}}\, d\zeta, \tag{3.77}$$

where $\zeta = (\omega/c)l$. If one chooses $n_1 - n_2 \propto (q^2 + 1)^{1/2}$, the quantity σ varies linearly with the coordinate l along a ray. The corresponding function $q(l)$ describes the so-called Epstein layer (see, e.g., [142]). Note that the parameter p is reduced to G_0, G_∞, $G_\perp$, and $G_\parallel$, if parameters v, u, α satisfy the criteria (3.68)-(3.71).

The solution of the equation (3.55) with the function $G(\eta)$ given by (2.79) is expressed in terms of hypergeometric functions [142]. The transformation coefficient

$$Q = [1 + \exp(2^{-3/2}\pi p)]^{-1} \tag{3.78}$$

in a layer with a transition from $q_{\rm inc} \gg 1$ to $q_{\rm out} \ll 1$ is entirely determined by the coupling parameter $G \equiv p$. In full accord with the previous qualitative analysis, $Q \ll 1$ at $p \gg 1$, $Q \sim 1$ at $p \sim 1$, and $Q = 1/2$ at $p \ll 1$.

Note that the last result has a clear physical meaning. Indeed, the polarization of radiation propagating in a plasma layer with $G \ll 1$ remains unchanged (see the discussion of the "isotropic approximation" basing upon the formula (3.57)). If linearly polarized radiation at the ray origin $q_{\rm inc} \gg 1$ corresponds to a certain mode (say, the ordinary one), the radiation emerges at $q_{\rm out} \ll 1$ in two circularly polarized waves, ordinary and extraordinary, of equal amplitudes. According to the definition (3.49), this corresponds to $Q = 1/2$.

3.2.4. PROPAGATION OF WAVES THROUGH A REGION WITH TRANSVERSE MAGNETIC FIELD

It is quite possible that a wave propagating through an inhomogeneous magnetic fields passes two adjacent layers with $q^2 \sim 1$. This case is realized, for example, if rays cross a layer with a transverse magnetic field (Fig. 3.7) or a neutral current sheet with zero magnetic field (Fig. 3.8). The axis z, directed from the left to the right in these figures, denotes the direction of wave propagation, and the points at both sides of zero or transverse magnetic field mark the values $q^2 = 1$. In both cases the z-projection of the magnetic field $\boldsymbol{B}_0$ changes the sign. According to the above analysis, polarization of waves passed through such regions depend on the values of the parameters G_0 or $G_\perp$ in the layers $q^2 \sim 1$.

Polarization of waves propagating through a region of quasi-transverse magnetic field varies in the following way [131]. According to (3.74), the parameter $G_\perp \propto \omega^{-4}$ in the frequency range $\omega^2 \gg \omega_{\rm L}^2$. Therefore, $G_\perp^2 \gg 1$ in the layers $q^2 \sim 1$ if the frequency is much less than the critical one,

$$\omega_t = \frac{1}{m_e c}\left(\frac{2\sqrt{2}\pi e^5 N B^3}{|d\alpha/dz|}\right)^{1/4}. \tag{3.79}$$

So, geometrical optics approximation is valid in the transition layers from quasi-transverse to quasi-longitudinal propagation (or vice versa), and the modes, ordinary and extraordinary, propagate independently. Each mode becomes linearly polarized in the region of quasi-transverse magnetic field. When escaping this region, modes become circularly polarized again. However, the reversal of the longitudinal (along the wave vector $\boldsymbol{k}$) component of the magnetic field changes the sign of rotation as compared to the initial one. The corresponding transformation coefficient $Q \ll 1$.

As frequency increases, geometrical optics approximation in the regions $q^2 \sim 1$ is violated, so at $\omega \gg \omega_t$, where $G_\perp^2 \ll 1$, wave propagation in the

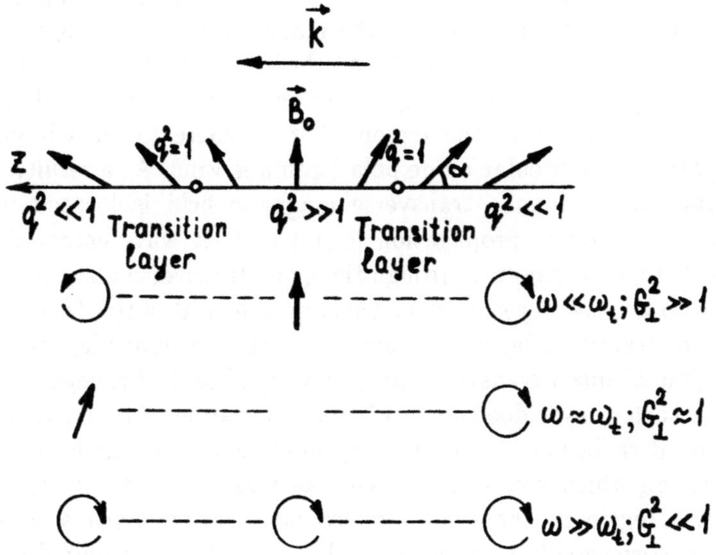

Figure 3.7. Variation of wave polarization in case of propagation through a region of quasi-transverse magnetic field.

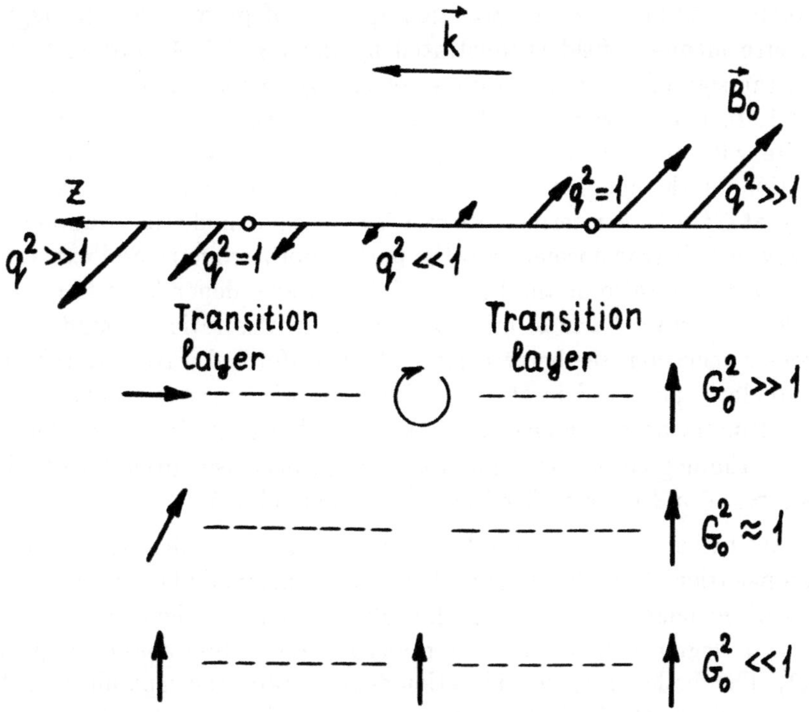

Figure 3.8. Variation of wave polarization in case of propagation through a neutral current sheet.

layers $q^2 \sim 1$ is similar to the case of vacuum or isotropic medium, and wave polarization is not influenced by the magnetic field inhomogeneity. More detailed analysis shows that at $G_\perp^2 \ll 1$ the layer $|G(\eta)| \simeq 1$, the conventional boundary which separate the regions of geometrical optics and "isotropic" propagation, shifts into the region $q^2 \ll 1$ where the quasi-longitudinal propagation with circular mode polarization is valid. As a result, the entire transition layer of quasi-transverse magnetic field is located inside the region of "isotropic" propagation $|G(\eta)| \ll 1$. A wave enters this region at $q^2 \ll 1$, then propagates through the quasi-transverse magnetic field, and escapes to another layer $q^2 \ll 1$. Thus, geometrical optics fails to describe the entire transition layer, including both the boundary layers $q^2 \sim 1$ and the region of quasi-transverse propagation $q^2 \gg 1$. Propagation of wave through such a layer does not lead to the variation of its polarization. For example, if radiation in the ordinary mode enters a region of "isotropic" propagation which contains a quasi-transverse magnetic field, it emerges from this region in the same circular polarization. Such a polarization, however, corresponds to another mode, since the longitudinal component of $\boldsymbol{B}_0$ has reversed. The transformation coefficient Q in this case is close to unity.

Possible variations of polarization in case of propagation through a transverse magnetic field is illustrated by the Fig. 3.7. According to this figure, the sign of polarization does not change in case of strong coupling $G_\perp^2 \ll 1$. In case of weak coupling $G_\perp^2 \gg 1$ the sign of rotation becomes opposite with respect to the initial one, in accord with the sign of the projection of the magnetic field $\boldsymbol{B}_0$ onto the direction of propagation. Finally, at $G_\perp^2 \approx 1$ the intermediate case is realized: a mode (for example, the ordinary one) is transformed into the pair of both ordinary and extraordinary waves. The ratio of amplitudes of these waves depends on frequency; it is close to unity if $\omega \simeq \omega_t$ (the exact value of this ratio is found by the rigorous calculation; see below). Since the transformed waves are coherent and circularly polarized in the opposite sense with respect to each other, the resulting polarization near $\omega \simeq \omega_t$ is purely linear. As frequency detunes from ω_t, the degree of linear polarization, ζ_l, decreases from 1 to 0. The cases $\zeta_l \approx 0$, $\zeta_c \approx \pm 1$ are realized at $G_\perp^2 \gg 1$ and $G_\perp^2 \ll 1$.

Quantitative analysis of the linear mode coupling which occurs due to the propagation through a region of transverse magnetic field is started by the use of the phase integral method. It allows one to get the transformation coefficient, degree of linear or circular polarizations as functions of frequency ω [141]. The forthcoming consideration demonstrates the capabilities of the phase integral method which, in more complicated form, is also used in this chapter to calculate transformation coefficients in the region $\omega_{\mathrm{L}} \simeq \omega$ located in inhomogeneous magnetoactive plasma with a homogeneous magnetic

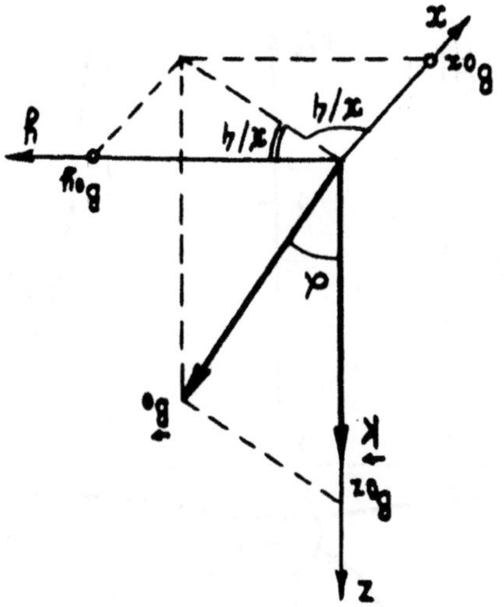

Figure 3.9. The frame.

field. As has already been mentioned above, the equations (3.40) of the quasi-isotropic approximation are inapplicable in this case.

The phase integral method, which the analysis of linear coupling in this case is based upon, yields the relation between the wave amplitudes far from the coupling region using only the geometrical-optical solution and some properties (for example, analyticity in the complex plane z) of the exact solution of the equations (3.37), but not calculating this solution.[14]

Following [141], consider propagation of plane electromagnetic waves along z axis in a plasma with inhomogeneous magnetic field $\boldsymbol{B}_0(z)$. Assume that the transverse component of the field is z-independent, while the longitudinal one, B_{0z} is monotonous function which nullifies at $z=0$. This assumption, on the one hand, realizes the case of propagation through a transverse magnetic field which we are interested in. On the other hand, the problem remains one-dimensional which simplifies its solution.

Strictly speaking, such a field $\boldsymbol{B}_0$ can not exist in plasma since it does not obey the Maxwell equation div $\boldsymbol{B}_0=0$. A magnetic field of the adopted field line pattern (Fig. 3.7) can only satisfy this equation if $\boldsymbol{B}_0$ depends on two coordinates. Analysis of the coupling in such a system on the basis of the equation of the quasi-isotropic approximation (3.40) has shown that the one-dimensional model describes fairly well the real picture of wave propagation (see below).

[14]The phase integral method is sometimes considered in a wider sense, so it is identified with the method of geometrical optics. One can get information on the method of phase integrals using the book [143]. This method in the form used below has been developed in [144] to solve some problems of quantum mechanics. Later on it has been applied to the theory of wave propagation in inhomogeneous plasma [145, 146].

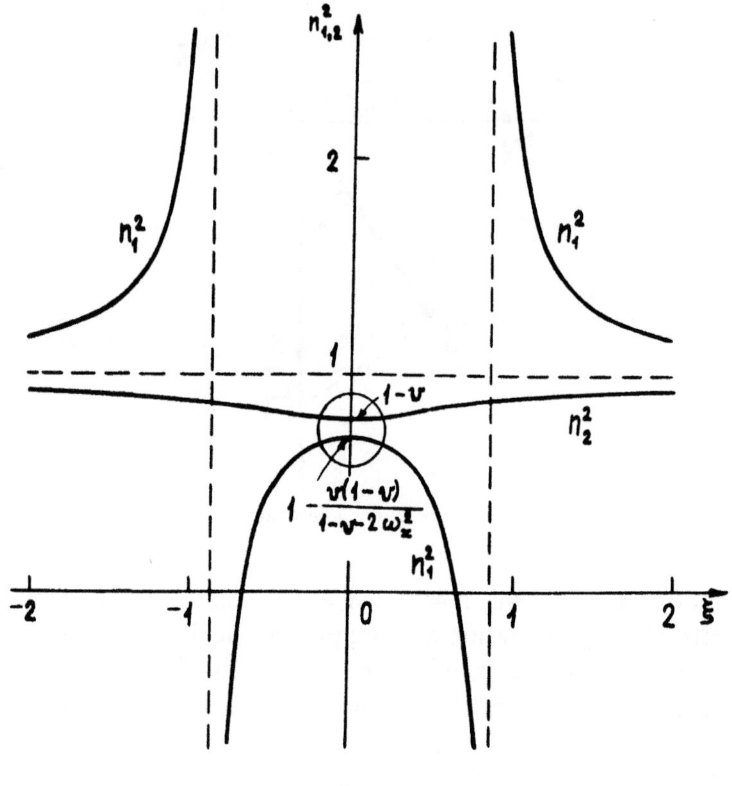

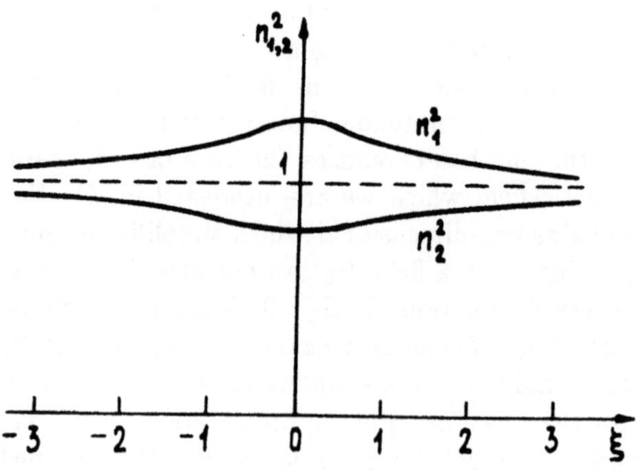

Figure 3.10. The dependence of the refractive index squared, n_j^2, on $\xi \equiv \omega_z$ in plasma with an inhomogeneous magnetic field of the kind (3.82): *a*, $u \sin^2 \alpha < 1 - v$; *b*, $u \sin^2 \alpha > 1 - v$.

Investigation of linear coupling of electromagnetic waves in a region of transverse magnetic field is based upon the system of equations (2.19)–(2.22) with the tensor $\epsilon_{ij}(\omega)$ in the form (2.70). If parameters of an inhomogeneous medium depend on the coordinate z only, then the fields $\boldsymbol{E}$ and $\boldsymbol{B}$ of a plane wave which propagates along z-axis should also depend on this coordinate. Therefore, the above system of equations takes the following form [141]:

$$\frac{d^2F_+}{dz^2} + \frac{\omega^2}{c^2}(C-B)F_+ = -i\frac{\omega^2}{c^2}AF_-,$$
$$\frac{d^2F_-}{dz^2} + \frac{\omega^2}{c^2}(C+B)F_- = i\frac{\omega^2}{c^2}AF_+, \tag{3.80}$$

where

$$F_\pm = E_x \pm iE_y,$$

$$A = -\frac{v\omega_x^2}{2\omega_x^2 - (1-\omega_z^2)(1-v)},$$
$$B = \frac{v(1-v)\omega_z}{2\omega_x^2 - (1-\omega_z^2)(1-v)}, \tag{3.81}$$
$$C = 1 - \frac{v(\omega_x^2 + v - 1)}{2\omega_x^2 - (1-\omega_z^2)(1-v)},$$

$$\omega_x = \frac{eB_{0x}}{mc\omega}, \qquad \omega_z = \frac{eB_{0z}}{mc\omega} = \sqrt{u}\cos\alpha.$$

The equations (3.80) are written in the frame in which z-axis is identical to that of the one in Fig. 2.1,b, whereas the angles between the corresponding x- and y-axes of these frames equal $\pi/4$. As seen from the Fig. 3.9, the projections of the field $\boldsymbol{B}_0$ onto x- and y-axis are equal in the new frame.

Coefficients A, B and C depend on z via the parameter ω_z. If $B_{0z}(z)$ is a linear function, this parameter can be written as

$$\omega_z = \Lambda z, \qquad \Lambda = \text{const.} \tag{3.82}$$

Thus, the magnetic field is perpendicular to the wave vector at $z=0$. Let new variable be $\xi \equiv \Lambda z = \omega_z$, then

$$\frac{d^2F_+}{d\xi^2} + \rho_B^2(C-B)F_+ = -i\rho_B^2 AF_-,$$
$$\frac{d^2F_-}{d\xi^2} + \rho_B^2(C+B)F_- = i\rho_B^2 AF_+, \tag{3.83}$$

where $\rho_B = \omega/\Lambda c$. This system can be reduced to a single, fourth-order differential equation with varied coefficients for the unknown function F_-.

Note that the characteristic scale of inhomogeneity of the magnetic field is much larger than the wavelength, and the parameter $\rho_B \gg 1$. Therefore, the solution of the above equation for F_- can be found in the geometrical optics approximation:

$$F_- = \exp\left\{\rho_B[S_0(\xi) + \rho_B^{-1}S_1(\xi) + \rho_B^{-2}S_2(\xi) + \ldots]\right\}. \tag{3.84}$$

This solution yields two possible values, n_1^2 and n_2^2, for $(dS(\xi)/d\xi)^2$. These quantities are identical to the refractive indices of homogeneous plasma determined by the expressions (2.121). They make possible the propagation of two modes, ordinary and extraordinary. The function $S_1(\xi)$, in its turn, obeys the equation

$$\left(\frac{dS_1(\xi)}{d\xi}\right)_{1,2} = -\frac{1}{2}\frac{(dS_0/d\xi)_{1,2}}{(S_0)_{1,2}} \mp \frac{1}{2}\frac{d\sigma/d\xi}{\sqrt{1+\sigma^2}} - \frac{1}{2}\frac{\sigma d\sigma/d\xi}{1+\sigma^2}, \tag{3.85}$$

where

$$\sigma = -\frac{B}{A} = (1-v)\frac{\omega_z}{\omega_x^2} = \frac{\xi}{\xi_0}, \qquad \xi_0 = \frac{\omega_x^2}{1-v}. \tag{3.86}$$

Substituting (3.85) into (3.84) and taking into account (2.121), we write the desired solution in the following form:

$$F_-(\xi) = \exp\left\{-\frac{1}{2}\frac{\sigma d\sigma}{1+\sigma^2}\left[\left(c_1 e^{i\rho_B\int n_1\, d\xi} + d_1 e^{-i\rho_B\int n_1\, d\xi}\right)\right.\right.$$

$$\times e^{-\frac{1}{2}\int n_1^{-1}dn_1 - \frac{1}{2}\int(1+\sigma^2)^{-1/2}d\sigma} + \left(c_2 e^{i\rho_B\int n_2 d\xi}\right.$$

$$\left.\left.\left. + d_2 e^{-i\rho_B\int n_2 d\xi}\right) e^{-\frac{1}{2}\int n_2^{-1}dn_2 + \frac{1}{2}\int(1+\sigma^2)^{-1/2}d\sigma}\right]\right\}. \tag{3.87}$$

As has been pointed out above, the counter waves are independent in the considered case when no reflection regions $n_{1,2}^2 \approx 0$ are taken into account.[15] Being included in the analysis, these waves make the solution quite complicated but yields no new results. So, in what follows only waves which propagate to the direction of positive ξ will be considered. The geometrical optics solution for such waves can be written in the form

$$F_-(\xi) = e^{i\rho_B\int\frac{1}{2}(n_1+n_2)d\xi - \frac{1}{2}\int\frac{\sigma\, d\sigma}{1+\sigma^2}}$$

$$\times\left(\frac{c_1}{\sqrt{n_1}}e^{-i\rho_B\int\frac{1}{2}(n_2-n_1)d\xi - \frac{1}{2}\int\frac{d\sigma}{\sqrt{1+\sigma^2}}}\right.$$

$$\left. + \frac{c_2}{\sqrt{n_2}}e^{i\rho_B\int\frac{1}{2}(n_2-n_1)d\xi + \frac{1}{2}\int\frac{d\sigma}{\sqrt{1+\sigma^2}}}\right). \tag{3.88}$$

[15] In other words, the constants c_1, c_2 and d_1, d_2 in (3.87) are not related to each other. As a result, linear coupling occurs between the waves of different types propagating in the same direction.

This expression at arbitrary c_1 and c_2 describes independent waves of two types, ordinary and extraordinary. The formula (3.88) is the asymptotic solution valid at $\rho_B \gg 1$ outside the regions of reflection or linear coupling. The location of the latter ones shall be deduced using (3.88) and recalling that the geometrical optics fails if the spatial scales of wave amplitudes are comparable to the wavelength. By analysing the formula (3.88) it may be found that geometrical optics may be violated at large ρ_B near the points $n_1 = 0$, $n_2 = 0$, or $n_1 = n_2$. The latter condition corresponds to the points $\sigma = \xi/\xi_0 = \pm i$ on a complex plane ξ.

Fig. 3.10 shows the squares of refractive indices as functions of the parameter $\xi \equiv \omega_z$. The forms of the dispersion curves at $u \sin^2\alpha = 2\omega_x^2 < 1 - v$ and $u \sin^2\alpha = 2\omega_x^2 > 1 - v$ are essentially different (cf. Fig. 3.10 *a* and *b*). If $u \sin^2\alpha < 1 - v$, i.e., $\xi_0 < 1/2$, then the case of the quasi-perpendicular propagation which is realized at small ξ is characterized by the close approach of the dispersion curves to each other. The smaller the product $v\omega_x$, the closer lie the curves. Therefore, radiation may pass from one dispersion curve to another. Such a "transition" is facilitated if the plasma density or magnetic field is small enough, and the effective linear coupling of the modes can occur. On the contrary, at $u \sin^2\alpha > 1 - v$ the dipersion curves in the region of quasi-transverse propagation are well separated from each other.

The following consideration is focused on the case $u \sin^2\alpha < 1 - v$. Note that the nulls and poles of the functions $n_{1,2}^2(\xi)$ should not be taken into account, since the linear approximation for the magnetic field is only valid at sufficiently small $\omega_z = \sqrt{u}\cos\alpha$. We suppose that at larger values of $\sqrt{u}\cos\alpha$ the field varies in such a way that n_1^2 does not reach the pole. Thus, normal waves leave freely the region of quasi-transverse magnetic field.

Since wave propagation through the coupling region marked by a circle at the Fig. 3.10, *a* is accompanied by their effective coupling and transformation into each other, the constants c_1 and c_2 in the geometrical optics solution can be different at different sides of the coupling region. Indeed, an exact solution of the system (3.83) is analytical function which is surely univalent in the coupling region and nearby. On the other hand, the solution (3.88) for the function $F_-(\xi)$ is a combination of multivalent functions. Thus, the latter can only approximate the former if the coefficients c_1 and c_2 are different at different parts of the complex plane. The solution of the problem of linear mode coupling in plasma is thus reduced to the deriving of the relations between the constants which characterize normal wave amplitudes at different sides of the coupling region.

It is known [143] that the coefficients c_1, c_2 in the solutions (3.88) are constant in the regions bounded by the so-called Stokes lines, while crossing such a line causes step-wise variations of these coefficients. The Stokes lines are determined by the condition

$$\arg\left(\pm i\rho_B \int_{\pm i\xi_0}^{\xi} \frac{1}{2}(n_2 - n_1)\, d\xi\right) = \pi + 2\pi l, \tag{3.89}$$

where $l = 0, 1, 2, \ldots$ The functions

$$\exp\left(\pm i\rho_B \int_{\pm i\xi_0}^{\xi} \frac{1}{2}(n_2 - n_1)\, d\xi\right)$$

which enter the geometrical optics expression (3.88) take infinitely large or small real values on these lines far from the points $\pm\xi_0$, while the ratio of such functions becomes extremal.

Crossing a Stokes line causes a step-wise variation of one of two constants which precedes the decaying exponent in the solution. The value of the step, $\Delta c = c_1' - c_1$, is proportional to the coefficient c_2 of the growing exponent:[16] $c_1' = c_1 + \alpha c_2$, $c_2' = c_2$. Analogous relations hold for the Stokes line at which the exponent corresponding to the coefficient c_1 becomes increasing: $c_2' = c_2 + \beta c_1$, $c_1' = c_1$.

To draw the Stokes lines, we use the expression (2.121) which yields

$$n_2 - n_1 \equiv \frac{n_2^2 - n_1^2}{n_2 + n_1} = \Xi(\xi)\sqrt{\xi^2 + \xi_0^2}, \tag{3.90}$$

where $\Xi(\xi)$ is an analytical function which does not nullify in the vicinity of the points $\xi = \pm\xi_0$. Fig. 3.11 shows configuration of the Stokes lines at the plane ξ cut along the line between the points $i\xi_0$ and $-\xi_0$ for the case $n_2 - n_1 = \sqrt{\xi^2 + \xi_0^2}$. Appearance of the function $\Xi(\xi)$ in the difference $n_2 - n_1$ leads to a continuous deformation of the Stokes lines but does not alter their general behavior. It is not necessary to know the exact configuration of the Stokes lines when dealing with coupling problems.

The relation between the constants c_1, c_2 at the point A and the corresponding values C_1, C_2 at the point B is definite if the steps of these coefficients caused by crossing of a Stokes line are known. The desired steps can be found by demanding that the initial values of the constants at A are identical to the corresponding ones in (3.88) obtained after passing a

[16] Indeed, since the initial system of equation is linear, the coefficients at different sides of a Stokes line should be linked via linear relation of the form $c_1' = \tilde{\alpha} c_1 + \alpha c_2$. However, to ensure the validity of the asymptotic approximation at $c_2 = 0$, it is necessary that $\tilde{\alpha} = 0$, i.e., the step is absent. At the same time, difference of α from zero (or, equivalently, the appearance of a step of the exponentially decaying term) does not affect much the accuracy of the geometrical optics approximation for the exact solution, since at a Stokes line the error brought by the exponentially growing term is of about the modulus of the exponentially decaying one.

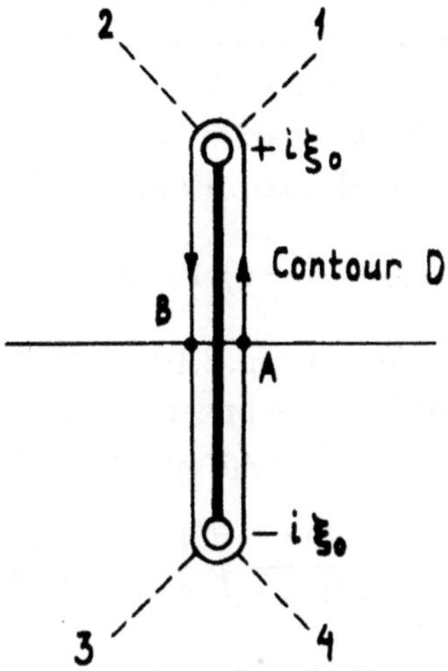

Figure 3.11. Location of the Stokes lines (marked by the dashed lines 1–4) on the complex plane for $n_2 - n_1 = \sqrt{\xi^2 + \xi_0^2}$.

complete cycle along the contour D which encloses the branch points $\pm i\xi_0$ of the function $n_2 - n_1$ (see Fig. 3.11).

Omitting the details of the above procedure which is described in [3,141], we just give here the resulting relations between the constants at the points A and B:

$$c_1 = \sqrt{1 - e^{-2\delta_0}}\, C_1 + e^{-\delta_0} C_2,$$
$$c_2 = -e^{-\delta_0} C_1 + \sqrt{1 - e^{-2\delta_0}}\, C_2. \tag{3.91}$$

Note that we consider the waves propagating to the direction of positive value of ξ. Therefore, the constants C_1 and C_2 characterize the amplitudes of extraordinary and ordinary mode, respectively, before the coupling region (the point B) whereas the constants c_1 and c_2 relate to the mode amplitudes after this region (the point A). The squares of the constants describe the intensities of the corresponding waves. Using (3.91) it is easy to check that the total intensity in both modes at the boundaries of the coupling region are equal:

$$|c_1|^2 + |c_2|^2 = |C_1|^2 + |C_2|^2.$$

It is clear from (3.91) that the relation between ordinary and extraordinary waves passing through a region of quasi-transverse magnetic field is characterized by the real parameter

$$2\delta_0 = -i\rho_B \oint \frac{1}{2}(n_2 - n_1)d\xi \tag{3.92}$$

in which the integral is taken over the closed contour D. The coupling effect is most pronounced in the limit $\delta_0 \ll 1$. Here a wave from certain dispersion curves in Fig. 3.10, *a* is completely transformed into the one corresponding to another dispersion curve: $C_2 = c_1$, $C_1 = c_2$. On the contrary, linear coupling is negligible if $\delta_0 \gg 1$ when $C_1 = c_1$ and $C_2 = c_2$.

The expression (3.92) for the parameter δ_0 comprises a contour integral which in general is quite complicated. However, the problem is simplified if

$$2v\omega_x^2 \ll (1-v)(1-v+2\omega_x^2). \tag{3.93}$$

Fortunately, dispersion curves in Fig. 3.10, *a* get closer in this case, so the above condition favors the effective linear coupling. Since $n_1 \approx n_2 \approx \sqrt{1-v}$ along the integration contour, the expression (3.92) is reduced to

$$2\delta_0 \approx -\frac{i\rho_B}{\sqrt{1-v}} \oint \frac{1}{4}(n_2^2 - n_1^2)d\xi.$$

Substituting the expressions (2.121) for $n_{1,2}^2$, we obtain [141]

$$2\delta_0 \approx \frac{i\rho_B v}{2\sqrt{1-v}} \oint \frac{\sqrt{\xi^2+\xi_0^2}}{\xi^2 + 2\xi_0^2 - 1} d\xi$$
$$= -\pi\rho_B \frac{v}{\sqrt{1-v}} \frac{\sqrt{1-2\xi_0} + \xi_0 - 1}{\sqrt{1-2\xi_0}}. \tag{3.94}$$

If $2\xi_0 \approx 2\omega_x(1-v)^{-1} \ll 1$, the coupling parameter takes the form [139, 140]

$$2\delta_0 \simeq \frac{\pi\rho_B}{2}\frac{v\omega_x^4}{(1-v)^{5/2}} = \frac{\pi\omega v u^{3/2}}{8c(1-v)^{5/2}|d\alpha/dz|}. \tag{3.95}$$

To come to the last expression, it should be taken into account that $2\omega_x^2$ equals $u = \omega_B^2/\omega^2$ at $z = 0$, while the parameter $\rho_B = \omega/c\Lambda$, where $\Lambda = d\omega_x/dz = \sqrt{u}|d\alpha/dz|$ (see (3.82)). But if $v \ll 1$, then [141]

$$2\delta_0 \simeq \frac{\pi\omega v u^{3/2}}{8c|d\alpha/dz|} = \frac{\pi\omega_L^2\omega_B^3}{8c\omega^4|d\alpha/dz|} = 2.5\cdot 10^{20}\frac{NB_0^3}{\omega^4|d\alpha/dz|}. \tag{3.96}$$

According to the expression (3.95), $2\delta_0 \equiv (\pi\sqrt{2}/8)G_\perp$, where the coupling parameter $G_\perp$ is determined by the formula (3.74) above. Note that both parameters are identical up to the factor of about unity.

Using (3.91), (3.96) and choosing the characteristics of radiation which enter the coupling region, it is not difficult to derive the frequency profiles of degrees of linear and circular polarizations of the radiation escaping from the coupling region into the region of quasi-longitudinal propagation.

As is known (see section 1.2), the degree of circular polarization is defined by the expression $\zeta_c = (I_l - I_r)/(I_l + I_r)$, where I_l and I_r are the

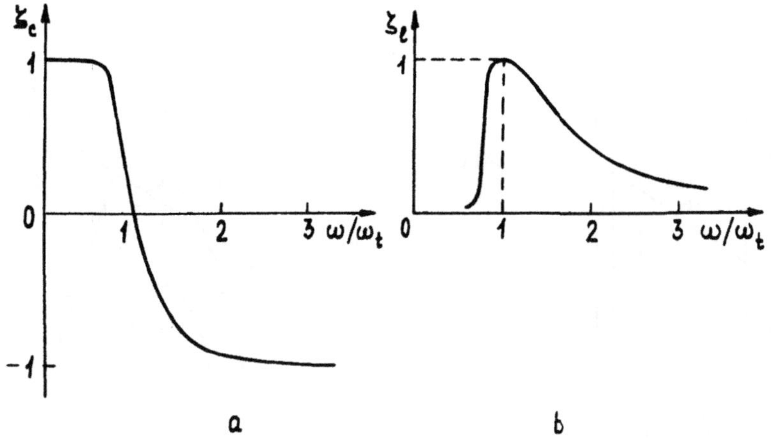

Figure 3.12. Frequency profiles of the degrees of circular (ζ_c) and linear (ζ_l) polarizations.

intensities of left- and right-hand components. In the case considered these intensities are proportional to $|c_2|^2$ and $|c_1|^2$, respectively. The degree of linear polarization, in its turn, is determined from the relation $\zeta_c^2 + \zeta_l^2 = \zeta^2$, where $\zeta = \text{const}$ is the total degree of polarization.

If radiation enters the coupling region in a single mode (for example, the ordinary one) with the intensity $|C_2|^2 = 1$, then, according to (3.91), the outgoing radiation on the other side will comprise both the ordinary component with the intensity $1 - e^{-2\delta_0}$ and the extraordinary one of the intensity $e^{-2\delta_0}$. Therefore, the transformation coefficient equals

$$Q = e^{-2\delta_0} = \exp\left(-\frac{\pi}{4\sqrt{2}} G_\perp\right). \tag{3.97}$$

Since $\zeta = 1$, the degrees of circular and linear polarization are given by the formulas

$$\zeta_c = 1 - 2e^{-2\delta_0}, \qquad \zeta_l = \sqrt{1 - \zeta_c^2} = 2e^{-\delta_0}\sqrt{1 - e^{-2\delta_0}}. \tag{3.98}$$

The sign of ζ_c and, respectively, the rotation over the resulting polarization ellipse reverse at $2\delta_0 = \ln 2 \simeq 0.7$. This value corresponds to the frequency

$$\omega_t = \left(\frac{\pi^2 e^5 N B_0^3}{2\ln 2\, m_e^4 c^4 |d\alpha/dz|}\right)^{1/4} \simeq 2\pi\left(3 \cdot 10^{17} \frac{N B_0^3}{|d\alpha/dz|}\right)^{1/4} \tag{3.99}$$

By the way, this value has been already pointed out above in (3.79), thought with less accuracy. The plots of ζ_c and ζ_l versus ω/ω_t are shown in Fig. 3.12. It is clearly seen that the sign of the circular polarization alters at the frequency $\omega = \omega_t$. At the same time, the intensity of the linearly polarized

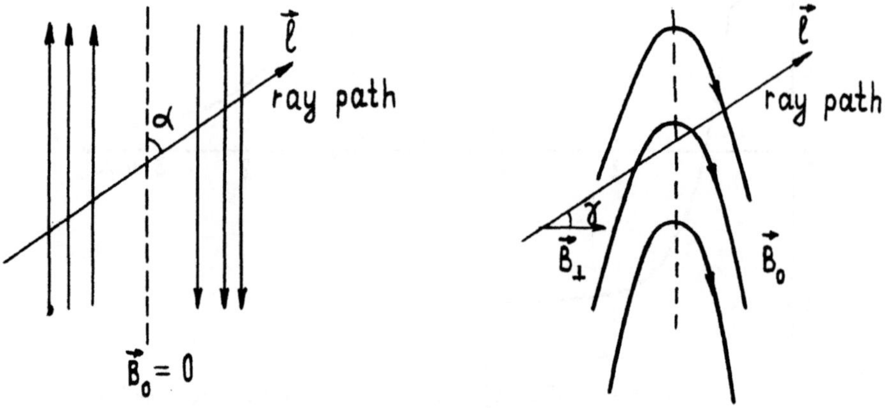

Figure 3.13. Field line configuration of the magnetic field $\mathbf{B}_0$ for the case of: (*a*) neutral current sheet containing a surface $\mathbf{B}_0 = 0$; (*b*) superposition of the fields by neutral current sheet, *a*, and a constant magnetic field $\mathbf{B}_\perp$ perpendicular to the zero-field surface.

component of the radiation escaping the region of quasi-transverse magnetic field is maximal at this frequency [141].

Note that the formulas (3.95), (3.98), (3.99) can be obtained using the analytical solution of the equations (3.40) of quasi-isotropic approximation, for example, in the form (3.55) [134]. If $\cos\alpha = l/\Lambda$ at $u =$ const and $v =$ const, then the function (3.64) in the case (3.66) takes the form:

$$G(\eta) = -\frac{vu^{3/2}}{4(1-v)^{5/2}}\frac{\omega}{c}\frac{\Lambda}{\cos^3 2\eta}. \tag{3.100}$$

Then an exact analytical solution of the equation (3.55) may be expressed in terms of the functions of parabolic cylinder. This solution yields the formulas (3.98) for degrees of linear and circular polarization and (3.99) for the critical frequency.

3.2.5. WAVE PROPAGATION THROUGH A ZERO MAGNETIC FIELD SURFACE

A surface at which the total magnetic field vanishes, $\boldsymbol{B}_0 = 0$, can exist in natural conditions inside neutral current sheets (see Fig. 3.13, *a*).

Consider propagation of waves through a neutral current sheet which is essentially determined by its parameters: characteristic width $L_{\rm sh}$, magnetic field strength $B_{\rm sh}$ at the edges, and the angle α between the ray and the magnetic field. The process of wave propagation depends crucially on the coupling parameter (3.72) which can be written in the form

$$G_0 = 32\sqrt{2}\pi e \frac{N L_{\rm sh}}{B_{\rm sh}} |\cos^3\alpha| \simeq 7\cdot 10^{-8}\frac{N L_{\rm sh}}{B_{\rm sh}}|\cos^3\alpha|. \qquad (3.101)$$

If $q^2 \ll 1$ at the edges of the sheet, this inequality should also hold inside it at $v \ll 1$. Therefore, quasi-longitudinal propagation is valid along a ray passing through the current sheet. According to the above qualitative analysis of the linear mode coupling, the effect is absent in this case. But if the parameter $q^2 \gg 1$ at the sheet edges, wave propagation takes place in the following way. Let a wave of a certain type (say, the ordinary one) encounter a neutral current sheet. This wave is linearly polarized, the electric field vector $\boldsymbol{E}$ lying in the $\boldsymbol{k} - \boldsymbol{B}_0$ plane. While propagating toward the surface of zero magnetic field, the wave passes successively through the regions $q^2 \sim 1$, $q^2 \ll 1$, and again $q^2 \sim 1$. Afterwards, it escapes to the region of quasi-transverse propagation at the other side of the current sheet. If the coupling parameter $G_0^2 \ll 1$, the geometrical optics is violated in the region $q^2 \sim 1$. If the modulus of the function $G(\eta)$ (see (3.52)) is much less than unity at $q^2 \lesssim 1$, the geometrical optics fails in the entire transition region $q^2 \lesssim 1$, while the "isotropic" approximation is valid there. Therefore, wave polarization does not change in this region. The considered incoming wave passing through the layer $q^2 \lesssim 1$ and escaping to the region $q^2 \gg 1$ remains linearly polarized in the $\boldsymbol{k}$–$\boldsymbol{B}_0$ plane. Such a polarization of the outgoing radiation corresponds to the ordinary mode. Therefore, strong coupling in this case prevents variation of the type of normal wave and the plane of polarization in the course of propagation of a wave through a current sheet between the regions $q^2 \gg 1$.

The considered case ($G_0^2 \ll 1$) of wave propagation through a current sheet is realized for the angles α close to $\pi/2$ (see (3.72)). But if α is quite different from $\pi/2$, the parameter G_0^2 is much larger than unity, and geometrical optics approximation remains valid in the region $q^2 \gtrsim 1$. It is easy to see that this approximation holds even in the nearest vicinity of the surface $\boldsymbol{B}_0 = 0$, at $q^2 \ll 1$. Thus, at $G_0^2 \gg 1$ the propagation of waves through the neutral current sheet between two regions $q^2 \gg 1$ of quasi-transverse propagation is well described by the geometrical optics approximation. Variations of the wave refractive index n_j^2 and polarization K_j in this case are given by the formulas (2.121) and (2.136) which have been derived for modes in a homogeneous medium. Inhomogeneity is taken into account by assuming that the parameter $u = u(z)$ varies along the direction of wave propagation. The function $n_j^2(\sqrt{u})$ is illustrated by Fig. 3.14. If $\sqrt{u} = al$, the dispersion curves shown in this figure describe, with appropriately chosen scales, the dependence $n_j^2(l)$ in an inhomogeneous magnetic field with a

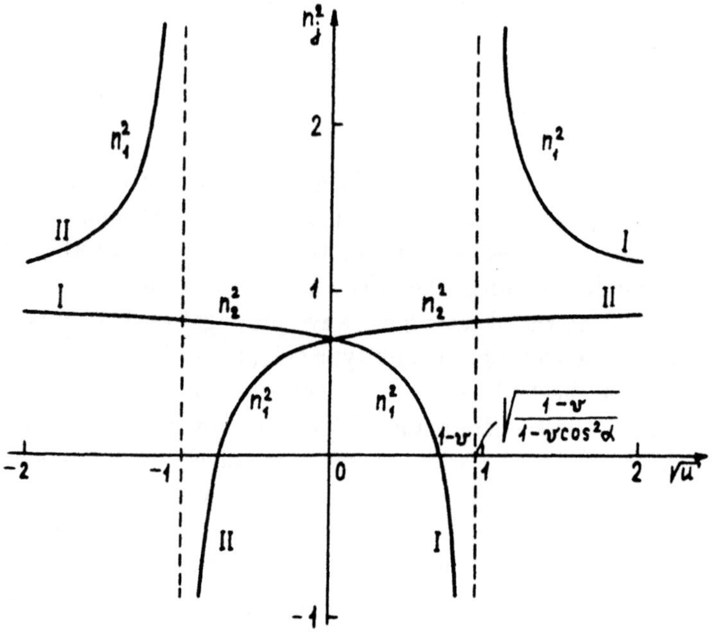

Figure 3.14. The plot of the square of the refractive index n_j^2 on $\sqrt{u}$ at $v=$ const and $\alpha \neq \pi/2$.

current sheet. Magnetic field reversal in the current sheet is taken into account by changing the sign of $\sqrt{u}$ at constant α.

If a wave of certain type (for example, the ordinary wave with the refractive index n_2), exists at the right-hand part of the Fig. 3.14, the curve *II* describes variation of the refractive index during propagation in an inhomogeneous magnetic field.[17] The curve *II* corresponds to the unified geometrical optics solution of the system of equations (3.37) which does not depend on the direction of the magnetic field. The second, linearly independent solution, is represented by the curve *I*. As a result, a wave which propagates along the dispersion curve *I* cannot pass to the curve *II* and vice versa — as long as the geometrical optics is applicable in an inhomogeneous magnetic field. On the contrary, the case $G_0^2 \ll 1$ of strong coupling corresponds to nearly complete transition of a wave from one dispersion curve to the other.

[17]Alternation of the mode name with direction of the magnetic field (i.e. in case of a transition from the region $\sqrt{u}>0$ to $\sqrt{u}<0$) is irrelevant here. According to the generally adopted convention, branches of a dispersion curve are called ordinary or extraordinary in such a way that the value n_j^2 for a given mode are the same for opposite directions of the magnetic field $\mathbf{B}_0 \to -\mathbf{B}_0$. This rule has set the notations of the dispersion branches in Fig. 3.14, in the formulas (2.128) for $n_{1,2}^2$ which comprise the expressions $|\sqrt{u}\cos\alpha|$, etc.

The above analysis allows one to follow the variation of wave polarization in a neutral current sheet at $G_0^2 \gg 1$. If a linearly polarized wave with vector $\boldsymbol{E}$ parallel to the $\boldsymbol{k}$-$\boldsymbol{B}_0$ plane propagates from the the right-hand region of quasi-transverse propagation ($q^2 \gg 1$) in Figs. 3.8, 3.14, then in the region of the quasi-longitudinal propagation ($q^2 \ll 1$) the polarization of this wave becomes circular. The sense of polarization does not reverse after crossing the layer $\sqrt{u}=0$, and then, in the left-hand region $q^2 \gg 1$, the wave should be linearly polarized with the vector $\boldsymbol{E}$ perpendicular to $\boldsymbol{k}$ and $\boldsymbol{B}_0$, i.e. become extraordinary. Thus, the geometrical optics propagation $G_0^2 \gg 1$ through a neutral current sheet alters the mode type and the orientation of the plane of linear polarization. On the contrary, orientation of the polarization plane is constant in the case of strong coupling $G_0^2 \ll 1$.

The above noted invariance of the sign of circular polarization after crossing the layer $\sqrt{u} = 0$ along a certain dispersion curve (I or II) in Fig. 3.14 clearly follows from the formula (2.136) for the polarization coefficient $K_j \equiv -iE_y/E_x$. This coefficient in the quasi-longitudinal approximation is given by the formula: $K_j \approx \sqrt{u}\cos\alpha(\pm|\sqrt{u}\cos\alpha|)^{-1}$. Alternation of the sign of $\sqrt{u}\cos\alpha$ changes the sign of the denominator (as well as the type of the mode which corresponds to the curve II). Hence, the value of K_j is unchanged.

The conclusion on the invariance of the sense of polarization of a wave which propagates through a zero-field region with negligible mode coupling is opposite to the one resulting from the analysis of the case of wave propagation through a transverse magnetic field. In the latter case the absence of the coupling (i.e., validity of the geometrical optics) results in polarization reversal.

The above qualitative description of the linear mode coupling in an inhomogeneous magnetic field yields the characteristic parameters of this effect. However, the approach used is not adequate to treat this problem quantitatively. Linear mode coupling in the region of zero magnetic field has been calculated numerically in [147]. The result is shown in Fig. 3.15 as a plot of the transformation coefficient Q versus the characteristic coupling parameter G_0 determined by the formula (3.49). Here the value Q is defined as the ratio of square amplitudes of outgoing to incoming ordinary waves. The similar ratio for the extraordinary waves entering and escaping the coupling region equals $1-Q_{\text{ord}}$. In the limiting cases of weak ($G_0 \gg 1$) and strong ($G_0 \ll 1$) coupling the analytical expressions for the coefficient of linear coupling have the following form [138]:

$$\begin{aligned} &Q \simeq \frac{\pi\sqrt{2}}{G_0}, && G_0 \gg 1, \\ &Q \simeq 1 - \frac{\pi}{2} G_0^{2/3} [\mathrm{Ai}'(0)]^2, && G_0 \ll 1. \end{aligned} \tag{3.102}$$

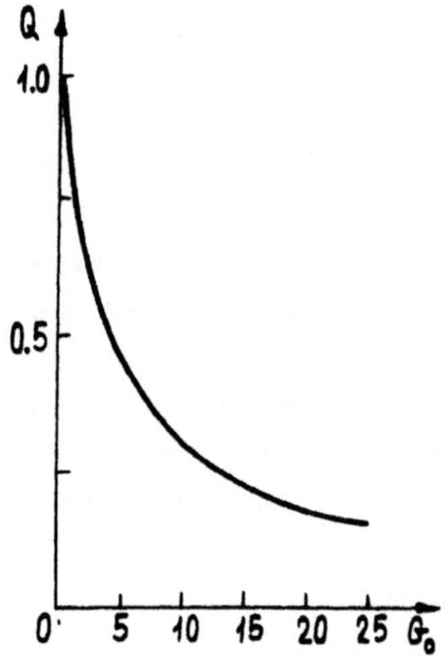

Figure 3.15. Transformation coefficient Q as a function of the interaction parameter G_0 in case of propagation through a region of zero magnetic field, when a wave passes from the dispersion curve II to the curve I.

Here $\mathrm{Ai}'(0) \approx -0.459$ is the Airy function derivative at zero argument. Finally, the solution of the relevant standard problem valid at arbitrary G_0 is given in [148].

If a neutral current sheet is permeated by a transverse magnetic field $B_\perp$ (see Fig. 3.13, *b*), the wave proparagion can drastically change. There is no zero-field surface $\boldsymbol{B}_0 = 0$ inside such a current sheet, though at $B_\perp \ll B_{\mathrm{sh}}$ the magnetic field in the middle of the sheet is much weaker than at its boundaries. The detalied analysis of the linear mode coupling in plasma current sheets has been made in [148]. To be definite, consider the superposition of a self-consistent neutral current sheet [149]:

$$B_{\mathrm{H}}^2 = B_{\mathrm{sh}}^2 \tanh^2(z/L_{\mathrm{sh}}), \qquad N = N_0 \cosh^{-2}(z/L_{\mathrm{sh}}), \tag{3.103}$$

and a homogeneous magnetic field $\boldsymbol{B}_\perp \perp \boldsymbol{B}_{\mathrm{sh}}$ orientated along z-axis. If the angle γ between the field $\boldsymbol{B}_\perp$ and the direction of the ray $\boldsymbol{l}$ is small enough:

$$\tan^2 \gamma < B_\perp^2 / B_{\mathrm{sh}}^2, \tag{3.104}$$

then the total field $\boldsymbol{B} = \boldsymbol{B}_{\mathrm{H}} + \boldsymbol{B}_\perp$ cannot become perpendicular to $\boldsymbol{l}$ at any point of the ray. The sign of the projection of $\boldsymbol{B}$ onto $\boldsymbol{l}$ is constant in this case. In the framework of the geometrical optics this means that the sign of rotation of the electric field vector $\boldsymbol{E}$ is constant, too. Moreover, calculations reveal that even if geometrical optics approximation fails in a domain $q^2 \sim 1$ along a ray, the transformation coefficient Q does not exceed $1/2$. Such a weak transformation cannot lead to the polarization reversal.

In the opposite case

$$\tan^2\gamma > B_\perp^2/B_{\rm sh} \tag{3.105}$$

there always exists a point $z_\perp$ at which the total magnetic field is perpendicular to the ray path. This point is determined by the formula

$$\tanh(z_\perp/L_s) = -B_\perp/B_{\rm sh}\tan\gamma. \tag{3.106}$$

In case of a well developed current sheet ($B_\perp^2 \ll B_{\rm sh}^2$) which we will focus on further, the point $z_\perp$ exists in a wide range of the angle γ excluding the narrow interval $\gamma^2 < B_\perp^2/B_\parallel^2 \ll 1$ in which the ray $\boldsymbol{l}$ is nearly perpendicular to the vector $\boldsymbol{B}_{\rm H}$. By passing the point $z_\perp$, the sign of the projection of $\boldsymbol{B}$ onto $\boldsymbol{l}$ is inversed, as well as the parameter $q(l)$ determined by (3.41) which crosses the point at infinity $\pm\infty$.

The picture of wave propagation through a current sheet is quite complicated. We only mention mention that, generally, there exist four regions along the ray in which $q^2 \simeq 1$ (see the details in [148]). However, the performed analysis has shown that only two of these at a time can have a significant effect on the resulting transformation coefficient. The variation of the sense of polarization due to the linear coupling in the domain of transverse magnetic field is determined, as before, by the parameter $G_\perp$, (3.74). It should be taken into account in this parameter that the magnetic field strength at the point $z_\perp$ is much weaker than $B_{\rm sh}$. Similarly, the value of $|d\alpha/dl|^{-1}$ is determined by the scale which is less than the thickness of the sheet $L_{\rm sh}$. All this leads to the decrease of $G_\perp$, so the conditions for the effective coupling in a quasi-transverse magnetic field, which conserves the sign of polarization, can be more easily realized in the current sheet considered. The resulting coupling parameter for the current sheet has the form

$$G_{\perp\rm sh} = \frac{\omega_{\rm L}^2\omega_{B\rm sh}^2 L_{\rm sh}}{\sqrt{2}c\omega^4\tan^5\gamma}\frac{B_\perp^4}{B_{\rm sh}^4}, \tag{3.107}$$

where $\omega_{B\rm sh} = eB_{\rm sh}/m_e c$. In the interval (3.105) and at $B_\perp^2 \ll B_0^2$ linear coupling is negligible if $G_{\perp\rm sh} \gg 1$ while polarization inversion occurs under the weaker condition $G_{\perp\rm sh} > 1$. It follows from (3.107) that the latter inequality is equivalent to the following criterion

$$\tan^{5/4}\gamma < \frac{B_\perp}{B_{\rm sh}}\left(\frac{\sqrt{2}c\omega^4}{\omega_{\rm L}^2\omega_{B\rm sh}^3 L_{\rm sh}}\right)^{1/4}. \tag{3.108}$$

If the opposite inequality is realized, linear coupling becomes strong and polarization conserves in the current sheet.

3.2.6. ON THE ROLE OF THE LINEAR MODE COUPLING IN COSMIC PLASMAS WITH AN INHOMOGENEOUS MAGNETIC FIELD

Propagation through the regions of transverse or zero magnetic field can significantly affect the polarization of radiation. The features of such a variation are determined by the above defined parameters $G_\perp$ (or $2\delta_0$) and G_0. For example, the case of linear coupling in a quasi-transverse magnetic field is realized easily for radiation propagating in the interstellar plasma. Indeed, in a medium with electron density $N \sim 0.1\,\mathrm{cm}^{-3}$, magnetic field $B_0 \sim 3 \cdot 10^{-6}\,\mathrm{G}$ and characteristic scale of the field inhomogeneity $L_\alpha = |d\alpha/dz|^{-1} \sim 3 \cdot 10^{20}\,\mathrm{cm} \equiv 100\,\mathrm{pc}$ the critical frequency (3.99) is about $7 \cdot 10^5\,\mathrm{s}^{-1}$, so that $f_t = \omega_t/2\pi \sim 0.1\,\mathrm{MHz}$. It is thus clear that $\omega \gg \omega_t$ over the entire range of frequencies which are explored in radio astronomy, so the condition of strong coupling $2\delta_0 \ll 1$ is satisfied. As a result, the sign of circular polarization of radiation propagating in inhomogeneous Galactic magnetic field is unchanged, in spite of multiple flips of the projection of the magnetic field $\boldsymbol{B}_0$ onto the line of sight.

The effect of linear mode coupling in inhomogeneous magnetic field can probably influence the formation of polarization pattern of pulsar radio emission. For example, chaotic variation of the value of ω_t in neutron star magnetosphere or nearby may cause strong pulse-to-pulse variability of polarization observed from some pulsars [150]. However, one must be careful in using the numerical estimates of the parameters which determine the coupling effect and the conditions for the escape of radiation from pulsar magnetospheres, since plasmas there may be relativistic. Unfortunately, the problem of linear mode coupling in relativistic plasmas, though topical, has not been investigated up to now.

The conditions of electromagnetic wave propagation through the regions of quasi-transverse magnetic field in the solar corona predetermine the observed polarization features of solar radio emission. As an example, one may refer to the effect of polarization reversal observed in microwave bursts at frequencies $f \sim 3 \cdot 10^3\,\mathrm{MHz}$ [151]. The explanation of this fact is based on the assumption that radio emission passes through the region of quasi-transverse magnetic field situated in the corona somewhere along the ray path from the source to the observer, the combination of the parameters N, B_0 and L_α making the critical frequency equal to $\omega_t \sim 6\pi \cdot 10^3\,\mathrm{MHz}$ [152]. According to (3.99), the latter requires that $N B_0^3 L_\alpha \sim 4 \cdot 10^{20}\,\mathrm{G}^3\mathrm{cm}^{-2}$. Assuming that $N \sim 3 \cdot 10^8\,\mathrm{cm}^{-3}$ and $L_\alpha \sim 10^{10}\,\mathrm{cm}$ in the coupling region, it is possible to obtain an estimate for the magnetic field strength there: $B_0 \simeq 5\,\mathrm{G}$. The sense of polarization in such a region is varied according to Fig. 3.12, *a*.

It is clear from the above that the measurement of the frequency ω_t when studying the polarization reversal in solar radio emission is a quite

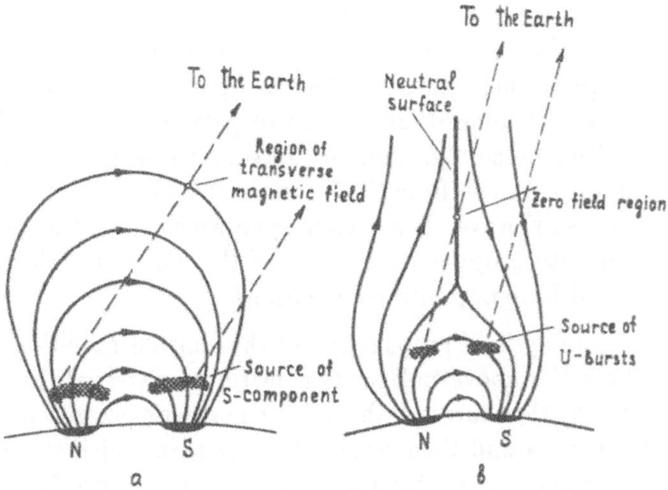

Figure 3.16. Escape of radio waves emitted in a solar active region above a bipolar group of sunspots: *a*, the case of propagation through a region of transverse magnetic field; *b*, the case of propagation through a zero-field surface.

convenient way to estimate the value of the magnetic field in the corona. In fact, the location of the coupling domain in the corona can be determined by adopting a certain configuration of the magnetic field lines in the coronal active region. Next, specifying the distribution of the electron density in the active region, it is possible to determine the product NL_α in the region of the quasi-transverse magnetic field. The estimate for the magnetic field B_0 which corresponds to the observed value of ω_t can be found using the formula (3.99). According to this formula, the deduced value of the magnetic field B_0 is set primarily by the frequency ω_t and depends very weakly on the parameters N and L_α.

Note that radiation at frequencies close to ω_t which escapes from the region of quasi-transverse magnetic field should exhibit strong linear polarization (see Fig. 3.12, *b*). However, strong dispersion of Faraday rotation angles in the corona above the coupling region prevents the reliable detection of such a polarization. This feature can only be observed with a specialized narrow-band facility (see section 4.2 and [3]).

Taking account of the polarization effects due to wave propagation in an inhomogeneous magnetic field is especially important for the interpretation of polarimetric observations of radio emission from local sources above bipolar groups of sunspots. As an example, we mention here the regions of generation of the S-component of solar radio emission at centimeter wavelengths (see section 6.2 and [3]). Sources of this type can also be found in the region of generation of noise storms, U-bursts etc. [18]

A typical feature observed in many such events is different signs of polarization of radio waves emitted by the sources with anti-parallel projections of the magnetic field onto the line of sight [3, 153, 154]). At first, it seems quite natural. Indeed, generation of polarized radiation leads to predominance of one mode over another (for example, extraordinary waves are dominant in the emission from the S-sources, whereas ordinary waves in U-bursts). As a result, waves of a certain type which leave the generation regions with opposite projections $B_0 \cos\alpha$ of the magnetic field onto the line of sight should have opposite polarizations.

However, the analysis of propagation of the emitted radiation through the corona makes the above polarization feature non-trivial. Indeed, the radiation emitted by the sources with opposite magnetic polarity propagates up through the corona and then reaches the layers in which the magnetic field is more homogeneous, so that the projection of $\boldsymbol{B}_0$ onto the directions of propagation becomes the same for both sources. It means that the radiation from one of these sources should inevitably pass through the region in which the projection $B_0 \cos\alpha$ changes direction. Such a flip may occur in either zero or transverse magnetic field. Propagation of radiation from the sources of opposite magnetic polarity to the outer corona in the latter case is illustrated by the Fig. 32 a which shows the structure of the magnetic field above a bipolar group of sunspots in which the S-component of solar radio emission is generated. If geometrical optics is valid in the region of transverse magnetic field, the observed polarizations of both sources should be the same. Since observations yield the opposite, we conclude that propagation through the region of transverse magnetic field does not lead to polarization inversion, i.e., takes place under the conditions of strong linear coupling (see Fig. 3.7). The case of strong coupling is realized if $G_\perp \ll 1$. The above considered example of polarization reversal of solar microwave bursts have shown that this inequality can be easily realized for centimeter waves in solar corona conditions. On the contrary, the sign of polarization in some cases does not correspond to the polarity of the magnetic field in the source [155]. Therefore, the opposite inequality $G_\perp \gg 1$ can also be realized for the S-component propagating through the corona.

According to [154] and a number of previous publications, polarization of radio waves in the meter band corresponds to the polarity of the sunspot

[18] U-bursts are a subclass of fast-drifting type III bursts discussed in section 7.2. According to the contemporary point of view, U-bursts are generated by electron beams moving along a magnetic arc with the foots basing on photospheric regions of opposite magnetic polarity.

above which the radiation source is located. Such a coincidence cannot be explained by the effect of linear coupling in a transverse magnetic field [156]. Indeed, for $\omega \sim 2\pi \cdot 10^8\,\mathrm{s}^{-1}$, $B_0 \sim 1\,\mathrm{G}$, and $L_\alpha \sim 10^{10}\,\mathrm{cm}$ the criterion $G_\perp \ll 1$, at which the sign of polarization does not change along the ray path through the transverse field region, can only be met if the electron density $N \ll 5{\cdot}10^4\,\mathrm{cm}^{-3}$. Thus, the region of transverse magnetic field should be situated high in the corona, at $R > 2.5\,R_\odot$, where the magnetic field is quasi-radial. On the other hand, there are no rationales for the assumption that the magnetic field B_0 in the coupling region is much less than 1 G, the characterictic value for the global magnetic field of the Sun (such an assumption would lead to the conclusion that the transverse field region is located closer to the solar surface in more dense plasma).

Since the sources of meter-wave bursts are situated high enough in the corona, it is quite probable that the projection $B_0 \cos\alpha$ for this radiation changes direction in neutral current sheets which contain surfaces of zero magnetic field $\boldsymbol{B}_0 = 0$ [147]. An example of such a field configuration is shown at Fig. 3.16, *b*.

It has been shown previously in this section that the sign of polarization is unchanged in case of quasi-longitudinal propagation in an inhomogeneous magnetic field, including the propagation through a neutral current sheet. The observed agreement between the direction of the magnetic field in the source and the sign of polarization of the meter-band radio emission is realized if the criterion of quasi-longitudinal propagation $q^2 \ll 1$ is satisfied over the entire ray path from the source to the point where it escapes from the corona into the interplanetary medium. The invariance of the sense of polarization in the course of propagation through neutral current sheets in the solar corona may also influence the observed polarization characteristics of microwave radio emission.

Note finally that for the directions at which the quasi-transverse approximation $q^2 \gg 1$ is valid at the boundaries of a neutral current sheet the resulting polarization of the escaping radiation depends on the value of the coupling parameter G_0 (see (3.101)). For $N \sim 10^7\,\mathrm{cm}^{-3}$, $B_{\mathrm{sh}} \sim 10\,\mathrm{G}$ and $L_{\mathrm{sh}} \sim 10^8 - 10^9\,\mathrm{cm}$ this value can be estimated as

$$G_0 \simeq (0.7 - 7) \cdot 10^7 |\cos^3\gamma|.$$

Hence, the effective coupling $G_0 \lesssim 1$ leading to the polarization inversion is only possible for $\gamma \lesssim 0^\circ.5$.

Therefore, in a wide interval of the angles γ at which the quasi-longitudinal approximation $q^2 \ll 1$ is realized over the entire ray from the source to the exit from the corona (neutral current sheets included), both extraordinary and ordinary waves remain unchanged (circularly polarized). But if the quasi-transverse propagation $q^2 \gg 1$ is realized at the current sheet boundaries, the sign of polarization inverts only for the values of γ within the narrow angular interval of about a degree.

It should be noticed that the long-term existence of neutral current sheets is quite improbable because of the tearing instability. A weak magnetic field $\boldsymbol{B}_\perp$ perpendicular to the neutral current sheet can stabilize this instability [157, 158]. The resulting structure called plasma current sheet is sketched in Fig. 3.14, *b*. According to the preceding paragraph, the sign of polarization of radiation passing through such a structure is conserved under the condition opposite to (3.108). According to [159], at $\omega \simeq 2\pi \cdot 160\,\mathrm{MHz}$, $N \simeq 10^7\,\mathrm{cm}^{-3}$, $B_{\mathrm{sh}} \simeq 10\,\mathrm{G}$, and $L_{\mathrm{sh}} \simeq 10^8\,\mathrm{cm}$ the quantity in the parentheses equals $3 \cdot 10^2$. However, the small exponent $-1/4$ leads to the quantity of the order of unity ($\simeq 0.25$). Hence, polarization reversal is only possible in the narrow angular interval

$$\tan^{5/4}\beta < B_\perp / B_{\mathrm{sh}} \ll 1.$$

On the contrary, for β from the wide region outside this interval the sign of polarization is unchanged. So, polarization of radiation escaping a well-developed current sheet ($B_\perp \ll B_0$) is the same as in the case of a neutral current sheet ($B_\perp = 0$) in spite of the radically different pictures of linear coupling and geometrical-optical propagation in these two cases.

3.2.7. COUPLING OF ELECTROMAGNETIC WAVES AND THE PROBLEM OF ESCAPE OF RADIO EMISSION FROM DENSE PLASMA

The problem of escape of radiation from plasma has been discussed earlier in this section concerning the formation of its polarization characteristics. There is however another important aspect of this problem: the question on the principal possibility for radiation to escape the plasma rather than to be reflected and, finally, absorbed there. In fact, analysis of the condition for the escape of radiation in such an aspect has been started in section 3.1 in the framework of the geometrical optics. It has been clarified there that only the waves corresponding to the branches of the dispersion curves $n_j^2(R)$ between the values $n_j^2 = 0$ in plasma and $n_j^2 = 1$ outside it (see Fig. 3.1–3.3) can escape a dense plasma. The condition $n_j^2(R) = 0$ determines the lower boundary of the domain from which radio waves escape freely from the plasma. Waves corresponding to other branches of dispersion curves cannot leave the plasma. Thus, the problem arises to find the conditions which favor the "transfer" of radiation from these branches onto the ones corresponding to escaping waves and to estimate the efficiency of such a "transition".

One of the possible solutions to the escape problem is yielded by linear coupling and transformation of different modes at $v \sim 1$ which, under certain conditions, may arise both in magnetoactive and in isotropic, regularly inhomogeneous plasma. Another possibility is connected to the scattering of waves by particles or waves in plasma which also results in the wave transformation (conversion) from one type to another. This effect will be considered in detail in chapter 8, while this section deals only with the discussion of the role of linear coupling in the problem of the escape of radiation from dense plasma and the process of the conversion of plasma waves into electromagnetic ones.

The problem mentioned is of special importance for the "plasma hypothesis" of the origin of sporadic radio emission from the Sun. Such a hypothesis was put forward in the first years of the radio astronomical era [160,161] and then also applied to radio emission by Jupiter, pulsar, and quasars [162–164,345,346]. According to this hypothesis, the observed radio emission originated due to plasma (Langmuir) waves (oscillations) excited by some "agent" (say, a particle beam) at frequencies $\omega \sim \omega_L$. Plasma waves, however, cannot propagate in vacuum. Being generated near the region $\omega \approx \omega_L$, these waves are completely absorbed at more tenuous layers of the plasma due to Landau damping (see section 6.1). Hence, consistent application of the above hypothesis to the origin of solar radio emission depends on the possibility of the effective transformation of plasma to electromagnetic waves under solar corona conditions.

The question of the transformation of plasma waves into electromagnetic ones requires special analysis since in a homogeneous plasma these modes propagate independently. The situation is however changed in the case of an oblique propagation with respect to the density gradient in an inhomogeneous plasma. Electromagnetic waves, linearly polarized in the plane of incidence, are coupled to plasma ones near the point of plasma wave reflection, both modes having the same projection of the wave vector onto the plane perpendicular to ∇N, since in this region the geometrical optics approximation is violated for both waves. As a result, part of the plasma wave energy is transformed into electromagnetic radiation which leaves the plasma region. The criterion of isotropic plasma (3.10) is very restrictive, so, as a rule, the influence of the magnetic field $\boldsymbol{B}_0$ on the linear mode coupling cannot be neglected. Hence, the effect of conversion of plasma waves into electromagnetic ones in isotropic plasma is not discussed here. The reader can get acquainted with it using [3,114,115].

Investigation of mode coupling in plasma with a constant magnetic field $\boldsymbol{B}_0$ is quite a complicated problem. So, in what follows there will only be considered the simplest cases of linear coupling in the region $v = \omega_L^2/\omega^2 \approx 1$ in which the magnetic field is homogeneous, plasma is plane-parallel, and waves propagate along the density gradient.

Linear coupling of electromagnetic and plasma waves has been studied in detail in [145, 146, 165, 166] by the method of phase integrals which has been used above in the analysis of mode coupling in a transverse magnetic field. So, only the final results are given below whereas the details of the calculation procedure which can be found in [3, 71] will be omitted.

Refer, first of all, to Fig. 2.5 which characterizes the distribution of the refractive indices squared of different modes $n_j^2(v)$, where $v = \omega_L^2/\omega^2$, not only in a homogeneous (ω_L = const) plasma at various frequencies but also in an inhomogeneous plasma at a fixed frequency. According to these figures, there exists the single coupling region at $u < 1$ in a regularly inhomogeneous plasma. Another coupling region in which $n_j^2 < 0$ is usually not reached by the waves since these decay exponentially as approaching it. This fact simplifies considerably the whole picture of wave coupling. It is also important that the effect of thermal motion of plasma particles at $u < 1$ can be neglected since the dispersion curve profiles near $v \approx 1$ is practically unchanged at $v_T^2/c^2 \ll 1$.

The solution of the problem under consideration is similar to the one used in the case of the coupling in a transverse magnetic field. The characteristic coupling parameter is given by

$$2\delta_{01} = -i\rho_N \oint \frac{n_2 - n_1}{2} d\epsilon. \tag{3.109}$$

The integration contour surrounds the "branching" points $n_1 = n_2$ on the complex plane $\epsilon = 1 - v$ (cf. (3.92)). Calculation of $2\delta_{01}$ at sufficiently small angles α between $\boldsymbol{B}_0$ and the direction of wave propagation yields

$$2\delta_{01} \approx \frac{\pi}{2} \frac{\omega}{c|\nabla\epsilon|} \frac{\alpha^2}{(1 + \omega/\omega_B)^{3/2}}. \tag{3.110}$$

To calculate the effect of linear coupling in a strong magnetic field $u \cos^2 \alpha > 1$, both coupling regions, which correspond to $n_j^2 > 0$ should be taken into account (Fig. 2.5). The detailed analysis [165, 166] shows that mode coupling in the lower domain in Fig. 2.8 in which the dispersion curves II and III are close to each other is determined by the same formulas as in the case $u < 1$ with the characteristic parameter δ_{01}, (3.110). The upper coupling domain in which the dispersion curves I and III approach each other is characterized by another parameter. This parameter at small angles is given by

$$2\delta_{02} \approx \frac{\pi}{2} \frac{\omega}{c|\nabla\epsilon|} \frac{\alpha^2}{(1 - \omega/\omega_B)^{3/2}}. \tag{3.111}$$

Different cases of linear coupling in the region $v \approx 1$ are sketched at Figs. 3.17, 3.18. In particular, according to Fig. 3.17, *a*, a plasma wave of unit amplitude which enters the region $v \approx 1$ from the left (i.e., from the

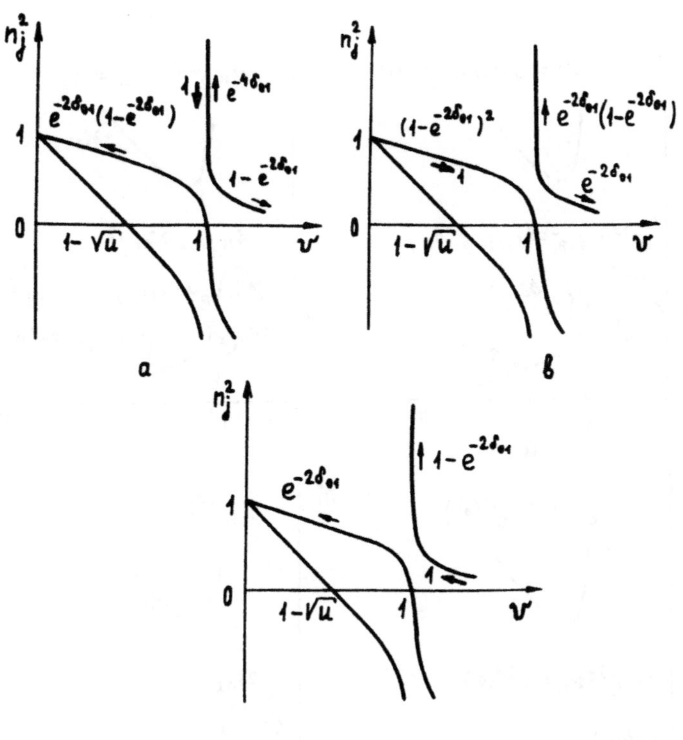

Figure 3.17. Different cases (a, b and c) of linear wave coupling at $u = \omega_B^2/\omega^2 < 1$. Thick arrows mark the incident wave of unit amplitude, and thin ones correspond to the outgoing waves which appear due to the coupling. The expressions near thin arrows give the modulus squared of the outgoing wave amplitudes.

region $v < 1$), gives rise to the ordinary electromagnetic wave which escapes to the region of small v. The modulus squared of the amplitude of this wave called transformation coefficient looks like

$$Q_{\rm ord} = e^{-2\delta_{01}}(1 - e^{-2\delta_{01}}). \tag{3.112}$$

Note that this formula does not account for the reflection of the ordinary radiation from the point $v = 1 + \sqrt{u}$. This effect is negligible due to, for example, strong absorption between the reflection point and the coupling region. In cases shown in Fig. 3.2 the reflection layer does not exist in the solar corona. Next, if $u\cos^2\alpha > 0$, then both ordinary and extraordinary waves with the amplitudes

$$\begin{aligned} Q_{\rm ord} &= e^{-2\delta_{01}-2\delta_{02}}(1 - e^{-2\delta_{01}}), \\ Q_{\rm ext} &= e^{-4\delta_{01}-2\delta_{02}}(1 - e^{-2\delta_{02}}). \end{aligned} \tag{3.113}$$

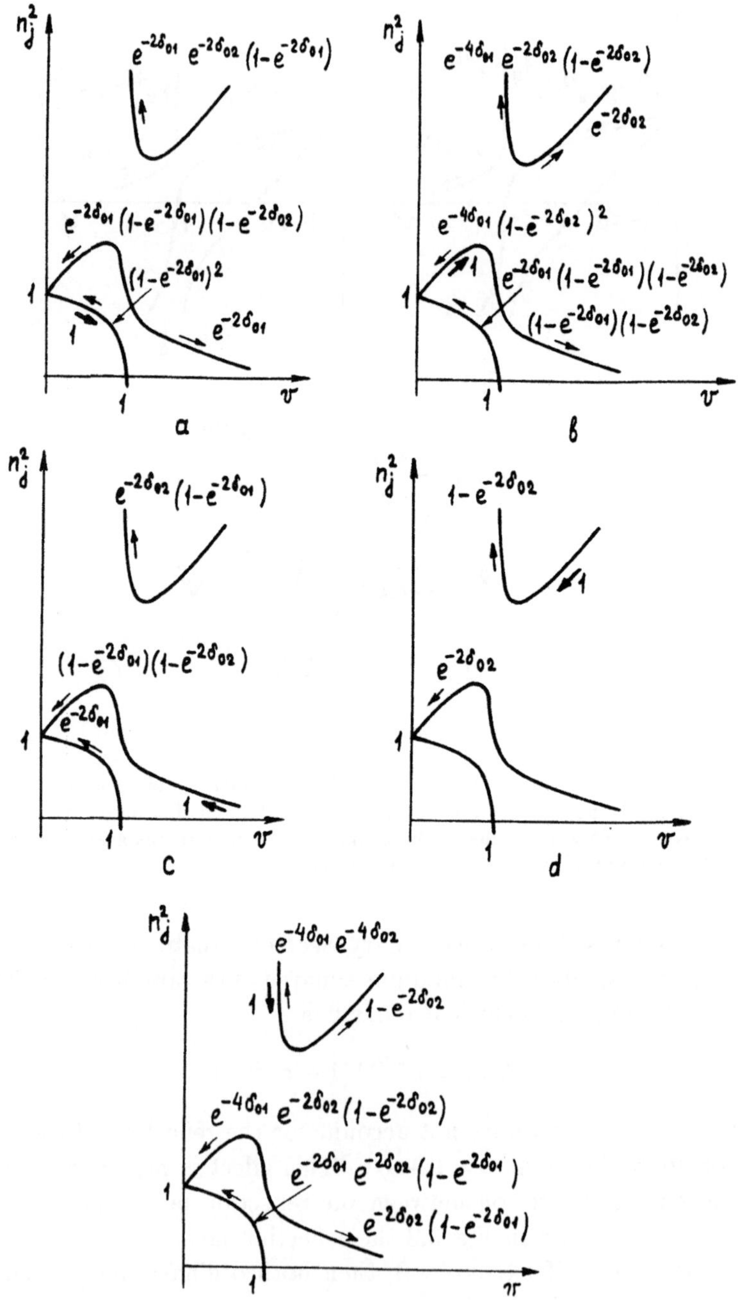

Figure 3.18. Different cases of linear wave coupling at $u\cos^2\alpha > 1$. The notations are the same as that of the previous figure.

escape to the region of small v as a result of the conversion of the incident plasma wave (Fig. 3.18, b). The relative intensities (3.113) determine the efficiency of the transformation.

The above transformation coefficients Q_{ord}, Q_{ext} are of the order of unity if $\delta_{01,02} \sim 1$. Analysis shows that the considered effects of linear wave coupling in the region $v \approx 1$ are negligible under solar corona conditions, since the values $\delta_{01,02} \sim 1$ are realized only at very small angles α between the direction of wave propagation and the magnetic field. The efficiency of the coupling may nevertheless increase if large gradients of electron density or shock discontonuities appear in the region $v \approx 1$.

3.2.8. LINEAR MODE COUPLING IN A SYSTEM "PLASMA + BIREFRINGENT VACUUM".

It has been noted in section 2.2 that vacuum in a superstrong magnetic field behaves as an uniaxial anisotropic, non-gyrotropic medium. Linear mode coupling in such media has been considered in [167] using the example of liquid cristals. It has been shown that the transformation can only occur if the direction of the anisotropy axis varies in space. In the case of birefringent vacuum this effect is connected with the shear of the magnetic field (i.e., rotation of the projection of the magnetic field onto the plane perpendicular to the direction of wave propagation), while it is absent in non-sheared magnetic fields. However, the coupling can occur in a system comprising both magnetized vacuum and plasma [168]. This case is considered below following the analysis made in [169]. Assume that plasma is rarefied enough, $v \ll 1$, yet thermal motion is not important (in particular, the criterion of the "cold plasma" approximation (2.138) is satisfied).

In a system comprising cold, classical plasma and birefringent vacuum there are two modes, ordinary and extraordinary, with the refraction indices

$$n_{1,2} = \epsilon \pm C\sqrt{1 + q^2} \tag{3.114}$$

and polarization coefficients $K_{1,2}$ determined by the formulas (3.43). Here, as usual, the upper sign and subscript 1 correspond to the extraordinary mode, while the lower sign and subscript 2, to the ordinary one. The following notations are adopted in the above formulas:

$$q = \frac{\sqrt{u}\sin^2\alpha}{2\cos\alpha}\left[1 + \frac{3a(1-u)}{uv}\right], \tag{3.115}$$

$$\begin{gathered} \epsilon = 1 + \frac{11}{2}a\sin^2\alpha - \frac{v(1-u\sin^2\alpha)}{2(1-u)}. \\ C = \frac{v\sqrt{u}\cos\alpha}{u-1}, \end{gathered} \tag{3.116}$$

The parameters u, v, a, and α has been defined in section 2.2.

According to the above qualitative analysis of the linear mode coupling, the efficiency of transformation is characterized by the parameter (3.63), where $\Lambda_q = |q^{-1}dq/dl|^{-1}_{q^2=1}$ is the characteristic scale of the quantity q along a ray. The case of strong coupling is realized if $G \ll 1$ in the region $q^2 \sim 1$. On the other hand, if the wave does not cross the region $q^2 \sim 1$ or if $G \gg 1$, the coupling is weak.

It has been noted in section 2.2 that the effect of vacuum birefringence can dominate the wave propagation only in very strong magnetic fields of neutron stars (up to 10^{13} G). At the same time, strong gravity field in the vicinity of such stars may provide inhomogeneous plasma distributions which mainly contribute the scale Λ_q, whereas the influence of the inhomogeneity of the magnetic field is relatively unimportant. This case will be considered in detail below. Other possible cases are discussed in [169].

Let quasi-transverse approximation $q^2 = u \sin^4 \alpha / 4 \cos^2 \alpha \gg 1$ be valid deep in a neutron star where $uv \gg 3a|1-u|$ and the influence of plasma on wave propagation prevails over that of vacuum. Polarizations of the modes in this domain are close to linear. In the frame shown in Fig. 2.1 *b* the electric field vector $\boldsymbol{E}$ of the extraordinary wave is directed along the x-axis, while in the ordinary one this vector is parallel to the y-axis. If plasma density decreases along the ray path, the parameter q (3.115) at $u > 1$ decreases, nullifies at the point

$$1 + \frac{3a(1-u)}{uv} = 0, \tag{3.117}$$

and then changes the sign and increases in modulus up to the values $|q| \gg 1$. Wave polarization in the latter region, which becomes linear again, is determined by birefringent vacuum. However, the vector $\boldsymbol{E}$ of the ordinary wave is now parallel to the x-axis and the extraordinary mode is polarized along the y-axis (see section 2.2). The effective coupling occurs if properties of the system "plasma + vacuum" undergo such strong variations in the region $q^2 \sim 1$ separating the domains of different orientation of mode polarizations that a wave passing through this region is unable to change the polarization in accord with the geometrical optics approximation. This is possible at $G \ll 1$. In the considered case of $\alpha =$ const, $u =$ const > 1 and varying v the coupling parameter (3.63) takes the form

$$G_v^{\perp} = 12\sqrt{2}\frac{\omega}{c}\frac{a}{u}\cot^2\alpha \left(\frac{v}{|dv/dl|}\right)_{q^2=1} \tag{3.118}$$

It follows from this expression that the effect of linear coupling becomes stronger as the direction of propagation approaches the transverse one $\alpha = \pi/2$.

If modes are circularly polarized ($q^2 \ll 1$) at plasma interiors where the vacuum birefringence is negligible, the form of the coupling parameter is different. As radiation propagates to the region of smaller plasma densities, the quantity q increases, so mode polarization becomes elliptical and then linear in the region where $q^2 \gg 1$ in which the effect of vacuum birefringence is dominant. If the condition

$$\left|1 + \frac{3a(1-u)}{uv}\right| \gg 1, \tag{3.119}$$

is satisfied for the transition layer, then the coupling parameter takes the form

$$G_v^{\parallel} = 3\sqrt{2}\frac{\omega}{c} a \sin^2\alpha \left(\frac{v}{|dv/dl|}\right)_{q^2=1}. \tag{3.120}$$

The coupling in this case is stronger at the longitudinal propagation of a wave: $\alpha \to 0$.

Consider now the transformation coefficients for the case characterized by the coupling parameter (3.118). The equations (3.40) of the quasi-isotropic approximation for a linear domain of the medium "plasma + birefringent vacuum", $v = v_0 - bl$ (where the value v_0 corresponds to the point (3.117) at which $q = 0$), are reduced to the Weber equation. The solution of this equation is expressed in terms of the functions of parabolic cylinder. The corresponding transformation coefficient (3.49) here takes the form:

$$Q = \exp\left(-\frac{\pi}{4\sqrt{2}} G_v^{\perp}\right). \tag{3.121}$$

The obtained solution is similar to the above considered case of linear mode coupling in plasma with a transverse magnetic field (cf. 3.121 and 3.97), though the expressions for the coupling parameters are drastically different.

If linear mode coupling in the system "plasma + birefringent vacuum" is characterized by the parameter (3.120), the expression for the transformation coefficient Q can be obtained using the Riccatti equation (3.51). One should take into account in this equation that the function $G(\eta)$ (see (3.52)), as before, is expressed via the difference $n_1 - n_2$ and the function $\Psi(q)$. However, $n_1 - n_2$ and q are determined now by the formulas (3.114) and (3.115), respectively. If plasma density on a star follows the barometric law $v \propto \exp(-l/H)$, the function $G(\eta)$ can be written as

$$G(\eta) = 3\frac{\omega}{c} aH \sin^2\alpha (\sin 2\eta \cos^2 2\eta)^{-1}, \tag{3.122}$$

where $\eta = \pi/4 + (1/2)\arctan q$ and the formula (3.119) has been taken into account. The solution of the Riccatti equation is expressed in terms of hypergeometrical functions, similar to the case of linear coupling in a transition layer [130]. The transformation coefficient is given by

$$Q = \left[1 + \exp\left(\frac{\pi}{\sqrt{2}} G_v^{||}\right)\right]^{-1/2} \tag{3.123}$$

where $G_v^{||}$ is determined by (3.120).

Examine now the role of the linear mode coupling in the formation of polarization of radiation from accreting neutron stars – X-ray pulsars. Assume for the estimate that the scale height of a plasma atmosphere of a neutron star $L \simeq H \simeq 10^2$ cm and the magnetic field strength $B_0 \simeq 5 \cdot 10^{12}$ G.

If the coupling is characterized by the parameter $G_v^{||}$, the linear coupling becomes sufficiently strong at $G_v^{||} \lesssim 1$. According to (3.120), the latter requires that $\hbar\omega \sin^2 \alpha \lesssim 4 \cdot 10^{-5}$ keV. Therefore, even for soft X-ray photons with energies $\hbar\omega \simeq 1$ keV the effective coupling can occur in a small angular interval of about several tens of arc minutes, so it is practically negligible.

In the other case when the coupling is characterized by the parameter $G_v^{\perp}$, the condition of strong coupling $G_v^{\perp} \lesssim 1$ is satisfied in the wide angular interval $\alpha \sim 1$ if $\hbar\omega \lesssim 0.4$ keV (see the formula (3.118)). At the same time, plasma density in the transition should ensure the validity of the equality (3.117) which yields $N_0 \simeq 10^{20}$ cm^{-3}. Since the effect is realized in the wide interval over α, it may become dominant as far as the formation of the polarization characteristics of X-ray radiation in the region $\hbar\omega \lesssim$ 1 keV is concerned. It should be stressed, however, that this effect can only influence polarized radiation, whereas unpolarized radiation remains the same after passing through a coupling region. Of course, this fact following from the identity of the transformation coefficients (3.49) and (3.50) should be accounted for in all the possible cases of linear mode coupling which have been considered above in this chapter.

Transfer of radiation in astrophysical plasmas

In the previous section 3.1 the so-called "ray treatment" of the geometrical optics approximation has been considered. According to this approach, a set of waves with wave vectors $\boldsymbol{k}_j$ from a phase space element d^3k, or wave packet, propagates along the rays, the curves tangential to the group velocity vector $\boldsymbol{v}_{gr}$. This vector in isotropic plasma is collinear with energy flux, or Poynting, vector. In magnetoactive plasma the time-average of the Poynting vector is parallel to $\boldsymbol{v}_{gr}$. In astrophysics a wave packet propagating along a ray is called "radiation" and characterized by the specific intensity I_ω. This value has been introduced and defined in the section 1.2. As soon as the ray pattern in a medium is known,[1] the problem arise to find the radiation intensity at any point along a ray. Of course, I_ω is constant along a rectilinear ray in a homogeneous, non-emitting, and non-absorbing medium. An inhomogeneity leads to the variation of the radiation intensity with the coordinate l along the ray. This variation which, generally, can also be caused by emission, absorption and scattering of radiation in the medium, is described by the so-called radiation transfer equation. The form of this equation and its solutions are given in the section 4.1.

Transfer of radiation in a magnetoactive medium is accompanied by variation of its polarization. "Polarization transfer problem" dealing with variation of polarization due to wave propagation through a medium is of special interest for radio astronomy, in particular for the studies of polarization of diffuse cosmic radio emission, radio emission from discrete sources, the Sun etc. Polarimetric observations are widely used in optical astronomy, as well. Hence, in addition to the transfer equation for radiation intensity, the polarization transfer equation which describes variation of the polarization parameters along a ray (see the section 1.2) plays an important role in the description of radiation characteristics in astrophysical plasmas. This equation and its solutions are discussed in section 4.2 of this chapter.

---

[1]The shape of a ray in an inhomogeneous isotropic medium is determined by the equation (3.11).

4.1. Transfer of radiation intensity

4.1.1. RADIATION TRANSFER EQUATION IN ISOTROPIC PLASMA

Transfer of radiation intensity along the direction of the vector $\boldsymbol{v}_{\text{gr}}$ in a homogeneous medium can generally be described by the equation

$$\frac{dI_\omega}{dl} = \mathcal{F}(I_\omega), \tag{4.1}$$

where I_ω is the specific intensity of radiation (see section 1.2) and dl is a line element along $\boldsymbol{v}_{\text{gr}}$ (along a ray). Since time does not enter the equation (4.1), it can only be used to describe a steady-state spatial distribution of the intensity I_ω in a medium with time-independent parameters. This equation can be generalized to account for non-stationary media with time-dependent parameters (see the end of this paragraph).

If the intensity I_ω is small enough, the functional $\mathcal{F}(I_\omega)$ can be expanded in powers of I_ω and the first two terms of this expansion can be used in the right-hand side of the transfer equation (4.1):

$$\frac{dI_\omega}{dl} = a_\omega - \mu_j I_\omega. \tag{4.2}$$

Here the first term a_ω is called emissivity. It does not depend on I_ω. The factor μ_j in the second term is called absorption coefficient.[2]

The form of the equation (4.2) clearly shows that the emissivity a_ω describes an increase of radiation intensity at the point l of a ray which occurs even if $I_\omega(l) = 0$. Therefore, a_ω characterizes the contribution of radiation emitted by plasma. By definition, radiation intensity equals the energy flux through a unit surface area into a unit solid angle per unit intervals of time and frequency. Since the emissivity a_ω stands for the increase of specific intensity over a ray element of unit length, it is identical to the spectral power of radiation emitted by a unit volume of a medium (see the definition of the spectral power of a source $\mathcal{P}_{\omega\Omega}$ in section 1.3). On the other hand, if no emission by a medium is taken into account: $a_\omega = 0$, the absorption coefficient μ_j in the equation (4.2) yields the relative decrease of the intensity per unit length of a ray: $\mu_j = -I_\omega^{-1}(dI_\omega/dl)$.

Since a_ω does not depend on radiation intensity I_ω, it characterizes spontaneous processes which take place without any respect to presence or absence of radiation along the ray considered. On the contrary, the term

[2]Similar to the subscript j in the notation n_j for the refractive index, the one in the absorption coefficient μ_j points out the mode of the j-th type, electromagnetic or plasma, which the corresponding value is related to. Of course, a_ω and I_ω are also related to the same mode j. To simplify formulas, this subscript is dropped in the notations for these values, as well as the subscript Ω which points out that they are related to unit solid angle.

$\mu_j I_\omega$ vanishes if $I_\omega \to 0$, since it accounts for induced (stimulated) processes of emission or absorption which can only occur under the action of the transferred radiation. The form of radiation transfer coefficients is found further on in this book: plasma emissivity is found in section 5.2, while the chapters 6, 7 deal with calculations of absorption coefficient μ_j. If the transfer equation (4.2) accounts for the scattering of radiation in plasma, both a_ω and μ_j for radiation propagating in a given direction depend on the intensity of radiation transferred in the opposite direction along the same ray and along other rays. The simplest example of a radiation transfer equation of scattering type and its solution are considered further on in this chapter. The more detailed analysis is postponed to the chapter 8 where the radiation transfer coefficients a_ω and μ_j are also given for various specific types of scattering by waves and particles in plasma.

Note that the very form of the expansion (4.1) of the functional $\mathcal{F}(I_\omega)$ ensures that the coefficients a_ω and μ_j are independent on the intensity I_ω along the ray for which the equations (4.1), (4.2) are written. But if radiation intensity is large enough, the non-linearity of $\mathcal{F}(I_\omega)$ should be taken into account. This is done by keeping the higher-order terms, non-linear in I_ω, in the expansion at the right-hand side of the transfer equation. Alternatively, one can assume that the absorption coefficient μ_j is a function of the intensity I_ω. An example of such a description used to investigate the quasi-linear relaxation of a beam–plasma system under the action of the excited plasma waves can be found in section 7.2.

It was noted earlier that the transfer equation (4.2) is only valid for a homogeneous medium. As for an inhomogeneous one with $n_j(\boldsymbol{r})$, refraction results in variation of the solid angle element $d\Omega$ corresponding to the intensity of radiation considered. The relevant transfer equation should take into account that in such a medium I_ω is not constant along a ray, even if $a_\omega = 0$ and $\mu = 0$. The transfer equation for $I_\omega(l)$ in an inhomogeneous medium can be derived using a continuity equation for radiation energy density in phase space $(\boldsymbol{r}, \boldsymbol{k})$. Such a procedure has been performed in [170, 171]. However, we postpone its description to the following section where the case of inhomogeneous, magnetoactive plasma is considered. Here the more illustrative and simple case of plane-parallel, isotropic plasma in which the refractive index n_j depends only on one Cartesian coordinate z is analyzed. The refraction can be accounted for in this case quite easily.

Let x, y, z be the Cartesian frame with z axis directed along grad N in a plane-parallel medium. Consider a ray lying in a plane comprising z-axis, ϑ being the angle between this plane and the coordinate plane yz (Fig. 4.1). The shape of this ray is determined by the equation (3.14) which may be written in the following form:

$$n_j \sin\phi = n_j' \sin\phi', \qquad \vartheta = \vartheta', \tag{4.3}$$

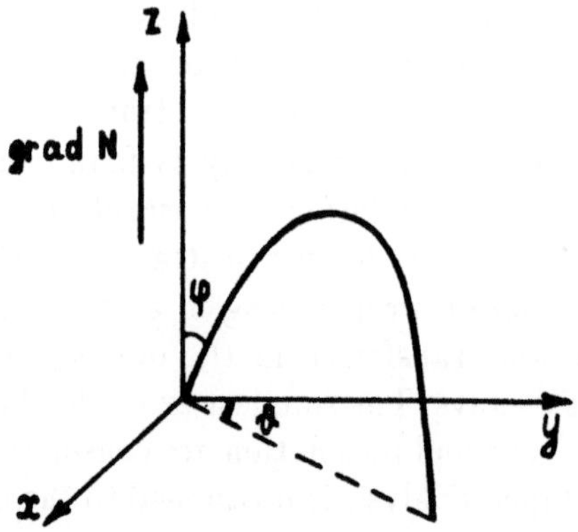

Figure 4.1. On accounting for refraction into a radiation transfer equation for a plane-parallel medium.

where non-primed quantities correspond to the coordinate z, while primed ones, to z'. Next, consider a pair of rays at z, one characterized by the angles ϕ, ϑ and the other one, by $\phi + d\phi$ and $\vartheta + d\vartheta$. Evidently, the relations

$$n_j \sin(\phi + d\phi) = n_j' \sin(\phi' + d\phi'), \qquad \vartheta + d\vartheta = \vartheta' + d\vartheta' \tag{4.4}$$

hold for the latter ray. With (4.3) taken into account, the above formulas for infinitesimal $d\phi$ and $d\vartheta$ can be transformed to

$$n_j \cos\phi d\phi = n_j' \cos\phi' d\phi', \qquad d\vartheta = d\vartheta' \tag{4.5}$$

Multiplying term by term the first equation in (4.3) by both in (4.5) and recognizing the solid angle elements $d\Omega = \sin\phi d\phi d\vartheta$ and $d\Omega' = \sin\phi' d\phi' d\vartheta'$ which comprise the rays from the intervals $d\phi d\vartheta$ and $d\phi' d\vartheta'$, we obtain:

$$n_j^2 \cos\phi d\Omega = n_j'^2 \cos\phi' d\Omega'. \tag{4.6}$$

Energy conservation demands that energy fluxes through equal areas $dS = dS'$ crossed by a given ray at the planes corresponding to the coordinates z and z' should be equal for a stationary intensity distribution in a plane-parallel medium with $a_\omega = 0$ and $\mu_j = 0$. Since energy fluxes through dS equals $I_\omega \cos\phi d\Omega dS$, this requirement yields the following relation:

$$I_\omega \cos\phi d\Omega = I_\omega' \cos\phi' d\Omega', \tag{4.7}$$

in which I_ω and I_ω' are the intensities along a ray at points z and z', respectively. Dividing (4.7) by (4.6), we get $I_\omega n_j^{-2} = I_\omega' n_j'^{-2}$. Since z and z' are arbitrary, we conclude that the invariant

$$\frac{I_\omega}{n_j^2} = \text{const} \tag{4.8}$$

holds along a ray in a transparent ($\mu_j = 0$) and non-emitting ($a_\omega = 0$) plane-parallel medium with stationary distribution of radiation intensity $I_\omega(l)$. The same invariant holds also for an isotropic medium with arbitrary but smooth inhomogeneity.

The refractive index of the solar corona plasma $n_{el}^2 < 1$ and increases with distance R from the solar center (see section 3.1). So, applying (4.8) to the case of wave propagation in the solar corona we find that, due to refraction, the intensity of radio waves along a ray increases with increasing distance from the Sun. Note that in vacuum ($n_{el} = 1$) the intensity of radiation from the Sun or any other extended source is constant along a ray. This seems quite natural if one recalls that both radiation flux density through the unit area F_ω determined by (1.31) and the solid angle Ω comprising the radiation at any point on this area decrease with distance from a source as R^{-2}. Therefore, the intensity $I_\omega = F_\omega/\Omega = \text{const}$.

It clear from the above that radiation transfer equations considered in this section, in particular, the invariant (4.8), are valid if radiation at any spatial point has an angular pattern of finite width. Only in this case can the radiation intensity be defined as radiation flux density per unit solid angle along group or phase velocity. But if angular spectrum of radiation at an arbitrary spatial point is characterized by definite $\boldsymbol{v}_{\text{ph}}$ and $\boldsymbol{v}_{\text{gr}}$ (for example, unscattered radiation from a point source in vacuum), it is impossible to apply the above definition or use the invariant $I_\omega/n_j^2 = \text{const}$.

Consider now the effect of the invariant (4.8) on the form of radiation transfer equation in an inhomogeneous medium. According to the formula (4.8), refraction leads to the following variation of the intensity per unit ray length: $dI_\omega/dl = (2I_\omega/n_j)(dn_j/dl)$. This term should be added to the right-hand side of the transfer equation (4.2). As a result, this equation takes the following form [172]:

$$n_j^2 \frac{d}{dl}\left(\frac{I_\omega}{n_j^2}\right) = a_\omega - \mu_j I_\omega. \tag{4.9}$$

If radiation intensity depends not only on the coordinate along a ray but also on time:[3] $I_\omega = I_\omega(l, t)$, the corresponding transfer equation should be written in the following form [171]:

$$\frac{1}{v_{\text{gr}}} \frac{\partial I_\omega}{\partial t} + n_j^2 \frac{\partial}{\partial l}\left(\frac{I_\omega}{n_j^2}\right) = a_\omega - \mu_j I_\omega. \tag{4.10}$$

[3] Such a time dependence may result from temporal variation of either radiation transfer coefficients a_ω, μ_j or boundary conditions for I_ω.

Finally, if the refraction index gradually varies in time, $n_j = n_j(\boldsymbol{r}, t)$, the radiation transfer equation is generalized the following way:

$$\frac{1}{v_{\text{gr}}} \frac{\partial}{\partial t}\left(\frac{I_\omega}{n_j^2}\right) + \frac{3}{c}\frac{\partial n_j}{\partial t}\frac{I_\omega}{n_j^2} + \frac{\partial}{\partial l}\left(\frac{I_\omega}{n_j^2}\right) = \frac{a_\omega}{n_j^2} - \mu_j \frac{I_\omega}{n_j^2}. \tag{4.11}$$

This equation describes the radiation transfer in a non-stationary, smoothly inhomogeneous medium. It has been derived in [173].

A radiation transfer equation which accounts for scattering can be written in the form

$$\frac{dI_\omega}{dl} = a_\omega - \mu_j I_\omega + \frac{1}{4\pi}\oint_{4\pi} \mu_j^{\text{sc}}(\boldsymbol{\Omega}') I_\omega(\boldsymbol{\Omega}') d\Omega' - \mu_j^{\text{sc}}(\boldsymbol{\Omega}) I_\omega. \tag{4.12}$$

Here μ_j^{sc} is the scattering coefficient,[4] $\boldsymbol{\Omega}$ is the unit vector along the ray, and $\boldsymbol{\Omega}'$ corresponds to a direction from which radiation of the intensity $I_\omega(\boldsymbol{\Omega}')$ is scattered into $\boldsymbol{\Omega}$. The third, integral, term in the right-hand side of the equation (4.12) plays a role of emissivity in the direction $\boldsymbol{\Omega}$, while the last one accounts for the decrease of the intensity $I_\omega(\boldsymbol{\Omega})$ due to scattering of radiation from $\boldsymbol{\Omega}$ to other directions, $\boldsymbol{\Omega}'$. Scattering transfer equation (4.12) is valid for a stationary medium in which the refractive index n_j is close to unity. In a plane-parallel medium all the parameters depend on one coordinate h, and this equation takes the form

$$\cos\Theta \frac{dI_\omega}{dh} = a_\omega - \mu_j I_\omega + \frac{\mu_j^{\text{sc}}}{2}\int_0^\pi I_\omega(\Theta')\sin\Theta'\, d\Theta' - \mu_j^{\text{sc}} I_\omega, \tag{4.13}$$

where Θ is the angle between $\boldsymbol{\Omega}$ and the h-axis. General information on radiation transfer equations in scattering media can be found in textbooks and monographs [174–177].

All the transfer equations considered above can be only used if the cosmological effects of the expanding Universe are negligible. In fact, the cosmological expansion leads to variation of radiation energy density and radiation intensity even in the case when, according to (4.11), they should remain constant. For example, (4.11) yields $\partial I_\omega/\partial t = 0$ for homogeneous intensity distribution in vacuum, though I_ω and the energy density W_ω do vary in the course of the cosmological expansion. This is caused by the variation of the volume occupied by radiation and its frequency in the non-stationary Universe. As shown below, the above effects can be neglected if characteristic times of radiation transfer are much shorter than the characteristic time of the expansion ($\sim 10^{10}$ yr). Thus, the emission and propagation of radiation observed from vast majority of the known astrophysical objects are not influenced by the cosmological expansion.

[4] In case of isotropic scattering μ_j^{sc} does not depend on $\boldsymbol{\Omega}$.

However, when dealing with cosmological sources, especially cosmic background radiation which comes from distances of the redshifts $z_c \gtrsim 10^3$, cosmological effects are of leading importance.[5]

According to the hot model of the Universe [178,179], cosmic background radiation originated during the pre-stellar epoch of the early Universe when the nearly homogeneous primordial plasma was heated up to the temperature of $5{\cdot}10^3$ K. After the cooling caused by the cosmological expansion this plasma (which was predominantly hydrogen) recombined to the neutral gas. Since free-free emission and absorption by neutral hydrogen is rather small, the efficiency of interaction between matter and radiation was greatly reduced at that moment. Therefore, since this recombination era the Universe has been filled by the "relict", or background radiation decoupled from the matter. Nowadays this radiation has the temperature of 2.7 K only. Cosmic background radiation is a unique source of information on physical conditions and properties of the primordial plasma. To extract such information one should study variations of the background radiation during the cosmological expansion using the appropriate radiation transfer equation.

Any characteristics of radiation or a source in the non-stationary model of homogeneous, isotropic Universe may depend on time t, not spatial coordinates. In particular, the spectral energy density of radiation $W_\omega = W_\omega(\omega, t)$, so the transfer equation should comprise the terms proportional to $\partial W_\omega/\partial t$ which describe temporal evolution of radiation, its "temporal transfer".[6]

As was said above, variations of W_ω in the course of the cosmological expansion are caused by two reasons: first, the increase of the volume occupied by the radiation and, second, variation of its frequency. The local distance between any two spatial points increases according to the Hubble law:

$$v = Hr, \tag{4.14}$$

where v is the relative velocity of two points at a distance r. The Hubble law is valid at distances r which correspond to $v \ll c$. The parameter H is called Hubble constant, H^{-1} in the present era being estimated as $(1-2)\cdot 10^9$ yr.

For the model considered it is convenient to begin with a derivation of the transfer equation for the spectral density of photons N_ω and, after that, transform it to the equation for the spectral energy density W_ω (see [179]). The above mentioned N_ω is defined as the number of photons per unit volume and unit range of frequency ω. This definition means that the

[5] The parameter $z_c = (\omega_0 - \omega)/\omega$, where ω_0 and ω are the frequencies of monochromatic radiation in source and observer frames, respectively, is introduced to characterize the relative redshift.

[6] The definition of W_ω is given in section 1.2.

spectral density of photons N_ω relates to the W_ω as

$$W_\omega = \hbar\omega N_\omega, \tag{4.15}$$

where $\hbar\omega$ is photon energy.

Note that the total number of photons in volume $V = (4/3)\pi r^3$ is conserved during cosmological expansion:

$$\frac{4\pi}{3} r^3 N_\omega = \text{const.}$$

Taking the time derivative of this equation we find

$$\frac{\partial N_\omega}{\partial t} r + 3N_\omega \frac{dr}{dt} = 0,$$

which, together with (4.14), yields the following:

$$\frac{\partial N_\omega}{\partial t} = -3HN_\omega. \tag{4.16}$$

Consider now the effect of cosmological expansion on photon frequency. The wavelength $\lambda = 2\pi c/\omega$ which is inversely proportional to the photon momentum $\hbar k$ varies similarly to the distance r: $(1/\lambda)d\lambda/dt = (1/r)dr/dt = H$, where the law (4.14) has been accounted for. This means that

$$\frac{d\omega}{dt} = -\frac{\omega}{\lambda}\frac{d\lambda}{dt} = -H\omega. \tag{4.17}$$

The above derivative characterizes the rate of photon drift along the ω-axis, while $N_\omega d\omega/dt$ equals the corresponding photon flux along this axis. Now it is easy to see that the variation of the spectral density N_ω due to such a frequency drift of photons is described by

$$\frac{\partial N_\omega}{\partial t} = -\frac{\partial}{\partial\omega}\left(\frac{d\omega}{dt}N_\omega\right) = H\frac{\partial}{\partial\omega}(\omega N_\omega). \tag{4.18}$$

Thus, the cosmological expansion leads to the decrease of the photon density (4.16) and to photon redistribution to the lower-frequency part of the spectrum (4.18).

The total variation of N_ω due to cosmological expansion results from both of the above effects, so

$$\frac{\partial N_\omega}{\partial t} = -3HN_\omega + H\frac{\partial}{\partial\omega}(\omega N_\omega). \tag{4.19}$$

Now, taking into account the relation (4.15) between N_ω and W_ω, one can easily arrive at the following transfer equation for spectral density of radiation which accounts for the cosmological expansion of the homogeneous and isotropic Universe [179]:

$$\frac{\partial W_\omega}{\partial t} = -3HW_\omega + H\omega\frac{\partial W_\omega}{\partial \omega} + \alpha_\omega - \beta W_\omega. \tag{4.20}$$

The terms describing emission (α_ω) and absorption (β) of radiation are added to the right-hand side of this equation. Since the refraction index of the matter was assumed to be equal to unity (i.e., $v_{gr} = c$), these are the only terms taking into account the presence of the matter in the Universe. Comparison of these terms with the ones in the right-hand side of the equation (4.10) for the intensity $I_\omega = v_{gr}W_\omega = cW_\omega$ shows that $\alpha_\omega = a_\omega$ and $\beta = \mu_j c$.

4.1.2. TRANSFER EQUATION FOR MAGNETOACTIVE PLASMA

The equations describing radiation transfer in vacuum or isotropic medium has been formulated in the previous section. The problem of radiation transfer in inhomogeneous magnetoactive plasma is also of interest and importance to plasma physics and, especially, to the theory of generation and propagation of radio waves in cosmic plasmas.

The method used to derive the relevant transfer equation [180] is analogous to the one applied in the case of an isotropic medium [170]. The key difference is arise from the fact that the dispersion relation for waves of the j-th type in magnetoactive plasma has the form

$$\omega = \omega_j(\boldsymbol{k}_j, \boldsymbol{r}),$$

i.e. the frequency ω depends not only on the value of the wave vector $\boldsymbol{k}_j$ (as in the case of an isotropic medium) but also on its direction.

If l is the coordinate along a ray, then $\boldsymbol{k}_j = \boldsymbol{k}_j(l)$ and $\boldsymbol{r} = \boldsymbol{r}(l)$, so that

$$\omega = \omega_j(\boldsymbol{k}_j(l), \boldsymbol{r}(l)) \tag{4.21}$$

However, in a stationary medium the frequency $\omega(l)$ is constant, contrary to varying wave k_j and wavelength $\lambda = 2\pi/k_j$. Therefore, taking the derivative of (4.21) with respect to l, we get

$$\frac{\partial \omega}{\partial \boldsymbol{k}_j}\frac{d\boldsymbol{k}_j}{dl} + \frac{\partial \omega}{\partial \boldsymbol{r}}\frac{d\boldsymbol{r}}{dl} = 0. \tag{4.22}$$

Since $d\boldsymbol{r}/dl = \boldsymbol{l}$ is the unit vector along a ray and $\partial\omega/\partial\boldsymbol{k}_j = \boldsymbol{v}_{gr} = v_{gr}\boldsymbol{l}$, the relation (4.22) may we written in the form

$$\boldsymbol{l}\left(v_{gr}\frac{d\boldsymbol{k}_j}{dl} + \frac{\partial \omega}{\partial \boldsymbol{r}}\right) = 0.$$

The obtained equation should be valid for an arbitrary ray crossing a given point $\boldsymbol{r}$. This is only possible if

$$v_{\text{gr}}\frac{d\boldsymbol{k}_j}{dl} = -\frac{\partial\omega}{\partial\boldsymbol{r}} \tag{4.23}$$

The rate of radiation transfer along a ray is given by the group velocity v_{gr}, so $v_{\text{gr}} = dl/dt$ and thus the relation (4.23) takes the form:

$$\frac{d\boldsymbol{k}_j}{dt} = \dot{\boldsymbol{k}}_j = -\frac{\partial\omega}{\partial\boldsymbol{r}}. \tag{4.24}$$

In what follows we will also use the relation

$$\frac{d\boldsymbol{r}}{dt} \equiv \dot{\boldsymbol{r}} = -\frac{\partial\omega}{\partial\boldsymbol{k}_j}, \tag{4.25}$$

which follows from the definition of group velocity $\boldsymbol{v}_{\text{gr}} = \partial\omega/\partial\boldsymbol{k}_j$.

Energy of radiation in the mode j within the phase space volume $d^3rd^3k_j$ is equal to $W_{\mathbf{k}}d^3rd^3k_j$, where $d^3r = dV$ and d^3k_j are volume elements in coordinate space and wave vector space, respectively, $W_{\mathbf{k}}$ is the spectral energy density defined in section 1.2.[7] Energy conservation demands that $W_{\mathbf{k}}$ obeys the following continuity equation:

$$\frac{\partial W_{\mathbf{k}}}{\partial t} + \frac{\partial}{\partial\boldsymbol{r}}(\dot{\boldsymbol{r}}W_{\mathbf{k}}) + \frac{\partial}{\partial\boldsymbol{k}_j}(\dot{\boldsymbol{k}}_jW_{\mathbf{k}}) = \alpha_{\mathbf{k}} - \beta_jW_{\mathbf{k}}. \tag{4.26}$$

Note the formal analogy between the equation (4.26) and the continuity equation for charge density ρ_e in coordinate space:

$$\frac{\partial\rho_e}{\partial t} + \nabla\cdot\boldsymbol{j} = 0,$$

where $\boldsymbol{j} = \rho_e\boldsymbol{v}$ is current density. The equation (4.26) however is written for the radiation energy density, so the divergence operator is applied to the intensity $I_{\mathbf{k}} = \boldsymbol{v}_{\text{gr}}W_{\mathbf{k}} = \dot{\boldsymbol{r}}W_{\mathbf{k}}$ which characterizes energy flux density per unit solid angle (see (1.32)). Since $W_{\mathbf{k}}$ depends also on $\boldsymbol{k}_j$, the above continuity equation comprises the divergence of energy flux density in wave vector space: $\nabla_{\boldsymbol{k}_j}\cdot(\dot{\boldsymbol{k}}_jW_{\mathbf{k}}) = \partial(\dot{\boldsymbol{k}}_jW_{\mathbf{k}})/\partial\boldsymbol{k}_j$. Finally, the terms $\alpha_{\mathbf{k}}$ and $\beta_jW_{\mathbf{k}}$ in the right-hand side of the equation (4.26) stand for the effect of spontaneous and induced processes in magnetoactive plasma, respectively.

Using (4.24) and (4.25), the equation (4.26) may be transformed to

$$\frac{\partial W_{\mathbf{k}}}{\partial t} + \dot{\boldsymbol{r}}\frac{\partial W_{\mathbf{k}}}{\partial\boldsymbol{r}} + \dot{\boldsymbol{k}}_j\frac{\partial W_{\mathbf{k}}}{\partial\boldsymbol{k}_j} = \alpha_{\mathbf{k}} - \beta_jW_{\mathbf{k}}. \tag{4.27}$$

[7] For the sake of simplicity we drop the subscript j in the notations $W_{\mathbf{k}}$, $I_{\mathbf{k}}$ and I_ω.

Taking into account the relations $\dot{\boldsymbol{r}} = v_{\rm gr}\boldsymbol{l} = v_{\rm gr} d\boldsymbol{r}/dl$, $\dot{\boldsymbol{k}}_j = v_{\rm gr} d\boldsymbol{k}_j/dl$ and the equality

$$\frac{d\boldsymbol{r}}{dl}\frac{\partial}{\partial \boldsymbol{r}} + \frac{d\boldsymbol{k}_j}{dl}\frac{\partial}{\partial \boldsymbol{k}_j} = \frac{d}{dl}$$

which determines the rule for taking the full derivative of the function $W_{\mathbf{k}}(\boldsymbol{r}(l), \boldsymbol{k}_j(l))$ with respect to l, we get the following transfer equation for radiation energy density:

$$\frac{\partial W_{\mathbf{k}}}{\partial t} + v_{\rm gr}\frac{dW_{\mathbf{k}}}{dl} = \alpha_{\mathbf{k}} - \beta_j W_{\mathbf{k}}. \tag{4.28}$$

The above considerations leading to this equation clearly show that (4.28) is valid for stationary[8] inhomogeneous magnetoactive plasma.

In addition to (4.28), the transfer equation for the specific intensity

$$I_\omega = \frac{k_j^2}{|\cos\vartheta_j|} W_{\mathbf{k}} \tag{4.29}$$

is frequently used. The relation (4.29), where ϑ_j is the angle between $\boldsymbol{k}_j$ and $d\omega/d\boldsymbol{k}_j$ which, generally, is non-zero in magnetoactive plasma, follows from the formulas (1.25) and (1.34). According to (4.28) and (4.29), the desired transfer equation for radiation intensity has the following form [180]:

$$\frac{1}{v_{\rm gr}}\frac{\partial I_\omega}{\partial t} + \frac{n_j^2}{|\cos\vartheta_j|}\frac{d}{dl}\left(\frac{I_\omega|\cos\vartheta_j|}{n_j^2}\right) = a_\omega - \mu_j I_\omega. \tag{4.30}$$

Here n_j denotes the refraction index of j-th mode with frequency ω constant along a ray, $a_\omega = k_j^2 \alpha_{\mathbf{k}} (|\cos\vartheta_j| v_{\rm gr})^{-1}$ and $\mu_j = \beta_j v_{\rm gr}^{-1}$ are the emissivity and absorption coefficient along a ray, respectively, for this mode. If parameters of the medium and radiation sources, as well as the intensity I_ω itself, are time-independent, the equation (4.30) is reduced to [3]

$$\frac{n_j^2}{|\cos\vartheta_j|}\frac{d}{dl}\left(\frac{I_\omega|\cos\vartheta_j|}{n_j^2}\right) = a_\omega - \mu_j I_\omega. \tag{4.31}$$

As follows from this equation, the invariant

$$\frac{I_\omega|\cos\vartheta_j|}{n_j^2} = \text{const} \tag{4.32}$$

[8]Similar to the case of isotropic plasma described by the equation (4.10), this requirement demands that the refraction index $n_j(\mathbf{r})$ and, thus, the group velocity $\mathbf{v}_{\rm gr}$ and the ray pattern in the medium are time-independent. The parameters $\alpha_{\mathbf{k}}$ and β_j in this case may be functions of time.

holds along a ray in a transparent ($\mu_j = 0$), non-emitting ($a_\omega = 0$) plasma.

The form (4.30) of the transfer equation is closely linked with the definition of I_ω in magnetoactive plasma adopted since section 1.2. This definition relates radiation intensity to the unit solid angle of wave vectors, not group velocities. The radiation transfer equation for the specific intensity defined another way has been derived, for example, in [181]. It is however more complex, the fact testifying to the convenience of the definition which is adopted in this book.

It should be stressed again that transfer equations considered are only valid within geometrical optics approximation, when the ray treatment of wave propagation is possible. In particular, the reflection and significant linear coupling of different modes in a smoothly inhomogeneous medium must be absent. Of course, radiation transfer equation can be applied both to equilibrium and to non-equilibrium media, the cases of unstable plasma and negative absorption coefficient μ_j included. The conditions which ensure the inequality $\mu_j < 0$ and lead to the amplification of radiation in a medium are discussed in chapter 7.

4.1.3. SOLUTIONS OF RADIATION TRANSFER EQUATIONS

This section describes the solution of radiation transfer equation for a stationary distribution of intensity $I_\omega(l)$ in isotropic and magnetoactive astrophysical plasma, and solution of the non-stationary transfer equation for radiation energy density $W_\omega(t)$ in homogeneous and isotropic Universe.

Since the equations (4.9), (4.31) are ordinary differential equations of the first order, their solutions may be reduced to quadratures. The solution of the equation (4.2) in a homogeneous medium with coordinate-independent transfer coefficients a_ω and μ_j looks like

$$I_\omega(l) = I_\omega(l_0)e^{-\mu_j(l-l_0)} + \frac{a_\omega}{\mu_j}(1 - e^{-\mu_j(l-l_0)}). \tag{4.33}$$

In an inhomogeneous medium with $a_\omega = a_\omega(l)$ and $\mu_j = \mu_j(l)$ the corresponding solution takes the following form:

$$I_\omega(l) = I_\omega(l_0)e^{-\tau_j} + \int_0^{\tau_j} \frac{a_\omega}{\mu_j} e^{\xi-\tau_j}\, d\xi. \tag{4.34}$$

The optical depth

$$\tau_j(l) = \int_{l_0}^{l} \mu_j(l)\, dl \tag{4.35}$$

between the points l_0 and l along a ray and the intensity of radiation $I_\omega(l_0)$ at l_0 enter the above formulas. The ratio $S(l) = a_\omega/\mu_j$ is called source function. The form of these solutions is simplified for equilibrium plasma (i.e., plasma with equilibrium distribution of particles over energies or momenta). In case of total thermodynamic equilibrium both plasma and radiation are equilibrium, radiation intensity $I_\omega^{(0)}$ being determined by the formula (1.37). This formula clearly shows that the left-hand sides of the equations (4.9), (4.31) are zero for $I_\omega = I_\omega^{(0)}$. Thus, the radiation transfer equation is reduced to the Kirchhoff law

$$a_\omega = \mu_j I_\omega^{(0)} \tag{4.36}$$

which relates emissivity a_ω and absorption coefficient μ_j in the case of thermodynamic equilibrium.

The Kirchhoff law can also be applied to describe transfer of non-equilibrium radiation in equilibrium plasma provided energy density of the former is not very high. The influence of radiation on the matter in this case is relatively weak, and radiation transfer coefficients a_ω and μ_j are only determined by the particle distribution over velocities, not by radiation intensity which may be non-equilibrium. Therefore, even if thermodynamic equilibrium between matter and radiation is not realized, the coefficients a_ω and μ_j in equilibrium plasma with kinetic temperature T may still be related via the Kirchhoff law (4.36) in which the equilibrium intensity $I_\omega^{(0)}$ corresponds to the plasma temperature T.

Since a_ω and $I_\omega^{(0)}$ are essentially positive, the absorption coefficient in equilibrium plasma cannot be negative. As a result, the intensity of radiation escaping a region of equilibrium plasma does not exceed the total radiation intensity of individual particles in this region. According to the definition given in section 1.3, this means that an equilibrium plasma may radiate via incoherent mechanism only.

In equilibrium magnetoactive plasma, in which, generally, $I_\omega \neq I_\omega^{(0)}$ but a_ω and μ_j are related through the Kirchhoff law (4.36), radiation transfer equation (4.31) takes the form

$$\frac{n_j^2}{|\cos\vartheta_j|}\frac{d}{dl}\left(\frac{I_\omega|\cos\vartheta_j|}{n_j^2}\right) = \mu_j(I_\omega^{(0)} - I_\omega) \tag{4.37}$$

or, equivalently,

$$\frac{d}{d\tau_j}\left(\frac{I_\omega|\cos\vartheta_j|}{n_j^2}\right) = \frac{\omega^2\kappa T}{8\pi^3c^2} - \frac{I_\omega|\cos\vartheta_j|}{n_j^2}. \tag{4.38}$$

The expression (1.37) for the equilibrium intensity is taken into account here. The form of the transfer equation (4.38) may be further simplified by the use of the effective radiation temperature T_b determined by (1.38) instead of the intensity I_ω:

$$\frac{dT_b}{d\tau_j} = T - T_b. \tag{4.39}$$

This equation clearly shows that ray pattern, distribution of the kinetic temperature along a ray $T = T(l)$, and the relation (4.35) between the optical depth τ_j and the coordinate l along a ray are entirely determined by the properties of equilibrium plasma (isotropic or magnetoactive): the form of its inhomogeneity and distribution of radiation sources.

The equation (4.39) can be easily solved by the method of integrating factor. As a result, the brightness radiation temperature at point l on a ray is found:

$$T_b(l) = e^{-\tau_j} \int_0^{\tau_j} T(\xi) e^{\xi}\, d\xi + e^{-\tau_j} T_b(l_0). \tag{4.40}$$

Here $T_b(l_0)$ is the brightness temperature of radiation at the point l_0, τ_j is the optical depth of the ray from l_0 to l, and $\xi(l) = \int_{l_0}^{l} \mu_j\, dl$ is the running value of the optical depth measured from the point l_0. The first term on the right-hand side of the equation (4.40) describes the contribution to the intensity at l by the radiation emitted at the considered part of the ray between the points l_0 and l. The second term stands for the residual of the radiation which enters the ray at the point l_0 and is partially extincted (attenuated by the factor of e^{τ_j}) when passing the interval $l - l_0$. Alternatively, the expression (4.40) may be written in terms of $\zeta = \tau_j - \xi$, the optical depth measured from the point l:

$$T_b(l) = \int_0^{\tau_j} T e^{-\zeta}\, d\zeta + e^{-\tau_j} T_b(l_0). \tag{4.41}$$

In case of a thermally homogeneous region in which the kinetic temperature is constant along a ray: $T = \text{const}$, the above solution takes the following form:

$$T_b(l) = T(1 - e^{-\tau_j}) + e^{-\tau_j} T_b(0). \tag{4.42}$$

Note that the medium may be inhomogeneous in this case: $n_j = n_j(l)$. According to (4.42), the temperature of radiation escaping a thermally homogeneous, optically thick region is equal to the kinetic temperature of this region:

$$T_b \approx T. \tag{4.43}$$

In other words, the intensity of radiation from such a region reaches the equilibrium intensity $I_\omega^{(0)}$ at the frequency ω considered. On the contrary, the brightness temperature of radiation from the optically thin region $\tau_j \ll 1$:

$$T_b \approx T\tau_j + T_b(l_0)(1 - \tau_j) \tag{4.44}$$

is much less than the kinetic temperature T.

If temperature distribution $T(l)$ is inhomogeneous, the brightness temperature of radiation by the optically thick region is given by (cf. (4.41))

$$T_b(l) = \int_0^{\tau_j} T(\zeta) e^{-\zeta} \, d\zeta. \tag{4.45}$$

The estimate of T_b in this case may be obtained from the following considerations. Regarding $T_b(l)$ as a function of l_0, we see from (4.45) that the main contribution to this value comes from those l_0^* for which the optical depth $\tau_j = \int_{l_0^*}^{l} \mu_j \, dl \sim 1$. If, moreover, the characteristic scale of T is large compared to $l - l_0^*$, we may put $T_b \sim T(l_0^*)$.

The applications of the above transfer equations and their solutions will be considered in sections 6.1, 6.2 following the ones in which various absorption coefficients $\mu_j(\omega)$ of astrophysical plasmas are derived. These coefficients are necessary to find the optical depth along a ray, $\tau_j(\omega)$. The last, in its turn, is needed to yield, relaying on the transfer equations considered, definite conclusions on the physical conditions inside radiation sources.

If absorption along a ray is negligible (or, equivalently, if optical depth of the region considered is small: $\tau_j \ll 1$), the time-dependent transfer equation (4.10) for radiation intensity is reduced to

$$\frac{1}{c}\frac{\partial I_\omega}{\partial t} + \frac{\partial I_\omega}{\partial l} = a_\omega(l, t). \tag{4.46}$$

The influence of the medium on the propagation of radiation is also neglected in the above equation by putting $n_j = 1$ and $v_{gr} = c$. The equation (4.46) may also be written in the form

$$\frac{\partial I_\omega}{\partial l} = a_\omega(l, t - l/c). \tag{4.47}$$

The solution of the last equation is

$$I_\omega(l, t) = \int_l^\infty a_\omega\left(l, t - \frac{l}{c}\right) dl. \tag{4.48}$$

If emissivity is time-independent: $a_\omega = a_\omega(l)$, the observed intensity equals the radiative losses of the system. On the contrary, such a relation is violated for non-stationary distribution of a_ω along a ray. For example, if a source of radiation moves as a whole along a ray with the velocity v, the emissivity $a_\omega = a_\omega(l - vt)$. Thus, the observed intensity, in full accord with the conclusion of the section 1.3 on the energy flux by a moving source (see (1.90)),[9] is given by the formula

$$I_\omega(l,t) = \int_l^\infty a_\omega\left(l - v\left(t - \frac{l}{c}\right)\right) dl = \frac{\int a_\omega(\xi)\, d\xi}{1 + v/c}. \tag{4.49}$$

Now we can find possible non-stationary solutions of the radiation transfer equation (4.20) with cosmological expansion taken into account. Parameters of the primordial plasma are changed in the course of the evolution of the Universe, so the coefficients α_ω and β which characterize proper emission and absorption in an expanding medium become time-dependent. In a simplified analysis given below these coefficients are assumed to be large enough to ensure that any radiation emitted is absorbed on a time scale much less than the characteristic time of the cosmological variation of matter properties.

In other words, the expansion stage considered is well described by the quasi-stationary solution of the equation (4.20) which is obtained by dropping all the terms except the first two in the right-hand side: $\alpha_\omega \approx \beta W_\omega$. According to the Kirchhoff law (4.36), however, the ratio α_ω/β equals

$$W_\omega^{(0)} = \frac{\hbar\omega^3}{8\pi^3 c^3 (e^{\hbar\omega/\kappa T} - 1)}, \tag{4.50}$$

the energy density of the equilibrium radiation. This can be immediately seen from (4.36) if one recalls that $\alpha_\omega = a_\omega$, $\beta = \mu_j c$, and $I_\omega = W_\omega c$. Therefore, the quasi-stationary solution for radiation in the primordial plasma looks like $W_\omega(t) = W_\omega^{(0)}(\omega, T)$ where $T(t)$ is the kinetic temperature of the plasma.

The above equilibrium of matter and radiation holds until $t \sim t_{\rm rec}$, when the temperature of hydrogen plasma falls to the value $T \sim T_{\rm rec} \simeq 5 \cdot 10^3\,{\rm K}$ at which electrons and protons recombine effectively to neutral hydrogen. This leads to the significant drop of α_ω and β at $t > t_{\rm rec}$. As a result, the radiation in the "post-recombination era" is decoupled from the neutral gas. Temporal evolution of radiation energy density is now described by the equation

$$\frac{\partial W_\omega}{\partial t} = -3HW_\omega + H\omega\frac{\partial W_\omega}{\partial \omega} \tag{4.51}$$

to be solved under the initial condition that $W_\omega(t_{\rm rec})$ is determined by the Planck law (4.50) with $T = T_{\rm rec}$.

[9]The difference in signs of the velocity in formulas (1.90) and (4.49) arises from the fact that the positive value of v in (4.49) corresponds to the source moving away from the observer, whereas in (1.90) the value $\beta\cos\theta > 0$ if a source is approaching the observer.

Substitution of (4.50) with $T = T(t)$ into (4.51) leads to the equation [179]

$$\frac{dT}{dt} = -HT. \tag{4.52}$$

Thus, radiation remains equilibrium during the cosmological expansion, and its temperature decreases as

$$T = T_{\mathrm{p}} e^{-Ht}. \tag{4.53}$$

According to (4.50), the maximum in the spectrum of the equilibrium radiation appears at the frequency $\omega_{\mathbf{max}} \approx 2.8\kappa T/\hbar$ which evolves similarly to the temperature (4.53):

$$\omega_{\mathbf{max}} = \omega_{\mathbf{max,p}} e^{-Ht}. \tag{4.54}$$

This seems quite natural since both the temperature T and the frequency ω in the expanding Universe obey similar equations, (4.52) and (4.17).

The temperature of the cosmic background radiation at the present epoch is $T \approx 2.7$ K, while it has been as high as $T_{\mathrm{p}} \sim 5 \cdot 10^3$ K at the time $t \sim t_{\mathrm{rec}}$ of the decoupling of matter and radiation. The relative decrease of the temperature $T/T_{\mathrm{rec}} \sim 2 \cdot 10^3$. As follows from (4.53), (4.54), the same factor, which equals the redshift $z_{\mathrm{c}} = (\omega_0 - \omega)/\omega$, describes the decrease of the frequency of the maximum of the cosmic background spectrum. Such a variation of the frequency can be treated as a result of the Doppler effect when observing the radiation emitted by sources in the primordial plasma which move apart with a velocity close to the speed of light. [10]

Concluding this section, we obtain some simplified solutions of radiation transfer equation in a scattering medium by the method of Schwarzschild–Shuster (see, for example, [174,177]). According to this method, the integro-differential transfer equation (4.13) for the angle-dependent intensity $I_\omega(\Omega)$ is approximated by the pair of differential equations for the angle-averaged values

$$I_1 = \int_0^{\pi/2} I_\omega(\Theta) \sin\Theta \, d\Theta, \qquad I_2 = \int_{\pi/2}^{\pi} I_\omega(\Theta) \sin\Theta \, d\Theta. \tag{4.55}$$

If the variable h in the equation (4.13) is treated as height, I_1 and I_2 characterize the average intensities of radiation propagating upward and downward, respectively, in the directions within the upper and lower hemisphere. Multiply (4.13) by $\sin\Theta$ and integrate it over Θ from 0 to $\pi/2$. Then approximate $\cos\Theta$ under the integral on the left-hand side by $1/2$, its average over the integration interval. As a result, the approximate equation

[10] On the electromagnetic background (relict) radiation see also [182]

for the intensity I_1 is found. To obtain the corresponding equation for I_2, one should perform the similar procedure for the interval $\pi/2 \le \Theta \le \pi$ at which the average value of $\cos\Theta$ equals $-1/2$. It is the feature of the Schwarzschild-Schuster method that the angular integral of the transfer operator on the left-hand side of the transfer equation is approximated with the average value of the comprising angular functions over the integration interval. Introducing the variables

$$x = \frac{I_1 + I_2}{I_\omega^{(0)}}, \qquad y = \frac{I_1 - I_2}{I_\omega^{(0)}}, \tag{4.56}$$

we arrive at the following system of equations which describe radiation transfer in a layer $h_0 - \Delta h < h < h_0$ of homogeneous, equilibrium plasma (for more details see, e.g., [183]):

$$\frac{dx}{d\tau} = 2(1+p)y, \qquad \frac{dy}{d\tau} = 2p(x-2). \tag{4.57}$$

Here the parameter $p = \mu_j/\mu_{\mathrm{sc}}$ and $\tau = \mu_j^{\mathrm{sc}}(h_0 - h)$ is the scattering optical depth of the region between the point considered, h, and the upper boundary of the layer, h_0. To come to (4.57) from (4.13), the Kirchhoff law (4.36) should be taken into account.

Let us now find the solution of the system (4.56) under the boundary conditions

$$x = y, \qquad \text{at} \quad h = h_0,$$
$$x + y = 2I^{\mathrm{inc}}/I_\omega^{(0)}, \qquad \text{at} \quad h = h_0 - \Delta h.$$

The first condition means that no radiation is incident in the upper boundary of the layer from the outside, while the second one takes into account that the radiation of the average intensity $I_1 = I^{\mathrm{inc}}$ enters the lower boundary of the layer.

The effect of scattering on radiation transfer is most pronounced if the parameter $p \ll 1$. In this case it is quite useful to distinguish between two opposite limiting cases. In case of a "thin" layer,

$$2\sqrt{p}\,\tau_j = 2\sqrt{\mu_j \mu^{\mathrm{sc}}}\,\Delta h \ll 1, \tag{4.58}$$

the outgoing intensity is given by

$$I_1 = I^{\mathrm{inc}}(1 + \tau^{\mathrm{sc}})^{-1} + 2I_\omega^{(0)}\tau_j, \tag{4.59}$$

where $\tau_j = \mu_j \Delta h$ and $\tau^{\mathrm{sc}} = \mu^{\mathrm{sc}}\Delta h$ are absorption and scattering optical depth, respectively. In the opposite case of a "thick" layer,

$$2\sqrt{p}\,\tau_j \gg 1, \tag{4.60}$$

this intensity is determined by the formula

$$I_1 = 4\sqrt{p}\, I^{\mathrm{inc}} e^{-2\sqrt{p\tau_j}} + 2\sqrt{p}\, I_\omega^{(0)} \tag{4.61}$$

The first term in both (4.59) and (4.61) corresponds to the fraction of the incoming radiation which has passed through the entire layer and escaped from the upper boundary, whereas the second one describes radiation emitted in the layer. The limiting transition $\Delta h \to \infty$ or $\tau_j \to \infty$ in the formula (4.61) yields the expression for the intensity escaping from the semi-infinite layer into the region $h > h_0$:

$$I_1 = 2\sqrt{\frac{\mu_j}{\mu^{\mathrm{sc}}}}\, I_\omega^{(0)}. \tag{4.62}$$

Since $p \ll 1$, the above outgoing intensity is small compared to the equilibrium intensity $I_\omega^{(0)}$, contrary to the case of a pure absorbing, optically thick layer (4.43).

Another example important for applications is the solution of a radiation transfer equation for an inhomogeneous medium (for example, the photosphere of a star) at hydrostatic equilibrium in a uniform gravity field. The plasma density in such a medium varies with height according to the barometric law

$$N = N_0 \exp(-h/H) \tag{4.63}$$

in which the scale height H equals $2\kappa T/m_{\mathrm{p}} g$ for a pure hydrogen plasma (m_{p} is proton mass and g is the free-fall acceleration).

Consider the solution of the transfer equation (4.13) in which the absorption is provided by particle collisions. Since the corresponding absorption coefficient $\mu_j \propto N^2$ (see the formulas (6.8), (6.12)) and the coefficient of scattering by plasma particles $\mu^{\mathrm{sc}} = \sigma^{\mathrm{sc}} N$, where σ^{sc} is the scattering cross section (see section 8.1), the transfer equation can be written in the form

$$\cos\Theta \frac{dI_\omega}{d\tau} = (1 + M\tau) I_\omega - M\tau I_\omega^{(0)} - \frac{1}{2}\int_0^\pi I_\omega(\Theta') \sin\Theta'\, d\Theta'. \tag{4.64}$$

Here the scattering optical depth is defined by the formula

$$\tau = \int_h^\infty \mu^{\mathrm{sc}}\, dh, \tag{4.65}$$

and the parameter

$$M = \frac{\mu_j}{H(\mu^{\mathrm{sc}})^2} = \tau_j (\tau^{\mathrm{sc}})^{-2}, \tag{4.66}$$

where $\tau_j = \int_h^\infty \mu_j \, dh$, is independent of h. The above described procedure of angular averaging relevant to the Schwarzschild–Shuster approximation leads to the following system of equations for average intensities (4.55):

$$\begin{aligned} \frac{1}{2}\frac{dI_1}{d\tau} &= (1+M\tau)I_1 - M\tau I_\omega^{(0)} - \frac{1}{2}(I_1+I_2), \\ -\frac{1}{2}\frac{dI_2}{d\tau} &= (1+M\tau)I_2 - M\tau I_\omega^{(0)} - \frac{1}{2}(I_1+I_2). \end{aligned} \tag{4.67}$$

In terms of the variables (4.56), this system can be written as:

$$\frac{dx}{d\tau} = 2(1+M\tau)y, \qquad \frac{dy}{d\tau} = 2M\tau(x-2). \tag{4.68}$$

Consider the solution of this system at sufficiently large heights at which

$$M\tau \ll 1 \tag{4.69}$$

and thus the scattering is of dominant importance compared to pure absorption. Here the term $M\tau$ is negligible in the first equation, and the resulting solution looks like [184]:[11]

$$x = 2 + C_1\mathrm{Ai}(\xi) + C_2\mathrm{Bi}(\xi), \qquad y = \frac{1}{2}\frac{dx}{d\tau}, \tag{4.70}$$

where $\mathrm{Ai}(\xi)$ and $\mathrm{Bi}(\xi)$ are Airy functions,

$$\xi = (4M)^{1/3}\tau, \tag{4.71}$$

and the constants C_1 and C_2 are determined by the boundary conditions of the problem. Since I_1 and I_2 must be finite deep in the photosphere, at large τ, the constant C_2 should be equal to zero to eliminate the term containing the function $\mathrm{Bi}(\xi)$ which grows to infinity as $\xi \to \infty$. Then, no radiation enters the photosphere from the outside, so $I_2 = 0$, or $x = y$, at $\tau = 0$. Using the first formula in (4.70), it is easy to find

$$C_1 = -2\,\frac{3^{4/3}\Gamma(2/3)\Gamma(4/3)}{3^{2/3}\Gamma(4/3) + (M/2)^{1/3}\Gamma(2/3)}. \tag{4.72}$$

Taking into account (4.70) and (4.56), we obtain that

$$\begin{aligned} I_1 + I_2 &= [2 + C_1\mathrm{Ai}(\xi)]I_\omega^{(0)}, \\ I_1 - I_2 &= (M/2)^{1/3}C_1 I_\omega^{(0)}\frac{d\mathrm{Ai}(\xi)}{d\xi}, \end{aligned} \tag{4.73}$$

[11] The similar solution for a layer with linear density profile has been obtained in [185].

where the constant C_1 is given by (4.72). In a strongly scattering medium where the parameter $M^{1/3} \ll 1$ the difference $I_1 - I_2$ is also small compared to the equilibrium intensity $I_\omega^{(0)}$. Deep in the photosphere, as $\xi \to \infty$, radiation becomes equilibrium: $I_1 \simeq I_2 \simeq I_\omega^{(0)}$. On the contrary, the outgoing intensity

$$I_1^{\text{out}} = \left(\frac{2}{3}\right)^{2/3} \frac{\Gamma(2/3)}{\Gamma(4/3)} M^{1/3} I_\omega^{(0)} \tag{4.74}$$

is much less than $I_\omega^{(0)}$. Comparison of (4.74) and (4.62) clearly reveals that the intensity of radiation escaping a scattering layer is strongly dependent on the electron density profile $N(h)$. Transition from the equilibrium intensity $I_1 \simeq I_\omega^{(0)}$ to the reduced one (4.74) takes place at the region of the photosphere where

$$\xi = (4\tau_j \tau)^{1/3} \simeq 1. \tag{4.75}$$

Since $\tau \gg \tau_j$ in case of strong scattering, the condition (4.75) corresponds to the heights at which $\tau \gg 1$ but $\tau_j \ll 1$.

The solution of the transfer equation (4.64) has been obtained under the condition (4.69). To ensure that the formulas (4.73) describe properly the spatial distributions of the intensities I_1, I_2, this condition must be valid up to $\xi \lesssim 1$, i.e., for $\tau \lesssim (4M)^{-1/3}$. Since $M\tau \lesssim (M/2)^{2/3} \ll 1$ in this interval, the above analysis is quite self-consistent.

Simplicity is the certain advantage of the approximate Schwarzschild-Shuster method of solution of radiation transfer problems, as well as the similar Eddington method (see [174]). The cost to be paid is the low accuracy of the resulting solution which, generally, is about 10 %. However, in many cases such an accuracy is quite enough to perform the qualitative analysis and to get approximate, the so-called "semi-qualitative" results. To achieve higher accuracy, one should use more rigorous techniques for analytical or numerical solution of radiation transfer problems in specific scattering media.

The above considered solutions of the radiation transfer equation in which strong scattering is taken into account will be used in chapter 8 when dealing with cyclotron effects in strongly magnetized plasma on white dwarfs and neutron stars.

4.2. Transfer of polarization

4.2.1. FARADAY AND COTTON–MOUTON EFFECTS IN PLASMA

Electromagnetic radiation in plasma may be either polarized (fully or partially) or not. Unpolarized, or chaotically polarized, radiation is a superposition of incoherent waves of various types. In its turn, fully polarized radiation which has well defined ellipse of polarization (with fixed excentricity and orientation) can be represented as the superposition of mutually coherent waves of various types. Of course, fully polarized radiation can be comprised of waves of a single type (either ordinary or extraordinary). Then the resulting polarization is the same as the one of the corresponding mode. The appearance of this or that polarization pattern of cosmic plasma radiation results from the features of the source and specific conditions along the radiation path, including the effects of linear mode coupling considered in section 3.2.

When waves of a certain type propagate in a magnetoactive plasma, the orientation of the polarization ellipse in the plane orthogonal to the direction of the wave vector $\boldsymbol{k}$ depends entirely on the orientation of the projection of the magnetic field $\boldsymbol{B}_0$ onto this plane. So, in a smoothly inhomogeneous medium, in which the geometrical optics approximation is valid, the ellipse of polarization holds the fixed spatial orientation or slowly rotates in accord with the shear of the magnetic field — variation of the considered projection of the vector $\boldsymbol{B}_0$. On the other hand, coherent ordinary and extraordinary waves with different refraction indeces (or, equivalently, phase velocities) propagate in another way. Since orientation of the polarization ellipse in this case is determined not only by the shape and orientation of the comprising ellipses of ordinary and extraordinary mode but also on the phase shift between them, the variation of the latter along the direction of propagation leads, under certain conditions, to the rotation of the resulting polarization ellipse, the phenomenon called Faraday effect.

Consider in more detail the influence of the phase shift between modes on the shape and orientation of the polarization ellipse. Let polarization ellipse of the propagating radiation differ from those of the normal waves in plasma. According to the above, this radiation can be represented as the superposition of two coherent waves, ordinary and extraordinary. The projections of the electric fields of these waves onto xy-plane perpendicular to the wave vector $\boldsymbol{k}$ have the components (see Fig. 1.4)

$$E_x = a\sin\omega t, \qquad E_y = ap\cos\omega t; \tag{4.76}$$

$$E_x = bp\sin(\omega t - \psi), \qquad E_y = -b\cos(\omega t - \psi). \tag{4.77}$$

Here a and b stand for amplitudes of the waves with counter-rotating electric field vectors $\boldsymbol{E}$, and ψ is phase shift between these waves. Since these waves are coherent, the value ψ is time-independent over a fixed plane xy. Finally, p describes the ellipticity E_y/E_x of the wave (4.76). The sign of this parameter indicates the direction of rotation of the vector $\boldsymbol{E}$. Namely, if $p < 0$, electric field vector of the wave (4.76) rotates clockwise (as seen along the direction of the vector $\boldsymbol{k}$) and that of the wave (4.77), counter-clockwise. If p equals K_1 from (2.136), the formulas (4.76) describe the extraordinary mode. The ordinary one is then described by (4.77).

It follows from (4.76), (4.77) that the components of the considered polarized radiation which has the frequency ω vary in time as

$$E_x = a \sin \omega t + bp \sin(\omega t - \psi) = A \sin(\omega t - \Psi),$$
$$E_y = ap \cos \omega t - b \cos(\omega t - \psi) = B \sin(\omega t - \Phi), \tag{4.78}$$

where

$$A^2 = a^2 + b^2 p^2 + 2abp \cos \psi,$$
$$B^2 = a^2 p^2 + b^2 - 2abp \cos \psi, \tag{4.79}$$

$$\tan \Psi = \frac{bp \sin \psi}{a + bp \cos \psi}, \qquad \tan \Phi = \frac{ap - b \cos \psi}{b \sin \psi}. \tag{4.80}$$

The ellipse determined by the relations (4.78)-(4.80) is orientated at the angle χ with respect to the x-axis and has eccentricity $\tan \sigma$ where

$$\tan(2\chi) = \frac{2AB \cos(\Psi - \Phi)}{A^2 - B^2}, \tag{4.81}$$

$$\tan^2 \sigma = \frac{A^2 + B^2 + \sqrt{(A^2 - B^2)^2 + 4A^2 B^2 \cos^2(\Psi - \Phi)}}{A^2 + B^2 - \sqrt{(A^2 - B^2)^2 + 4A^2 B^2 \cos^2(\Psi - \Phi)}}. \tag{4.82}$$

Substituting into these formulas the expressions (4.79) and (4.80) for A, B, Ψ and Φ, we find [186]

$$\tan 2\chi = \pm 2 \frac{1 + p^2}{p} \frac{\sin \psi}{(p^{-1} - p)(ab^{-1} - ba^{-1}) + 4 \cos \psi}, \tag{4.83}$$

$$\sin 2\sigma = \pm 2 \frac{b^2 - a^2 + ab(p^{-1} - p) \cos \psi}{(a^2 + b^2)(p^{-1} + p)}. \tag{4.84}$$

The additional phase shift between ordinary and extraordinary components gained at a distance L along $\boldsymbol{k}$ is equal to

$$\Delta\psi = \frac{\omega}{c} \int_L (n_2 - n_1)\, dl, \tag{4.85}$$

where the refractive indices n_1 and n_2 of extraordinary and ordinary waves are determined by (2.121). Propagation through the region with definite sign of the difference $n_2 - n_1$ leads to the monotonous increase of the value $\psi = \psi_0 + \Delta\psi$, where ψ_0 is the initial phase shift. Therefore, orientation and eccentricity of the resulting ellipse of polarization vary according to (4.83),

(4.84). If the angle χ_0 characterizes the initial position of the polarization ellipse, then the corresponding angle at a distance L equals

$$\chi = \chi_0 + \Delta\chi. \tag{4.86}$$

In the simplest and frequently appearing case of the quasi-longitudinal propagation, polarizations of both extraordinary and ordinary waves are close to circular ($p \approx -1$). According to (4.84)), the ellipticity of radiation is constant here, while the Faraday rotation of the resulting polarization ellipse is described by the relation

$$\Delta\chi = \frac{\Delta\psi}{2} = \frac{\omega}{2c}\int_L (n_2 - n_1)\, dl. \tag{4.87}$$

The first equality follows from the relation $\tan 2\chi = \pm\tan\psi$ which (4.83) is reduced to in the case considered. To come to the second one, the expression (4.85) has been taken into account.

To relate directly the value of the rotation $\Delta\chi$ to plasma parameters and radiation frequency, we take into account that, according to (2.128), the difference of the refractive indices at sufficiently high frequencies at which both the inequalities $v = \omega_L^2/\omega^2 \ll 1$, $\sqrt{u} = \omega_B/\omega \ll 1$ and the criterion (2.126) of the quasi-longitudinal propagation are satisfied, is given by

$$n_2 - n_1 \approx \frac{n_2^2 - n_1^2}{2} \approx v|\sqrt{u}\cos\alpha|. \tag{4.88}$$

This expression is valid the latter ensuring that both modes are circularly polarized. As follows from (4.87) and (4.88),

$$\Delta\chi \approx \frac{\omega}{2c}\int_L v\sqrt{u}\cos\alpha\, dl \simeq \frac{2.35\cdot 10^4}{f^2}\int_L N B_0 \cos\alpha\, dl, \tag{4.89}$$

where f is measured in Hz, N in cm^{-3}, B_0 in G, and l in cm. This formula clearly shows that the Faraday rotation $\Delta\chi$ is inversely proportional to the square of frequency and directly proportional to the "rotation measure" $\mathrm{RM} = \int N B_0 \cos\alpha\, dl$. If the amplitudes of circularly polarized waves are equal, the resulting radiation becomes linearly polarized. The Faraday effect appears in this case as the uniform rotation of the plane of polarization with the increase of the rotation measure RM. But if the amplitudes of the comprising modes are different, the radiation is elliptically polarized and Faraday effect leads to the uniform rotation of the axes of polarization ellipse with increasing RM (see Fig. 4.2, a).

It should be specially emphasized that the formula (4.89) has been derived for the case when $\sqrt{u}\cos\alpha > 0$, i.e. the angle between $\boldsymbol{k}$ and $\boldsymbol{B}_0$ is acute. However, one can easily comprehend that (4.89) also remains valid for $\sqrt{u}\cos\alpha < 0$ when rotations of ordinary and extraordinary waves are inverse compared to the ones in the above considered case. This changes the sign of the refractive index difference between waves with right- and left-hand polarizations and, respectively, the sign of rotation of the resulting polarization ellipse. This is taken into account in (4.89) by using $\sqrt{u}\cos\alpha$ instead of $|\sqrt{u}\cos\alpha|$ in the formula (4.88).

In case of quasi-transverse propagation (2.127) the modes are linearly polarized. The phase shift $\Delta\chi$ results in variation of polarization which is drastically different from the case considered above. Since $K_1 \to 0$ and $K_2 \to -\infty$ for the quasi-transverse propagation, the ellipticity $p \to 0$. The formulas (4.83), (4.84) then yield $\tan 2\psi = \pm\infty$, or $\chi = \pm\pi/4$, for equal amplitudes of the modes. One of these values is realized at the interval where the phase ψ increases from $2\pi m$ to $2\pi m + \pi$, the other one corresponding to $2\pi m + \pi \le \psi \le 2\pi m + 2\pi$, where m is an integer. Resulting polarization varies periodically from linear with $\chi = \pi/4$ through circular to another linear with $\chi = -\pi/4$ and further to circular again but with an opposite sign of rotation (see Fig. 4.2, b). Linear polarization occurs at $\psi = \pi m \pm \pi/2$, $2\pi m + 3\pi/2$, the circular one at $\psi = \pi m$. The described variation of polarization is called Cotton–Mouton effect. The term "QT Faraday effect" is sometimes used, contrary to the case of the quasi-longitudinal propagation (and the corresponding "QL Faraday effect") which is described by the formulas (4.87), (4.89).

The complete cycle of variation of polarization ellipse for Cotton–Mouton effect corresponds to the interval $\Delta\psi = 2\pi$. In case of quasi-transverse propagation (2.127) of radiation in a tenuous plasma ($v \ll 1$) the difference of the refractive indices (2.129) equals

$$n_2 - n_1 \approx \frac{n_2^2 - n_1^2}{2} \approx \frac{uv\sin^2\alpha}{2(1 - u\sin^2\alpha)}, \tag{4.90}$$

and the phase shift (4.85) between two linearly polarized modes is given by

$$\delta\psi = \frac{\omega}{2c}\int_L \frac{uv\sin^2\alpha}{1 - u\sin^2\alpha}\,dl. \tag{4.91}$$

In plasma with both $v \ll 1$ and $u \ll 1$ the quasi-perpendicular propagation is only realized in the narrow angular interval near $\alpha = \pi/2$. The formula (4.91) is simplified in this case:

$$\delta\psi = \frac{\omega}{2c}\int_L uv\,dl, \tag{4.92}$$

or, using different notations,

$$\delta\psi = \frac{1}{2\omega^3 c}\int_L \omega_L^2\omega_B^2\,dl \approx \frac{6.5\cdot 10^{10}}{f^3}\int_L N B_0^2\,dl. \tag{4.93}$$

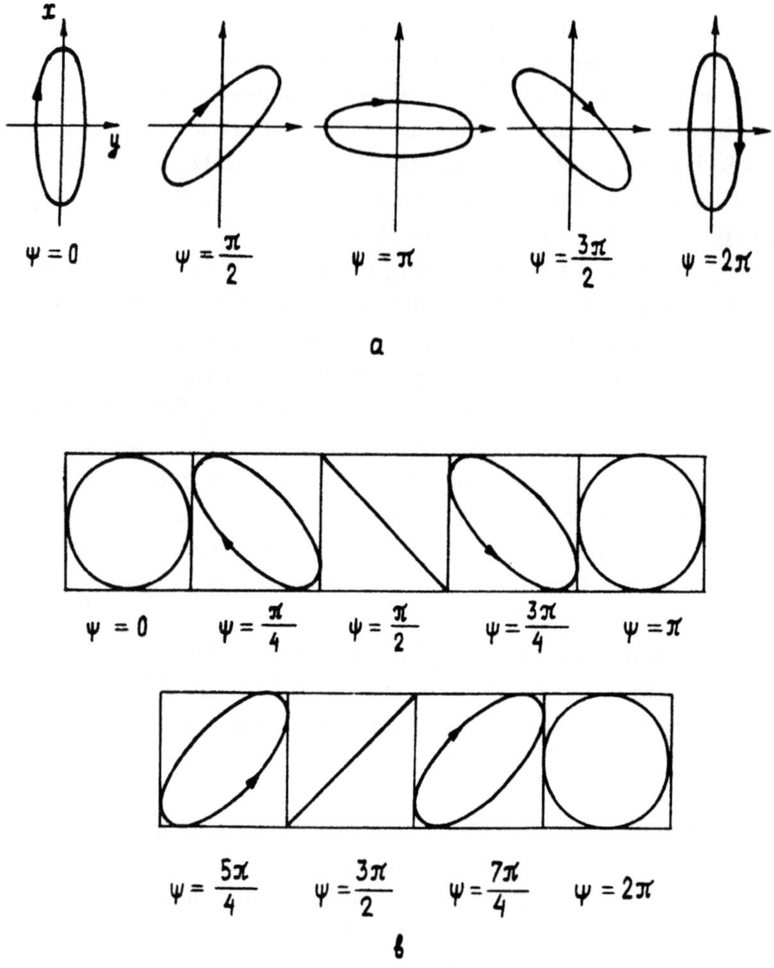

Figure 4.2. Variation of polarization with increasing phase shift ψ: *a*, Faraday effect ($a\!=\!b$, $p\!=\!1$); *b*, Cotton–Mouton effect ($a\!=\!b$, $p\!=\!0$).

The integral PM $= \int_L NB_0^2\,dl$ entering (4.93) may be called "pulsation measure" in the case of Cotton–Mouton effect. An increase of PM leads to the decrease of the distance over which the full cycle of the polarization ellipse variation shown in Fig. 4.2, b takes place.

In the general case of elliptical polarization with an arbitrary ratio of extraordinary and ordinary components the complicated formulas (4.83)-(4.85) with refractive indices (2.121) should be used instead of simpler ones (4.87), (4.89) and (4.93). Analysis of (4.83) shows that the Faraday rotation becomes non-uniform though (4.87) properly describes the mean variation of $\Delta\chi$ over the interval $\Delta\psi \gg 2\pi$. Finally, under the condition

$$|p^{-1} - p| \cdot |ab^{-1} - ba^{-1}| > 4, \tag{4.94}$$

when the ratio of normal wave amplitudes is far from unity and the resulting polarization ellipse differs significantly from a circle, the rotation of polarization ellipse is replaced by its oscillations over χ with amplitude not larger than $\pi/4$. This follows immediately from (4.83). Indeed, under the condition (4.94) the right-hand side of the formula (4.83) remains finite for any ψ, contrary to the case of rotation (monotonous variation of χ) in which both sides of this equality should go to infinity at $\psi = -\pi/4, \pi/4, \ldots$ Next, for

$$|p^{-1} - p| \cdot |ab^{-1} - ba^{-1}| \gg \max[4; 2p^{-1}(1+p^2)] \tag{4.95}$$

the right-hand side of (4.83) is much less than unity. Then

$$\chi = \pm \frac{1+p^2}{1-p^2} \frac{\sin \Psi}{ab^{-1} - ba^{-1}} + \pi m/2. \tag{4.96}$$

The fixed orientation of polarization ellipse in the case (4.95) seems quite natural since the ellipse of a small-amplitude wave added to the ellipse of a large-amplitude one cannot produce any significant variation of the resulting polarization ellipse.

Faraday and Cotton–Mouton effects play quite significant roles in cosmic plasmas. Since these effects, especially the former one, may radically change polarization of radiation sources, they should be accounted for when dealing with the polarimetric observations of extra-terrestrial radio emission.

For example, Faraday rotation $\Delta\chi$ in interstellar medium with $N \sim 0.1\,\text{cm}^{-3}$ and $B_0 \sim 10^{-6}\,\text{G}$ over a distance $L \sim 100\,\text{pc} \sim 3 \cdot 10^{20}\,\text{cm}$ equals 70 and 0.7 rad at the frequencies $f \sim 10^8\,\text{Hz}$ ($\lambda \sim 3\,\text{m}$) and $f \sim 10^9\,\text{Hz}$ ($\lambda \sim 30\,\text{cm}$) , respectively. Note that in the chaotic model of the Galactic magnetic field the above distance $L \sim 100\,\text{pc}$ (or, sometimes, $L \sim 10\,\text{pc}$) is usually assumed to be the typical size of "magnetic clouds", the regions of quasi-homogeneous field. Since the value of the angle α along a ray is uniformly distributed over the interval $[0, 2\pi]$, the ray-averaged value $\overline{\cos\alpha} = 0$. Therefore, the average value of Faraday rotation $\overline{\Delta\chi} = 0$ at constant N and B_0 in an interstellar medium (see the formula (4.89)). The corresponding variance of the rotation angle, $\sqrt{\overline{(\Delta\chi)^2}} \sim \Delta\chi_{\text{cl}}\sqrt{n}$, may be quite significant, since it depends on the number of the clouds crossed by the ray ($\Delta\chi_{\text{cl}}$ stands for the typical value of Faraday rotation in a single cloud, the numerical estimates of this value have been given above).

Propagation of linearly polarized radiation through a medium with chaotically orientated magnetic field leads to effective compensation of Faraday rotations produced by individual clouds. Note that such a compensation is significantly reduced or absent for rays lying within the Galactic disk where the magnetic field is more homogeneous (see section 1.1).

The estimates based on the formula (4.89) show that the Faraday effect is also very strong for radio waves in the solar corona. Indeed, even for very moderate values of electron density $N \sim 10^8\,\mathrm{cm}^{-3}$, magnetic field $B_0 \sim 4\,\mathrm{G}$ and the ray length $L \sim 10^{10}\,\mathrm{cm}$ the angle $\Delta\chi$ in the corona is about $10^{23} f^{-2}$ rad (to compare, the same angle in the Earth's ionosphere is a million times less). This means that in coronal conditions $\Delta\chi \lesssim 1$ only for wavelengths $\lambda \lesssim 0.1\,\mathrm{cm}$. The value of Faraday rotation angle sharply increases with increasing wavelength up to 10^6 rad at $\lambda \sim 1\,\mathrm{m}$. The actual value of the rotation can even exceed the above estimates, in particular, above active regions with increased N and B_0.

The loss of information on the position of the polarization plane inside a radiation source is not the only consequence of Faraday rotation occurring either inside the source itself or on the way to the observer. Another consequence is depolarization, the decrease of the degree of linear polarization ζ_l of the observed radiation. Such an effect is observed due to the finite size of the emitting region and finite bandwidth Δf_{rec} of the receiver. The role of the former factor is discussed in the last paragraph of this section where the role of the source in the formation of the polarization characteristics of the observed radiation is analyzed using polarization transfer equation. Now the effect of the latter factor, the finite bandwidth of the receiver, is estimated.

It is clear from the expressions (4.86), (4.87) that the angle χ is the function of radiation frequency f, even if χ_0, the initial orientation of polarization ellipse, does not depend on f. The function $\chi(f)$ may be linearized in the following way:

$$\chi(f) = \chi(f_0) + \left.\frac{\partial(\Delta\chi)}{\partial f}\right|_{f=f_0} (f - f_0) \tag{4.97}$$

within a narrow frequency interval, for example, within the bandwidth $\Delta f_{\mathrm{rec}} \ll f_0$ of a polarimeter with "working" frequency f_0. Then the variation of the orientation of the polarization ellipse over the bandwidth equals

$$\delta\chi \approx \left.\frac{\partial(\Delta\chi)}{\partial f}\right|_{f=f_0} \Delta f_{\mathrm{rec}}. \tag{4.98}$$

In particular, $\Delta\chi \propto f^{-2}$ for quasi-longitudinal propagation in a region $n_1 \approx n_2 \approx 1$ (see(4.89)), so the derivative $\partial(\Delta\chi)/\partial f|_{f=f_0} = -2 f_0^{-1} \Delta\chi|_{f=f_0}$, and (4.98) becomes

$$\delta\chi \approx 2 f_0^{-1} \Delta\chi|_{f=f_0} \Delta f_{\mathrm{rec}}. \tag{4.99}$$

The degree of linear polarization of the observed radio emission ζ_l is decreased compared to the initial one due to the difference in position of polarization ellipses at different frequencies. The observed value of ζ_l sharply drops if $\delta\chi \gg 1$.

The above effect is quite pronounced in solar corona conditions due to the strong rotation of the polarization plane. This demands the accurate account of the effects connected with the dispersion of Faraday rotation in the corona (see in this respect [3, 187]). For example, if $\Delta\chi \sim 10^6$ rad at wavelengths $\lambda \sim 1$ m, then, according to (4.99), the dispersion $\delta\chi$ can be less than unity only for $\Delta f_{\text{rec}} < 150$ Hz which is less than the typical bandwidth of polarimetric devices. Thus, valuable linear polarization of the solar radio emission can only be detected with specialized narrow-band facilities. Recently, it has been observed by radiometer with a bandwidth of about 20 kHz [188].

Depolarization due to finite bandwidth of the polarimeter may also be used in polarization measurements of non-thermal radio emission from an interstellar medium. Modulation of the bandwidth Δf_{rec} and registration of the resulting variation of ζ_l allows one to detect a weak linear polarization of cosmic radio emission against numerous interfering effects [189].

Faraday effect in cosmic plasmas in not the only cause of depolarization of the observed radio emission from cosmic sources. Such a depolarization may also appear due to the difference in orientation of the plane of linear polarization of radiation from several sources covered by antenna pattern [190]. Synchrotron radiation of relativistic electrons in chaotic magnetic field can be a specific example of such an effect. Polarization plane of such radiation is perpendicular to the projection of the magnetic field on the plane perpendicular to the line of sight (see sections 5.1, 5.2). Polarizations produced by different regions of quasi-homogeneous field are chaotically orientated with respect to each other. When taken into account in the Galactic model with chaotic magnetic field, this effect leads to a strong depolarization, so that the degree of linear polarization falls below the observable level. Detection of polarization of cosmic radio emission is the evidence for the existence of regions of regular magnetic fields in our Galaxy.

4.2.2. EQUATIONS OF POLARIZATION TRANSFER IN MAGNETOACTIVE PLASMA

The problem of the formation of polarization characteristics is not completely solved by taking into account the phase shifts between different modes which lead to Faraday or Cotton–Mouton effects. Polarization of radiation also depends on polarization of radiating sources and on absorption of radiation in magnetoactive plasma. It can be found using the equation of polarization transfer which accounts for all these effects. Such an equation is derived below.

It has been noted in section 1.2 that polarization of radiation at any point of the medium is completely determined by four Stokes parameters I, Q, U, and V or four components of polarization tensor $I_{\alpha\beta} = \overline{D_\alpha D_\beta^*}$ (see (1.51) or (1.55), respectively), where D_α, D_β are the components of electric induction vector $\boldsymbol{D}$ of the electromagnetic radiation. So, the polarization transfer equation should describe spatial variation of either polarization tensor or Stokes parameters.

We now derive the simplest and most illustrative equations of polarization transfer assuming that magnetoactive plasma is homogeneous and stationary, radiation characteristics are time-independent, and mode polarizations are orthogonal. In doing this, the influence of phase shifts between modes on polarization tensor $I_{\alpha\beta}$ is taken into account using the procedure proposed in [191]. Moreover, this procedure is extended to the whole equation of polarization transfer [57]. Possible generalizations and other forms of polarization transfer equations will then be discussed.

Consider a component of electromagnetic radiation in a medium with wave vectors orientated along z-axis and frequencies in the interval $\Delta\omega$. Electric induction vector $\boldsymbol{D}$ of this radiation lies in xy-plane, so α-component of this vector ($\alpha = x, y$) is given by

$$D_\alpha(z,t) = \int_{\Delta\omega} D_1 e_{1\alpha} \exp(-i\omega t + ik_1 z + i\psi_1)\, d\omega + \int_{\Delta\omega} D_2 e_{2\alpha} \exp(-i\omega t + ik_2 z + i\psi_2)\, d\omega. \tag{4.100}$$

Here the first and the second integrals, respectively, correspond to extraordinary (subscript 1) and ordinary (subscript 2) waves with amplitudes D_1 and D_2, wave vectors k_1, k_2, and phases ψ_1, ψ_2, while $e_{1\alpha}$ and $e_{2\alpha}$ denote the components of unit polarization vectors $\boldsymbol{e}_1$ and $\boldsymbol{e}_2$ of the modes. Evidently,

$$\boldsymbol{e}_1 \boldsymbol{e}_1^* = 1, \qquad \boldsymbol{e}_2 \boldsymbol{e}_2^* = 1. \tag{4.101}$$

Generally, all the values introduced above depend on the frequency ω.

If a medium considered is absorbing (or amplifying), the wave vectors are complex:

$$k_1 = k_1' + ik_1'', \qquad k_2 = k_2' + ik_2'', \tag{4.102}$$

where

$$k_{1,2}' = \frac{\omega}{c} n_{1,2}, \qquad k_{1,2}'' = \frac{\omega}{c} \eta_{1,2}, \tag{4.103}$$

$n_{1,2}$ and $\eta_{1,2}$ are refractive and absorption indeces of different modes which have been introduced in section 2.1.

If mode polarizations are orthogonal, the following relations are also valid for the vectors e_1 and e_2:

$$e_1 e_2^* = 0 \qquad e_1^* e_2 = 0 \tag{4.104}$$

In case of non-orthogonal polarizations of the modes these equations should be substituted with

$$e_1 e_2^* = \overline{\alpha}, \qquad e_1^* e_2 = \overline{\alpha}^*, \tag{4.105}$$

which define the degree of non-orthogonality $\overline{\alpha}$. In what follows it is assumed that $\overline{\alpha}=0$ and thus the relations (4.104) are valid. Possible limitations on the validity range of the resulting polarization transfer equation will be discussed below.

Writing down the component $D_\beta(z,t)$ (where $\beta = x, y$) in the form (4.100), multiplying the complex conjugate of it by $D_\alpha(z,t)$ and taking the derivative with respect to z, we find

$$\begin{aligned}\frac{\partial}{\partial z}(D_\alpha D_\beta^*) = &\int_{\Delta\omega}\int_{\Delta\omega} D_1 \tilde{D}_1 e_{1\alpha} \tilde{e}_{1\beta}^* (ik_1 - i\tilde{k}_1) \\ &\times \exp(-i\omega t + i\tilde{\omega} t + i\psi_1 - i\tilde{\psi}_1 + ik_1 z - i\tilde{k}_1 z)\, d\omega d\tilde{\omega} + \\ &\int_{\Delta\omega}\int_{\Delta\omega} D_2 \tilde{D}_2 e_{2\alpha} \tilde{e}_{2\beta}^* (ik_2 - i\tilde{k}_2) \\ &\times \exp(-i\omega t + i\tilde{\omega} t + i\psi_2 - i\tilde{\psi}_2 + ik_2 z - i\tilde{k}_2 z)\, d\omega d\tilde{\omega} + \\ &\int_{\Delta\omega}\int_{\Delta\omega} D_1 \tilde{D}_2 e_{1\alpha} \tilde{e}_{2\beta}^* (ik_1 - i\tilde{k}_2) \\ &\times \exp(-i\omega t + i\tilde{\omega} t + i\psi_1 - i\tilde{\psi}_2 + ik_1 z - i\tilde{k}_2 z)\, d\omega d\tilde{\omega} + \\ &\int_{\Delta\omega}\int_{\Delta\omega} D_2 \tilde{D}_1 e_{2\alpha} \tilde{e}_{1\beta}^* (ik_2 - i\tilde{k}_1) \\ &\times \exp(-i\omega t + i\tilde{\omega} t + i\psi_2 - i\tilde{\psi}_1 + ik_2 z - i\tilde{k}_1 z)\, d\omega d\tilde{\omega}\end{aligned} \tag{4.106}$$

Here tilde marks the values which enter the expression for D_β^*.

Assuming that the conditions

$$\begin{aligned}|k_1' - \tilde{k}_1'| &\ll |k_1'' + \tilde{k}_1''| \approx 2|k_1''|, \\ |k_2' - \tilde{k}_2'| &\ll |k_2'' + \tilde{k}_2''| \approx 2|k_2''|,\end{aligned} \tag{4.107}$$

are valid within narrow frequency band $\Delta\omega$, we neglect the differences $k_1' - \tilde{k}_1'$ and $k_2' - \tilde{k}_2'$ in pre-exponential factors in (4.106) and take out of the integral the values of the components of the vectors $e_{1,2}$ and of the wave vectors $k_{1,2}$ corresponding to the middle of the band $\Delta\omega$. The remaining integrals in (4.106) reduce to

$$\begin{aligned}
&\int_{\Delta\omega}\int_{\Delta\omega} D_1\tilde{D}_1 \exp(-i\omega t + i\tilde{\omega}t + i\psi_1 - i\tilde{\psi}_1 \\
&\quad + ik_1 z - i\tilde{k}_1 z)\, d\omega d\tilde{\omega} = e^*_{1\gamma}e_{1\delta}D_\gamma D^*_\delta, \\
&\int_{\Delta\omega}\int_{\Delta\omega} D_2\tilde{D}_2 \exp(-i\omega t + i\tilde{\omega}t + i\psi_2 - i\tilde{\psi}_2 \\
&\quad + ik_2 z - i\tilde{k}_2 z)\, d\omega d\tilde{\omega} = e^*_{2\gamma}e_{2\delta}D_\gamma D^*_\delta, \\
&\int_{\Delta\omega}\int_{\Delta\omega} D_1\tilde{D}_2 \exp(-i\omega t + i\tilde{\omega}t + i\psi_1 - i\tilde{\psi}_2 \\
&\quad + ik_1 z - i\tilde{k}_2 z)\, d\omega d\tilde{\omega} = e^*_{1\gamma}e_{2\delta}D_\gamma D^*_\delta, \\
&\int_{\Delta\omega}\int_{\Delta\omega} D_2\tilde{D}_1 \exp(-i\omega t + i\tilde{\omega}t + i\psi_2 - i\tilde{\psi}_1 \\
&\quad + ik_2 z - i\tilde{k}_1 z)\, d\omega d\tilde{\omega} = e^*_{2\gamma}e_{1\delta}D_\gamma D^*_\delta.
\end{aligned} \tag{4.108}$$

The validity of these formulas can be checked using (4.101), (4.104) and writing down the formulas similar to (4.100) for D_γ and D_δ. Substituting (4.108) into (4.106) and averaging over time, we arrive at the following polarization transfer equation [57]:

$$\frac{dI_{\alpha\beta}}{dz} = S_{\alpha\beta} + T_{\alpha\beta\gamma\delta}I_{\gamma\delta}, \tag{4.109}$$

where $T_{\alpha\beta\gamma\delta}$ is called polarization transfer tensor. This tensor can be written in the form

$$T_{\alpha\beta\gamma\delta} = R_{\alpha\beta\gamma\delta} + K_{\alpha\beta\gamma\delta}, \tag{4.110}$$

where $R_{\alpha\beta\gamma\delta}$ describes the change of polarization due to phase shift, and $K_{\alpha\beta\gamma\delta}$, due to different absorption of the modes in magnetoactive plasma:

$$\begin{aligned}
R_{\alpha\beta\gamma\delta} &= i(k_1' - k_2')(e_{1\alpha}e^*_{2\beta}e^*_{1\gamma}e_{2\delta} - e_{2\alpha}e^*_{1\beta}e^*_{2\gamma}e_{1\delta}), \\
K_{\alpha\beta\gamma\delta} &= (k_1'' + k_2'')(e_{1\alpha}e^*_{2\beta}e^*_{1\gamma}e_{2\delta} + e_{2\alpha}e^*_{1\beta}e^*_{2\gamma}e_{1\delta}) \\
&\quad + 2k_1''e_{1\alpha}e^*_{1\beta}e^*_{1\gamma}e_{1\delta} + 2k_2''e_{2\alpha}e^*_{2\beta}e^*_{2\gamma}e_{2\delta}.
\end{aligned} \tag{4.111}$$

The right-hand side of the equation (4.109) also comprises the polarization tensor $S_{\alpha\beta}$ of radiation sources in plasma which describes variation of $I_{\alpha\beta}$ due to proper emission by the medium.

If radiation is comprised of a single mode, for example, $D_1 \neq 0$, $D_2 = 0$, then, according to (4.108) and (1.55), $e^*_{2\gamma}e_{2\delta}I_{\gamma\delta} = 0$, $e^*_{1\gamma}e_{2\delta}I_{\gamma\delta} = 0$ and $e^*_{2\gamma}e_{1\delta}I_{\gamma\delta} = 0$. Therefore, the terms related to the phase shift between modes are cancelled in the equation (4.109): $R_{\alpha\beta\gamma\delta}I_{\gamma\delta} = 0$, whereas the ones describing the effect of absorption are essentially simplified:[12]

$$K_{\alpha\beta\gamma\delta}I_{\gamma\delta} = 2k_1''e_{1\alpha}e^*_{1\beta}e^*_{1\gamma}e_{1\delta}I_{\gamma\delta} = 2k_1''I_{\alpha\beta}. \tag{4.112}$$

[12] To come to the last equality in (4.112), one should use the first formula in (4.108) and the expression $D_\alpha D^*_\beta$ written for the case $D_1 \neq 0$, $D_2 = 0$.

If proper emission of the medium is absent ($S_{\alpha\beta}=0$), we get the transfer equation

$$\frac{dI_{\alpha\beta}}{dz}=-2k_1''I_{\alpha\beta}, \tag{4.113}$$

which just reflects the trivial fact that the components of the vector $\boldsymbol{D}$ in the extraordinary wave behave as $e^{-k_1''z}$ due to the effect of absorption. Evidently, only the first Stokes parameter, radiation intensity $I_1=I_{xx}+I_{yy}$, varies in this case according to the equation

$$\frac{dI_1}{dz}=-2k_1''I_1. \tag{4.114}$$

The remaining polarization parameters ζ, p, χ, determined by the ratios of the components $I_{\alpha\beta}$ through the formulas (1.52) and (1.54), are constant in this case. These parameters are also constant if radiation sources described by the tensor $S_{\alpha\beta}$ contribute to extraordinary radiation. Since mode polarizations are orthogonal in the case considered, the intensity of arbitrary polarized radiation can be presented as the sum of two components corresponding to different modes: $I=I_1+I_2$. It can be shown (see the next section) that radiation intensity in each mode satisfies the corresponding transfer equation similar to (4.114).

The assumption on orthogonality of the mode polarizations (4.114) has been essentially used to derive the equation (4.109) with polarization transfer tensor in the form (4.110), (4.111). Generally speaking, such an orthogonality is violated due to absorption (or amplification) of waves in plasma when $k_1'', k_2''\neq 0$. However, the equations (4.109)-(4.111) can also be applied to a absorbing medium if mode non-orthogonality is weak: $\overline{\alpha\alpha^*}\ll 1$.[13]

In case of arbitrary (not weak) non-orthogonality of the modes the polarization transfer tensor becomes more complicated [192]:

$$\begin{aligned}T_{\alpha\beta\gamma\delta}&=(1-\overline{\alpha}^2)^{-2}\\&\times\Big\{-2k_1''e_{1\alpha}e_{1\beta}^*[e_{1\gamma}^*e_{1\delta}-\overline{\alpha}(e_{1\gamma}^*e_{2\delta}+e_{2\gamma}^*e_{1\delta})+\overline{\alpha}^2e_{2\gamma}^*e_{2\delta}]\\&\quad-2k_2''e_{2\alpha}e_{2\beta}^*[e_{2\gamma}^*e_{2\delta}-\overline{\alpha}(e_{1\gamma}^*e_{2\delta}+e_{2\gamma}^*e_{1\delta})+\overline{\alpha}^2e_{1\gamma}^*e_{1\delta}]\\&\quad+[i(k_1'-k_2')-(k_1''+k_2'')]e_{1\alpha}e_{2\beta}^*\\&\quad\times[e_{1\gamma}^*e_{2\delta}-\overline{\alpha}(e_{1\gamma}^*e_{1\delta}+e_{2\gamma}^*e_{2\delta})+\overline{\alpha}^2e_{2\gamma}^*e_{1\delta}]\\&\quad+[-i(k_1'-k_2')-(k_1''+k_2'')]e_{2\alpha}e_{1\beta}^*\end{aligned}$$

[13]The strong non-orthogonality of the modes in plasma ($\overline{\alpha\alpha^*}\sim 1$) is realized if the Hermitian part of dielectric tensor, $\epsilon_{\alpha\beta}+\epsilon_{\beta\alpha}^*$, is comparable to anti-Hermitian one, $\epsilon_{\alpha\beta}-\epsilon_{\beta\alpha}^*$. But if the dielectric tensor $\epsilon_{\alpha\beta}$ is close to pure Hermitian or anti-Hermitian, the non-orthogonality becomes weak. This justifies the use of the tensors (4.111) to study the polarization transfer in this case [192].

$$\times \left[e_{2\gamma}^* e_{1\delta} - \overline{\alpha}(e_{1\gamma}^* e_{1\delta} + e_{2\gamma}^* e_{2\delta}) + \overline{\alpha}^2 e_{1\gamma}^* e_{2\delta}\right]\Big\} \quad (4.115)$$

The above expression can be derived in a similar way as (4.111) but using (4.105) instead of (4.104). Note that, for the sake of simplicity, the expression (4.115) is written for the case of real $\overline{\alpha}$. In particular, this condition is met in plasma with isotropic momentum distribution of particles.

Radiation comprised of single mode waves is described by the equations (4.113), (4.114) even in the case of non-orthogonal mode polarizations. Such radiation holds the initial polarization while propagating through the plasma unless another mode is present. But if two modes propagate in such a medium, the radiation intensity is not equal to the sum of the normal wave intensities: $I \neq I_1 + I_2$, and the transfer equation in the form (4.114) is no longer valid in this case.[14] Nevertheless, it is possible to select a part of the radiation which obeys the equation of the type (4.114) [195, 196]. This is done by introducing "transfer polarization vectors" $\boldsymbol{t}_1$, $\boldsymbol{t}_2$ which are orthogonal to the normal wave polarizations $\boldsymbol{e}_1$, $\boldsymbol{e}_2$:

$$\boldsymbol{e}_1 \boldsymbol{t}_1^* = 0 \qquad \boldsymbol{e}_2 \boldsymbol{t}_2^* = 0 \quad (4.116)$$

The intensities I_{t_1} and I_{t_2} of radiation with "transfer polarizations" $\boldsymbol{t}_1$ and $\boldsymbol{t}_2$ vary according to the equations

$$\frac{dI_{t_1}}{dz} = -2k_2'' I_{t_1} \quad (4.117)$$

$$\frac{dI_{t_2}}{dz} = -2k_1'' I_{t_2}. \quad (4.118)$$

respectively.

These equations can be rigorously derived from the transfer equations (4.109). They become quite clear if one takes into account that waves with polarization vector $\boldsymbol{e}_2$ do not enter the radiation with transfer polarization $\boldsymbol{t}_2$. This means that only extraordinary radiation contributes to the intensity I_{t_2}. So, the identity of the transfer equation (4.118) for I_{t_2} and (4.114) which describes the intensity of extraordinary radiation seems to be quite natural. Of course, the same is valid for the equation (4.117). Since the transfer polarizations are non-orthogonal, $\overline{\alpha} \neq 0$, the total radiation intensity does not equal $I_{t_1} + I_{t_2}$.

The case of a rarefied, weakly anisotropic medium is quite important for cosmic plasmas, in particular, for an interstellar medium or discrete source. This case is realized in the low-density plasma $v = \omega_L^2/\omega^2 \ll 1$ in a weak

[14] The total intensity in this case is oscillating function of z with the spatial period $2\pi/(k_1' - k_2')$ [192–194].

magnetic field $u=\omega_B^2/\omega^2 \ll 1$ when the components of the dielectric tensor satisfy the criterion

$$|\epsilon_{\alpha\beta} - \delta_{\alpha\beta}| \ll 1, \tag{4.119}$$

where $\delta_{\alpha\beta}$ is Kronecker delta. This leads to the substantial simplification of the expression (4.115) for the polarization transfer tensor $T_{\alpha\beta\gamma\delta}$ which can be written as [192]

$$T_{\alpha\beta\gamma\delta} = i\frac{\omega}{c}(\epsilon_{\alpha\gamma}\delta_{\beta\delta} - \epsilon^*_{\beta\delta}\delta_{\alpha\gamma}). \tag{4.120}$$

If dielectric tensor $\epsilon_{\alpha\beta}$ depends explicitly on the wave vector $\boldsymbol{k}$, one should put $k = \omega/c$ when substituting the expression for $\epsilon_{\alpha\beta}$ in (4.120). Since Greek subscripts may take two values, x and y, the elements of the tensor $T_{\alpha\beta\gamma\delta}$ in (4.120) are determined by the components of $\epsilon_{\alpha\beta}$ in the xy-plane.

Transfer equation (4.118) with tensor $T_{\alpha\beta\gamma\delta}$ in the form (4.120) has been derived from Maxwell equations in [197]. The expressions relating the components of the tensor to that of $\epsilon_{\alpha\beta}$ are rather simple. This fact makes the expression (4.120) essentially different from (4.110), (4.111) and (4.115) which relate $T_{\alpha\beta\gamma\delta}$ directly to the mode parameters: polarization vectors, refractive and absorption indeces. The choice between two representations of the polarization transfer tensor should be made according to the type of the problem considered. To a great extent, the choice is based upon convenience. Of course, one should bear in mind the validity ranges of various forms of polarization transfer tensor.

The above discussion has left apart the important problem of calculation of the source polarization tensor $S_{\alpha\beta}$. This problem is quite complicated. In general, tensor $S_{\alpha\beta}$ can be written in terms of the integrals of the product of Fourrier components of the external current in the source. Such expressions for weakly anisotropic plasma can be found in [197]. This paper gives also the formulas for these expressions corresponding to synchrotron radiation sources of specific types. However, the problem is simplified for thermal radiation, i.e., radiation by a particle system with equilibrium energy or momentum distribution. In this case the tensor $S_{\alpha\beta}$ can be found in terms of polarization transfer tensor $T_{\alpha\beta\gamma\delta}$ and polarization tensor of equilibrium radiation $I^{(0)}_{\alpha\beta}$, the same as emissivity a_ω is related to the equilibrium intensity $I^{(0)}_\omega$ and absorption coefficient μ_j by the Kirchhoff law (4.36). Indeed, in case of total thermodynamic equilibrium between radiation and a homogeneous medium $dI_{\alpha\beta}/dz=0$, so (see (4.109))

$$S_{\alpha\beta} = -T_{\alpha\beta\gamma\delta}I^{(0)}_{\gamma\delta}. \tag{4.121}$$

Since $S_{\alpha\beta}$ is determined by the properties of the source (by the distribution of external currents) and does not depend on the radiation around it (as long as its intensity is not too large), it is clear that the relation (4.121) holds even if the matter and radiation are not in strict thermodynamic equilibrium. For this relation to be valid it is sufficient that plasma particles which determine the dielectric tensor have equilibrium distribution.

The tensor $S_{\alpha\beta}$ can be most simply derived with (4.121) in case of weakly anisotropic rarefied plasma since, at a first approximation, the tensor $I_{\alpha\beta}^{(0)}$ can be put equal to the vacuum one:

$$I_{\alpha\beta}^{(0)} = I_{\omega}^{(0)} \delta_{\alpha\beta}. \tag{4.122}$$

This polarization tensor is normalized so that $I_{\alpha\alpha} = I_{\omega}^{(0)}$ equals energy flux per unit frequency range and unit solid angle interval. Otherwise, at $\alpha \neq \beta$, the components of the tensor (4.122) are zero. This reflects the absence of polarization of equilibrium radiation in vacuum. Taking into account the expression (4.120) for $T_{\alpha\beta\gamma\delta}$, we find from (4.121) and (4.122) that

$$S_{\alpha\beta} = -i\frac{\omega}{2c} I_{\omega}^{(0)} (\epsilon_{\alpha\beta} - \epsilon_{\beta\alpha}^{*}) \tag{4.123}$$

in a weakly anisotropic rarefied equilibrium plasma [195, 196].[15] The equilibrium intensity $I_{\omega}^{(0)}$ in this case is given by (1.38) in which $n = 1$ and $\cos\vartheta = 1$. According to (4.123), the source polarization tensor in an equilibrium plasma is directly related to the anti-Hermitian part of the dielectric tensor $\epsilon_{\alpha\beta}$.

The analog of the Kirchhoff law, the equation (4.121), has been used in [198] to derive the tensor $S_{\alpha\beta}$ in the more complex case of equilibrium magnetoactive plasma with arbitrary anisotropy but assuming that polarizations of the modes are orthogonal. The sources of thermal radiation in plasma have been found to emit ordinary and extraordinary waves which are incoherent with respect to each other. The polarization of the thermal emission in magnetoactive plasma should therefore coincide with one of the mode polarizations. For example, under the conditions of quasi-longitudinal propagation (2.126) the modes are circularly polarized. In addition to the naturally polarized radiation, thermal sources can therefore emit only a circularly polarized one. Linear polarization of radiation in such a medium cannot originate from its proper thermal sources. Respectively, in case of quasi-transverse propagation with linearly polarized modes thermal sources contribute only to the linearly polarized radiation and are unable to emit a circularly polarized one.

The above considered polarization transfer equations (4.109) is valid for a homogeneous medium. The attempt to generalize these equation for an

---

[15] The formula (4.121) clearly shows that the effect of the small corrections to $I_{\alpha\beta}^{(0)}$ due to finite anisotropy and plasma density on $S_{\alpha\beta}$ is negligible.

arbitrary inhomogeneous medium encounters significant difficulties, since the difference of ray trajectories for extraordinary and ordinary radiation should be taken into account.[16] However, if an inhomogeneous medium is rarefied enough, $n_j \approx 1$, the refraction becomes negligible. Transfer of polarization can be described in this case by the above equations with the components of source polarization tensor $S_{\alpha\beta}$ and polarization transfer tensor $T_{\alpha\beta\gamma\delta}$ treated as functions of spatial coordinates. In the so-called quasi-isotropic approximation $|n_1 - n_2| \ll |n_1 + n_2|$ which has been considered in section 3.2 both ray trajectories coincide. Here the polarization transfer equation takes the following form [137]:

$$\epsilon \frac{d}{dl}\left(\epsilon^{-1} I_{\alpha\beta}\right) = T_{\alpha\beta\gamma\delta} I_{\gamma\delta}. \tag{4.124}$$

This equation can be derived from Budden–Kravtsov equations (3.40). The right-hand side of (4.124) differs from that of (4.109) only by the dependence of the tensor components on spatial coordinates, whereas the dielectric constant $\epsilon = n^2$ of isotropic plasma enters the left-hand side of this equation, similar to the stationary transfer equation for radiation intensity (4.10). Note that the equation (4.124) is valid beyond the geometrical optics approximation and thus takes into account the effects of linear mode coupling. Therefore, the equation of polarization transfer (4.124) can be successfully used instead of the Badden–Kravtsov equation (see section 3.2) when dealing with such problems.

4.2.3. SOLUTIONS OF POLARIZATION TRANSFER EQUATIONS

To analyze polarization transfer equation with the tensor $T_{\alpha\beta\gamma\delta}$ in a general form is quite a complex problem, since the corresponding solutions are cumbersome and inconvenient. This inevitably leads to the extensive use of various approximations, quasi-longitudinal and quasi-transverse ones playing the leading role among them. In the first limit both modes, extraordinary and ordinary, are circularly polarized, while in the second one the mode polarization is linear. Due to its wide validity range, quasi-longitudinal approximation is applied to various cosmic plasmas: solar corona, supernova envelopes, interstellar medium etc. Another case of linearly polarized modes is of interest in some special cases of radio waves propagation in solar corona, Jovian ionosphere and in neutron star magnetospheres. Both approximations lead to the significant simplification of the polarization transfer equation (4.109) with tensor $T_{\alpha\beta\gamma\delta}$ in the form (4.110), (4.111).

[16] In the approximation of small-angle scattering this effect is taken into account in the transfer equation derived in [199] for a chaotically inhomogeneous medium.

Indeed, polarization vectors of circularly polarized modes are equal to

$$e_1 = \frac{1}{\sqrt{2}}(\boldsymbol{x}_0 + i\boldsymbol{y}_0), \qquad e_2 = \frac{1}{\sqrt{2}}(i\boldsymbol{x}_0 + \boldsymbol{y}_0), \tag{4.125}$$

where $\boldsymbol{x}_0$ and $\boldsymbol{y}_0$ are unit vectors along x- and y-axis, respectively, which determine a plane perpendicular to the wave vector $\boldsymbol{k}$. Substituting the expressions (4.125) into (4.111), we write the polarization transfer equations (4.109) in terms of the Stokes parameters (1.54):

$$\begin{aligned} \frac{dI}{dz} &= S_I - (k_1'' + k_2'')I + (k_1'' - k_2'')V, \\ \frac{dV}{dz} &= S_V - (k_1'' + k_2'')V + (k_1'' - k_2'')I, \end{aligned} \tag{4.126}$$

$$\begin{aligned} \frac{dQ}{dz} &= S_Q - (k_1'' + k_2'')Q + (k_1' - k_2')U, \\ \frac{dU}{dz} &= S_U - (k_1'' + k_2'')U + (k_1' - k_2')Q. \end{aligned} \tag{4.127}$$

Here

$$\begin{aligned} S_I &= S_{xx} + S_{yy}, \\ S_V &= S_{xx} - S_{yy}, \\ S_Q &= i(S_{yx} - S_{xy}), \\ S_U &= S_{yx} + S_{xy}. \end{aligned}$$

In case of quasi-perpendicular propagation the polarization vectors of extraordinary and ordinary modes equal

$$e_1 = \boldsymbol{x}_0, \qquad e_2 = \boldsymbol{y}_0. \tag{4.128}$$

It is assumed here that the magnetic field $\boldsymbol{B}_0$ is directed along $\boldsymbol{y}_0$. The equations of polarization transfer, (4.109)–(4.111), take the following form:

$$\begin{aligned} \frac{dI}{dz} &= S_I - (k_1'' + k_2'')I - (k_1'' - k_2'')Q, \\ \frac{dQ}{dz} &= S_Q - (k_1'' + k_2'')Q - (k_1'' - k_2'')I, \\ \frac{dU}{dz} &= S_U - (k_1'' + k_2'')U - (k_1' - k_2')V, \\ \frac{dV}{dz} &= S_V - (k_1'' + k_2'')V + (k_1' - k_2')U, \end{aligned} \tag{4.129}$$

where

$$\begin{aligned} S_I &= S_{xx} + S_{yy}, \\ S_Q &= S_{xx} - S_{yy}, \\ S_V &= i(S_{yx} - S_{xy}), \\ S_U &= S_{yx} + S_{xy}. \end{aligned}$$

Since the case of quasi-longitudinal propagation is of greater interest for applications, we will consider polarization transfer in this case in more detail [57]. The analysis of the corresponding polarization transfer equation in quasi-perpendicular approximation can be done in a similar way (for example, see in this respect [193, 194]).

So, turning to the system (4.126), (4.127), we note that the former equations are not influenced by the Faraday effect which is determined by $k_1' - k_2'$, the phase shift between the normal waves. This is quite natural since the rotation of the polarization ellipse does not change the total radiation intensity I and the degree of circular polarization $\zeta_c = V/I$. However, as seen from (4.127), this effect may alter the degree of linear polarization $\zeta_l = \sqrt{Q^2 + U^2}/I$ and, according to the relation $\tan 2\chi = U/Q$, the orientation of the polarization ellipse χ (see section 1.2).

The equations (4.126) can be transformed to

$$\frac{dI_1}{dz} = S_1 - 2k_1'' I_1, \qquad \frac{dI_2}{dz} = S_2 - 2k_2'' I_2, \tag{4.130}$$

where $S_I = S_1 + S_2$ and $S_V = S_2 - S_1$, while I_1 and I_2 are the intensities of extraordinary and ordinary radiation which relate to the Stokes parameters as follows (see (1.61)):

$$I = I_1 + I_2, \qquad V = I_2 - I_1. \tag{4.131}$$

The formulas (4.130) illustrate the above noted fact that radiation in each of the orthogonal modes (for example, the extraordinary one) is described by the intensity transfer equation of the type (4.114) which is independent of radiation in another mode.

It is not difficult to find the general solution of the system (4.126), (4.127) of linear differential equations with constant coefficients. The integration constants in such a solution can be related to the values of Stokes parameters at the layer boundary $z = 0$ (see [57]). If the effect of radiation sources inside the layer is negligible: $S_I = S_V = S_Q = S_U = 0$, these solutions yield the following expressions for the degree of circular[17]

$$\zeta_c = \frac{V}{I} = \frac{1 + \zeta_c^0 - (1 - \zeta_c^0)e^{-2(k_1'' - k_2'')z}}{1 + \zeta_c^0 + (1 - \zeta_c^0)e^{-2(k_1'' - k_2'')z}} \tag{4.132}$$

[17] Here and thereafter the superscript 0 stands for the value at $z = 0$.

and linear polarization

$$\zeta_l = \frac{\sqrt{Q^2+U^2}}{I} = \frac{2\zeta_l^0}{(1-\zeta_c^0)e^{-(k_1''-k_2'')z} + (1+\zeta_c^0)e^{-(k_1''-k_2'')z}}. \qquad (4.133)$$

Orientation of polarization ellipse χ is determined by the relation

$$\tan 2\chi = \frac{U}{Q} = \tan[2\chi^0 - (k_1' - k_2')z].$$

According to this formula, Faraday effect is described in the following way (cf. (4.86), (4.87))

$$\Delta\chi = \chi - \chi^0 = \frac{1}{2}(k_2' - k_1')z = \frac{\omega}{2c}(n_2 - n_1)z. \qquad (4.134)$$

As seen from the above formulas, polarization ellipse of radiation propagating through magnetoactive, non-emitting plasmas rotates uniformly, the degree of linear polarization ζ_l becoming exponentially small under the condition

$$|k_1'' - k_2''|z \gg 1. \qquad (4.135)$$

On the contrary, the degree of circular polarization ζ_c is close to ± 1 even if the incoming radiation at $z=0$ has no circular polarization ($\zeta_c^0=0$). The latter is caused by the difference in absorption of modes with right-hand and left-hand polarizations. The sign of ζ_c is the same as the difference $k_1'' - k_2''$.

Consider now the variation of polarization with proper emission by magnetoactive plasma taken into account. Firstly, we analyze the case $I^0 = V^0 = Q^0 = U^0 = 0$ when no radiation enters the layer boundary. If the layer is thin enough to neglect the effect of absorption:

$$|k_{1,2}''z| \ll 1, \qquad (4.136)$$

a general solution of the system (4.126)–(4.127) yields $I = S_I z$ and $V = S_V z$. Therefore, the degree of circular polarization $\zeta_c = V/I =$const. The degree of linear polarization depends on the size z of the homogeneous layer [189]:

$$\zeta_1 = \sqrt{U^2+Q^2}/I = \zeta_1^S \left| \frac{\sin\left[\frac{1}{2}(k_1' - k_2')z\right]}{\frac{1}{2}(k_1' - k_2')z} \right|. \qquad (4.137)$$

Here $\zeta_l^S = \sqrt{S_U^2 + S_Q^2}/S_I$ is the corresponding degree of linear polarization in the layer. It is seen from (4.137) that the degree of linear polarization ζ_l is much less than ζ_l^S if

$$\frac{1}{2}(k_1' - k_2')z = \frac{\omega}{2c}(n_1 - n_2)z \gg 1. \qquad (4.138)$$

Next,

$$\frac{U}{Q} = \tan\left[2\chi^S - \frac{1}{2}(k_1' - k_2')z\right],$$

where $\tan 2\chi^S = S_U/S_Q$, so the orientation of polarization ellipse varies as

$$\chi = \chi^S + \frac{1}{4}(k_2' - k_1')z + \frac{\pi}{2}m. \tag{4.139}$$

The integer m increases by 1 at the points z where the degree of linear polarization (4.137) goes to zero, i.e., $(k_1' - k_2')z = 2\pi m$. If z approaches these values, the angle χ goes to $\chi^S + \pi/4$ or $\chi^S - \pi/4$, depending on the sign of the difference $k_1' - k_2'$. At the same time, variation of χ beyond these points always starts from the value χ^S given by the radiation sources in the layer. The corresponding rate of variation of the angle χ is two times less than the one for radiation entering the layer from the outside (see the formulas (4.139) and (4.134)). The slower rotation of the polarization ellipse is caused by the fact that radiation from an elementary source undergoing the usual Faraday rotation (4.134) during its propagation through the plasma is added to the radiation from other sources which is characterized by the fixed orientation of the polarization ellipse [57].

When analyzing the case (4.136) of an optically thin source, special attention should be drawn to the condition (4.138) leading to the strong depolarization, the decrease of the observed value of ζ_l compared to ζ_l^S. Such an effect makes impossible the registration of linear polarization even in case of high values of ζ_l^S. Comparison of (4.138) and (4.134) shows that depolarization occurs in case of the strong Faraday rotation $\Delta\chi \gg 1$ of radiation within the generation region. Due to this fact orientations of polarization ellipses of radiation which has been emitted in the source at different points of a ray and then propagated to an observer are spread over the wide angular interval $\Delta\chi \gg 1$. The elliptically polarized source radiation is thus observed as the circular polarized one, while linearly polarized source radiation appears as naturally polarized. The above depolarization is absent only if $|\Delta\chi| \ll 1$. As can be easily shown, this condition poses serious constraints on the size of the generation region of linearly or elliptically polarized radiation. Indeed, according to (4.89), the condition $|\Delta\chi| \ll 1$ can only be met for an emitting source of the size L if

$$L \ll f^2/2.35 \cdot 10^4 N B_0 |\cos\alpha|. \tag{4.140}$$

In the solar corona with $N \sim 10^8\,\mathrm{cm}^{-3}$ and $B_0|\cos\alpha| \sim 10\,\mathrm{G}$ the value of L should be less than $2 \cdot 10^6\,\mathrm{cm}$ for $f \sim 10^{10}\,\mathrm{Hz}$. The corresponding value at the meter band should be 10^4 times less. Since the usual size of the sources exceeds the above values, the radiation in the source region will be completely depolarized due to Faraday effect. The appearance of linear or elliptical polarization of solar radio emission is only possible under specific conditions in the corona, for example due to the effect of linear mode coupling which has been discussed in section 3.2.

Consider now the polarization of radiation by a homogeneous, optically thick source in which $2k''_{1,2}z \gg 1$. Since absorption in such a source is very strong, the influence of the incoming radiation may be neglected. When applied to this case, the solution of the system (4.126)–(4.127) shows that the degrees of linear and circular polarization, ζ_l and ζ_c, are independent of the coordinate z and determined by the combined effect of source polarization (ζ_l^S and ζ_c^S) and absorption ($k''_{1,2}$). In spite of the Faraday effect, the polarization ellipse does not rotate, since χ reaches the limiting value at distances where $2k''_{1,2}z \gg 1$. If the Faraday effect is weak compared to absorption: $|k'_1 - k'_2| \ll k''_1 + k''_2$, the angle χ is close to χ^S. On the contrary, $\chi \approx \chi^S \pm \pi/4$ in the opposite case of the strong Faraday effect $|k'_1 - k'_2| \gg k''_1 + k''_2$, the sign being the same as the one of the difference $k'_2 - k'_1$. Such results can be easily understood by recalling that radiation in each mode at a point z which is far enough from the layer boundary $(2k''_{1,2}z \gg 1)$ is emitted by the sources of equal sizes $\Delta z_{1,2} \sim (2k''_{1,2})^{-1}$.

It was noted in the previous paragraph that thermal radiation sources in an equilibrium plasma have circular polarization in the quasi-longitudinal approximation: $S_U = S_Q = 0$ and $\zeta_l^S = 0$. As follows from (4.137), the radiation emitted by the plasma layer does not contain a linearly polarized component [198]. The same conclusion is valid for the source of arbitrary optical depth. The appearance of linearly polarized radio emission from the regions in which the conditions of quasi-longitudinal propagation are realized (e.g., in the interstellar medium of supernovae envelopes) is strong evidence for the non-thermal origin of such radiation, unless the observed polarization is formed along the path from a source, say, due to the effects of linear mode coupling.

The identity of the polarization of thermal radiation and the mode polarization in plasma means that extraordinary and ordinary components comprising such radiation are incoherent with respect to each other. Of course, this fact does not depend on the angle between $\boldsymbol{k}_{1,2}$ and $\boldsymbol{B}_0$. Therefore, polarization of sources of thermal radiation coincides with mode polarization for an arbitrary angle between $\boldsymbol{k}_{1,2}$ and $\boldsymbol{B}_0$ (of course, if the modes themselves are orthogonal).

If the conditions $k''_{1,2} < 0$ favoring amplification of radiation in a medium are realized (see chapter 7 for more details on these conditions), polarization of the resulting radiation is sharply different from that considered above. The degree of circular polarization in the region $-2k''_{1,2}z \gg 1$ is constant. If $2|k''_1 - k''_2|z \ll 1$, the value of ζ_c is determined by the polarization of the sources and incident radiation. However, at sufficiently large z satisfying the opposite criterion, $\zeta_c(z) \to \pm 1$ independently of the properties of the source and the incident radiation. The sign of ζ_c in this case is determined by that of the difference $k_2'' - k_1''$. Since the total degree of polarization $\sqrt{\zeta_l + \zeta_c} \leq 1$, the degree of linear polarization obviously goes to zero: $\zeta_l \to 0$.

The above example of a medium with circular polarization of extraordinary and ordinary waves illustrates the general statement on the possible polarization characteristics of radiation escaping cosmic sources in which wave amplification occurs. Namely, the radiation escaping a generation region of the size L with sufficiently high $-2k_{1,2}''L \gg 1$ but strongly different $2|k_1'' - k_2''|L \gg 1$ mode gains will be polarized similar to the mode with greater gain. The resulting polarization is uniquely determined by the polarization of this mode and does not depend at all on polarization of radiation emitted by distinct particles in the source.

We have now finished the restricted and incomplete, due to the unavoidable requirement to keep the size of the book reasonable, analysis of the solutions of polarization transfer equations. In conclusion, we mention some references [66, 195–197, 200–203] from the copious literature on polarization transfer of various types of radiation, especially the synchrotron one. Equations of polarization transfer proved to be useful for the analysis of the influence of Zeeman effect on the polarization of spectral line radiation from the Sun and magnetic stars [204]. This problem, however, extends beyond the scope of this book. The problems of propagation and polarization of radiation in astrophysical plasmas are also considered in the monograph [205].

Emission by astrophysical plasmas

In this chapter various processes of emission by charged particles are considered which contribute to the emissivity of astrophysical plasmas. Particle emission depends on its energy, the type of the motion in electric and/or magnetic fields, and on the properties of the surrounding medium. The major contributions to the emissivity of astrophysical plasma come from bremsstrahlung, Čerenkov, and magnetic bremsstrahlung emissions which are the main subject of this chapter. These processes influence the formation of the spectra of the observed extra-terrestrial radiation either directly or via scattering of waves which is considered after in chapter 8. Of course, the above mentioned emission processes do not cover the whole variety of those realized in astrophysical conditions. For example, a lot of spectral lines are registered in the spectra of cosmic sources which originate due to electron transitions between discrete energy levels of atoms and molecules. The analysis of spectral lines by interstellar HI, OH, etc. appearing at radio frequencies is the powerful tool for the studies of the interstellar medium. Since plasma processes do not play a significant role in spectral line formation, the latter is not considered in this book.[1]

5.1. Emission by a charged particle

5.1.1. BREMSSTRAHLUNG

This term is used to denote radiation emitted by a charged particle moving near another one (colliding with it) and remaining free (unbound to the collision center) after this event. Thus, quantum transitions resulting in emission of bremsstrahlung quanta are usually called free-free in astrophysics, contrary to free-bound (recombination) ones leading to electron capture by the ion.

When analyzing a collision by particles with essentially different masses (e.g., electron and ion) it is possible to assume the heavier particle to be motionless. Then bremsstrahlung emission occurs due to electron energy losses. The energy of the emitted quantum $\hbar\omega$ is certainly not larger than the initial electron kinetic energy $\mathcal{E}_{\rm kin}$. Therefore, a wide, continuous brems-

[1] See, nevertheless, the works [206,207] where the effect of induced scattering in plasma on the spectral line characteristics is discussed.

strahlung spectrum is limited by the maximal frequency $\omega_{\max}$:

$$\omega \leq \omega_{\max} = \frac{\mathcal{E}_{\text{kin}}}{\hbar}. \tag{5.1}$$

Maximal variation of electron kinetic energy in the Coulomb electric field of an ion, $\Delta\mathcal{E}$, corresponds to the nearest approach by colliding particles up to the distance $r_{\min} \sim q$, where q is the impact parameter of the collision. [2] The potential energy of an electron in the ion Coulomb field at a distance $r_{\min}$ equals $Ze^2/r_{\min}$, where Ze is the ion charge. Thus, the values

$$q \sim r_{\min} \sim \frac{Ze^2}{\mathcal{E}_{\text{kin}}} \sim 2\frac{Ze^2}{m_e v^2} \tag{5.2}$$

correspond to an "effective" collision in which $\Delta\mathcal{E} \sim Ze^2/r_{\min}$ is comparable to $\mathcal{E}_{\text{kin}}$. The latter condition is valid for a non-relativistic electron with velocity v.

Electron motion in the electric field of an ion can be described classically if $\bar{\lambda}_B \ll r_{\min}$, where $\bar{\lambda}_B = \hbar/m_e v$ is de-Broglie wavelength of a particle with the velocity v. With (5.2) taken into account, this inequality for a non-relativistic electron is reduced to

$$v \ll \frac{2Ze^2}{\hbar}. \tag{5.3}$$

Therefore, bremsstrahlung emission at low frequencies $\omega \ll \omega_{\max}$ can be described classically, whereas this process (as well as the particle motion influenced by the "recoil" effect originating due to emission of bremsstrahlung quanta) exhibits quantum features at high frequencies $\omega \sim \omega_{\max}$.

Calculation of bremsstrahlung by a non-relativistic electron which moves along a "classical" trajectory with the velocity v satisfying the criterion (5.3) is considerably simplified at sufficiently lower frequencies

$$\omega \ll \frac{m_e v^2 c}{2Ze^2}. \tag{5.4}$$

The radiation becomes dipole in this case: the corresponding wavelength is much larger than the size of the system "electron–ion" at the moment of collision $r \sim r_{\min}$. Therefore, the energy radiated into the unit frequency interval in a single collision is given by (see, e.g., [56])

$$\mathcal{E}_\omega = \frac{8\pi}{3c^2}|\ddot{\boldsymbol{d}}_\omega|^2. \tag{5.5}$$

[2]The impact parameter is the minimal distance the colliding particles would approach if their trajectories have been rectilinear.

Here $\boldsymbol{d}_\omega$ is the component of Fourier expansion of the dipole moment $\boldsymbol{d} = -e\boldsymbol{r}_e(t) + Ze\boldsymbol{r}_i$ of the system "electron–ion". To make the formula (5.5) valid for an isotropic medium, one should supplement its right-hand side with the refractive index $n_{em}(\omega)$. The expressions for $\mathcal{E}_\omega$ get more complicated in a magnetoactive plasma.

As follows from (5.5), no dipole bremsstrahlung occurs in a collision of two identical non-relativistic particles (say, two electrons).[3] Only quadrupole emission occurs in this case, the corresponding radiation energy being $(v/c)^2$ times less than that of the dipole emission due to electron collision with a massive particle at rest. That is why the bremsstrahlung of non-relativistic particles is contributed to mainly by the emission due to electron collisions with ions or nuclei. At the same time, bremsstrahlung by heavy particle collisions is negligible due to the small accelerations experienced by the colliding particles.

Spectral energy of bremsstrahlung emission is strongly dependent on the impact parameter q. At the same time, emission by a particle system colliding at various q, not by a single colliding pair, is usually of interest, as far as specific problems of radiation of cosmic plasmas are concerned. So, for applications it is useful to know the effective spectral power of radiation

$$\kappa_\omega = 2\pi \int_0^\infty \mathcal{E}_\omega q \, dq, \tag{5.6}$$

the average of $\mathcal{E}_\omega$ over the impact parameter q. It is easy to see that κ_ω is the spectral power of radiation by a particle stream of a unit flux density due to the collisions with a single particle at rest.[4] In addition to κ_ω, the effective cross-section of bremsstrahlung emission, $\sigma_\omega = \kappa_\omega/\hbar\omega$, is also used in the theory of bremsstrahlung radiation. This value characterizes the number of photons from the unit frequency range emitted in a unit time by unit particle flux colliding with a single particle.

If a Coulomb center holds motionless, then

$$\kappa_\omega = \frac{16e^6Z^2}{3v^2c^3m_e^2} \ln\left(\frac{2m_ev^3}{\delta\omega e^2 Z}\right), \quad \omega \ll \frac{m_ev^3}{Ze^2}, \tag{5.7}$$

$$\kappa_\omega = \frac{16\pi e^6Z^2}{3\sqrt{3}v^2c^3m_e^2}, \quad \omega \gg \frac{m_ev^3}{Ze^2}. \tag{5.8}$$

Here $\delta = e^C \approx 1.781$, where $C \approx 0.577$ is the Euler constant. Note that the approximation of a motionless collision center is quite reasonable for

[3]The dipole moment of a pair of electrons equals $\mathbf{d} = e(\mathbf{r}_1 + \mathbf{r}_2)$, where $\mathbf{r}_1 + \mathbf{r}_2 = \mathbf{R}$ is the radius-vector of the center of mass of the system. If no external force acts upon the particles, then $\ddot{\mathbf{R}} = 0$ and $\ddot{\mathbf{d}} = 0$.

[4]The unit particle flux density corresponds to a particle flow in which one particle crosses a unit area in unit time.

electron–ion collisions. The procedure of deriving (5.6) from (5.5) and then transforming the former to (5.7), (5.8) is described in [56]. Note that both of the above formulas are obtained in the dipole approximation (5.4), the expression (5.7) being valid both for oppositely and similarly charged particles, while (5.8) is applicable in the former case only. The expression $\kappa_\omega \propto \exp(-2\pi\omega e^2 Z/m_e v^3)$ should be used instead of (5.8) if charge signs of colliding particles are the same.

The total bremsstrahlung energy losses by a particle penetrating a medium with ion density N_+ is evidently expressed as follows:[5]

$$-\left(\frac{d\mathcal{E}}{dt}\right)_{\rm b} = \int_0^{\omega_{\rm max}} \kappa_\omega \, d\omega N_+ v. \tag{5.9}$$

This value can be estimated using the expression (5.8) for κ_ω:

$$-\left(\frac{d\mathcal{E}}{dt}\right)_{\rm b} \sim \frac{8\pi e^6 Z^2 N_+ v}{3\sqrt{3} m_e c^3 \hbar} \tag{5.10}$$

(see, e.g., [208]). The errors in the formula (5.10) arise due to extrapolation of the expression (5.8) to the lower-energy region beyond its validity range and, to a greater extent, to higher frequencies up to $\omega = \mathcal{E}_{\rm kin}/\hbar$.

If the condition (5.3) is violated (for ions with $Z \sim 1$ it may occur even for non-relativistic electron velocities $v/c \gtrsim (e^2/\hbar c)Z = Z/137$), the collision process acquires quantum features. The corresponding formulas for bremsstrahlung emission are given in [210]. In the case opposite to (5.3) but still at $v \ll c$ the value

$$\kappa_\omega = \frac{16 e^6 Z^2}{3 v^2 c^3 m_e^2} \ln\left(\frac{2mv^2}{\hbar\omega}\right). \tag{5.11}$$

The formula (5.11) is valid at frequencies $\omega \ll \omega_{\rm max}$ at which the relative variation of electron velocity due to emission in a single collision remains small. To derive this formula, one should use the Born approximation for the probability of a free-free collisional transition and the dipole one for the emission process.

The total (frequency-integrated) energy loss (5.9) of a non-relativistic particle moving in a medium with the velocity $v \gg 2e^2 Z/\hbar$ is determined by the expression [55]

$$-\left(\frac{d\mathcal{E}}{dt}\right)_{\rm b} = \frac{16 e^6 Z^2 N_+ v}{3 m_e c^3 \hbar}. \tag{5.12}$$

[5]The value $P_\omega = \kappa_\omega N_+ v = \hbar\omega\sigma_\omega N_+ v$ is the spectral power of bremsstrahlung emission by a particle, i.e., the energy emitted per unit time and into unit frequency range.

Then, the effective power of bremsstrahlung radiation by a relativistic electron equals [208]

$$\kappa_\omega = \frac{16e^6Z^2}{3m_e^2c^5}\left[\ln\left(\frac{2\mathcal{E}^2}{m_ec^2\hbar\omega}\right) - \frac{1}{2}\right]. \tag{5.13}$$

The time-averaged bremsstrahlung energy losses of an electron equal [211]

$$-\left(\frac{d\mathcal{E}}{dt}\right)_b = \frac{4e^6Z^2N_+\mathcal{E}}{m_e^2c^4\hbar}\left[\ln\left(\frac{2\mathcal{E}}{m_ec^2}\right) - \frac{1}{3}\right]. \tag{5.14}$$

The major contribution to this value comes, however, from the emission of photons with energy $\hbar\omega \sim \hbar\omega_{\max} \sim \mathcal{E}$. The real decrease of electron energy at a given instant can strongly differ from (5.14).

It should be stressed again that the influence of the surrounding medium (plasma) on bremsstrahlung radiation has been completely neglected in the above formulas. The presence of plasma, on the one hand, alters the electrostatic field. That of an ion differs dramatically from the Coulomb one and sharply drops at distances $r > D$ from the charge considered, the effect called Debye screening (see the formula (1.5)). Such a screening, evidently, does not influence the above results if $D \gg r_{\min}$, where $r_{\min}$ is determined by (5.2). In the opposite case $r_{\min} > D$ the value κ_ω can be estimated by putting equal to D the upper limit in the integral (5.6) over the impact parameter. The accurate calculation of bremsstrahlung emission in this case requires re-derivation of $\mathcal{E}_\omega$ using the Debye form (1.5) of the charged particle potential in plasma. On the other hand, the bremsstrahlung spectrum is altered due to the difference of the refractive index from unity. The change of the spectrum can be easily taken into account in case of isotropic plasma and dipole bremsstrahlung: the spectrum is cut at the low-frequency side $(\omega > \omega_L)$ [6], while the radiation power κ_ω equals the vacuum value multiplied by the refractive index of electromagnetic waves $n_{em}(\omega)$ given by (2.104). In the case of bremsstrahlung the effect of the magnetoactive medium on the radiation power is more complicated. In addition, bremsstrahlung emission of plasma waves becomes possible if the thermal motion of plasma particles is taken into account. Further information on bremsstrahlung in plasma can be found in [212–215].

Bremsstrahlung emission of radio waves by astrophysical plasmas will be considered in section 6.1. In concluding this section, we just mention that the above formulas are widely used to calculate X-ray emission by solar and stellar coronas and by the sources in X-ray binaries, and also to estimate the radiation losses of energetic particles moving in an interstellar medium.

[6] Electromagnetic waves with $\omega < \omega_L$ cannot propagate in plasma since $n_{em}^2 < 0$ at these frequencies (see section 2.2).

5.1.2. ČERENKOV RADIATION

To radiate in vacuum, a charged particle must move non-uniformly. As an example, we may recall the formula (5.7) for dipole emission. In a medium, however, certain conditions allow the uniformly moving charge to radiate. Such an effect in a homogeneous medium is named after Vavilov and Čerenkov. In an inhomogeneous medium the so-called transition radiation occurs in addition to the Čerenkov one. We will not consider the transition radiation and simply refer the reader to [208,209] and the bibliography therein.

The conditions of Čerenkov radiation can be found by solving the classical problem of electromagnetic fields which are generated in plasma by a uniformly moving charge. However, these conditions can also be derived in a much simpler way by applying the energy and momentum conservation laws in the quantum form to the system "electron + radiation" [216]. This is an example of the fruitfulness of the elementary quantum approach even in case of the essentially classical problem when the Planck constant does not enter the resulting solution.

So, let a non-relativistic electron move with a velocity $\boldsymbol{v}$ in a medium with the refraction index $n_j(\omega)$. If such an electron emits a photon with energy $\hbar\omega$ and momentum $\hbar\boldsymbol{k}_j$, its initial energy $m_e v^2/2$ and momentum $m_e\boldsymbol{v}$ change to the values $m_e v_1^2/2$ and $m_e\boldsymbol{v}_1$. The above emission of a photon by a single electron is only possible if the following energy and momentum conservation laws are valid:

$$\frac{1}{2}m_e v^2 = \frac{1}{2}m_e v_1^2 + \hbar\omega, \tag{5.15}$$

$$m_e\boldsymbol{v} = m_e\boldsymbol{v}_1 + \hbar\boldsymbol{k}_j. \tag{5.16}$$

Putting $\boldsymbol{v}_1 = \boldsymbol{v} + \Delta\boldsymbol{v}$, and substituting this into (5.15), (5.16), we find

$$m_e\boldsymbol{v}\Delta\boldsymbol{v} + \frac{1}{2}m_e(\Delta\boldsymbol{v})^2 + \hbar\omega = 0, \tag{5.17}$$

$$m_e\Delta\boldsymbol{v} + \hbar\boldsymbol{k}_j = 0. \tag{5.18}$$

Assuming that the velocity change due to photon emission is small: $|\Delta\boldsymbol{v}| \ll |\boldsymbol{v}|$, or, equivalently, $m_e v^2/2 \gg \hbar\omega$, we neglect the second term in (5.17) compared to the first one.[7] Eliminating $\Delta\boldsymbol{v}$ from the equations (5.17), (5.18), we arrive at the following simple condition for the Vavilov–Čerenkov effect:

$$\omega = \boldsymbol{k}\boldsymbol{v}. \tag{5.19}$$

---

[7]This means that the "recoil" due to photon emission is neglected. Such an approximation leads to the classical limit when deriving the conditions of Vavilov–Čerenkov emission.

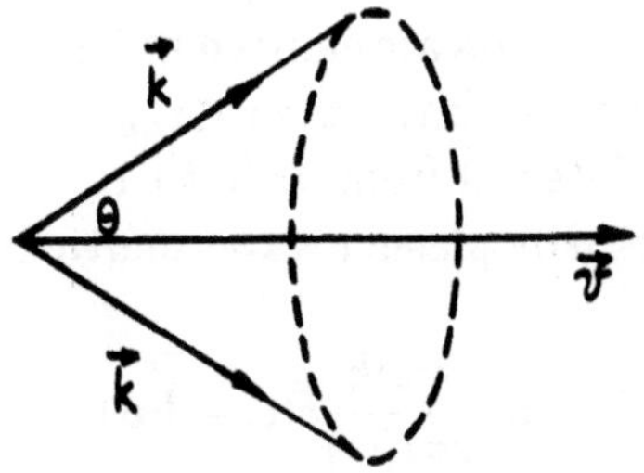

Figure 5.1. Wave vector cone in case of the Vavilov–Čerenkov effect.

Though it was supposed for the simplicity that $v \ll c$, the criterion (5.19) holds for arbitrary electron velocity.

The above criterion can be written in a different form by substituting $k_j = n_j \omega / c$, $v = \beta c$ and introducing the angle θ between $\boldsymbol{k}$ and $\boldsymbol{v}$:

$$\beta n_j \cos\theta = 1. \qquad (5.20)$$

The third form of the criterion considered,

$$v_{\text{ph}} = v \cos\theta, \qquad (5.21)$$

results from (5.20) by taking into account the formula for the phase velocity of the waves: $v_{\text{ph}} = c/n_j$. At fixed v and v_{ph} the above criteria determine the so-called "Čerenkov cone" with the axis directed along $\boldsymbol{v}$ and opening angle 2θ (see Fig. 5.1). Radiation with wave vectors along this conic surface is generated as a result of Čerenkov emission.

Since the velocity of radiating particle $v < c$ and $\cos\theta \le 1$, the conditions (5.19)–(5.21) can only be met in a medium for the waves with phase velocities $v_{\text{ph}} < c$ or $n_j > 1$. According to (2.104), the refractive index of electromagnetic waves in isotropic plasma $n_j = n_{\text{em}} < 1$. This prevents Čerenkov emission of such waves. On the contrary, as seen from (2.116) and Fig. 2.2, the refractive index of longitudinal plasma waves can exceed unity at frequencies not very close to the plasma frequency ω_{L}. These are the waves which can be emitted in an isotropic plasma due to the Vavilov–Čerenkov effect. It follows from (5.21) that an electron moving in plasma can emit waves with phase velocities $v_{\text{ph}} = \omega/k < v$. On the other hand, it will be shown in section 6.1 that plasma waves are strongly damped if v_{ph} is comparable to the electron thermal velocity in plasma, $v_T = \sqrt{\kappa T / m_{\text{e}}}$. Thus the phase velocities of plasma waves emitted due to the Čerenkov effect range from v to v_T (the condition $v \gg v_T$ is assumed). But if the electron velocity $v \lesssim v_T$, it practically cannot emit Čerenkov plasma waves. The field induced by the motion of such an electron in plasma is similar to the electrostatic one (1.5). Next, since $n_{\text{pl}} = n_{\text{pl}}(\omega)$, a given frequency corresponds to the distinct value of the angle θ. Note that $n_{\text{pl}}(\omega)$ increases with ω, so higher-frequency waves are emitted at larger angles θ with respect to the velocity $\boldsymbol{v}$. For the above interval $v_T < v_{\text{ph}} < v$ of phase velocities the values of $\cos\theta$ fall in the interval from v_T/v to 1.

Spectral power of Čerenkov plasma waves emitted by an electron equals (see, e.g., [89])[8]

$$P_{\omega\Omega} = \frac{e^2\omega k}{2\pi\epsilon(\omega)}\delta(\omega - \boldsymbol{k}\boldsymbol{v}). \tag{5.22}$$

Dirac delta-function $\delta(\omega - \boldsymbol{k}\boldsymbol{v})$ in the above expression accounts for the criterion (5.19) which determines the frequencies and directions of Čerenkov radiation. Multiplying (5.22) by $2\pi\sin\theta$ and integrating over θ from 0 to π with the delta function taken into account, we calculate the integral of $P_{\omega\Omega}$ over solid angles. This integral is nothing but the spectral power of Čerenkov radiation [217, 218]:

$$P_\omega = \int P_{\omega\Omega} d\Omega = \frac{e^2\omega}{v\epsilon(\omega)}. \tag{5.23}$$

Here $\epsilon(\omega)$ is the dielectric permeability of isotropic plasma determined by (2.73).

To find the energy loss rate of a particle due to the Čerenkov emission of plasma waves, one should integrate (5.23) over all the frequencies emitted:

$$-\left(\frac{d\mathcal{E}}{dt}\right)_{\text{Čer}} = \int_{\omega_1}^{\omega_2} P_\omega d\omega \approx \frac{e^2\omega_{\text{L}}^2}{2v}\ln\left(\frac{2}{3}\frac{v^2}{v_T^2}\right). \tag{5.24}$$

The lower limit in this integral is determined by the condition $\beta n_{\text{pl}}(\omega) = 1$ and the upper by the following one: $v_{\text{ph}} \sim v_T$ (or its equivalent $kD \approx kv_T/\omega_{\text{L}} \sim 1$, where D is the Debye radius given by (1.4)). Note that Čerenkov losses by a particle with $v \gg v_T$ are comparable to collisional energy losses and greatly exceed bremsstrahlung losses.

Up to now, the Vavilov–Čerenkov effect in isotropic plasma has been considered. The presence of a static magnetic field makes plasma magnetoactive which complicates considerably the characteristics of emission by charged particles. The criteria of Čerenkov emission (5.19)-(5.21) remain valid for a rectilinear motion of a particle in a magnetoactive plasma (say, along the magnetic field $\boldsymbol{B}_0$). In addition, however, electromagnetic waves for which $n_{1,2}^2 > 1$ can now be emitted. As seen from Fig. 2.3-2.5, such value of refractive indices correspond primarily to frequencies at which the parameter $u = \omega_B^2/\omega^2 > 1$. In this case extraordinary waves are emitted

[8] The spectral power of radiation $P_{\omega\Omega}$ and the radiation power per unit frequency range P_ω which is used below are defined in section 1.3.

parameter $u = \omega_B^2/\omega^2 > 1$. In this case extraordinary waves are emitted in the interval $0 < v = \omega_L^2/\omega^2 < 1$ and whistlers in the region $v > 1$. Corresponding expressions for the spectral power of Čerenkov radiation can be found in [219,220].

Čerenkov emission of plasma waves by fast particle beams plays an important role in the plasma mechanisms of radio emission which are widely used to explain some types of radio emission of the Sun, pulsars etc. (see section 7.2). It is quite important here that the branches of dispersion curves at which Čerenkov emission in plasma occurs correspond to the waves which cannot directly escape from plasma to vacuum with weak magnetic field. This can be easily seen from Fig. 3.1-3.4. Thus, when interpreting observed radiation such as Čerenkov emission of charged particles in astrophysical plasma, one should bear in mind that the emitted waves can only escape the source due to some transformation process, either linear coupling of the modes (section 3.2) or scattering effects (sections 8.1,8.4).

5.1.3. MAGNETIC BREMSSTRAHLUNG

The trajectory of an electron in a magnetic field $\boldsymbol{B}_0$ is, generally, a helical line with the axis along the field. Projection of this motion onto the plane perpendicular to $\boldsymbol{B}_0$ is a circle. Angular velocity of electron rotation over this circle equals relativistic gyrofrequency Ω_B. The spiral motion radically changes the electron emission as compared to the case of a rectilinear one. This results both in variation of the properties of Čerenkov radiation and in the appearance of magnetic bremsstrahlung which can occur not only in a medium but also in vacuum. It is shown further that an electron which moves helically along a classical trajectory in a homogeneous magnetic field radiates at the infinite number of harmonics $s = \pm 1, \pm 2, \pm 3, \ldots$ set by the Doppler formula (5.34). The zeroth harmonic corresponds to modified Čerenkov radiation, and its frequency is determined by the equation (5.32).

Magnetic bremsstrahlung of a non-relativistic (weakly relativistic) particle with $\mathcal{E}_{\rm kin} \ll m_e c^2$ is called cyclotron radiation. As will be shown below, this radiation is of dipole type. On the contrary, radiation by a relativistic electron ($\mathcal{E}_{\rm kin} \gg m_e c^2$) is called synchrotron. To be more correct, the term "synchrotron" is used to describe radiation of an electron with relatively large pitch angle $\vartheta \gg m_e c^2/\mathcal{E}$ between the velocity $\boldsymbol{v}$ and the magnetic field $\boldsymbol{B}_0$. On the contrary, radiation by particles with small pitch angles $\vartheta \ll m_e c^2/\mathcal{E}$ is called relativistic dipole radiation. The features of the last case arise from the fact that the radiating particle is non-relativistic in the moving frame A' in which the velocity component along the field $\boldsymbol{B}_0$ vanishes. In the frame A' the dipole radiation does not differ from the cyclotron one. Moreover, relativistic dipole radiation is nothing but cyclotron radiation in the frame A' which is transformed to the frame A moving relative to A' at a relativistic velocity. Emission by mildly relativistic electrons with energy $\mathcal{E} \sim m_e c^2$ is called gyrosynchrotron.

The above consideration corresponds to the case of a homogeneous magnetic field. However, the so-called "curvature radiation" by an electron moving along the curved magnetic field lines can also be regarded as radiation of the magnetic bremsstrahlung type.

To determine the frequency spectrum of radiation by an electron in a homogeneous magnetic field, apply again the energy and momentum conservation laws to the emission of a photon with energy $\hbar\omega$ and longitudinal momentum $\hbar \boldsymbol{k}_{\|} = \hbar(\omega/c)n_j \cos\alpha$, where $\boldsymbol{k}_{\|}$ is the projection of the wave vector $\boldsymbol{k}$ onto the magnetic field $\boldsymbol{B}_0$ and α is the angle between $\boldsymbol{k}$ and $\boldsymbol{B}_0$. These laws in the relativistic form can be written as follows [79,221]:

$$\begin{aligned} \mathcal{E}_m - \mathcal{E}_n &= \sqrt{m_e^2 c^4 + p_m^2 c^2} - \sqrt{m_e^2 c^4 + p_n^2 c^2} = \hbar\omega, \\ p_{m\|} - p_{n\|} &= \frac{\hbar\omega}{c} n_j \cos\alpha, \end{aligned} \tag{5.25}$$

where $\mathcal{E}_m$, $\mathcal{E}_n$ and p_m, p_n are electron energies and momenta before and after the emission of the photon (the states m and n, respectively). Since, contrary to $p_\perp^2$ and $p_{\|}$, the components of electron transverse momentum $p_\perp$ in a quantum state with fixed energy $\mathcal{E}$ are indefinite, only the longitudinal components enter the momentum conservation law.

The square of electron transverse momentum is quantized, its values corresponding to the so-called Landau levels [222]:

$$\frac{p_\perp^2}{2m_e} = \left(n + \frac{1}{2}\right)\hbar\omega_B. \tag{5.26}$$

Here n is an integer, and $\omega_B = eB/m_e c$ is cyclotron, or gyrofrequency.[9] In accord with this formula,

$$\begin{aligned} (p_{m\perp}^2 - p_{n\perp}^2)/2m_e &= s\hbar\omega_B, \\ s = m - n &= 0, \pm1, \pm2, \ldots, \end{aligned} \tag{5.27}$$

where s is called the harmonic number, and the values $p_{m,n\perp}^2/2m_e$ have the meaning of the energy of transverse motion in the case of a non-relativistic electron. It follows from (5.25), (5.27) that

$$\hbar\omega = \frac{m_e c^2}{\sqrt{1-\beta^2}} \frac{1 - n_j \beta_{\|} \cos\alpha}{1 - n_j^2 \cos^2\alpha}$$

[9] Note that the non-relativistic gyrofrequency enters (5.26), though this formula is valid for relativistic particles.

$$\times\left[1\pm\sqrt{1-2s\hbar\omega_B\,\frac{1-\beta^2}{m_ec^2}\,\frac{1-n_j^2\cos^2\alpha}{(1-n_j\beta_\parallel\cos\alpha)^2}}\,\right], \tag{5.28}$$

where $\beta_\parallel = v_\parallel/c$, and $v_\parallel$ is the longitudinal velocity of the electron before emission. If

$$2\hbar\omega_B\,\frac{1-\beta^2}{m_ec^2}\,\frac{|1-n_j^2\cos^2\alpha|}{(1-n_j\beta_\parallel\cos\alpha)^2}\,|s|\ll 1, \tag{5.29}$$

the solution corresponding to the plus sign in (5.28) can be approximately written as

$$1-n_j\beta_\parallel\cos\alpha\simeq\hbar\omega_B\,\frac{\sqrt{1-\beta^2}}{2m_ec^2}\,(1-n_j^2\cos^2\alpha), \tag{5.30}$$

while the one with minus sign, as

$$\omega\simeq\frac{s\omega_B\sqrt{1-\beta^2}}{1-n_j\beta_\parallel\cos\alpha}+\hbar\left(\frac{s\omega_B\sqrt{1-\beta^2}}{1-n_j\beta_\parallel\cos\alpha}\right)^2\frac{\sqrt{1-\beta^2}(1-n_j^2\cos^2\alpha)}{2m_ec^2(1-n_j\beta_\parallel\cos\alpha)} \tag{5.31}$$

The formula (5.30) is accurate for $s=0$. In the classical limit $\hbar\to 0$ for arbitrary s it is reduced to the well-known condition for Čerenkov radiation by a helically moving electron in a magnetic field:

$$1-n_j(\omega)\beta_\parallel\cos\alpha=0. \tag{5.32}$$

The solution (5.30) points out that the frequency of Čerenkov radiation in magnetic field is weakly dependent on the transitions between the Landau levels. The condition (5.32) is apparently different from the criterion (5.20) of Vavilov–Čerenkov effect in isotropic medium and cannot be reduced to the latter by the limiting transition $B_0\to 0$ since $\beta_\parallel\neq\beta$ and $\alpha\neq\vartheta$ for non-zero values of electron pitch angle ϑ. The frequency of the modified Čerenkov emission in a dispersive medium (plasma) is determined by the equation (5.32).

The solution of (5.31) at $s=0$ corresponds to zero frequency. This means that the electron state is unchanged, and photons are not emitted. But if $s\neq 0$, the solution yields a sequence of the harmonics of magnetic bremsstrahlung radiation. The frequencies of magnetic bremsstrahlung by a non-relativistic electron ($\beta^2\ll 1$) in vacuum ($n_j=1$) may be found from the equation

$$\omega\simeq\frac{s\omega_B+(\hbar\omega^2/2m_ec^2)\sin^2\alpha}{1-\beta_\parallel\cos\alpha}. \tag{5.33}$$

The term $(\hbar\omega^2/2m_ec^2)\sin^2\alpha$ describes the so-called "recoil effect": the frequency of the emitted photon shifts due to variation of longitudinal momentum and transverse energy of the radiating electron. This effect vanishes in case of emission along the magnetic field in vacuum and becomes strongest if $\alpha = \pi/2$. Since energy $\hbar\omega$ of the photon emitted by a non-relativistic electron should be much less than $m_e c^2$, the quantum correction due to the recoil is small compared to $s\omega_B$.

The role of the recoil effect under neutron star conditions has been discussed in [223]. For example, $\hbar\omega/m_e c^2 \sim 0.1$ for the frequency $\omega \simeq \omega_B \simeq 7 \cdot 10^{19}\,\mathrm{s}^{-1}$ which corresponds to the magnetic field $B \simeq 4 \cdot 10^{12}\,\mathrm{G}$ in the X-ray pulsar Her X-1. Thus, quantum recoil correction of the frequency of the cyclotron line observed from this object is similar to the gravitational redshift on the neutron star. Both effects are partically compensated, since gravitational redshift leads to the decrease of the frequency of the observed radiation, whereas recoil increases it.

The recoil effect vanishes in the classical limit $\hbar \to 0$. The formula (5.33) in this limit is transfromed into the following expression for the frequency of the magnetic bremsstrahlung radiation:

$$\omega = \frac{s\omega_B\sqrt{1-\beta^2}}{1 - n_j\beta_{||}\cos\alpha}. \tag{5.34}$$

Here $\omega_B\sqrt{1-\beta^2}$ is the frequency of electron gyration in the magnetic field, and s is called the harmonic number in the spectrum of magnetic bremsstrahlung. In quantum description s equals the difference between the quantum numbers, or Landau level numbers, m and n corresponding to the states before and after photon emission (see (5.27)). Generally, (5.34) is relativistic the Doppler formula for the frequencies of magnetic bremsstrahlung radiation of an electron. If $n_j = n_j(\omega)$, this formula should be treated as the equation for the frequency ω.

Note that the Planck constant $\hbar$ does not enter the Doppler formula (5.34) which is therefore of classical nature. At the same time, one should bear in mind that the frequency ω deduced from (5.34) cannot be negative since it determines the photon energy $\hbar\omega$. This leads to important conclusions on the possible variation of the state (on the direction of transitions) of an electron emitting magnetic bremsstrahlung photon.

Indeed, in the region of normal Doppler effect,

$$n_j\beta_{||}\cos\alpha < 1, \tag{5.35}$$

the positive frequency in (5.34) corresponds to $s = m - n > 0$. Therefore, by emitting the quantum $\hbar\omega$, an electron transits from a state m with the "transverse energy" $\epsilon_m = p_\perp^2/2m_e$ to the lower one n which is characterized by the smaller value ϵ_n. Otherwise, in the region of anomalous Doppler effect,

$$n_j\beta_{||}\cos\alpha > 1, \tag{5.36}$$

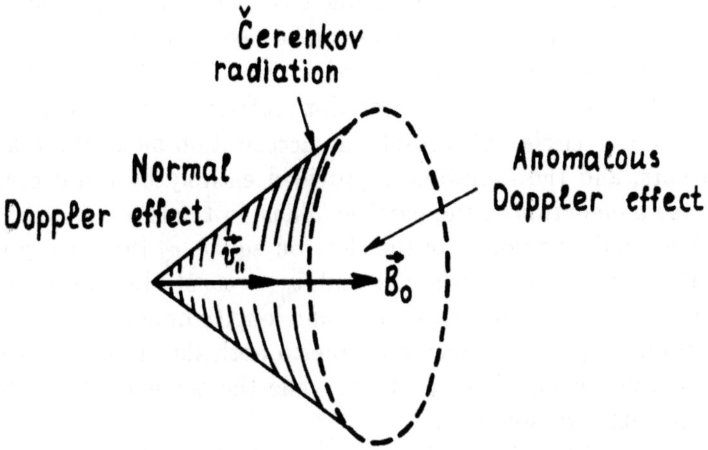

Figure 5.2. The regions of anomalous (A) and normal (N) Doppler effect in case of magnetic bremsstrahlung of an electron in a medium. The conical surface corresponds to the Čerenkov emission.

the relation (5.34) can only be satisfied by posivite ω if $s < 0$. Thus, the difference $\epsilon_m - \epsilon_n$ can be negative in this case. So, emission of a photon ω results in the increase of $\epsilon = p_\perp^2/2m_e$, and the required energy is gained from the longitudinal motion.[10] The inequality (5.36) shows that the anomalous Doppler effect can only be realized in a medium with $n_j(\omega) > 1$ and for acute angles between $\boldsymbol{k}$ and the electron longitudinal velocity, when both $\beta_\parallel$ and $\cos\alpha$ are of the same sign. The angles relevant to anomalous and normal Doppler effect correspond, respectively to the regions inside and outside the Čerenkov cone defined by the equation

$$n_j \beta_\parallel \cos\alpha = 1 \tag{5.37}$$

(see Fig. 5.2). Since the directions of radiative transitions in the cases of normal and anomalous Doppler effect are opposite, the question arises on the resulting change of $p_\perp^2$. The corresponding analysis for a moving oscillator has shown that radiation in an isotropic medium results in the damping of the "transverse" oscillations, whereas in an anisotropic medium, particularly, in plasma, the transverse dergee of freedom may be excited due to photon emission in the directions which correspond to the anomalous Doppler effect [226].

In vacuum Čerenkov emission is impossible, so magnetic bremsstrahlung takes place under the conditons of normal Doppler effect, so the electron velocity varies under the action of the force from a radiation field in the

[10] Specific features of electron emission in the region of anomalous Doppler effect have be pointed out in [225] in which the motion of a two-level system in a medium has been considered.

following way. As $t \to \infty$, the transverse velocity $v_\perp$ goes to zero, whereas the longitudinal one $v_\parallel$ keeps constant. This can be easily seen by transition into the inertial frame A' moving with the velocity $v_\parallel$. The electron trajectory is circular in this frame, and the radiation pattern is symmetric relative to the plane of this circle. As a result, no acceleration along the magnetic field $\boldsymbol{B}_0$ occurs, and the radiation is provided entirely by the decrease of the transverse momentum of the particle. Thus, in the laboratory frame in which the longitudinal velocity of the electron equals $v_\parallel$ the corresponding longitudinal acceleration $dv_\parallel/dt = 0$ and $v_\parallel = \text{const}$. In this frame the emission process changes both transverse and longitudinal electron momentum. The former, $p_\perp = mv_\perp$, decreases due to both the transverse velocity and the relativistic mass of the particle, while the decrease of $p_\parallel = mv_\parallel$ is caused by the latter reason only.

The above consideration allows one to point out the universal one-dimensional momentum distribution $f(\boldsymbol{p}, t \to \infty)$ of independent radiating particles in a magnetic field which is formed as a result of their relaxation due to magnetic bremsstrahlung.[11] The final distribution $f(\boldsymbol{p}, t \to \infty)$ does not depend on the initial one $f_0(\boldsymbol{p})$ as long as the latter is relativistic, $p \gg m_e c$, and isotropic $f_0(\boldsymbol{p}) = f_0(p)$. Indeed, the initial distribution of the relativistic isotropic system in the momentum space looks like

$$F(v) = Q\delta(v - c),$$

where Q is the normalizing factor defined by the condition that the particle density $N = \int F(v)d^3v = 4\pi c^2 Q$. Then the distribution over longitudinal velocities equals

$$F(v_\parallel) = 2\pi \int_0^c F(v)v_\perp dv_\perp = \frac{N}{2c}$$

in the interval $0 \le |v_\parallel| \le c$. Since the longitudinal velocity of a radiating particle is constant, the above expression for longitudinal distribution is valid during the entire relaxation process, the final state included. In this state, however, $v_\perp = 0$ and $p_\parallel = m_e v_\parallel / \sqrt{1 - v_\parallel^2/c^2}$. Thus the resulting particle distribution has the form

$$f(p_\parallel, t \to \infty) = F(v_\parallel)\frac{dv_\parallel}{dp_\parallel} = \frac{N}{2m_e c}\left(1 + \frac{p_\parallel^2}{m_e^2 c^2}\right)^{-3/2}.$$

Since particle distribution becomes one-dimensional, the delta-function $\delta(p_\perp)$ (to be more correct, the factor $\delta(p_\perp)/2\pi p_\perp$ accounting for the above

[11] An example of such a system is optically thin ($\tau = \int_L \mu_j dl \ll 1$) relativistic plasma with the linear size L. The re-absorption in such a source is negligible, so any particle radiates and loses momentum independently of others.

normalizing condition) should enter the expression for $f(\mathbf{p}, t \to \infty)$. Finally, we arrive at [227]

$$f(\mathbf{p}, t \to \infty) = \frac{N}{m_e c}\left(1 + \frac{p_{\|}^2}{m_e^2 c^2}\right)^{-3/2} \frac{\delta(p_\perp)}{2\pi p_\perp}. \tag{5.38}$$

This universal distribution has been used in section 2.2 to analyze the characteristics of normal waves in a relativistic plasma. The energy loss rate due to emission is proportional to B_0^2 (see the formula (5.99) below), so such a distribution can be quickly established due to radiation losses by an isotropic particle system in a strong magnetic field (for example, in the vicinity of a neutron star).

Returning to the analysis of the magnetic bremsstrahlung of a single particle, we give below the general formulas for radiation power for the case of helically moving electron in a cold magnetoactive plasma with dielectric tensor $\epsilon_{\alpha\beta}$ determined by (2.67), (2.68). These formulas have been derived in [228,229].

Radiation power at harmonics $s \neq 0$ in a unit solid angle is described by the following expression:

$$P_{\Omega j} = \frac{n_j e^2 \omega^2 T_j^2 \{\beta_\perp J_s'(\zeta) + [g_j s \beta_\perp \zeta^{-1} + h_j \beta_{\|}] J_s(\zeta)\}^2}{2\pi c |1 - \beta_{\|} \cos\alpha (n_j + \omega dn_j/d\omega)|}, \tag{5.39}$$

where $J_s(\zeta)$ and $J_s'(\zeta)$ are Bessel function of the order s and its derivative, respectively,

$$\zeta = k_j r_B \sin\alpha = \frac{s n_j \beta_\perp \sin\alpha}{1 - \beta_{\|} n_j \cos\alpha} \tag{5.40}$$

is the argument of these functions, $r_B = v_\perp/\Omega_B$ is electron gyroradius in the magnetic field, $\beta_\perp = v_\perp/c$, and $k_j = (\omega/c)n_j$ is the wave number of the j-th normal wave ($j = 1$ for the extraordinary wave and $j = 2$ for the ordinary one). Radiation frequency ω is determined by the Doppler equation (5.34). The parameters g_j, h_j are given by the formulas (2.140), (2.136), the refractive index n_j by (2.121), and the factor

$$T_j^2 = (1 + K_j^2)^{-1} \tag{5.41}$$

is expressed via the coefficients K_j defined by (2.136).

The above formula for the magnetic bremsstrahlung radiation power is quite cumbersome and inconvenient for further analysis. The simpler expressions are obtained for an electron which moves helically in isotropic plasma. The approximation of isotropic plasma is valid if $\sqrt{u} = \omega_B/\omega \ll 1$ and $v = \omega_L^2/\omega^2$ is not too close to unity (see section 2.2). The radiation power then can be found from (5.39) by the limiting transition $\sqrt{u}\to 0$ at $r_B=$ const. In this case $n_j^2=1-v$, $T_j^2=1/2$, $g_j=\pm\cos\alpha$, $h_j=\mp\sin\alpha$, so the total power of radiation into both modes equals

$$P_\Omega = P_{\Omega 1}+P_{\Omega 2} = \frac{n_j e^2\omega^2[\beta_\perp^2 J_s'^2(\zeta)+(n_j\sin\alpha)^{-2}(\cos\alpha - n_j\beta_\parallel)^2 J_s^2(\zeta)]}{2\pi c(n_j-\beta_\parallel\cos\alpha)}. \quad (5.42)$$

This formula is reduced to the following one:

$$P_\Omega = \frac{e^2\omega^2}{2\pi c(1-\beta_\parallel\cos\alpha)}\left[\beta_\perp^2 J_s'^2(\zeta)+\sin^{-2}\alpha(\cos\alpha-\beta_\parallel)^2 J_s^2(\zeta)\right] \quad (5.43)$$

for an electron which moves in vacuum ($v\equiv\omega_{\rm L}^2/\omega^2\to 0$); the argument of the Bessel functions equals $\zeta=s\beta_\perp\sin\alpha(1-\beta_\parallel\cos\alpha)^{-1}$ in this case. Next, the famous Schott formula which describes radiation of a circularly moving electron in vacuum follows from (5.43) at $v\to 0$ and $\beta_\parallel=0$:

$$P_\Omega = \frac{e^2\omega^2}{2\pi c}\left[\beta^2 J_s'^2(\zeta)+\cot^2\alpha\, J_s^2(\zeta)\right]. \quad (5.44)$$

Here $\omega = s\omega_B/\sqrt{1-\beta^2}$ and $\zeta = s\beta\sin\alpha$.

In addition to the formula (5.42) for the radiation power per unit solid angle, the following expression for radiation power related to unit frequency and solid angle ranges is widely used:

$$P_{\omega\Omega} = \frac{e^2\omega^2}{2\pi c}\sum_{s=1}^{\infty}\left[\left(\frac{\cos\alpha-\beta_\parallel}{\sin\alpha}\right)^2 J_s^2(\zeta)+\beta_\perp^2 J_s'^2(\zeta)\right]\delta(y) \quad (5.45)$$

(see, for example, [85]). Here the delta-function of the argument $y=(1-\beta^2)^{1/2}s\omega_B-\omega(1-\beta_\parallel\cos\alpha)$ takes into account the equation (5.34) which determines the frequencies of the magnetic bremsstrahlung radiation at $n_j=1$. The formula (5.45) is valid under the same conditions as (5.42).

The above formulas characterize the magnetic bremsstrahlung of an electron at harmonics $s\neq 1$. The following expression gives the power of Čerenkov radiation ($s=0$) into unit frequency range due to the helical motion of an electron in cold magnetoactive plasma:

$$P_{\omega j} = \frac{T_j^2 e^2\omega}{2v_\parallel}\frac{[\beta_\perp J_0'(\xi)+h_j\beta_\parallel J_0(\xi)]^2}{|1-(\partial n_j/\partial\alpha)/n_j\sqrt{n_j^2\beta_\parallel^2-1}|}, \quad (5.46)$$

where $n_j\beta_\parallel\geq 1$, $J_0(\xi)$ and $J_0'(\xi)$ are the zeroth-order Bessel function and its derivative, respectively, and $\xi=(\omega r_B/v_\parallel)\sqrt{n_j^2\beta_\parallel^2-1}$. Radiation frequency is determined by the condition (5.37). The formula (5.46) is simplified if an electron moves rectilinearly along $\boldsymbol{B}_0$, so that $v_\perp = 0$ and $v_\| = v$:

$$P_{\omega j} = \frac{T_j^2 e^2 \omega}{2c^2} \frac{h_j^2 v}{|1 - (\partial n_j/\partial \alpha)/n_j \sqrt{n_j^2 \beta_\|^2 - 1}|}. \tag{5.47}$$

The corresponding expression for plasma wave emission by a charge moving in a magnetoactive plasma along a helical trajectory are obtained in [220] with thermal motion taken into account.

5.1.4. CYCLOTRON RADIATION

Consider now in more detail the features of the magnetic bremsstrahlung by a weakly relativistic electron. Let $\beta = v/c$ be small enough for the following conditions

$$\beta^2 \ll 1, \qquad |n_j \beta_\|| \ll 1, \qquad \left|\beta_\| \omega \frac{\partial n_j}{\partial \omega}\right| \ll 1, \tag{5.48}$$

$$|s n_j \beta_\perp \sin\alpha| \ll 1 \tag{5.49}$$

to be valid. The above collection of criteria ensures the cyclotron character of radiation by an electron in magnetoactive plasma. As seen from (5.34), frequencies of cyclotron radiation are close to $s\omega_B$. If, in addition to (5.48), the following condition is also valid:

$$|2 n_j \beta_\| \cos\alpha| \gg \beta^2, \tag{5.50}$$

the difference $\omega - s\omega_B$ can be written in the form

$$\omega - s\omega_B \approx s\omega_B n_j \beta_\| \cos\alpha, \tag{5.51}$$

where $s = 1, 2, \ldots$ As follows from the second inequality in (5.48), cyclotron radiation is considered here in the region of normal Doppler effect (5.35).

Next, since electron gyroradius $r_B = v_\perp/\omega_B$ and cyclotron radiation frequency $\omega \approx s\omega_B$, the inequality (5.49) demands that the size of the radiating system along the wave vector of the emitted radiation, $r_B \sin\alpha$, is much less than the wavelength $\lambda\!\!\!^{-} \equiv \lambda/2\pi = c\omega n_j$. Therefore, it ensures the dipole character of cyclotron radiation. Since the parameter $|\zeta|$ is much less than unity under condition (5.49), one can approximate the Bessel functions $J_s(\zeta)$ and $J_s'(\zeta)$ which enter the formula (5.39) for magnetic bremsstrahlung radiation power by the first terms of the corresponding expansions in powers of ζ. Taking into account also the inequalities (5.48) and putting $\omega \approx s\omega_B$ and

$$1 + g_j + n_j h_j \beta_\| \sin\alpha \approx (1 + g_j)_{\omega = s\omega_B}, \tag{5.52}$$

we finally get the cyclotron radiation power in a magnetoactive plasma [230]:

$$P_{\Omega j} = \frac{e^2\omega_B^2\beta_\perp^2}{2\pi c}\left[\frac{n_j}{1+K_j^2}(1+g_j)^2(sn_j\beta_\perp \sin\alpha)^{2s-2}\right]_{\omega=s\omega_B} 2^{-2s}s^4(s!)^{-2}. \tag{5.53}$$

The second of the conditions (5.48) makes this approximation quite good at harmonics $s \geq 2$. The exception is the case of transverse propagation ($\alpha=\pi/2$) of the ordinary radiation ($j=2$) in which the parameter $h_j \to \infty$. The formula (5.53) at $v \ll 1$ takes the following form:

$$P_{\Omega j} = \frac{e^2\omega_B^2\beta_\perp^2}{2\pi c}\frac{(1+K_j\cos\alpha)^2}{1+K_j^2}2^{-2s}s^6(s!)^{-2}(s\beta_\perp \sin\alpha)^{2s-2}. \tag{5.54}$$

The coefficients K_j determined by the formulas (2.137) may be used here in the form

$$K_j = \frac{2s\cos\alpha}{\sin^2\alpha \pm \sqrt{\sin^4\alpha + 4s^2\cos^2\alpha}}. \tag{5.55}$$

The situation is radically different at the first harmonic $s = 1$, since $(1+g_j)_{\omega=s\omega_B} = 0$. The latter is explained by the peculiar polarization of normal waves at gyrofrequency in a cold magnetoactive plasma, the fact noticed in section 2.2. As a result, at $s=1$ the factor

$$1 + g_j + n_j h_j \beta_\parallel \sin\alpha \approx \beta_\parallel \cos\alpha\left[n_j\left(\omega\frac{\partial g_j}{\partial\omega} + h_j\tan\alpha\right)\right]_{\omega=s\omega_B}, \tag{5.56}$$

and the power of cyclotron radiation near the gyrofrequency ω_B equals

$$P_{\Omega j} = \frac{e^2\omega_B^2\beta_\perp^2}{8\pi c}\beta_\parallel^2\cos^2\alpha\left[\frac{n_j^3}{1+K_j^2}\left(\omega\frac{\partial g_j}{\partial\omega} + h_j\tan\alpha\right)^2\right]_{\omega=s\omega_B}. \tag{5.57}$$

The inequality $|2n_j\beta_\parallel\cos\alpha| \gg \beta^2$ makes it possible to use the approximate formula (5.51) for the radiated frequency instead of (5.34) in the expression (5.56). Thus, it restricts the validity of the formula (5.57) at $\alpha \approx \pi/2$ and $n_j \approx 0$. In other cases the formula (5.57) is only applicable to sufficiently fast particles with velocity $v_\parallel \gg v_T$, where v_T is the thermal velocity of plasma particles. The last restriction arises due to the fact that at frequencies at which radiation is emitted by electrons with $v_\parallel \sim v_T$ the expression for $P_{\Omega j}$ is drastically changed due to thermal corrections to the polarization coefficients of ordinary and extraordinary waves.

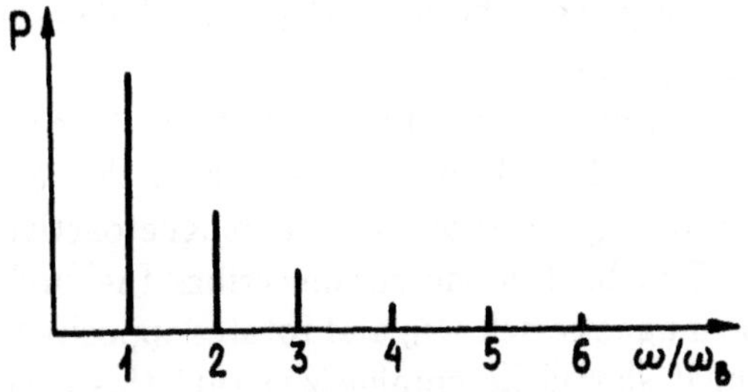

Figure 5.3. Frequency spectrum of cyclotron radiation by an electron in vacuum.

Finally, the following expression describes the radiation power of an electron moving along a helical line in vacuum:

$$P_\omega = \frac{e^2\omega_B^2\beta_\perp^2}{2\pi c}(1+\cos^2\alpha)2^{-2s}s^6(s!)^{-2}(s\beta_\perp \sin\alpha)^{2s-2}. \quad (5.58)$$

It follows from (5.44) at $\beta \ll 1$ and $s\beta_\perp \sin\alpha \ll 1$.

It is clear from (5.58) that only radiation at the cyclotron fundamental $s=1$ is emitted along the magnetic field ($\alpha = 0$), whereas higher harmonics are not. The emitted radiation is circularly polarized, the electric field vector $\boldsymbol{E}$ rotating in the same direction as an electron in a magnetic field. Next, elliptically polarized waves at all the harmonics are emitted at angles $\alpha \neq 0$. At $\alpha = \pi/2$ only linearly polarized waves are emitted by electrons with $\beta_\| = 0$. The vector $\boldsymbol{E}$ in such a radiation lies in the plane of electron gyration.

The expression (5.58) is simple enough to be integrated over solid angles. This yields the following formula for the total power of cyclotron radiation at s-th harmonic in vacuum:

$$P = \int (P_{\Omega 1} + P_{\Omega 2}) d\Omega = \frac{2e^2\omega_B^2\beta_\perp^{2s}}{c}\frac{s^{2s+1}(s+1)}{(2s+1)!}. \quad (5.59)$$

This value decreases with the harmonic number mainly as $\beta_\perp^{2s}$ (see Fig. 5.3). This is explained by the dipole nature of cyclotron radiation.

Comparison of (5.57) and (5.58) shows that the radiation power by an electron in a magnetoactive plasma at $s = 1$ is much less compared to the corresponding vacuum value (roughly, by a factor of $\beta_\|^2$, apart from the contribution by the quantities n_j, T_j, g_j, h_j etc). At the same time, the radiation power at higher harmonics $s \geq 2$ is practically unchanged (cf. (5.53) and (5.58)). So, cyclotron radiation power in a magnetoactive plasma decreases as $\beta_\perp^{2s}$ starting from $s \geq 2$, contrary to $s=1$ for the vacuum case. Due to the presence of the small factor $\beta_\|^2$ in the expression (5.57) for the first harmonic radiation power, the ratio $P_\Omega(s=2)/P_\Omega(s=1) \sim \beta_\perp^2/\beta_\parallel^2$ is of the order of unity at $\beta_\perp \sim \beta_\parallel$.

The effect of depression of cyclotron radiation at the first harmonic has been pointed out in [231]. It originates due to the features of mode polarizations at electron gyrofrequency in a magnetoactive plasma. The electric field vector $\boldsymbol{E}$ in both mode circumscribes the circle in the plane perpendicular to the magnetic field $\boldsymbol{B}_0$ and rotates opposite to the direction of electron gyration. It should be emphasized that this effect has nothing in common with the well-known depression of synchrotron radiation in a medium (plasma) with the refractive index $n < 1$ (see the following section).

Depression of cyclotron radiation at gyrofrequency is absent in vacuum and in case of longitudinal propagation in plasma, since the factor $(1 + g_j)$ in (5.53) is non-zero at $\omega = \omega_B$ for either $v = 0$ or $\alpha = 0$. At the same time, $(1 + g_j)|_{\omega=\omega_B} = 0$ for $v \neq 0$ or $\alpha \neq 0$. This makes non-trivial the limiting transitions to the cases of vacuum or small propagation angles α. Such transitions can only be rigorously done taking into account the thermal motion by using the dielectric tensor in the form (2.99). The corresponding analysis [232] shows that the depression at the first cyclotron harmonic occurs in a sufficiently dense plasma in which the parameter

$$v = \omega_L^2/\omega^2 \gg \beta_T n_j |\cos\alpha / w(Z_j)|. \tag{5.60}$$

Here $w(Z_j)$ is the Cramp function (2.91) of the argument

$$Z_j = (1 - \omega_B/\omega)/\sqrt{2}\beta_T n_j \cos\alpha$$

and $\beta_T = v_T/c$. Alternatively, in the limit of rarefied plasma,

$$\omega_L^2/\omega^2 \ll \beta_T n_j |\cos\alpha / w(Z_j)| \tag{5.61}$$

the total power of cyclotron radiation in both modes is close to the vacuum value (5.58). Radiation power for a distinct mode at $s = 1$ is determined by the formula (5.53) in which the appropriate polarization coefficients at $\omega \approx \omega_B$ in rarefied plasma:

$$g_1 = \cos^2\alpha, \qquad h_1 = -\sin\alpha\cos\alpha, \qquad K_1 = \cos\alpha \tag{5.62}$$

for extraordinary waves and

$$g_2 = -1, \qquad h_2 = \tan\alpha, \qquad K_2 = -1/\cos\alpha \tag{5.63}$$

for the ordinary ones, should be used. With the adopted accuracy of the polarization coefficients g_j and h_j, the power of cyclotron radiation for the extraordinary mode at the first harmonic is described by (5.58), whereas this quantity for the ordinary mode is nullified. The above expressions for the polarization coefficients are valid up to the terms of the order of β_T. The omitted thermal corrections are negligible for the expressions (5.62). At the same time, taking them into account in (5.63) and then substituting the corrected polarization parameters into a more accurate formula than (5.53),[12] one may obtain the non-vanishing, though relatively weak, value for the radiation power into the ordinary mode: $P_{\Omega 2} \sim \beta_T^2 P_{\Omega 1}$. Such a weak emission (as well as absorption and scattering) of the ordinary waves in an extremely rarefied plasma at gyrofrequency is explained by the peculiar mode polarization. In the approximation (5.62), (5.63) the extraordinary mode polarization is identical to the polarization of the cyclotron emission at ω_B by a circularly moving electron in vacuum, whereas the ordinary mode polarization is orthogonal to it. Therefore, such an electron is able to emit (absorb) only extraordinary radiation. Weak emission and absorption of the ordinary component arise due to the electrons with velocities of about v_T which slightly change the polarization of ordinary waves as compared to (5.63). The resulting emission and absorption of the ordinary waves at $s=1$ are different by a factor β_T^2 from extraordinary ones and are comparable to the corresponding effects at the second harmonic $s=2$.

The above expressions for the power of cyclotron radiation by a classical electron in plasma are sufficient for the theory of cyclotron radiation from the Sun, Jupiter, or magnetic white dwarfs. But these are certainly not enough for the theory of cyclotron lines observed in the spectra of X-ray sources (for example, X-ray pulsars Her X-1 or 4U 0115+63) and cosmic gamma-ray bursts (if they are related to neutron stars).

First, strong magnetic fields lead to vacuum polarization and magnetization in the vicinity of neutron stars. The resulting vacuum birefringence affects propagation and emission of electromagnetic waves. This effect has already been considered in section 2.2 in which the criteria (2.187), (2.189) at which mode polarizations and refractive indices are dominated by the effect of vacuum birefringence has been given. Under these conditions the power of cyclotron radiation either at $s=1$ or at $s \geq 2$ is described by the formula (5.53) in which, in accord with mode polarization in birefringent vacuum, $n_j \simeq 1$, $\Gamma_j = 0$ and $K_j = \infty, 0$. The powers of cyclotron radiation into the extraordinary and ordinary mode equal

$$\begin{aligned} P_{\Omega 1} &= \frac{e^2\omega_B^2\beta_\perp^2}{2\pi c}\cos^2\alpha(s\beta_\perp \sin\alpha)^{2s-2}\frac{s^6}{2^{2s}(s!)^2}, \\ P_{\Omega 2} &= \frac{e^2\omega_B^2\beta_\perp^2}{2\pi c}(s\beta_\perp \sin\alpha)^{2s-2}\frac{s^6}{2^{2s}(s!)^2}. \end{aligned} \qquad (5.64)$$

[12] Such a formula can be derived from (5.39) in a similar way as (5.53) but without using the approximation (5.52). The expression for the polarization parameters with thermal corrections are given in section 6.2.

Note that the corresponding total power of cyclotron radiation into both modes is the same as in the case of non-birefringent vacuum (5.58).

Second, motion of an electron in a strong magnetic field becomes quantized, and this may inluence the properties of cyclotron radiation. In fact, the above formulas for $P_{\Omega j}$ are valid in the classical limit, at sufficiently high energy of electron transverse motion: $p_\perp^2/2m_e \gg \hbar\omega_B$. Most likely, this criterion is violated on neutron stars with $B_0 \sim 4 10^{12}$ G and $\hbar\omega_B/m_e c^2 \sim 0.1$. So, electrons with $p_\perp^2/2m_e \sim \hbar\omega_B$ occupy lower Landau levels, and the energy $\hbar\omega$ of the emitted cyclotron photon is comparable to the electron energy. Both the motion and radiation of such electrons acquire quantum features.

In addition to the principal quantum number n introduced by (5.26), quantum states of an electron in a magnetic field are characterized by orbital l and spin ν quantum numbers. These numbers are related as follows: $n = l + (\nu + 1)/2$. Note that radiation at frequency $\omega = \omega_B$ can be emitted due to magneto-dipole spin-flip transition $\nu = 1 \to \nu = -1$. This emission is much weaker than cyclotron radiation which corresponds to electro-dipole transitions at which the orbital number l is changed at constant ν. The following expression [223] characterizes the probability of a spontaneous cyclotron transition of a non-relativistic electron from the initial state (l, ν) to the final one $l' = l - s$, $\nu' = \nu$ resulting in emission of a photon at frequency $\omega \simeq \omega_B$

$$A_l^{l'} = \frac{2\pi^2 e^2 \omega_B}{m_e \omega} \frac{l!}{(l-s)!} \frac{|a_j|}{[(s-1)!]^2} \left[\frac{\hbar\omega^2 \sin^2\alpha}{2\omega_B m_e c^2}\right]^{s-1}$$
$$\times \delta\left\{\omega(1 - \beta_\parallel \cos\alpha) - s\omega_B - \frac{\hbar\omega^2}{2m_e c^2}\sin^2\alpha\right\}. \tag{5.65}$$

This probability is related to radiation in a phase space volume $d^3k/(2\pi)^3$. The last term in the argument of the delta-function accounts for the recoil effect (see (5.33)), and the factor a_j describes the influence of mode polarization. For example, in birefringent vacuum $a_1 = \cos^2\alpha$ and $a_2 = 1$.

Total probability of a cyclotron transition without a spin-flip in birefringent vacuum is obtained by integrating (5.65) over the wavevector space and summing up the results for both modes $j = 1$ and $j = 2$:

$$A_s = \frac{e^2\omega_B^2}{m_e c^3}\left(\frac{B}{2B_{\rm cr}}\right)^{s-1} \frac{(2s)^{2s}}{(2s+1)!} \frac{(s+1)l!}{(l-s)!}. \tag{5.66}$$

Thus, the ratio of radiation powers at the first two harmonics equals [13]

[13] The coefficient 2 appears due to the ratio of photon energies at the first and second harmonics.

$$\frac{P(s=2)}{P(s=1)} = 2\frac{A_2}{A_1} = \frac{24}{5}\frac{B}{B_{cr}}(l-1). \tag{5.67}$$

Note that the formulas (5.65)–(5.67) are valid in the dipole approximation which requires that $lB/B_{cr} \ll 1$.

5.1.5. SYNCHROTRON RADIATION

This and the following sections deal with characteristics of radiation by relativistic particles in a magnetic field. At first, consider synchrotron radiation. By definition, it is emitted by a helically moving relativistic particle with $\vartheta \gg m_e c^2/\mathcal{E}$, where ϑ is the pitch angle between the velocity $\boldsymbol{v}$ and the magnetic field $\boldsymbol{B}_0$.

Synchrotron radiation is very widespread in astrophysical conditions, and thus plays an exclusively important role in astrophysics. Synchrotron mechanism of radio emission works effectively in any source which contains relativistic electrons in a magnetic field: in the Jovian magnetosphere, solar corona, interstellar medium, supernova remnants etc. Moreover, this mechanism has provided an opportunity to relate the observed Galactic diffuse radio emission to cosmic rays in an interstellar medium [233, 234]. Finally, it is the first radio astronomical radiation mechanism which has been succesfully used by classical astrophysics. We mean the interpretation of optical emission by solar flares and continuum spectrum of Crab nebula in terms of synchrotron radiation [235,236]. Later on the synchrotron origin of the X-ray emission by this nebula and, probably, optical and X-ray emission by the Crab pulsar itself was also pointed out.

The angular and frequency spectrum of synchrotron radiation by a charged particle in vacuum can be found using the Lienart-Vichert potentials (see, e.g., [56]):

$$\boldsymbol{A} = \left[\frac{e\boldsymbol{v}}{c(R - \boldsymbol{R}\boldsymbol{v}/c)}\right]_{t-\frac{R}{c}}, \qquad \varphi = \left[\frac{e}{R - \boldsymbol{R}\boldsymbol{v}/c}\right]_{t-\frac{R}{c}}, \tag{5.68}$$

which determine the radiation field by a point charge, $\boldsymbol{E} = -(1/c)\partial\boldsymbol{A}/\partial t - \nabla\varphi$ and $\boldsymbol{H} = \mathrm{curl}\,\boldsymbol{A}$, using the radius-vector $\boldsymbol{R}(t')$ and velocity $\boldsymbol{v}(t')$ at the advanced instant $t' = t - R/c$. The value t' is set by the requirement that the wave emitted by the particle at the moment t' reaches the point $\boldsymbol{R}(t')$ at a time t.

If the direction of $\boldsymbol{R}$ is close to the instantaneous velocity $\boldsymbol{v}(t')$, the denominator in the expression (5.68) for the potentials $\boldsymbol{A}$ and φ can be written in the form

$$R - \frac{\boldsymbol{R}\boldsymbol{v}}{c} = R\left(1 - \frac{v}{c}\cos\theta\right) \approx R\left(1 - \beta + \beta\frac{\theta^2}{2}\right), \tag{5.69}$$

where θ is the angle between $\boldsymbol{R}$ and $\boldsymbol{v}$, and $\beta = v/c$. It is clear from (5.69) that the denominator increases at $\theta \gg \sqrt{2(1-\beta)} \approx \sqrt{1-\beta^2}$, whereas the potentials $\boldsymbol{A}$ and φ sharply decrease at these directions. Therefore, the radiation by a relativistic electron is beamed within the angular interval

$$\theta \lesssim \sqrt{1-\beta^2} = \frac{m_e c^2}{\mathcal{E}} \ll 1 \tag{5.70}$$

near the direction of $\boldsymbol{v}(t')$. In other words, the instantaneous radiation pattern by a relativistic electron has the width $\Delta\theta \approx 2m_e c^2/\mathcal{E}$. More details on the angular pattern of radiation by a relativistic electron can be found in [237].

The duration Δt of a pulse of radiation by an electron which emits into the narrow angular pattern (5.70) determines the characteristic width of the radiation spectrum. In fact, the procedure of calculation of the value Δt for an electron with $\beta_{\|}=0$ orbiting a circle in a magnetic field is analogous to the one considered in section 1.3 in connection with a source orbiting a star. It has been found there that, due to the source motion, the time Δt is less than the time during which the angular pattern of the radiating particle is directed toward the observer. The width 2θ of the source radiation pattern is related to the time Δt by the formula (1.94). In the case considered the rotation period $\tau = 2\pi/\Omega_B = 2\pi/\omega_B\sqrt{1-\beta^2}$ and the angle $\theta \approx \sqrt{1-\beta^2}$. Thus, provided $1-\beta \ll 1$, we obtain

$$\Delta t \sim \frac{1-\beta^2}{\omega_B} \approx \frac{1}{\omega_B}\left(\frac{m_e c^2}{\mathcal{E}}\right)^2. \tag{5.71}$$

The above radiation pulses with duration Δt follow periodically with the period $2\pi/\Omega_B$. Therefore, the fields $\boldsymbol{E}(t)$, $\boldsymbol{H}(t)$ at an observing point are time-periodic functions which can be written as a Fourier series over the harmonics

$$\omega = s\Omega_B = s\omega_B \frac{m_e c^2}{\mathcal{E}}, \tag{5.72}$$

determining the spectrum of synchrotron radiation. It is quite natural that this spectrum is identical to the one (5.34) of magnetic bremsstrahlung by an electron with $\beta_{\|}=0$. Since the duration of an individual pulse equals δt, the large-amplitude harmonics span over the frequencies near

$$\omega_{\max} \sim \frac{1}{\Delta t} \sim \omega_B \left(\frac{\mathcal{E}}{m_e c^2}\right)^2. \tag{5.73}$$

Amplitudes of the harmonics and the corresponding synchrotron radiation power drop both at $\omega \gg \omega_{\max}$ and $\omega \ll \omega_{\max}$. The latter is explained by the absence of the static component in the synchrotron radiation pulses:

$\int_{-\infty}^{\infty} \boldsymbol{E}(t)dt = 0$, which leads to the decrease of the amplitudes of lower harmonics.

Thus, the discrete synchrotron spectrum comprises the harmonics separated by the intervals $\Delta\omega = \omega_B m_e c^2/\mathcal{E}$. The maximum of this spectrum falls onto the frequencies $\omega \sim \omega_{\max}$ given by (5.73) which correspond to the harmonic numbers $s \sim (m_e c^2/\mathcal{E})^3$. The radiation spectrum drops at both sides of $\omega_{\max}$.

The above qualitative picture is well justified by the rigorous calculations given below. The radiation power of an electron at s-th harmonic in vacuum can be found by integrating the Schott formula (5.44) over solid angles:

$$P_s = \int P_\Omega d\Omega =$$
$$\frac{2e^2\omega_B^2}{c}\frac{1-\beta}{\beta}\left[s\beta^2 J'_{2s}(2s\beta) - s^2(1-\beta^2)\int_0^\beta J_{2s}(2s\xi)\,d\xi\right]. \quad (5.74)$$

For a relativistic electron $1-\beta \ll 1$ the Bessel functions at higher harmonics $s \gg 1$ can be substituted with the corresponding asymptotic expansions in terms of the McDonald function $\mathrm{K}_{5/3}$. As a result, the following expression for the value P_s is obtained (see, e.g., [238]):

$$P_s = \frac{\sqrt{3}}{2\pi}\frac{e^2\omega_B^2}{c}\frac{m_e c^2}{\mathcal{E}}\frac{\omega}{\omega_c}\int_{\omega/\omega_c}^{\infty} \mathrm{K}_{5/3}(x)\,dx. \quad (5.75)$$

Here ω_c denotes the "critical" frequency

$$\omega_c = \frac{3}{2}\omega_B\left(\frac{\mathcal{E}}{m_e c^2}\right)^2. \quad (5.76)$$

The quantity

$$P_\omega(\omega,\mathcal{E}) = P_s/\Delta\omega = \frac{\sqrt{3}}{2\pi}\frac{e^2\omega_B}{c}\frac{\omega}{\omega_c}\int_{\omega/\omega_c}^{\infty} \mathrm{K}_{5/3}(x)\,dx, \quad (5.77)$$

where $\Delta\omega = \omega_B m_e c^2/\mathcal{E}$ is the interval between distinct harmonics in the synchrotron spectrum, is frequently used in astrophysics instead of P_s [239]. The introduction of P_ω, the radiation power per unit frequency range, is justified by the fact that the synchrotron spectrum comprised of the very close spectral lines is apparently quasi-homogeneous. The function

$$F(\omega/\omega_c) = (\omega/\omega_c)\int_{\omega/\omega_c}^{\infty} \mathrm{K}_{5/3}(x)\,dx$$

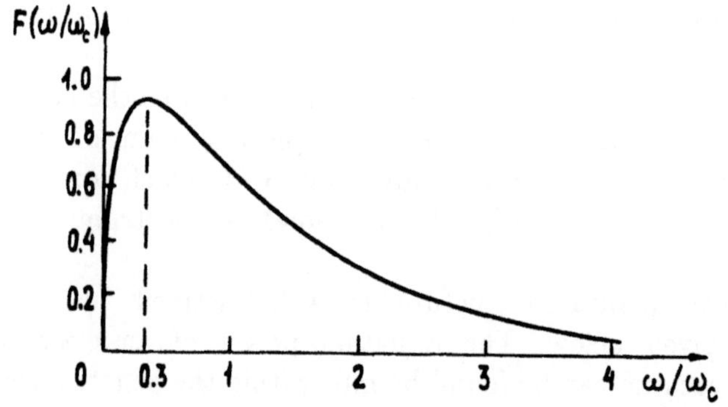

Figure 5.4. The frequency spectrum of synchrotron radiation by an electron in vacuum.

which describes the frequency spectrum of synchrotron radiation is plotted in Fig. 5.4.

At lower frequencies $\omega \ll \omega_c$ the synchrotron radiation power is proportional to $\omega^{1/3}$:

$$P_\omega \approx \frac{\sqrt{3}}{2\pi}\frac{e^2\omega_B}{c} 2^{2/3}\Gamma(2/3)\left(\frac{\omega}{\omega_c}\right)^{1/3}, \tag{5.78}$$

where $\Gamma(x)$ is the gamma-function. The value P_ω reaches the maximal value

$$P_\omega^{\max} \approx \frac{1.6}{2\pi}\frac{e^2\omega_B}{c} \tag{5.79}$$

at the frequency

$$\omega_{\max} \approx 0.3\,\omega_c \approx \frac{1}{2}\omega_B\left(\frac{\mathcal{E}}{m_e c^2}\right)^2 \tag{5.80}$$

and becomes exponentially small in the region $\omega \gg \omega_c$:

$$P_\omega \approx \frac{\sqrt{3}}{2\pi}\frac{e^2\omega_B}{c}\sqrt{\frac{\pi}{2}\frac{\omega}{\omega_c}}\,e^{-\omega/\omega_c}. \tag{5.81}$$

Note that the maximum of the synchrotron spectrum does not depend on particle energy and is entirely determined by the magnetic field strength. The particle energy influences the position $\omega_{\max}$ of the maximum in the frequency spectrum and the frequency width of the maximum, $\Delta\omega$. Both of these values are proportional to $\mathcal{E}^2$.

When analyzing the synchrotron reabsorption (see section 7.4) it is important to know the dependence of $P_\omega(\omega, \mathcal{E})$ on particle energy $\mathcal{E}$. The dependence of P_ω on $\mathcal{E}$ at fixed $\omega =$ const is the function which exhibits the maximum at

$$\mathcal{E}_{\max} \sim m_e c^2\left(\frac{\omega}{\omega_B}\right)^{1/2} \tag{5.82}$$

and drops at both sides of this value (see (5.73)). However, the decrease of $P_\omega(\omega, \mathcal{E})$ with increasing energy is weaker than with frequency (cf. (5.81)). In the region $\mathcal{E} \gg \mathcal{E}_{\max}$ the function $P_\omega(\omega, \mathcal{E})$ decreases more slowly than $\mathcal{E}^{-2}$, so the product $P_\omega \mathcal{E}^2$ increases monotonously with $\mathcal{E}$. Of course, when analyzing the dependence of P_ω on particle energy $\mathcal{E}$ one should always bear in mind that the considered frequency distribution of radiation power is "smoothed" over harmonics. In fact, the synchrotron radiation at a given frequency ω is emitted by particles with distinct energies (see (5.72))

$$\mathcal{E} = s \frac{\omega}{\omega_B} m_e c^2. \tag{5.83}$$

Now we briefly discuss the polarization of synchrotron radiation. The analysis of the synchrotron radiation field shows that radiation emitted along any given direction $\boldsymbol{k}$ at fixed frequency ω is, generally, elliptically polarized [240]. One of the ellipse axes is perpendicular to the magnetic field $\boldsymbol{B}_0$, and another one is parallel to the projection of $\boldsymbol{B}_0$ onto the plane orthogonal to the wave vector $\boldsymbol{k}$. Spectral powers of linearly polarized components of synchrotron radiation with electric field vectors $\boldsymbol{E}^{(1)}$ and $\boldsymbol{E}^{(2)}$ oriented along the polarization ellipse axes are given by

$$\begin{aligned} P_{\omega\Omega}^{(1)} &= \frac{3}{8\pi^3} \frac{e^2 \omega_B}{c} \left(\frac{\mathcal{E}}{m_e c^2}\right)^5 \left(\frac{\omega}{\omega_c}\right)^2 \left[\left(\frac{m_e c^2}{\mathcal{E}}\right)^2 + \Theta^2\right]^2 \mathrm{K}_{2/3}^2(g_\omega), \\ P_{\omega\Omega}^{(2)} &= \frac{3}{8\pi^3} \frac{e^2 \omega_B}{c} \left(\frac{\mathcal{E}}{m_e c^2}\right)^5 \left(\frac{\omega}{\omega_c}\right)^2 \left[\left(\frac{m_e c^2}{\mathcal{E}}\right)^2 + \Theta^2\right] \Theta^2 \mathrm{K}_{1/3}^2(g_\omega), \end{aligned} \tag{5.84}$$

where $\Theta = \pi/2 - \alpha$ is the angle between the wave vector $\boldsymbol{k}$ and the plane of the electron orbit (α is the angle between $\boldsymbol{k}$ and $\boldsymbol{B}_0$), and the McDonald function argument is given by

$$g_\omega = \frac{\omega}{2\omega_c} \left(\frac{\mathcal{E}}{m_e c^2}\right)^3 \left[\left(\frac{m_e c^2}{\mathcal{E}}\right)^2 + \Theta^2\right]^{3/2}. \tag{5.85}$$

The ratio of polarization ellipse axes (the ellipticity) equals

$$p = \sqrt{\frac{P_{\omega\Omega}^{(2)}}{P_{\omega\Omega}^{(1)}}} = \Theta \left[\left(\frac{m_e c^2}{\mathcal{E}}\right)^2 + \Theta^2\right]^{-1/2} \mathrm{K}_{1/3}(g_\omega) \mathrm{K}_{2/3}^{-1}(g_\omega). \tag{5.86}$$

It is seen from this formula that the polarization of synchrotron radiation varies over the angular pattern. In particular, it changes sign in accord with the quantity Θ. The synchrotron radiation emitted along the plane of electron orbit $\Theta = 0$ is linearly polarized. At $\Theta \gg m_e c^2/\mathcal{E}$ the polarization is close to circular, though radiation power is strongly decreased in this angular interval.

Next, intergation of the expressions (5.84) over solid angles yields the overall power of linearly polarized synchrotron components [240]:

$$\begin{aligned} P_{\omega 1} &= \frac{\sqrt{3}\, e^2 \omega_B}{2\pi c} \frac{\omega}{2\omega_c} \left[\int_{\omega/\omega_c}^{\infty} \mathrm{K}_{5/3}(x)\, dx + \mathrm{K}_{2/3}\left(\frac{\omega}{\omega_c}\right) \right], \\ P_{\omega 2} &= \frac{\sqrt{3}\, e^2 \omega_B}{2\pi c} \frac{\omega}{2\omega_c} \left[\int_{\omega/\omega_c}^{\infty} \mathrm{K}_{5/3}(x)\, dx - \mathrm{K}_{2/3}\left(\frac{\omega}{\omega_c}\right) \right]. \end{aligned} \tag{5.87}$$

Of course, the sum $P_{\omega 1} + P_{\omega 2}$ coincides with the total power of synchrotron radiation P_ω determined by (5.77).

A medium, even quite a rarefied one, can drastically change the characteristics of radiation by a relativistic electron. The influence of a medium can be qualitatively demonstrated using the Lienart-Vichert potential for a non-dispersive medium $n(\omega) = \text{const}$ [241]:

$$\boldsymbol{A} = \left[\frac{e\boldsymbol{v}}{c(R - n\boldsymbol{R}\boldsymbol{v}/c)} \right]_{t - \frac{R}{c} n}, \qquad \varphi = \left[\frac{e}{n^2 (R - n\boldsymbol{R}\boldsymbol{v}/c)} \right]_{t - \frac{R}{c} n}. \tag{5.88}$$

These formulas show that the ratio of particle velocity v to the phase velocity of the wave $v_{\mathrm{ph}} = c/n$ is the key parameter determining the radiation in a medium. Since the relativistic particle velocity, though close to c, cannot exceed this value, the influence of the medium is different in the cases $v_{\mathrm{ph}} > c$ $(n < 1)$ and $v_{\mathrm{ph}} < c$ $(n > 1)$.

If $n > 1$, the denominator in the expressions (5.88) gets small at the directions close to the Čerenkov cone $(v/c) n \cos\theta = 1$ at which $R - n\boldsymbol{R}\boldsymbol{v}/c = 0$, not along the particle velocity $\boldsymbol{v}$. Hence, the instantaneous radiation pattern peaks at the Čerenkov cone. The account of the circular motion of the electron leads to the redistribution of synchrotron radiation power. The synchrotron radiation in this case is emitted mainly into the angular interval $\beta n \cos\alpha \lesssim 1$, and the radiation pattern is of complicated, multi-lobe form [228, 229].

On the contrary, the synchrotron radiation in a medium with $n < 1$ but $1 - n \ll 1$ is considerably different. The denominator in (5.88) for small angles θ between $\boldsymbol{R}$ and $\boldsymbol{v}$ takes the form

$$R - \frac{\boldsymbol{R}\boldsymbol{v}}{c} n = R(1 - \beta n \cos\theta) \approx R\left(1 - \beta n + \beta n \frac{\theta}{2}\right)^2. \tag{5.89}$$

Thus, the characteristic width of synchrotron radiation pattern in such a medium is

$$\theta \sim \sqrt{1-\beta n}. \tag{5.90}$$

The influence of the medium is negligible and the above value is practically reduced to (5.70) if

$$1 - n^2 \ll \left(\frac{m_e c^2}{\mathcal{E}}\right)^2. \tag{5.91}$$

Alternatively, the width of radiation pattern

$$\theta \sim \sqrt{1-n} \tag{5.92}$$

is increased and entirely determined by the medium if the opposite condition,

$$1 - n^2 \gg \left(\frac{m_e c^2}{\mathcal{E}}\right)^2, \tag{5.93}$$

is satisfied.

It is clear from (5.88), (5.89) that the effect of the medium can be neglected at any angles θ within the radiation pattern unless the criterion (5.91) is met. Thus, the above criterion allows application of the "vacuum" formulas (5.75)–(5.82) for synchrotron radiation power to the corresponding medium. In the case (5.93), however, these formulas should dramatically change since, contrary to the vacuum case, the denominator of the potentials $\boldsymbol{A}$ and φ is limited by the finite value $R(1-n)$ for $\boldsymbol{R} \| \boldsymbol{v}$, even as $\beta \to 1$. This results in the sharp decrease (depression) of the synchrotron radiation power compared to the vacuum value.

The depression of synchrotron radiation pointed out in [224] appears in a dispersive medium as well. The Lienart-Vichert potentials in the form (5.88) cannot be applied to calculate the radiation fields in this case. However, the criteria (5.91), (5.93) of weak and strong influence of the medium together with the estimates (5.90), (5.92) of the width of synchrotron radiation pattern hold for sufficiently rarefied, isotropic plasma in which $\omega_L^2/\omega^2 \ll 1$ and $\omega_B^2/\omega^2 \ll 1$. When using the above mentioned formulas one should just substitute $n = n_{em}(\omega) = \sqrt{1 - \omega_L^2/\omega^2}$, so the medium affects primarily the radiation at sufficienty low frequencies.[14]

The synchrotron radiation power in an isotropic dispersive medium is determined by the following formula [245–247]:

$$P_\omega(\omega, \mathcal{E}) =$$

[14]The physical interpretation of the depression of synchrotron radiation in plasma has been given in [242–244].

$$\frac{\sqrt{3}\,e^2\omega_B}{2\pi\quad c}\left[1+(1-n^2)\left(\frac{\mathcal{E}}{m_ec^2}\right)^2\right]^{-1/2}\frac{\omega}{\omega_c'}\int_{\omega/\omega_c'}^{\infty}\mathrm{K}_{5/3}(x)\,dx$$

$$\omega_c'=\omega_c\left[1+(1-n^2)\left(\frac{\mathcal{E}}{m_ec^2}\right)^2\right]^{-3/2} \qquad (5.94)$$

To neglect the influence of the medium in the main interval of the frequencies emitted, $\omega\sim\omega_{\max}$, the following condition should be satisfied:

$$\frac{\omega_L^2}{\omega_B^2}\ll\left(\frac{\mathcal{E}}{m_ec^2}\right)^2. \qquad (5.95)$$

Such a medium is only able to depress the synchrotron radiation at frequencies $\omega\ll\omega_{\max}$. In the opposite case,

$$\frac{\omega_L^2}{\omega_B^2}\gg\left(\frac{\mathcal{E}}{m_ec^2}\right)^2, \qquad (5.96)$$

the depression covers the wider frequency range including the main region $\omega\sim\omega_{\max}$ corresponding to vacuum emission. This shifts the maximum of the synchrotron spectrum to the frequency

$$\omega_{\max}'\approx\sqrt{2}\,\omega_L\frac{\mathcal{E}}{m_ec^2}\gg\omega_{\max} \qquad (5.97)$$

and to the sharp decrease of synchrotron energy losses by a relativistic electron [248]:

$$-\left(\frac{d\mathcal{E}}{dt}\right)_{\text{sync}}\approx-\left(\frac{d\mathcal{E}}{dt}\right)_{\text{sync}}^{(0)}\frac{3}{4}\,\delta e^{-\delta}, \qquad (5.98)$$

where

$$-\left(\frac{d\mathcal{E}}{dt}\right)_{\text{sync}}^{(0)}=\frac{2e^2\omega_B^2}{3c}\left(\frac{\mathcal{E}}{m_ec^2}\right)^2 \qquad (5.99)$$

sharacterizes the synchrotron losses in vacuum [56] and the parameter

$$\delta=\sqrt{3}\,\frac{\omega_L}{\omega_B}\frac{m_ec^2}{\mathcal{E}}\gg1.$$

The medium also changes the dependence of P_ω at fixed frequency on electron energy $\mathcal{E}$. Contrary to the vacuum case in which $\mathcal{E}^2P_\omega$ increases monotonously with increasing $\mathcal{E}$, this value in the energy range

$$\mathcal{E}^2\gg\left(m_ec^2\frac{\omega}{\omega_L}\right)^2 \qquad (5.100)$$

derived from the criterion (5.93) reaches maximum at

$$\mathcal{E}'_{\max} \approx m_e c^2 \omega_B \frac{\omega^2}{\omega_L^3} \tag{5.101}$$

and then decreases exponentially with increasing $\mathcal{E}$. But if the value $\mathcal{E}'_{\max}$ does not fall into the interval (5.100), the function $\mathcal{E}^2 P_\omega$ decreases exponentially in the entire energy region affected by the medium [311]. It is this drop of $\mathcal{E}^2 P_\omega$ at high energies that makes possible the synchrotron instability in isotropic plasma.[15] The synchrotron instability is considered in detail in section 7.4.

To estimate the influence of astrophysical plasmas on the observed synchrotron radiation it is convenient to transform in some way the criterion (5.95) by taking into account that the radiation at a frequency ω is mainly emitted by electrons with energy $\mathcal{E} \sim \mathcal{E}_{\max}$, the latter being determined by (5.82). Substituting this value of $\mathcal{E}$ into (5.95) we find the following condition at which the deviation of the refraction index from unity does not influence the characteristics of synchrotron radiation [242,243,247]:

$$\omega \gg \frac{\omega_L^2}{2\omega_B}. \tag{5.102}$$

This condition can also be written in the equivalent form:

$$f = \frac{\omega}{2\pi} \gg \frac{eNc}{B_0} \approx 15\frac{N}{B_0}, \tag{5.103}$$

where f is measured in Hz, N in cm^{-3}, and B_0 in G.

Type IV solar radio emission can possibly be explained as synchrotron radiation though other mechanisms seem to be more probable. The following estimates hold for the corresponding generation regions in the solar corona: $N \sim 10^8$ cm^{-3} and $B_0 \sim 10$ G. So, the influence of the coronal plasma on radiation can only be neglected at frequencies above 150 MHz. Since type IV component is also observed at lower frequencies, the influence of the corona on synchrotron radiation cannot be ignored. The strong depression of synchrotron radiation in the case opposite to (5.102) means that the condition $f > 15N/B_0$ should be realized in the observed sources of intense type IV radio emission [248].

The role of the interstellar medium in the Galaxy on the synchrotron emission by relativistic electrons is, as a rule, insignificant. For example, at frequencies $f \sim 10$ MHz the condition (5.102) for a field $B_0 \sim 10^{-6}$ G can only be violated at $N > 1$ cm^{-3}. However, the electron density in

[15] See [249] on the influence of magnetoactive plasma on synchrotron radiation.

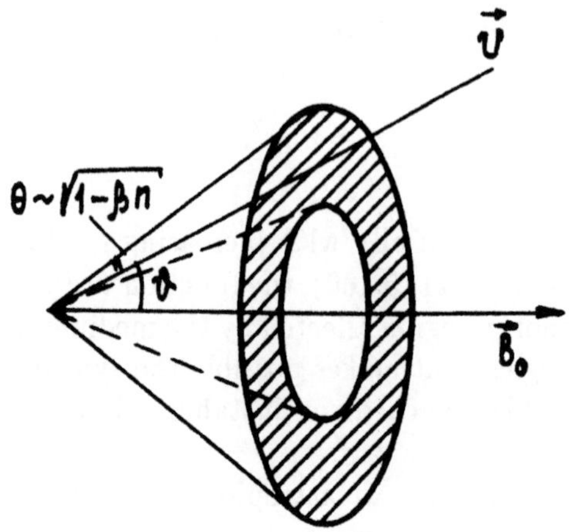

Figure 5.5. Angular pattern of synchrotron radiation. Radiation is emitted mainly into the hatched hollow cone.

an interstellar medium is 1-2 orders of magnitude less than this value. Thus, the depression of synchrotron radiation can appear only at lower frequencies of about hundreds kilohertz or inside discrete sources (for example, supernova remnants) with larger electron density. The effect considered leads to the strong depression in the low-frequency part of the observed spectrum.

The above analysis of synchrotron emission has been done for the case of plane (pure circular) motion of an electron when the velocity component along the magnetic field is absent ($\beta_{\|}=0$) and the electron trajectory is a circumference of the radius $r_B=c\Omega_B$. Synchrotron radiation by such a particle is emitted primarily into the narrow angular interval $\theta\lesssim\sqrt{1-\beta n}$ near the plane of motion (see (5.70), (5.90)). But if $\beta_{\|}\neq 0$, the resulting synchrotron radiation is emitted into the hollow cone near the one circumscribed by the velocity vector of the electron moving in the magnetic field (Fig. 5.5). The angular width $\Delta\theta$ of this hollow cone is again determined by the formulas (5.70), (5.90). If the pitch-angle of the electron ϑ is much larger than the above interval $\Delta\theta\sim m_ec^2/\mathcal{E}$ near $\boldsymbol{v}$ into which synchrotron radiation is emitted, all the formulas for synchrotron radiation (5.77)–(5.102) except (5.83) are made valid for the case $\beta_{\|}\neq 0$ by substituting $\omega_B\sin\vartheta$ instead of ω_B (in other words, the value $B_0\sin\vartheta$ of the projection of magnetic field onto the plane perpendicular to $\boldsymbol{v}$ is used instead of B_0). In the opposite case, $\vartheta\ll m_ec^2/\mathcal{E}$, the electron moves along the strongly stretched helical line with the transverse velocity $\beta_\perp\ll(m_ec^2/\mathcal{E})\beta_{\|}$. As a result, the magnetic bremsstrahlung in vacuum exhibits dipole features.

The power of relativistic dipole radiation per unit solid angle can readily be obtained from (5.39) by performing the limiting transition $v=\omega_{\rm L}^2/\omega^2\to 0$

and using the expansions of the Bessel function $J_s(\zeta)$ and its derivative $J'_s(\zeta)$ in the powers of the argument $\zeta = s\beta_\perp \sin\alpha(1 - \beta_\| \cos\alpha)^{-1}$. The function $\zeta(\alpha)$ reaches the maximal value $s\beta_\perp \mathcal{E}/m_e c^2$ at $\cos\alpha = \beta_\|$. Thus, $\zeta \ll 1$ for any α at harmonics $s \ll m_e c^2/\mathcal{E}\beta_\perp$. Keeping only the leading terms in the expansions of Bessel function and its derivative at these harmonics, we arrive at the following formula:

$$P_\Omega = P_{\Omega 1} + P_{\Omega 2} = \frac{e^2\omega^2}{2\pi c}\frac{\zeta^{2s-2}}{2^{2s}(s!)^2} \times \frac{s^2\beta_\perp^2(1+\cos^2\alpha) + \zeta^2\beta_\|^2 \sin^2\alpha - 2s\beta_\|\beta_\perp\zeta\sin\alpha\cos\alpha}{1-\beta_\|\cos\alpha}. \quad (5.104)$$

Since $\zeta \ll 1$, the radiation power decreases steeply (approximately as ζ^{2s-2}) with increasing harmonic number s. The angular pattern of relativistic dipole radiation is mainly determined by the factor $1 - \beta_\| \cos\alpha$ in the denominator of the expression (5.104). The radiation pattern is stretched along the magnetic field and has the characteristic width $\alpha \sim m_e c^2/\mathcal{E}$, whereas the pitch angle ϑ of the radiating particle is much less than this value.

Relativistic dipole radiation is realized in a medium if the parameter ζ determined by the formula (5.40) is much less than unity. The corresponding formula for $P_{\Omega j}$ in this case can also be obtained from (5.39) by keeping the first terms of the Bessel function power series in it.[16]

As has been shown above, the general formula (5.39) for magnetic bremsstrahlung is essentially simplified in the limiting cases of cyclotron radiation by a weakly relativistic electron and synchrotron radiation by an ultra-relativistic one. However, in the intermediate case of gyrosynchrotron emission by mildly relativistic particles with energies $\mathcal{E} \sim m_e c^2$ the simplifications are only possible in the expressions of gyrosynchrotron emissivity and the related absorption coefficient. The corresponding formulas are given in sections 5.2 and 6.2, respectively. The results of numerical calculations of angular and frequency spectra of gyrosynchrotron radiation by single particles may be found in [7, 251].

5.1.6. CURVATURE RADIATION

The above sections deal with the magnetic bremsstrahlung of a particle in a homogeneous magnetic field. In an inhomogeneous field the additional

[16] Various aspects of the theory of synchrotron radiation which are not touched upon in this section but nevertheless important for astrophysical applications are discussed in [66, 201, 208, 247]. The features of radiation by charged particles are also discussed in [208, 209].

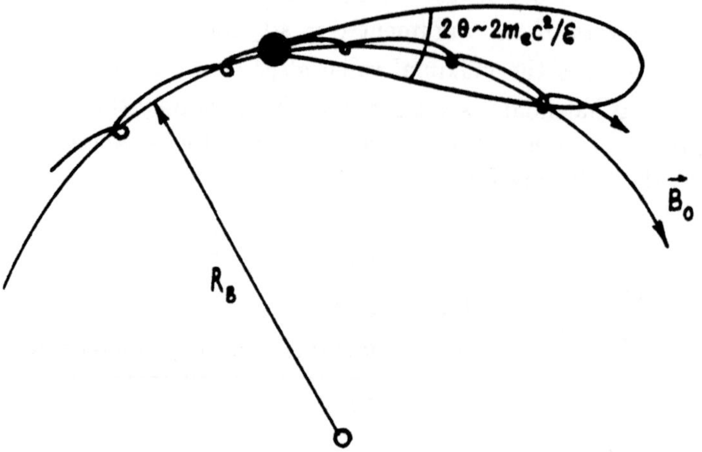

Figure 5.6. The instantaneous angular pattern of curvature radiation (R_B is the curvature radius of the magnetic field line).

acceleration is experienced by a charged particle which is caused by its motion along the curved field lines. This additional acceleration gives rise to the so-called curvature radiation. Of course, a charged particle can not move strictly along a magnetic field line, since the Lorenz force $(e/c)[\boldsymbol{v}\boldsymbol{B}_0]$ is nullified in this case. In fact, in addition to the drift velocity component along $\boldsymbol{B}_0$, there is also the one which is perpendicular to the plane tangential to the field line. It is this latter component that provides the Lorenz force necessary to fit the particle trajectory to the field line [252].

Curvature radiation by relativistic particles is quite similar to the synchrotron one. In particular, the angular pattern of this radiation has the same width, $\theta \sim m_e c^2/\mathcal{E}$, as the synchrotron angular pattern (Fig. 5.6). The differences arise from the fact that in case of magnetic curvature radiation the role of the gyration radius $r_B = v_\perp/\Omega_B$ of an electron in a magnetic field is played by the curvature radius R_B of a field line. Therefore, the observed pulse duration Δt is obtained from (5.71) by the ansatz $r_B = \mathcal{E}/\omega_B m_e c^2 \to R_B$:

$$\Delta t \approx \frac{R_B}{c}\left(\frac{m_e c^2}{\mathcal{E}}\right)^3 . \tag{5.105}$$

This yields the following estimates for the characteristic width of frequency spectrum, $\Delta\omega$, and the characteristic frequency, $\omega_{\max}$, of the curvature radiation:

$$\Delta\omega \sim \omega_{\max} \sim \frac{1}{\Delta t} \simeq \frac{c}{R_B}\left(\frac{\mathcal{E}}{m_e c^2}\right)^3 . \tag{5.106}$$

The same ansatz $r_B \to R_B$ allows to get the total energy loss due to curvature radiation using the corresponding formula (5.104) for synchrotron radiation:

$$-\left(\frac{d\mathcal{E}}{dt}\right)^{(0)}_{\text{curv}} = \frac{2e^2c}{3R_B^2}\left(\frac{\mathcal{E}}{m_ec^2}\right)^4 \tag{5.107}$$

Note the more rapid growth of $\omega_{\max}$ and energy loss with increasing particle energy $\mathcal{E}$ for magnetic curvature compared to synchrotron radiation. This is explained by the fact that the curvature radius of the trajectory of a particle moving along a magnetic field line does not depend on the energy $\mathcal{E}$ and is just equal to R_B. On the contrary, the electron gyroradius r_B in a magnetic field increases with increasing $\mathcal{E}$, so the trajectory is straightened.

Since motion of a particle along a magnetic field line is aperiodic, the frequency spectrum of the curvature radiation is continuous: the discrete lines typical for synchrotron spectrum overlap each other. However, the envelopes of the frequency spectra of both curvature and synchrotron radiation are similar.

If the curvature radius of the magnetic field is large enough: $R_B \gg r_B$, the characteristic frequency of magnetic curvature radiation is much less than of synchrotron radiation at the same particle energy. This makes possible the unified explanation of optical and radio emission of the Crab pulsar as radiation by electrons moving in a magnetic field near the light cylinder at a distance $R \sim 10^8$ cm from the neutron star [253]. For example, the characteristic frequency of curvature radiation by electrons with $\mathcal{E}/m_ec^2 \sim 10^2$ moving along field lines with curvature radius $R_B \sim 10^8$ cm falls into the radio band: $\omega_{\max} \sim 3 \cdot 10^8\,\text{s}^{-1}$. On the other hand, the same particles moving in a magnetic field $B_0 \sim 10^4 - 10^5$ G at larger pitch angles $\vartheta \sim 1$ give rise to synchrotron radiation with characteristic frequency (see (5.80)) $\omega \sim \omega_{\max} \sim 10^{15} - 10^{16}\,\text{s}^{-1}$ which corresponds to the optical band.

5.2. Plasma emissivity

Starting from this section, the analysis of radiation by astrophysical plasma is more or less connected with two values, emissivity a_ω and absorption (or scattering) coefficient μ_j which enter the radiation transfer equation and its solutions (see section 4.1). This seems to be quite natural since, on the one hand, these values finally determine the observed radiation intensity of an astrophysical source and, on the other hand, these radiation transfer coefficients themselves are determined by the nature of elementary radiative processes in the source and on plasma particle momentum distribution. Thus, the analysis of these coefficients allows one to judge on the relative role and importance of various emission processes in the formation of radiation in astrophysical plasmas, on the possibility for a maser emission mechanism to be realized etc.

The plasma emissivity a_ω is considered below. By definition, a_ω is the spectral power of radiation emitted by the unit volume of a medium into a unit solid angle. In this section the emissivities due to various emission processes: bremsstrahlung, Čerenkov, cyclotron, gyrosynchrotron, and synchrotron, are found. The analysis of contribution by scattering processes to a_ω is postponed to chapter 8. To calculate the emissivity a_ω, one should use the expressions for the spectral power of radiation by a single particle, $P_{\omega\Omega}$, derived in section 5.1. Then the desired a_ω can be found by summing up the contribution of all the plasma particles. Of course, in doing this one should account for their momentum distribution. The emissivity of a particle system (a medium) with a given distribution function $f_0(\boldsymbol{p}, \boldsymbol{r}, t)$ at a point $\boldsymbol{r}$ and instant t equals

$$a_\omega = \int P_{\omega\Omega} f_0(\boldsymbol{p}, \boldsymbol{r}, t) d^3p. \tag{5.108}$$

5.2.1. BREMSSTRAHLUNG EMISSIVITY

The formula (5.108) cannot be directly used to calculate a_ω for bremsstrahlung, since, instead of the required quantity $P_{\omega\Omega}$, the expressions for another value κ_ω, spectral power of radiation by a unit particle flux colliding with a single particle, have only been given in section 5.1. Since the flux of electrons with momenta $\boldsymbol{p}$ within the volume element d^3p in the momentum space equals $v f_0(\boldsymbol{p}) d^3p$, the desired emissivity is related to κ_ω in the following way:

$$\int a_\omega d\Omega = N_+ \int \kappa_\omega v f_0(\boldsymbol{p}) d^3p, \tag{5.109}$$

where N_+ is the density of particles which the considered electrons collide with. Since κ_ω comprises radiation from a unit frequency interval emitted to all directions, the integral over solid angles enters the left-hand side of (5.109). If particle distribution $f_0(\boldsymbol{p})$ is isotropic, the resulting emissivity is isotropic, too. Thus,

$$a_\omega = N_+ \int_{p_{\min}}^{\infty} \kappa_\omega v f_0(p) p^2 dp. \tag{5.110}$$

The lower limit $p_{\min}$ is set by the condition $\mathcal{E}_{\mathrm{kin}} = \hbar\omega$, since an electron with the kinetic energy $\mathcal{E}_{\mathrm{kin}} < \hbar\omega$ is unable to emit the quantum $\hbar\omega$. For non-relativistic particles the energy $\mathcal{E}_{\mathrm{kin}} = p^2/2m_e$, so $p_{\min} = \sqrt{2\hbar\omega m_e}$. To find the emissivity of a particle system with anisotropic distribution, one should know the angular dependence of bremsstrahlung radiation power (see e.g. [56,210,211]).

If non-relativistic electrons have Maxwell momentum distribution, the function $f_0(p) = m_e^{-3} F_0(v)$, where the velocity distribution $F_0(v)$ is determined by the formula (2.95). The emissivity could then be found by substituting (2.95) and the expressions (5.7), (5.8), (5.11) for κ_ω into (5.110). The following formulas result from this procedure.

In a non-relativistic plasma with the temperature[17]

$$T \ll \frac{\pi e^4 Z^2 m_e}{2\kappa\hbar^2} \sim 4 \cdot 10^5 Z^2 \ \mathrm{K} \tag{5.111}$$

the emissivity at frequencies

$$\omega \ll \frac{(8\kappa T)^{3/2}}{\pi^{3/2} m_e^{1/2} e^2 Z} \tag{5.112}$$

is given by ([71], S 37)

$$a_\omega \approx \frac{4}{3\pi}\sqrt{\frac{2}{\pi}}\frac{e^6 Z^2 N N_+}{(\kappa T)^{1/2} m_e^{3/2} c^3} \ln\left[\frac{2(2\kappa T)^{3/2}}{\delta^{5/2} e^2 Z m_e^{1/2}\omega}\right], \tag{5.113}$$

where $\delta = e^C$ and $C = 0.577$ is the Euler constant. The above expression is of interest mainly for the radio emission band where the inequality (5.112) is valid over the wide temperature range relevant to astrophysical plasmas (for example, for $T \gtrsim 10^2$ K at $\omega \ll 10^{11}\ \mathrm{s}^{-1}$). Next, at frequencies

$$\frac{(8\kappa T)^{3/2}}{\pi^{3/2} m_e^{1/2} e^2 Z} \ll \omega \ll \frac{4c\kappa T}{\pi e^2 Z} \tag{5.114}$$

the corresponding emissivity

$$a_\omega \approx \frac{4}{3}\sqrt{\frac{2}{3\pi}}\frac{e^6 Z^2 N N_+}{(\kappa T)^{1/2} m_e^{3/2} c^3}. \tag{5.115}$$

In plasma with temperature

$$T \gg \pi e^4 Z^2 m_e / 2\kappa\hbar^2 \tag{5.116}$$

we have

$$a_\omega \approx \frac{4}{3\pi}\sqrt{\frac{2}{\pi}}\frac{e^6 Z^2 N N_+}{(\kappa T)^{1/2} m_e^{3/2} c^3} \ln\left[\frac{4\kappa T}{\delta\hbar\omega}\right]. \tag{5.117}$$

[17] Here and thereafter the temperature constraints are obtained from the inequalities (5.3), (5.4), (5.7) and (5.8) by using the "arithmetic mean" velocity of electrons defined by the relation $\bar{v} = \int v F_0(v) d^3 v = \sqrt{8\kappa T/\pi m_e}$.

It should be stressed that the above formulas for bremsstrahlung emissivity account for the total emission into two orthogonal polarizations: $a_\omega = a_{\omega 1} + a_{\omega 2}$. Since the bremsstrahlung by an equilibrium particle system in the absence of a magnetic field $\boldsymbol{B}_0$ is unpolarized, the emissivity per single mode is two times less than a_ω. Note also that the formulas (5.113), (5.115), and (5.117) are valid in the Rayleigh-Jeans limit $\hbar\omega \ll \kappa T$ where $p_{\min} = m_e v_{\min} \ll m v_T$. This allow placing $p_{\min} = 0$ when integrating in (5.110). If $\hbar\omega \gtrsim \kappa T$, the additional term $g(\omega, T)e^{-\hbar\omega/\kappa T}$ appears in the expression for the emissivity a_ω (see e.g. [208]). The exponent is originated due to the difference of $p_{\min}$ from zero, and the so-called Gaunt factor $g(\omega, T)$ accounts for the quantum effects accompanying the emission by particles with $p \sim p_{\min}$ which significantly contribute to a_ω at $\hbar\omega \gtrsim \kappa T$ [254]. Though the Gaunt factor is usually of the order of unity, it should be taken into account in detailed calculations of bremsstrahlung.

In the Born approximation

$$v_T \gg \frac{2Ze^2}{\hbar} \tag{5.118}$$

the bremsstrahlung emissivity

$$a_\omega = \frac{4}{3}\frac{e^6 Z^2 N N_+}{m_e^2 c^3}\left(\frac{m_e}{2\pi\kappa T}\right)^{1/2} g(\omega, T)e^{-\hbar\omega/\kappa T}, \tag{5.119}$$

where the Gaunt factor

$$g(\omega, T) = \frac{\sqrt{3}}{\pi}\,\mathrm{K}_0(\hbar\omega/2\kappa T)e^{\hbar\omega/2\kappa T}. \tag{5.120}$$

Here K_0 is the modified Hankel function (or McDonald function) of the zero order. In the limit $\hbar\omega/2\kappa T \ll 1$ the expression (5.120) is reduced to

$$g(\omega, T) = \frac{\sqrt{3}}{\pi}\,e^{\hbar\omega/2\kappa T}\ln\left(\frac{4\kappa T}{\delta\hbar\omega}\right) \tag{5.121}$$

and the formula (5.119), to the emissivity (5.117). At higher frequencies $\hbar\omega/2\kappa T \gg 1$ the Gaunt factor equals

$$g(\omega, T) = \sqrt{\frac{3}{\pi}\frac{\kappa T}{\hbar\omega}}, \tag{5.122}$$

so the corresponding emissivity

$$a_\omega = \frac{4}{\sqrt{6}\pi}\frac{e^6 Z^2 N N_+}{m_e^2 c^4}\left(\frac{m_e c^2}{\hbar\omega}\right)^{1/2} e^{-\hbar\omega/\kappa T} \tag{5.123}$$

(see, e.g., [85]). This formula is widely used for calculation of X-ray emission from non-relativistic plasmas in the solar and stellar coronas and other X-ray sources.

The emissivity of relativistic plasmas is not given here. This is, first and foremost, due to uncertainty in the choice of the appropriate particle distribution function $f_0(\boldsymbol{p})$. It seems quite improbable that this function is equilibrium in astrophysical sources (for example, in neutron star magnetospheres). If necessary, the emissivity a_ω for any other distribution can be readily obtained using the formulas (5.13), (5.110) and the expressions for bremsstrahlung radiation power given in [1,208,255] where the problem of X-ray and gamma-ray emission by particles in astrophysical conditions is discussed.

5.2.2. ČERENKOV EMISSIVITY

Bearing in mind further applications to the problem of plasma wave absorption by equilibrium plasma (section 6.1) and amplification (Čerenkov instability) of these waves by particle beams (section 7.2), the emissivity a_ω of a particle system with the velocity distribution function

$$F_S(\boldsymbol{v}) = N_S\left(\frac{m_e}{2\pi\kappa T_S}\right)^{3/2}\exp[-m_e(\boldsymbol{v}-\boldsymbol{v}_S)^2/2\kappa T_S] \qquad (5.124)$$

is derived in this section. This function describes a particle beam with Maxwell distribution in the co-moving frame which moves at the bulk velocity $\boldsymbol{v}_S$. The function (5.124) is reduced to the electron distribution in equilibrium plasma (2.95) if $v_S=0$, $T_S=T$, and $N_S=N$. Let the particle beam penetrate isotropic plasma with the dielectric constant $\epsilon=1-\omega_L^2/\omega^2$. Any particle in this case emits Čerenkov radiation at frequencies $\omega=\boldsymbol{kv}$, the corresponding spectral power $P_{\omega\Omega}$ being determined by the expression (5.22). Substituting (5.22) and (5.124) into (5.108) we get (see the details in [217]):

$$a_\omega = \frac{e^2\omega N_S}{(2\pi)^{3/2}v_{T_S}\epsilon(\omega)}\exp\left[-\frac{m_e}{2\kappa T_S}(v_{ph}-v_S\cos\theta)^2\right], \qquad (5.125)$$

where $v_{T_S}=\sqrt{\kappa T_S/m_e}$, $v_{ph}=c/n_{pl}(\omega)$, and $n_{pl}(\omega)$ is determined by the formula (2.116). According to (5.125), phase velocities of the waves emitted by the beam at an angle θ to the velocity $\boldsymbol{v}_S$ lie within the interval $\Delta v_{ph}\sim 2\sqrt{2}v_{T_S}$ near $v_{ph}=v_S\cos\theta$. Čerenkov emissivity by electrons of equilibrium plasma without the beam,

$$a_\omega = \frac{e^2\omega N}{(2\pi)^{3/2}v_T\epsilon(\omega)}\exp\left[-\frac{m_e v_{ph}^2}{2\kappa T}\right], \qquad (5.126)$$

can be easily found from (5.125) by setting $v_S = 0$, $T_S = T$, $v_{T_S} = v_T$ and $N_S = N$.

Both formulas for a_ω have been found assuming that Čerenkov radiation is emitted in a non-absorbing medium, so they are valid unless $v_{ph} \gg v_T = \sqrt{\kappa T/m_e}$ and the Landau damping is weak (see section 6.1). Note also that the expression (5.22) for the spectral power $P_{\omega\Omega}$ does not account for the effect of beam particles on the dielectric constant of plasma. Therefore, for the formula (5.125) to be valid it is also necessary that the electron density in the beam is much less than that in the background plasma.

5.2.3. CYCLOTRON EMISSIVITY

To derive the cyclotron emissivity from the general formula (5.108), one should use the expression for the spectral power of cyclotron radiation $P_{\omega\Omega}$ at a given harmonic s. Since the direction of the emission (the angle α) and the emitted frequency are related through (5.51), the required value $P_{\omega\Omega}$ should comprise the delta-function $\delta[\omega - s\omega_B(1 + n_j\beta_\parallel \cos\alpha)]$, similar to the expression for $P_{\omega\Omega}$ in the case of Čerenkov radiation (5.22). However, the formulas (5.53), (5.57), (5.58) for another value, the cyclotron radiation power per mode into unit solid angle $P_{\Omega j}$, are given in section 5.1. These formulas can be used in the following way to find the desired emissivity.

Recalling that $P_\Omega = \int P_{\omega\Omega} d\omega$, we integrate both sides of (5.108) over frequency:

$$\int a_{\omega j} d\omega = \int P_{\Omega j} f_0(\mathbf{p}) d^3p.$$

Note, moreover, that a stationary electron distribution in a magnetic field should be independent on the azimuthal angle in the plane perpendicular to the field direction. Therefore, $f_0(\mathbf{p}) = f_0(p_\perp, p_\parallel)$ and $d^3p = 2\pi p_\perp^2 dp_\perp dp_\parallel$, where $p_\perp$ and $p_\parallel$ are projections of electron momentum onto this plane and the magnetic field direction, respectively. Next, the frequency ω of cyclotron radiation at s-th harmonic is related to $p_\perp$ and $p_\parallel$ by the Doppler formula (5.34). Thus, the ansatz $p_\parallel \to \omega$ leads to the following expression:

$$\int a_{\omega j} d\omega = 2\pi \int_\omega \int_0^\infty P_{\Omega j} f_0(p_\perp, p_\parallel) \left|\frac{\partial p_\parallel}{\partial \omega}\right| p_\perp dp_\perp d\omega. \tag{5.127}$$

Then the desired cyclotron emissivity into j-th mode is given by

$$a_{\omega j} = 2\pi \int_0^\infty P_{\Omega j} f_0(p_\perp, p_\parallel) \left|\frac{\partial p_\parallel}{\partial \omega}\right| p_\perp dp_\perp. \tag{5.128}$$

The approximate expression

$$\omega = s\omega_B \left(1 + \frac{n_j p_\parallel}{m_e c} \cos\alpha\right) \tag{5.129}$$

which follows from the Doppler condition (5.34) ensures the univalent relation $p_{\parallel} = p_{\parallel}(\omega)$ for cyclotron radiation. The formula (5.129) then yields

$$\left|\frac{\partial p_{\parallel}}{\partial \omega}\right| \approx \frac{m_e c}{\omega n_j |\cos\alpha|}. \qquad (5.130)$$

Next, the momentum distribution of equilibrium (Maxwellian) plasma is given by (cf. (2.87))

$$f_0(p_\perp, p_\parallel) = \frac{N_0}{(2\pi\kappa T m_e)^{3/2}} \exp\left[-\frac{p_\perp^2 + p_\parallel^2}{2m_e\kappa T}\right]. \qquad (5.131)$$

Substituting (5.130), (5.131) and (5.53) into the expression (5.128) we finally derive the cyclotron emissivity of an equilibrium plasma into mode j at s-th harmonic ($s \geq 2$) [230]:

$$a_{\omega j} \approx \frac{s^{2s}}{2^s (2\pi)^{3/2} s!} \frac{e^2 \omega N_0}{c^{2s}} v_T^{2s-1} \frac{\sin^{2s-2}\alpha}{|\cos\alpha|} \times [n_j^{s-1} T_j (1+g_j)]^2_{\omega = s\omega_B} \exp\left[-\frac{(\omega - s\omega_B)^2}{2k_\parallel^2 v_T^2}\right]. \qquad (5.132)$$

Here $v_T = \sqrt{\kappa T/m_e}$ and $k_\parallel = (\omega/c) n_j \cos\alpha$. The corresponding expression for radiation at the first harmonic is found by the substitution of $P_{\Omega j}$ in the form (5.57) into (5.128) [230]:

$$a_{\omega j} \approx \frac{1}{2(2\pi)^{3/2}} \frac{e^2 \omega N_0 v_T}{c^2 |\cos\alpha|} \left(\frac{\omega}{\omega_B} - 1\right)^2 \times \left[T_j \left(\omega \frac{\partial g_j}{\partial \omega} + h_j \tan\alpha\right)\right]^2_{\omega = \omega_B} \exp\left[-\frac{(\omega - \omega_B)^2}{2k_\parallel^2 v_T^2}\right]. \qquad (5.133)$$

The validity criteria of the above formulas for the cyclotron emissivity $a_{\omega j}$ are obtained by substituting β, $\beta_\parallel$ and $\beta_\perp$ with $\beta_T = v_T/c$ in the expressions (5.48)–(5.50) which ensure the validity of the formulas for cyclotron emission from section 5.1 used to derive (5.132), (5.133):

$$\beta_T^2 \ll 1, \qquad n_j \beta_T \ll 1, \qquad \beta_T \left|\omega \frac{\partial n_j}{\partial \omega}\right| \ll 1,$$
$$s n_j \beta_T |\sin\alpha| \ll 1, \qquad 2n_j |\cos\alpha| \gg \beta_T. \qquad (5.134)$$

The last criterium, in particular, indicates that the above expressions for $a_{\omega j}$ are invalid at $\alpha \approx \pi/2$. In addition, the formula (5.133) is only applicable at frequencies

$$(\omega - s\omega_B)^2 \gg 2k_\parallel^2 v_T^2 \qquad (5.135)$$

in the wings of the cyclotron line. The latter is explained by the fact that the formula (5.57) for $P_{\Omega j}$ is valid for particles with velocity $v_{\parallel} \gg v_T$ which radiate at frequencies (5.135). However, the entire profile of the cyclotron line can be obtained using the Kirchhoff law and the correct expression for the cyclotron absorption coeficient derived by the method of kinetic equation (see section 6.2).

Finally, for low-density plasma ($v \ll 1$) we obtain, according to (5.54), the following expression:

$$a_{\omega j} \approx \frac{s^{2s}}{2^s (2\pi)^{3/2} s!} \frac{e^2 \omega N_0 v_T^{2s-1}}{c^{2s}} \frac{\sin^{2s-2}\alpha}{|\cos\alpha|} \times \frac{(1 + K_j \cos\alpha)^2}{1 + K_j^2} \exp[-(\omega - s\omega_B)^2 / 2k_{\parallel}^2 v_T^2], \quad (5.136)$$

where K_j are defined by (5.55). Using (5.136) one can easily obtain the following total cyclotron emissivity:

$$a_\omega = a_{\omega 1} + a_{\omega 2} \approx \frac{s^{2s+1}}{2^{s-1}(2\pi)^{3/2}} \frac{e^2 \omega N_0 v_T^{2s-1}}{c^{2s}} \times \frac{\sin^{2s-2}\alpha}{|\cos\alpha|} (1 + \cos^2\alpha) \exp[-(\omega - s\omega_B)^2 / 2k_{\parallel}^2 v_T^2]. \quad (5.137)$$

When deriving the last formula the relation $K_1 K_2 = -1$ has been taken into account. The formulas (5.136), (5.137) are valid for $s \geq 2$, the limiting case of rarefied plasma included. The formula (5.136) can also be used to calculate the cyclotron emissivity for extraordinary waves in rarefied plasma at the first harmonic. The corresponding expression

$$a_{\omega 1} = \frac{1}{(2\pi)^3} \sqrt{\frac{\pi}{8}} \omega_B \omega_{\mathrm{L}}^2 \frac{m_e}{c} \beta_T \frac{1 + \cos^2\alpha}{|\cos\alpha|} \exp[-(\omega - \omega_B)^2 / 2k_{\parallel}^2 v_T^2] \quad (5.138)$$

is obtained by taking into account the different mode polarization in rarefied plasma in which $K_1 = \cos\alpha$ at $\omega = \omega_B$ (see (5.62)). The case of the ordinary mode is more complicated. According to (5.63), the coefficient $K_2 = 1/\cos\alpha$ for this mode at the first harmonic, so the emissivity $a_{\omega 2}$ found with (5.136) is equal to zero. This means that the value $a_{\omega 2}$ at $s = 1$ can only be derived by a more rigorous technique which accounts for the corrections to the polarization coefficients due to particle thermal motion in plasma. The resulting emissivity looks like [183]

$$a_{\omega 2} = \frac{\omega_B \omega_{\mathrm{L}}^2 m_e}{(2\pi)^{7/2} c} \beta_T^3 \times \frac{\sin^4\alpha \cos^4\alpha (3 + \tan^2\alpha)^2}{|W(Z_2)|^2 |\cos\alpha| (1 + \cos^2\alpha)^3} \exp[-(\omega - \omega_B)^2 / 2k_{\parallel}^2 v_T^2] \quad (5.139)$$

where $W(Z_2)$ is the Kramp function (2.91) of the argument $Z_2 = (\omega - \omega_B)/\sqrt{2}\omega\beta_T \cos\alpha$.

If the magnetic field is strong enough that the mode polarization is determined by the effect of vacuum birefringence, cyclotron emissivity at arbitrary harmonic is found from (5.136) by substituting the appropriate polarization coefficients $K_1 = \infty$ and $K_2 = 0$ (see section 2.2):

$$a_{\omega 1} = \frac{s^{2s}}{2^s(2\pi)^{3/2}s!}\frac{e^2\omega N_0 v_T^{2s-1}}{c^{2s}} \times \sin^{2s-2}\alpha|\cos\alpha|\exp[-(\omega - s\omega_B)^2/2k_{\|}^2v_T^2], \quad (5.140)$$

$$a_{\omega 2} = \frac{s^{2s}}{2^s(2\pi)^{3/2}s!}\frac{e^2\omega N_0 v_T^{2s-1}}{c^{2s}} \times \frac{\sin^{2s-2}\alpha}{|\cos\alpha|}\exp[-(\omega - s\omega_B)^2/2k_{\|}^2v_T^2], \quad (5.141)$$

Of course, the same result can be obtained if the expression for cyclotron radiation power in birefringent vacuum (5.64) is substituted into the formulas (5.128), (5.130).

For the analysis of the interaction of radiation with plasma in strong magnetic fields it is important to know the expressions for cyclotron emissivity, absorption or scattering coefficients (see sections 6.2,8.2) in plasma with temperature anisotropy. Electron momentum distribution in such a plasma is described by the function

$$f_0(p_\perp, p_\|) = \frac{N_0 \exp(-p_\perp^2/2m_e\kappa T_\perp - p_\|^2/2m_e\kappa T_\|)}{(2\pi m_e\kappa T_\perp)(2\pi m_e\kappa T_\|)^{1/2}}, \quad (5.142)$$

where $T_\perp$ and $T_\|$ are transverse and longitudinal temperatures with respect to the magnetic field $\boldsymbol{B}_0$. Since the formula (5.128) comprises only the integral over transverse momenta $p_\perp$, the factor $T_\|^{-1/2}\exp(-p_\|^2/2m_e\kappa T)$ from (5.142) will appear unchanged in the final expression for $a_{\omega j}$. The longitudinal momentum $p_\|$ is related to the frequency ω by the Doppler condition (5.129). This allows one to find the emissivity of plasma with temperature anisotropy using the above formulas for isotropic distribution with $T_\perp = T_\| = T$:

$$a_{\omega 1} = \frac{1}{(2\pi)^3}\sqrt{\frac{\pi}{8}}\,\omega_B\omega_L^2\frac{m_e}{c}\frac{\beta_{T_\perp}^2}{\beta_{T_\|}}\frac{1+\cos^2\alpha}{|\cos\alpha|} \times \exp[-(\omega - \omega_B)^2/2k_\|^2v_{T_\|}^2] \quad (s = 1), \quad (5.143)$$

$$a_{\omega 2} = \frac{\omega_B\omega_L^2 m_e}{(2\pi)^{7/2}c}\beta_{T_\perp}^2\beta_{T_\|}\frac{\sin^4\alpha\cos^4\alpha(3+\tan^2\alpha)^2}{|W(Z_2)|^2|\cos\alpha|(1+\cos^2\alpha)^3}$$

$$\times \exp[-(\omega - \omega_B)^2/2k_\parallel^2 v_{T_\parallel}^2] \quad (s=1), \quad (5.144)$$

$$a_{\omega j} = \frac{1}{(2\pi)^3}\sqrt{\frac{\pi}{2}}\,\omega_B\omega_{\mathrm{L}}^2\,\frac{m_{\mathrm{e}}}{c}\,\frac{\beta_{T_\perp}^{2s}}{\beta_{T_\parallel}}\,\frac{s^{2s+1}}{2^s s!}\,\frac{\sin^{2s-2}\alpha(1+K_j\cos\alpha)^2}{|\cos\alpha|(1+K_j^2)}$$

$$\times \exp[-(\omega - s\omega_B)^2/2k_\parallel^2 v_{T_\parallel}^2] \quad (s \geq 2). \quad (5.145)$$

Here $\beta_{T_\perp} = (\kappa T_\perp/m_e c^2)^{1/2}$, $\beta_{T_\parallel} = (\kappa T_\parallel/m_e c^2)^{1/2}$, $v_{T_\parallel} = (\kappa T_\perp/m_e)^{1/2}$, and the coefficients K_j are determined by (5.55). Finally, cyclotron emissivities at arbitrary harmonic in birefringent vacuum can be derived from the formula (5.145) by putting $K_1 = \infty$ and $K_2 = 0$.

Consider now the influence of various harmonics s on radiation at a given frequency ω. The expression (5.136) clearly shows the frequency width of the cyclotron line at the s-th harmonic: $\Delta\omega \sim \sqrt{2}|k_\parallel|v_T = \sqrt{2}s\omega_B\beta_T|\cos\alpha|$. Since the interval between the center frequencies of adjacent harmonics equals ω_B, the cyclotron radiation lines do not overlap for [18]

$$s < \frac{1}{\sqrt{2}\beta_T|\cos\alpha|}. \quad (5.146)$$

At the same time, the dipole nature of cyclotron radiation is ensured by the inequality (5.49) which takes the form $s\beta_T \sin\alpha \ll 1$ in case of homogeneous rarefied plasma with $n_j \approx 1$. Comparison of two criteria shows that for $\alpha \sim 1$ the cyclotron lines overlap at the same harmonics at which the dipole approximation fails. Thus, the radiation at frequencies corresponding to harmonics $s \gtrsim 1/\sqrt{2}\beta_T$ should be calculated taking into account the contribution by several harmonics to radiation at a given frequency ω. Such calculations have been made in [256, 257]. The radiation by X-ray pulsars Cen X-3 and Her X-1 in binary systems has been interpreted in [258] in terms of emission by weakly relativistic electrons at "overlapping" cyclotron harmonics. The observed radiation has been assumed to occur due to magnetic bremsstrahlung by high-temperature plasma at harmonics $s \gtrsim 5$. The plasma accretes to the magnetic poles of a neutron star, so the magnetic field $B_0 \sim 10^{10} - 10^{12}$ G shifts the cyclotron frequencies to the X-ray band.

5.2.4. GYROSYNCHROTRON EMISSIVITY

Consider now plasma emissivity in the more complicated case of mildly or semi-relativistic electrons with energies $\mathcal{E}_{\mathrm{kin}} \sim m_e c^2$. Magnetic bremsstrahlung by such electrons is called gyrosynchrotron radiation. The desired emissivity can be derived by substituting the total power of magnetic

[18] We repeat that the cyclotron radiation in a homogeneous magnetic field is considered here. If the magnetic field is inhomogeneous, the frequency intervals corresponding to different harmonics do not overlap under the additional conditions outlined in section 6.2.

bremsstrahlung P_Ω determined by (5.42) or (5.43) into the formula (5.128). However, it can also be obtained in a different way using the general expression (5.108) for plasma emissivity and the spectral power $P_{\omega\Omega}$ in the form (5.45) which comprises delta-function reflecting the resonance condition. Such a procedure has been performed in the paper [259] which we rely upon in the following consideration. The emissivity into both modes at $n_j^2 = 1$ can be written in the form:

$$a_\omega = a_{\omega 1} + a_{\omega 2} = \frac{e^2\omega^2}{(2\pi)^2 c \sin^2\alpha} \sum_{s=1}^{\infty} \int_1^{\infty} d\gamma \int_{-1}^{+1} d\mu\, f_0(\gamma,\mu)\delta(y)$$
$$\times \left\{ (\cos\alpha - \beta\mu)^2 J_s^2(\zeta) + (1-\beta\mu\cos\alpha)^2 \left[\frac{\zeta}{s} J_s'(\zeta)\right]^2 \right\}, \quad (5.147)$$

where

$$y = \frac{s\omega_B}{\gamma} - \omega(1-\beta\mu\cos\alpha), \qquad \zeta = \frac{\omega\gamma}{\omega_B}\beta\sin\alpha(1-\mu^2)^{1/2},$$

$\gamma = (1-\beta^2)^{-1/2} \equiv \mathcal{E}/m_e c^2$ is Lorenz-factor, $\mu = \cos\vartheta$ is the cosine of electron pitch angle, and $f_0(\gamma,\mu)$ denotes electron distribution over energies and pitch angles.

Contrary to the weakly relativistic limit $\gamma - 1 \ll 1$ at which the effect of higher harmonics is negligible compared to the radiation at the fundamental and few lower ones, the contribution by higher harmonics to the emission of mildly relativistic electrons is more significant. At frequencies $\omega \gg \omega_B$, which mainly contribute to the emissivity considered, the interval between adjacent harmonics $\Delta\omega \sim \omega_B \ll \omega$. Therefore, the sum over harmonic numbers can be approximated by the integral which, in its turn, can be easily calculated due to the presence of delta-function in the integrand. Moreover, the following asymptotic can be used for Bessel functions and their derivatives at $s \gg 1$ [260]:

$$J_s(\zeta) \simeq (2\pi s)^{-1/2}(1-z)^{-1/4} L^s,$$
$$\zeta J_s'(\zeta) \simeq (1-z^2)^{1/2} J_s(sz), \quad (5.148)$$

where

$$L = \frac{z\exp[(1-z^2)^{1/2}]}{1+(1-z^2)^{1/2}}, \qquad z = \frac{\zeta}{s} = \frac{\beta\sin\alpha(1-\mu^2)^{1/2}}{1-\beta\mu\cos\alpha}.$$

This approximation is valid if

$$\gamma^2 \ll \omega/\omega_B. \quad (5.149)$$

The above criterion is violated for radiation of ultra-relativistic particles $\gamma = \mathcal{E}/m_e c^2 \gg 1$ at frequencies $\omega \sim \omega_B \gamma^2$. The synchrotron limit of the magnetic bremsstrahlung is realized in this case. Synchrotron emissivity of relativistic electrons is calculated in the next section.

As for the gyrosynchrotron emission, the following expression for the emissivity can be obtained from (5.147) by account of the asymptotic (5.148):

$$a(\alpha) = \frac{e^2 \omega}{(2\pi)^2 c \sin^2 \alpha} \int_1^\infty d\gamma \int_{-1}^{+1} d\mu\, f_0(\gamma, \mu)\, Y L^{2s}, \tag{5.150}$$

where

$$Y = \frac{(\cos\alpha - \mu\beta)^2 + (1 - z^2)(1 - \beta\mu\cos\alpha)^2}{(1 - z^2)^{1/2}(1 - \beta\mu\cos\alpha)},$$
$$s = \frac{\omega\gamma}{\omega_B}(1 - \beta\mu\cos\alpha). \tag{5.151}$$

The function $Y(\mu)$ is slowly-varying as compared to L^{2s} which comprises an exponential factor. The latter, in its turn, decreases steeply from the maximal value at $\mu_{\max}$ with increasing $|\mu - \mu_{\max}|$. This fact is caused by relatively high directivity of the angular pattern of gyrosynchrotron radiation of an electron at each harmonic s. The maximum of the emissivity (5.150) at $\omega \gg \omega_B$ corresponds to the direction

$$\cos\alpha = \mu_{\max}/\beta. \tag{5.152}$$

If, moreover, the distribution $f_0(\gamma, \mu)$ is isotropic or slowly varying over the angular width of the radiation pattern: $d\ln f_0/d\mu \ll \omega/\omega_B$, the integral over μ in (5.150) can be calculated by the method of steepest descent. As a result,

$$a_\omega(\alpha) = \frac{e^2 \omega_B}{(2\pi)^2 c} \left(\frac{\pi\omega}{\omega_B}\right)^{1/2} \int_0^\infty d\gamma\, f_0(\gamma, \mu_{\max}) \gamma^{-1}$$
$$\times \left[(1 + 2\gamma^{-2}\cot^2\alpha)(1 - \beta^2\cos^2\alpha)^{1/4}\right] L_{\max}^{2s}, \tag{5.153}$$

where

$$L_{\max} = \frac{t \exp[(1 + t^2)^{-1/2}]}{1 + (1 + t^2)^{1/2}}, \quad s = \frac{\omega}{\gamma\omega_B}(1 + t^2), \quad t = \beta\gamma\sin\alpha. \tag{5.154}$$

Emissivity of electrons with isotropic mono-energetic distribution $f_0 \propto \delta(\gamma - \gamma_0)$ in the direction $\alpha = \pi/2$ is given by a simpler formula. Since in this case the expression in square brackets in (5.153) equals unity and $1+t^2=\gamma^2$, the emissivity is proportional to [256]

$$\left(\frac{\omega}{\omega_B}\right)^{1/2} e^{2\omega/\omega_B}\left(\frac{\gamma_0-1}{\gamma_0+1}\right)^{\gamma_0\omega/\omega_B}. \qquad (5.155)$$

Gyrosynchrotron emissivity of particles with wide energy distribution (for example, Maxwell or power-law one) can be calculated by the method of steepest descent. Note that in (5.153) the function L^{2s} increases as $(\gamma-1)^{2s}$ at $\gamma-1\ll 1$ and goes to unity as $\gamma\to 1$, the expression in square brackets is the slowly-varying function of γ, whereas particle distribution, as usual, decreases rather steeply with increasing γ. Therefore, the major contribution to the emissivity comes from the interval of γ near the maximum of the function $f_0 L^{2s}\gamma^{-1}$, and hence

$$a_\omega(\alpha) = \frac{e^2\omega}{4\pi c}\left(\frac{\omega}{\omega_B}\right)^{1/2}(1+2\gamma_0^{-2}\cot^2\alpha)(1-\beta_0^2\cos^2\alpha)^{1/4} \times f_0(\beta_0\cos\alpha,\gamma_0)L^{2s}_{\max}(t_0)X. \qquad (5.156)$$

Here $L_{\max}$ is given by (5.154), the value γ_0, which sets $\beta_0=(1+\gamma_0^{-2})^{1/2}$ and $t_0=\beta_0\gamma_0\sin\alpha$, is determined by the equation

$$\frac{1}{t^2(1+t^2)^{1/2}}+\left(1-\frac{\beta^2\cos^2\alpha}{t^2}\right)\ln L_{\max} = -\frac{\omega_B}{2\omega\sin^2\alpha}\frac{d\ln(f_0\gamma^{-1})}{d\gamma}, \qquad (5.157)$$

the function X is given by

$$X^{-2} = -\gamma_0^2\frac{d^2\ln(f_0\gamma^{-1})}{2d\gamma^2}\bigg|_{\gamma=\gamma_0} - \frac{d\ln(f_0\gamma^{-1})}{d\ln\gamma}\bigg|_{\gamma=\gamma_0}\left(\frac{3}{2}+\frac{1}{\gamma_0^2-1}\right), \qquad (5.158)$$

and $s=(\omega/\gamma\omega_B)(1+t_0^2)$.

The expression (5.156) for gyrosynchrotron emissivity is simplified in two limiting cases of large and small values of the parameter

$$\psi = \frac{\omega_B}{2\omega\sin^2\alpha}\frac{d\ln(f_0\gamma_0^{-1})}{d\gamma}. \qquad (5.159)$$

The corresponding formulas look like:

$$a_\omega(\alpha) = \frac{e^2\omega_B}{4\pi c}\left(\frac{\omega}{\omega_B}\right)^{1/2} f_0(\beta_0\cos\alpha,\gamma_0)X \times \begin{cases} \sin^{1/2}\alpha\,\exp\{d^2\ln(f_0\gamma^{-1})/d\ln\gamma^2|_{\gamma=\gamma_0}\} & \psi\ll 1, \\ \dfrac{1+\cos^2\alpha}{\sin^2\alpha}\left(\dfrac{e\,\beta_0\sin\alpha}{2}\right)^{2\omega/\omega_B} & \psi\gg 1. \end{cases} \qquad (5.160)$$

Here and thereafter the quantity $\mathrm{e}=2.718\ldots$ The limiting case $\psi \ll 1$ is realized at sufficiently large angles $\alpha \gg \alpha_c$, where the value α_c is set by the equation $t_0 \equiv \beta_0\gamma_0 \sin\alpha_c = 1$.

Gyrosynchrotron emission by plasma with the isotropic Maxwell distribution of electrons is widely applied to explain the origin of solar microwave bursts, radio emission from stellar coronas, emission from accreting columns of magnetic white dwarfs [261–263]. Such a distribution at an arbitrary ratio $\kappa T/m_e c^2$ is given by

$$f_0(\gamma,\mu) = NCe^{-\gamma m_e c^2/\kappa T}(\gamma^2-1)^{1/2}. \tag{5.161}$$

Here C is the normalizing factor which has the following form in the limit $\kappa T \ll m_e c^2$:

$$C = \frac{1}{2}\left[2\pi\left(\frac{\kappa T}{m_e c^2}\right)^3\right]^{-1/2}\left(1-\frac{15\kappa T}{8m_e c^2}\right)e^{m_e c^2/\kappa T}. \tag{5.162}$$

For mildly relativistic temperatures $m_e c^2/\kappa T \ll \omega/\omega_B$ the parameter ψ in (5.159) is small at any α except the values satisfying the condition $\sin\vartheta \ll \omega_B m_e c^2/\omega\kappa T$. Therefore,

$$\beta_0 \simeq 1, \qquad \gamma_0 \simeq \left(\frac{4\omega\kappa T}{3\omega_B m_e c^2 \sin\alpha}\right)^{1/3}, \qquad X \simeq \left(\frac{2\kappa T}{3\gamma_0 m_e c^2}\right)^{1/2},$$

so the emissivity is determined by the first expression in (5.160) and reduced to the formula

$$a_\omega(\alpha) \simeq \frac{e^2\omega_B NC}{3\sqrt{2}\pi c}\,\frac{\omega\kappa T}{\omega_B m_e c^2}\exp\left\{-\frac{\omega}{\omega_B}\left[\frac{4.5}{\sin\alpha}\left(\frac{\omega_B m_e c^2}{\omega\kappa T}\right)^2\right]^{1/3}\right\}. \tag{5.163}$$

For non-relativistic particles ($\kappa T \ll m_e c^2$) at frequencies $\omega/\omega_B \ll m_e c^2/\kappa T$ the parameter $\psi \gg 1$ and

$$\gamma_0 \simeq 1, \qquad \beta_0^2 \simeq \frac{2\omega\kappa T}{\omega_B m_e c^2}, \qquad X = \left(2\frac{\omega}{\omega_B}\right)^{1/2}.$$

Then the second expression in (5.160) yields

$$a_\omega(\alpha) = \frac{2e^2\omega_B NC}{\pi c\,\mathrm{e}^{3/2}} \times(1+\cos^2\alpha)(\sin\alpha)^{2\omega/\omega_B-2}\left(\frac{\mathrm{e}\,\omega\kappa T}{2\omega_B m_e c^2}\right)^{\omega/\omega_B+3/2}. \tag{5.164}$$

Note that gyrosynchrotron emission in the direction $\alpha=\pi/2$ of electrons with isotropic Maxwell distribution has been calculated in [256, 264]. The formulas (5.163), (5.164) are reduced to the ones from these papers as $\alpha \to \pi/2$. Numerical calculations of gyrosynchrotron emission [259,265,266] have revealed that the formulas (5.163), (5.164) provide an accuracy of a few per cent even at lower harmonics at which the method of steepest descent should work only poorly.

Concluding this section, consider the expressions for gyrosynchrotron emissivity of the particle system with the distribution

$$f_0(\gamma, \mu) = N g(\mu) C \left(1 + \frac{\gamma - 1}{\gamma_c - 1}\right)^{-\delta}, \tag{5.165}$$

where

$$C = \frac{\delta - 1}{\gamma_c - 1} \tag{5.166}$$

is the normalizing factor, the pitch-angle distribution $g(\mu)$ is normalized to unity: $\int_{-1}^{+1} g(\mu)\, d\mu = 1$, and the quantity $\gamma_c \neq 1$ ensures the convergence of the distribution at lower energies $\gamma \to 1$. The distribution (5.165) has been used in the above cited paper [259] which our following consideration is relied upon. It is reduced to power-law distribution $f_0 \propto \gamma^{-\delta}$ at $\gamma_c \gg |\gamma_c - 2|$, whereas $f_0 \simeq \mathrm{const}$ if $\gamma - 1 \ll \gamma_c - 1$. If angular anisotropy of particle spectrum is sufficiently weak:

$$\frac{d \ln g(\mu)}{d\mu} \ll \frac{\omega}{\omega_B}, \tag{5.167}$$

the emissivity is obtained in the same way as for the thermal distribution (5.161). At frequencies $\omega \gg \omega_B \delta$ the relativistic factor $\gamma_0 \gg 1$. Hence,

$$\beta_0 \simeq 1, \qquad \gamma_0^2 = \frac{4\omega}{3\omega_B (1 + \delta) \sin\alpha}, \qquad X = (1 + \delta^2)^{-1/2},$$

so the first expression in (5.160) yields

$$a_\omega(\alpha) \simeq \frac{\sqrt{3} e^2 \omega_B N C}{4\pi c} (\gamma_c - 1) g(\mu_0) \sin\alpha \, e^{-(\delta+1)/2} \times \left[\frac{3\omega_B (\gamma_c - 1)^2 (\delta + 1) \sin\alpha}{4\omega}\right]^{(\delta-1)/2}, \tag{5.168}$$

where $\mu_0 = \beta_0 \cos\alpha$. The obtained emissivity depends on the parameters ω, ω_B and α in a similar way as the synchrotron emissivities (5.185)–(5.187) for a power-law energy spectrum derived in the next section with a different technique. Such a coincidence up to the numerical factor of about 2 seems natural since the formula (5.168) is related to the ultra-relativistic limit $\gamma_0 \gg 1$ in which the distribution (5.165) is reduced to the power-law one.

If $\gamma_c - 1 \ll 1$, the emissivity at frequencies $\omega/\omega_B \ll \delta(\gamma_c - 1)^{-1}$ is determined by (5.164) in which the quantity $\kappa T/m_e c^2$ should be substituted with $(\gamma_c - 1)/\delta$:

$$a_\omega = \frac{2e^2\omega_B NC}{\pi c\, e^{3/2}}(1 + \cos^2\alpha)(\sin\alpha)^{2\omega/\omega_B - 2}\left[\frac{e\,\omega(\gamma_c - 1)}{2\omega_B\delta}\right]^{\omega/\omega_B + 3/2}. \quad (5.169)$$

It has been taken into account here that in the case considered the parameters

$$\gamma_0 \simeq 1, \qquad \beta_0^2 \simeq 2\omega(\gamma_c - 1)/\omega_B\delta.$$

In the intermediate case $\gamma_c - 1 \sim 1$ no simple analytical formulas are known for gyrosynchrotron emissivity.

Note that the above formulas for the emissivity a_ω are valid for weakly anisotropic distributions of electrons over pitch angles. On the contrary, if the velocities of radiating electrons are beamed within the narrow pitch-angle interval near $\mu_0 = \cos\vartheta_0$ and the distribution function has the form

$$f_0(\mu, \gamma) = \frac{2}{1 - \mu_0^2}\exp\left\{-\frac{1 - \mu^2}{1 - \mu_0^2}\right\} f_0(\gamma), \quad (5.170)$$

where $1 - \mu_0^2 \ll \omega_B/\omega$, then the main contribution to the emissivity comes from the particles with

$$\beta_0 = \cos\vartheta_0 \qquad \alpha \neq \pi/2,$$

$$\beta_0^2 = 2\left[\ln\frac{\omega_B}{\omega(1 - \mu_0^2)}\right]^{-1} \qquad \alpha = \pi/2.$$

The corresponding gyrosynchrotron emissivity then takes the form

$$a_\omega = \frac{e^2\omega N}{2\pi c} f_0(\gamma_0) A, \quad (5.171)$$

where the Lorenz factor $\gamma_0 = (1 - \beta_0^2)^{-1/2}$, the function

$$A = \cot\alpha\left[\ln\frac{4\omega_B\tan\alpha}{e\,\omega(1 - \mu_0^2)\cos\alpha}\right]^{-1/2}\left(\frac{4\omega_B\tan\alpha}{e\,\omega(1 - \mu_0^2)\cos\alpha}\right)^{-\omega\sin\alpha/\omega_B} \quad (5.172)$$

if $\alpha \neq \pi/2$ and

$$A = \sqrt{2}\left[\ln\frac{4\omega_B}{e\,\omega(1 - \mu_0^2)}\right]^{-1/2}\left(\frac{2\omega_B}{e\,\omega(1 - \mu_0^2)}\right)^{-\omega/\omega_B} \quad (5.173)$$

at $\alpha = \pi/2$. These formulas are valid for mildly relativistic particles. In the case of Maxwell distribution these correspond to the temperatures $\kappa T/m_e c^2 \gtrsim \omega_B/\omega$.

The analytical expressions for gyrosynchrotron emissivity given in this section are quite awkward. Simpler empirical formulas have been obtained in [266] as an interpolation of numerical results for isotropic pitch-angle distribution and power-law energy spectrum of the form

$$f_0(\mathcal{E}) = NC\mathcal{E}^{-\delta}, \qquad C = (\delta - 1)\mathcal{E}_0^{\delta-1}. \tag{5.174}$$

The corresponding gyrosynchrotron emissivity into extraordinary mode at $\omega/\omega_B \geq 10$, $\alpha \geq 20°$ and $2 \leq \delta \leq 7$ is fitted by the formula

$$a_{\omega 1} \approx \frac{B_0 N}{2\pi} \cdot 3.3 \cdot 10^{-24} 10^{-0.52\,\delta} (\sin\alpha)^{-0.43+0.65\,\delta} (\omega/\omega_B)^{1.22-0.90\,\delta} \tag{5.175}$$

with an accuracy not lower than 25 %. It is assumed here that plasma is rarefied enough ($\omega_{\mathrm{L}}^2 \ll \omega^2$), so that the mode polarizations are close to circular. This is possible only for quasi-longitudinal propagation which is violated if $|\pi/2 - \alpha| \lesssim \omega_B/2\omega$ (see section 2.2). The degree of circular polarization equals

$$\zeta_c = \frac{a_{\omega 1} - a_{\omega 2}}{a_{\omega 1} + a_{\omega 2}} \approx$$
$$0.20 \cdot 10^{0.05\,\delta} 10^{1.93\cos\alpha - 1.16\cos^2\alpha} (\omega/\omega_B)^{-0.21-0.37\sin\alpha}. \tag{5.176}$$

Analytical formulas for gyrosynchrotron emissivity into a distinct mode for arbitrary angle α and $\omega \gg \omega_{\mathrm{L}}, \omega_B$ have been obtained in [267]. Better approximations for Bessel functions and their derivatives [268] which have been extensively used in the paper [267] allowed authors to get the correct formulas valid both for gyrosynchrotron and synchrotron radiation which is emitted by ultra-relativistic particles. The formulas from [267] are reduced to the results of the papers [256,259,265] in the relevant limiting cases.

5.2.5. SYNCHROTRON EMISSIVITY. SYNCHROTRON MECHANISM OF COSMIC RADIO EMISSION

Note, first and foremost, that distribution function over energies $N(\mathcal{E})$ (or particle energy spectrum), along with the momentum distribution function $f_{\mathrm{S}}(\boldsymbol{p})$, is frequently used in the analysis of emission by relativistic particles. By definition, $N(\mathcal{E})d\mathcal{E}$ equals the number of particles in a unit volume of coordinate space and with energy in the interval from $\mathcal{E}$ to $\mathcal{E} + d\mathcal{E}$ in the unit solid angle of momentum space. Since $\mathcal{E} = pc$ in the relativistic limit, it is clear from the above definition that $N(\mathcal{E})d\mathcal{E} = f_{\mathrm{S}}(\boldsymbol{p})p^2dp$ and

$$N(\mathcal{E}) = \frac{f_{\mathrm{S}}(\boldsymbol{p})p^2}{c}. \tag{5.177}$$

Of course, the above definition does not account for possible angular dependence of the particle energy spectrum. The spectra exhibiting such a dependence are called anisotropic. For the sake of simplicity (and recalling, in particular, the observed isotropy of cosmic rays), the subsequent analysis is restricted to particle systems with isotropic spectrum. Particle density in such a system is determined by

$$N_S = 4\pi \int N(\mathcal{E})d\mathcal{E}. \tag{5.178}$$

Note that the formulas thereafter are also valid for anisotropic spectra with an angular scale much larger than the characteristic width of the synchrotron radiation pattern.

As has been mentioned above, it is quite improbable that the equilibrium distribution of relativistic particles is realized in astrophysical conditions. This makes quite useless the calculation of the emissivity for this case. On the other hand, the appearance of the mono-energetic spectra of the form

$$N(\mathcal{E}) = \frac{N_S}{4\pi}\delta(\mathcal{E} - \mathcal{E}_0) \tag{5.179}$$

seems rather doubtful, too. However, owing to its peerless simplicity, the above spectrum is frequently used for the initial estimates of the possibility to apply the synchrotron mechanism for the interpretation of the observed radiation by astrophysical sources.

The narrowness of the synchrotron radiation pattern beamed along the particle velocity $\boldsymbol{v}$ allows one to write the emissivity (5.108) in the form

$$a_\omega = \int P_\omega(\omega, \mathcal{E})N(\mathcal{E})d\mathcal{E} \tag{5.180}$$

It is taken into account here that the emissivity which is related to the unit solid angle is contributed by all the electrons with velocity in this solid angle. Since a_ω is also related to the unit frequency interval, the contribution of an individual electron equals the radiation power P_ω in this interval integrated over the narrow solid angle near $\boldsymbol{v}$ into which synchrotron radiation is emitted.

The emissivity of mono-energetic particle system (5.179) is determined by the formula

$$a_\omega = P_\omega(\omega, \mathcal{E}_0)\frac{N_S}{4\pi}, \tag{5.181}$$

where $P_\omega(\omega, \mathcal{E}_0)$ in vacuum is given by (5.77) and in isotropic plasma, by (5.94). The emissivities for distinct linearly polarized components of synchrotron radiation with electric field vectors $\boldsymbol{E}^{(1)}$ and $\boldsymbol{E}^{(2)}$ orientated along the axes of polarization ellipse[19] are described by the similar formulas:

$$a_{\omega 1} = \frac{P_{\omega 1} N_S}{4\pi}, \qquad a_{\omega 2} = \frac{P_{\omega 2} N_S}{4\pi} \tag{5.182}$$

Here $P_{\omega 1}$ and $P_{\omega 2}$ are the powers of synchrotron radiation by single particle into linearly polarized components 1 and 2, respectively, determined by (5.87). Using (5.182) and (5.87), it is possible to find the degree of linear polarization of the mono-energetic particle system [269]:

$$\zeta_l = \frac{a_{\omega 1} - a_{\omega 2}}{a_{\omega 1} + a_{\omega 2}} = \frac{\mathrm{K}_{2/3}(\omega/\omega_c)}{\int_{\omega/\omega_c}^{\infty} \mathrm{K}_{5/3}(x)\,dx}. \tag{5.183}$$

This formula yields $\zeta_l \approx 1/2$ at lower frequencies $\omega \ll \omega_c$ and $\zeta_l \approx 1$ at higher ones $\omega \gg \omega_c$. Thus, the radiation of the particle system considered exhibits the large degree of linear polarization.

Quite important for astrophysical applications is the synchrotron radiation of electrons with the power-law energy spectrum

$$N(\mathcal{E}) = \mathcal{K}\mathcal{E}^{-\gamma}, \qquad \mathcal{K} = \frac{N_S}{4\pi}\left[\int_{\mathcal{E}_1}^{\mathcal{E}_2} \mathcal{E}^{-\gamma}\, d\mathcal{E}\right]^{-1}, \tag{5.184}$$

where the normalizing factor $\mathcal{K}$ is expressed in terms of N_S, the number of particles in the energy band from $\mathcal{E}_1$ to $\mathcal{E}_2$. Such a specific role of the distribution (5.184) is explained by the fact that the observed spectra of cosmic ray particles, including the relativistic electrons producing the non-thermal diffuse radio emission of the Galaxy (see below) are power-law. On the other hand, the frequency spectra of a number of discrete sources are fitted well by a power law. Being interpreted in terms of synchrotron radiation mechanism, the observed power-law radiation spectra demand the power-law energy spectrum of the emitting particles.

The emissivity of relativistic particles with an isotropic energy spectrum in a uniform magnetic field can be found by substituting the expressions (5.77) and (5.184) into the formula (5.180). As a result, the frequency spectrum $a_\omega(\omega)$ in vacuum has the following form [1]:

$$a_\omega = \tilde{\mathcal{K}}\omega^{-\alpha}, \tag{5.185}$$

where

$$\alpha = \frac{\gamma - 1}{2} \tag{5.186}$$

[19]The orientation of these axes in case of synchrotron radiation is discussed in the comments preceding the formulas (5.84).

is spectral index, and the coefficient

$$\tilde{\mathcal{K}} = b(\gamma)\mathcal{K}\frac{e^3}{m_e c^2}\left(\frac{3e}{m_e^3 c^5}\right)^{\frac{\gamma-1}{2}}(B_0 \sin\alpha)^{\frac{\gamma+1}{2}}, \quad (5.187)$$

where

$$b(\gamma) \doteq \frac{\sqrt{3}}{2\pi(\gamma+1)}\Gamma\left(\frac{3\gamma-1}{12}\right)\Gamma\left(\frac{3\gamma+19}{12}\right)$$

and Γ denotes the gamma-function. The degree of linear polarization can be obtained using the emissivities $a_{\omega 1}$ and $a_{\omega 2}$ for the linearly polarized components oriented along the major axes of polarization ellipse. These are found by substitution of (5.87) into (5.180). The resulting degree of linear polarization,

$$\zeta_l = \frac{a_{\omega 1} - a_{\omega 2}}{a_{\omega 1} + a_{\omega 2}} = \frac{\gamma+1}{\gamma+7/3}, \quad (5.188)$$

depends entirely on the spectral index and does not vary with frequency [270].

If a relativistic electron moves in a chaotic magnetic field which can be considered quasi-homogeneous over a spatial scale l_0, the following emissivity can be introduced to characterize the synchrotron radiation of such a system:

$$\bar{a}_\omega = L^{-1}\int_L a_\omega dl. \quad (5.189)$$

This quantity results from the averaging of emissivity a_ω determined by (5.185)-(5.187) over the magnetic field direction (i.e., the angles α). The averaging interval should be much larger than l_0. The corresponding expression for the average emissivity $\bar{a}_\omega$ can be presented in the form (5.185), (5.186) with different $\tilde{\mathcal{K}}$ [1]:

$$\tilde{\mathcal{K}} = a(\gamma)\mathcal{K}\frac{e^3}{m_e c^2}\left(\frac{3e}{2m_e^3 c^5}\right)^{\frac{\gamma-1}{2}} B_0^{\frac{\gamma+1}{2}}, \quad (5.190)$$

Here $B_0^{\frac{\gamma+1}{2}}$ is the mean value of this value over the region considered, and

$$a(\gamma) = \frac{2^{\frac{\gamma-1}{2}}\sqrt{3}}{(\gamma+1)4\sqrt{\pi}}\Gamma\left(\frac{3\gamma-1}{12}\right)\Gamma\left(\frac{3\gamma+19}{12}\right)\Gamma\left(\frac{\gamma+5}{4}\right)\Gamma^{-1}\left(\frac{\gamma+7}{4}\right)$$

is the slowly-varying function of γ. The values of $a(\gamma)$ at various spectral indices are given in Table 5.1.

Linear polarization of synchrotron radiation evidently vanishes in a chaotic magnetic field.

TABLE 5.1.

γ	1	1.5	2	2.5	3	4	5
$a(\gamma)$	0.566	0.294	0.206	0.170	0.148	0.145	0.184

When deriving the formulas (5.185)–(5.190), the initial expression (5.180) for a_ω has been integrated over $\mathcal{E}$ from 0 to infinity. Since the synchrotron radiation at a frequency ω is mainly contributed by electrons with $\mathcal{E} \sim \mathcal{E}_{\max}$ (see (5.82)), such an extension of the integration interval does not lead to significant errors in the resulting emissivity a_ω at frequency ω if the particle spectrum deviates from the power law at "unimportant" energies [20]

$$\mathcal{E} \ll m_e c^2 \left(\frac{\omega}{\omega_B}\right)^{1/2} \quad \text{or} \quad \mathcal{E} \gg m_e c^2 \left(\frac{\omega}{\omega_B}\right)^{1/2} . \tag{5.191}$$

The above expressions for synchrotron emissivity clearly show that particles with a power-law energy spectrum emit synchrotron radiation whose frequency spectrum is power-law, too. The corresponding spectral indices are related by the formula (5.186) both for homogeneous and for chaotic magnetic field. This important relation pointed out in [271] is widely used for the interpretation of radiation by astrophysical sources in terms of synchrotron mechanism. Since the observed spectrum of many sources can be well approximated by a power law over a wide frequency interval, it allows deciphering of the energy spectrum of radiating particles by the use of the frequency spectrum of the observed radiation.

The formula (5.186) was obtained in a formal way as a result of the calculation of the function $a_\omega(\omega)$. It can also be explained by the following simple considerations. Since, as has been already said, the synchrotron radiation at frequency ω is emitted mainly by particles with energies $\mathcal{E} \sim \mathcal{E}_{\max} \sim m_e c^2 (\omega/\omega_B)^{1/2}$ in the interval $\Delta\mathcal{E} \sim \mathcal{E}_{\max}$, the emissivity at frequency ω can be estimated as $a_\omega \sim P_\omega(\omega, \mathcal{E}_{\max}) N(\mathcal{E}_{\max}) \Delta\mathcal{E}$. However, the value $P_\omega(\omega, \mathcal{E}_{\max})$ for synchrotron radiation depends only on the magnetic field B_0. [21] Taking into account that $N(\mathcal{E}_{\max}) \propto \mathcal{E}_{\max}^{-\gamma} \propto \omega^{-\gamma/2}$ for a power-law spectrum and that $\Delta\mathcal{E} \sim \mathcal{E}_{\max} \propto \omega^{1/2}$, we arrive at the frequency spectrum $a_\omega \propto \omega^{-(\gamma-1)/2}$.

[20] The more rigorous criteria are given in [247].

[21] According to (5.79), the maximum (5.79) of $P_\omega(\omega, \mathcal{E})$ as a function of ω does not depend on $\mathcal{E}$. Thus, $P_\omega(\omega, \mathcal{E})$ as a function of $\mathcal{E}$ reaches maximum at any given frequency ω.

All the above results neglect the influence of the medium on the synchrotron radiation. The effect of the medium becomes pronounced for a power-law electron spectrum if the inequality (5.102) fails, i.e., at $\omega \lesssim \omega_{\mathrm{L}}^2/2\omega_B$. Depression of synchrotron radiation in this interval alters the sign of the spectral index and leads to the exponential cutoff of a_ω with decreasing frequency [245,246]:

$$a_\omega \propto \omega^{1-\gamma} \exp\left(\frac{-\sqrt{3}\omega_{\mathrm{L}}^2}{\omega\omega_B}\right). \tag{5.192}$$

Such a behavior is explained by the presence of the exponential factor in the expression (5.98) for the synchrotron radiation power of an electron.

The formulas for the synchrotron emissivity of relativistic electrons in a magnetic field are the basis for the investigations of the synchrotron mechanism of cosmic radio emission. If reabsorption along a ray is negligible, the observed intensity of radiation by a stationary source is related to the emissivity as

$$I_\omega = \int a_\omega dl \tag{5.193}$$

(see section 4.1). Power-law electrons in a chaotic magnetic field generate the synchrotron radiation with the specific intensity

$$I_f \approx 8.57 \cdot 10^{-22} a(\gamma) L \mathcal{K} B_0^{\frac{\gamma+1}{2}} \left(\frac{6.47 \cdot 10^{18}}{f}\right)^{\frac{\gamma-1}{2}}, \quad \frac{\mathrm{erg}}{\mathrm{cm^2\,s\,sr\,Hz}}. \tag{5.194}$$

Here $I_f = 2\pi I_\omega$ is the intensity of both polarizations related to the unit interval of the frequency $f = \omega/2\pi$, the size of the source is measured in cm, B_0 in G, f in Hz, and the coefficient $\mathcal{K}$ determined by (5.184) accords to the energy measured in ergs. The formula (5.194) is valid under the conditions which allow one to neglect the influence of the medium. The features of synchrotron emission are also discussed in the monograph [208].

Apply the obtained formulas to the non-thermal component of Galactic diffuse radio emission.[22] Assume that this radiation is emitted by relativistic electrons moving in interstellar magnetic fields. The frequency spectrum of the non-thermal component at frequencies $f \sim 10$ MHz can be approximated by a power law with the index $\alpha \approx 0.3-0.4$ which slowly increases up to 0.8 at higher frequencies $f \gtrsim 300$ MHz. The observed intensity I_f of radio emission exhibits strong variation with Galactic altitude and longitude and is about $1.5 \cdot 10^{-18}$ erg/cm^2 s Hz sr at $f = 400$ MHz in the direction of the Galactic anti-center [272].

[22] The diffuse radio emission of the Galaxy also comprises the thermal component originated due to bremsstrahlung emission by H II regions; see section 6.1.

Assuming that the Galactic magnetic field is chaotic and neglecting absorption of radio waves over a distance of about the size of the Galaxy,[23] we use the formula (5.194) relevant for the adopted conditions. According to this formula, the observed power-law frequency spectrum with $\alpha \approx 0.8$ may originate due to synchrotron emission by particles with power-law energy spectrum which has the index $\gamma = 2\alpha + 1 \approx 2.6$. The obtained value agrees very well with the spectral index derived by the analysis of the electron component of primordial cosmic rays [272].

Next, with $\gamma \approx 2.6$, the expected value of radiation intensity is estimated using (5.194) in the following way:

$$I_f \approx 1.6 \cdot 10^{-7} L \mathcal{K} B_0^{1.8} f^{-0.8} \ \frac{\text{erg}}{\text{cm}^2\,\text{s}\,\text{sr}\,\text{Hz}}. \qquad (5.195)$$

The parameters L, $\mathcal{K}$ and B_0 in the above formula are known, at least by the order of magnitude. Indeed, the effective ray length in the Galaxy can be estimated as $5\,\text{kpc} = 1.5 \cdot 10^{22}\,\text{cm}$, one third of the disk radius. The magnetic field strength $B_0 \sim 5 \cdot 10^{-6}\,\text{G}$ is determined by the independent measurements (in particular, Zeeman splitting of the neutral hydrogen line $\lambda = 21\,\text{cm}$). Finally, the coefficient $\mathcal{K} = 1.2 \cdot 10^{-17}$ typical for the electron component of cosmic rays [272] is used for the following estimates.

Substituting the numerical values into (5.195), we find $I_f \sim 1.1 \cdot 10^{-18}$ erg/cm^2 s Hz sr at the frequency $f = 400\,\text{MHz}$. Taking into account the uncertainty of the value of L and the absence of the data on the mean values of B_0 and $\mathcal{K}$ over the Galaxy, the agreement of the obtained estimate with the observed value $I_f \sim 1.5 \cdot 10^{-18}\,\text{erg/cm}^2\,\text{s}\,\text{sr}\,\text{Hz}$ is rather good. The agreement of the observed and predicted values of I_f and the index γ of electron energy spectrum is the strong evidence in favor of the synchrotron theory of non-thermal component of diffuse radio emission of the Galaxy [233,234,271].

Thus, the formula (5.194) can help to check the hypothesis on the synchrotron origin of the non-thermal component of diffuse radio emission of the Galaxy using the independent measurements of the parameters which enter this formula. The application of (5.194) to discrete sources (supernova envelopes, radio galaxies, galactic jets etc.), which are also interpreted in terms of the synchrotron mechanism, yields rather modest results. Here the source size L can only be estimated, while the B_0 and $\mathcal{K}$ are unknown. Thus, the formula (5.194) can only yield the spectral index γ of relativistic particle spectrum and the combination $\mathcal{K} B_0^{\frac{\gamma+1}{2}}$ from the known spectral index of radiation, α, and the intensity I_f at a given frequency. One additional relation is required to separate $\mathcal{K}$ and B_0 in this case. Such a relation can

[23] The absorption can be neglected at $\lambda < 10\,\text{m}$ if the ray considered does not pass through the dense H II regions; see sections 6.1,7.4.

be obtained in the framework of the synchrotron model itself if the low-frequency depression is seen in the source spectrum which is caused either by reabsorption inside the source[24] or by the influence of the medium (plasma). In the latter case the known frequency $\omega_{cr} \sim \omega_L^2/2\omega_B$ of the edge of the depression band allows one to determine B_0 in terms of N, the electron density of the background plasma in the source. It is however a difficult problem to get the needed value of N. So, the additional relation between $\mathcal{K}$ and B_0 is usually obtained by setting the ratio of energy density of relativistic electrons, W_e, to the energy density W_B of the magnetic field in the source. Dynamics stability of a charged particle system in a magnetic field is possible if $W_e < W_B$. Bearing in mind that $W_e/W_B \sim 10^{-2}$ is determined for the Galaxy,[25] the value $W_e/W_B \sim 10^{-1} - 10^{-2}$ is usually used for the above estimates.

Recently, synchrotron mechanism has been used to explain the microwave radiation from the X-ray source 1E 1740.7-2942 in the Galactic Center region [273]. Combined with the data available on the X-ray emission in a continuum and in the 511 keV annihilation line, these results have imposed severe constraints on the parameters of plasma and relativistic particles (electrons and positrons) in the source.

It should be emphasized that the synchrotron mechanism is the first radio astronomical radiation mechanism which has been successfully used in optical and X-ray astronomy. We refer to the paper [274] in which the optical continuum emission from the Crab nebula, the remnant of 1054 Supernova, has been explained in terms of synchrotron emission of relativistic electrons moving in the magnetic field of the nebula. The discovery of linear polarization of the optical radiation from the Crab nebula, which has been predicted on the basis of the synchrotron mechanism in [235,275], has become a strong argument in favor of this interpretation.

---

[24] See section 7.4 on the effect of synchrotron reabsorption.

[25] See [208] on the problem to determine the radio W_e/W_B far from the Earth.

Absorption in thermal astrophysical plasmas

From the previous chapter it is clear that any elementary process of emission by distinct particles: bremsstrahlung, Čerenkov, magnetic bremsstrahlung, etc. contributes to plasma emissivity, so that $a_\omega = a_\omega^{\rm b} + a_\omega^{\rm \check{C}er} + a_\omega^{\rm magn} + \ldots$ On the other hand, the Kirchhoff law (4.36) relates emissivity a_ω and the absorption coefficient μ_j of equilibrium plasma: $a_{\omega j} = \mu_j I_\omega^{(0)}$. Therefore, each emission process is related to a specific type of absorption. Since the emissivities of various types are summed to yield the total value, the resulting absorption coefficient at a given frequency ω is the sum of bremsstrahlung, magnetic bremsstrahlung, and the others.

Of course, it is not always possible to distinguish different components of particle emission. For example, in a non-relativistic magnetoactive plasma for which $\omega_B \sim \nu_{\rm eff}$ ($\nu_{\rm eff}$ is the effective collision frequency of electrons) the gyration period $2\pi/\omega_B$ is comparable to the mean time between collisions $1/\nu_{\rm eff}$. Since collisions change particle trajectories dramatically, it is impossible to separate bremsstrahlung and cyclotron emissions in this case. Another example is synchrotron or curvature radiation of an electron in the case when the gyration radius r_B is comparable to the curvature radius of the field line R_B. Here, again, one cannot separate these two types of emission (i.e., their radiation powers and, thus, emissivities and absorption coefficients). However, such cases are quite rare in cosmic plasmas, where, as a rule, the values of parameters ensure the emissivity a_ω and absorption coefficient μ_j to be additive functions comprised of the terms which correspond to various elementary emission processes. Only these typical cases will be considered henceforth.

Together with the absorption coefficient μ_j entering the Kirchhoff law and radiation transfer equations, the absorption index η_j (see (2.55)) may also be introduced. If $\eta_j \neq 0$, the amplitude of j-th normal wave varies along the direction of the wave vector $\boldsymbol{k}_j$ as $\exp(-\omega\eta_j z/c)$. Since radiation intensity is proportional to the square of the amplitude, it behaves as $\exp(-2\omega\eta_j z/c)$. In an anisotropic medium, however, the vectors $\boldsymbol{v}_{\rm gr}$ and $\boldsymbol{k}_j$ are not collinear, the angle between them being equal to ϑ_j. Therefore, the intensity varies as $\exp(-2\omega\eta_j l\cos\vartheta_j/c)$ in the direction of $\boldsymbol{v}_{\rm gr}$, and its relative variation per unit ray length equals $2\omega\eta_j\cos\vartheta_j/c$. On the other hand, this variation is determined by the absorption coefficient μ_j, so

$$\mu_j = 2\frac{\omega}{c}\eta_j \cos\vartheta_j.$$

In addition to the studies of spatial absorption of radiation, in radio astronomy it is sometimes interesting to consider the temporal problem of the damping of some initial electromagnetic disturbance which comprises waves with real $\boldsymbol{k}_j$. This process is characterized by the decrement γ_j. The temporal decrease of the intensity is described by the exponent $e^{-2\gamma_j t}$, where the decrement γ_j is related to η_j via (2.57). This formula together with the above one for μ_j yields the relation

$$2\gamma_j = \mu_j v_{\mathrm{gr}}.$$

There are two possible ways to find μ_j and/or γ_j in an equilibrium medium. The first one is based on the Kirchhoff law: the desired absorption coefficient is found using the known emissivity a_ω and equilibrium intensity $I_\omega^{(0)}$. This way is simple and obvious. However, its application is restricted by the requirement that the absorption is small enough that it does not change the equilibrium intensity (1.37) (in particular, the refraction index n_j and the angle ϑ_j), as compared to the non-absorbing medium. Another way, more formal but valid in the wide region of parameters, is to derive η_j by the solution of the dispersion equation (2.41) for normal waves. In this case it is necessary to know the dielectric tensor $\epsilon_{\alpha\beta}(\omega, \boldsymbol{k})$. The latter is found, depending on the nature of the problem and the accuracy desired, in the framework of elementary theory, in quasi-hydrodynamic approximation, or by the method of kinetic equation (see section 2.1). Generally, both approaches are used in physics of cosmic plasma. The choice between them in the common region of validity is made by the demands of simplicity and convenience.

6.1. Absorption in isotropic plasma

6.1.1. ABSORPTION DUE TO PARTICLE COLLISIONS. BREMSSTRAHLUNG RADIATION MECHANISM

In the absence of a static magnetic field and, thus, magnetic bremsstrahlung, the dissipation of radiation in plasma is caused by free-free (bremsstrahlung) and Čerenkov absorptions. The latter, evidently, may occur for plasma waves only, since the Vavilov-Čerenkov effect is absent for electromagnetic waves in isotropic plasma (see section 5.1). Free-free absorption exists for both types of waves. The corresponding absorption coefficient may be found using the Kirchhoff law.

In a low-density, isotropic plasma with $v = \omega_L^2/\omega^2 \ll 1$ and temperatures $T \ll 4 \cdot 10^5$ K the bremsstrahlung absorption coefficient for electromagnetic waves is equal to [75]

$$\mu_{\rm em} = \frac{a_\omega}{I_\omega^{(0)}} \approx \frac{32\pi^2}{3\sqrt{2\pi}} \frac{e^6 N^2}{(\kappa T m_{\rm e})^{3/2} c\omega^2} \ln\left[\frac{2(2\kappa T)^{3/2}}{\delta^{5/2} e^2 m_{\rm e}^{1/2} \omega}\right], \tag{6.1}$$

where $\delta = 1.781$. The absorption coefficient for non-relativistic plasma with temperature $4 \cdot 10^5\,{\rm K} \ll T \ll 10^{10}$ K differs from (6.1) only by the form of the logarithmic factor:

$$\mu_{\rm em} \approx \frac{32\pi^2}{3\sqrt{2\pi}} \frac{e^6 N^2}{(\kappa T m_{\rm e})^{3/2} c\omega^2} \ln\left[\frac{4\kappa T}{\delta \hbar\omega}\right]. \tag{6.2}$$

The emissivity a_ω in these cases is given by (5.113) and (5.117), respectively. The above absorption coefficients are written for $N = N_+$ and $Z = 1$.[1] The equilibrium intensity $I_\omega^{(0)}$ is given by (1.37). In the case of rarefied isotropic plasma one should put $n = n_{\rm em} = 1$ and $\cos\vartheta_j = 1$. Since the bremsstrahlung emissivity a_ω accounts for the emission into both modes, the doubled equilibrium intensity $I_\omega^{(0)}$ should be used to get the expression for the absorption coefficient μ_j per one mode from the Kirchhoff law.

Bremsstrahlung emission in plasma occurs due to collisions of charged particles, hence the corresponding absorption coefficient may be found by taking into account the effect of collisions on the dielectric permeability of plasma. In the framework of elementary theory it is made in the following way.

Let $\nu_{\rm eff}$ be the effective frequency of collisions of an electron with other particles in which it changes the velocity significantly and loses the regular momentum $m_{\rm e}\boldsymbol{v}$ gained up to the instant of collision under the action of the wave field. Then the loss of momentum per unit time equals $m_{\rm e}\nu_{\rm eff}\boldsymbol{v}$. The influence of collisions may therefore be accounted for by adding the term $-m_{\rm e}\nu_{\rm eff}\boldsymbol{v}$ to the right-hand side of the equation of motion of an electron (2.62). The resulting transverse dielectric permeability has the following form:

$$\epsilon_\perp = \epsilon' = 1 - \frac{\omega_L^2}{\omega(\omega + i\nu_{\rm eff})}, \tag{6.3}$$

and the dispersion equation for electromagnetic waves (2.46) is written as

$$\omega^2 = \frac{\omega_L^2}{1 - i\nu_{\rm eff}/\omega} + c^2k^2. \tag{6.4}$$

[1] The electron density is thus equal to the density of single-charged ions. Since hydrogen is the dominant species of the astrophysical plasma, this assumption is usually very well satisfied.

The longitudinal dielectric permeability $\epsilon_{\|}(\omega, \boldsymbol{k})$ for collisionless plasma in the kinetic approximation is given by the formula (2.88). The effect of collisions may be taken into account by using the complex argument $Z' = (\omega + i\nu_{\text{eff}})/\sqrt{2}kv_T$ instead of the real one $Z = \omega/\sqrt{2}kv_T$ which enters the expression (2.88).[2] In the region $|Z'| \gg 1$ which, for nearly real ω, corresponds to large phase velocities $v_{\text{ph}} \approx \omega/k \gg v_T$ the dispersion equation (2.47) for plasma waves takes the form

$$(\omega + i\nu_{\text{eff}})^2 = \omega_{\text{L}}^2 + 3k^2v_T^2 \tag{6.5}$$

(cf. (2.115)).

In the high-frequency limit $\omega \gg \nu_{\text{eff}}$ which is mostly interesting for physics of cosmic plasma the refraction indices of electromagnetic and plasma waves, n_{em} and n_{pl}, found with dispersion equations (6.4) and (6.5) are close to the expressions (2.104) and (2.116), respectively. However, the absorption indices are no-zero now and given by

$$\eta_{\text{em}} \approx \frac{\omega_{\text{L}}^2}{2\omega^3 n_{\text{em}}}\nu_{\text{eff}} \tag{6.6}$$

for an electromagnetic wave and by

$$\eta_{\text{pl}} \approx \frac{1}{3\omega\beta_T^2 n_{\text{pl}}}\nu_{\text{eff}} \tag{6.7}$$

for plasma one. The corresponding absorption coefficients $\mu_j = 2\omega\eta_j/c$ look like:

$$\mu_{\text{em}} \approx \frac{\omega_{\text{L}}^2\nu_{\text{eff}}}{\omega^2 c\sqrt{\epsilon}}, \qquad \mu_{\text{pl}} \approx \frac{2\nu_{\text{eff}}}{\sqrt{3\epsilon}v_T} \simeq \frac{2v_{\text{ph}}\nu_{\text{eff}}}{3v_T^2}. \tag{6.8}$$

Here $\epsilon = 1 - \omega_{\text{L}}^2/\omega^2$ and the relation (2.117) for plasma waves is taken into account in the last expression for μ_{pl}.

The bremsstrahlung decrements $\gamma_j = \mu_j v_{\text{gr}}/2$ have the following form:

$$\gamma_{\text{em}} \approx \frac{\omega_{\text{L}}^2}{\omega^2}\frac{\nu_{\text{eff}}}{2}, \qquad \gamma_{\text{pl}} \approx \frac{\nu_{\text{eff}}}{2} \tag{6.9}$$

These values are close to each other at frequencies $\omega \approx \omega_{\text{L}}$. The formulas (6.9) are valid for $\omega \gg \nu_{\text{eff}}$ while (6.6), (6.7) and (6.8) are constrained by the more severe condition $\omega\epsilon \gg \nu_{\text{eff}}$ in the region $\epsilon > 0$. The effective frequency of collisions may be written in the form

$$\nu_{\text{eff}} = \nu_{ei} + \nu_{ea}, \tag{6.10}$$

[2] Such an ansatz is justified by the re-calculation of $\epsilon_{\|}(\omega, \mathbf{k})$ from the kinetic equation (2.79) with the integral of collisions in the form $J' = \nu_{\text{eff}} f'$ (see section 2.1).

where the first and the second terms describe electron collisions with ions and neutral atoms (molecules), respectively. The electron-electron collisions do not contribute to the bremsstrahlung since the total momentum, including the regular one gained under the action of radiation field, is not changed by a collision of identical particles.

The value ν_{ei} can be estimated from the following simple considerations. During an effective collision in which the electron velocity is drastically changed leading to the significant change of electron velocity the distance between colliding particles decreases to the value r_{min} at which electron potential energy e^2/r_{min} in the ion electric field is comparable to its kinetic energy κT. This yields the estimate $r_{\text{min}} \sim e^2/\kappa T$. The value $\pi r_{\text{min}}^2 \sim \pi e^4/(\kappa T)^2$ is called the effective cross-section of a collision. During a time interval Δt an electron moving with the velocity v collides with all ions inside the cylinder with the base πr_{min}^2 and the height $v\Delta t$. The number of such ions equals $\pi r_{\text{min}}^2 N v \Delta t$. Putting in this expression $v \sim v_T$ and dividing it by Δt we arrive at the estimate

$$\nu_{ei} \sim \pi \frac{e^4}{(\kappa T)^2} v_T N, \tag{6.11}$$

where the above value for the effective cross-section πr_{min}^2 is taken into account. Since $v_T = \sqrt{\kappa T/m_e}$, the effective collision frequency of electrons and ions is proportional to $T^{-3/2}$ and decreases with increasing temperature.

The accurate calculations [71] yield more complicated expression:

$$\nu_{ei} = \pi \frac{e^4}{(\kappa T)^2} \overline{v} N A \approx \frac{5.5\, N}{T^{3/2}} A, \tag{6.12}$$

where the logarithmic factor equals

$$A = \ln \left[0.37 \frac{\kappa T}{e^2 N^{1/3}} \right] \approx \ln \left[220 \frac{T}{N^{1/3}} \right] \tag{6.13}$$

at $T < 4 \cdot 10^5$ K and

$$A = \ln \left[\delta \left(\frac{m_e e^4}{\hbar^2 \kappa T} \right)^{1/3} \frac{\kappa T}{e^2 N^{1/3}} \right] \approx \ln \left[10^4 \frac{T^{2/3}}{N^{1/3}} \right] \tag{6.14}$$

otherwise. Both formulas yield similar values in the region $T \sim 4 \cdot 10^5$ K. It is assumed in the formulas (6.12)-(6.14) that the factor $\delta \sim 1$, the arithmetic mean velocity $\overline{v} = \sqrt{8\kappa T/\pi m_e}$, and $N_+ = N$. The possible difference of electron and ion temperatures leads only to the correction of the argument of the logarithm and thus provides minor changes to the value of ν_{ei}.

The expressions (6.12)–(6.14) have been derived using the kinetic equation with collision integral $J_{\rm st}$ which accounts for pair collisions only.[3] Such an approach is valid if the time of collision Δt is small compared to the period of high-frequency field $1/\omega$. In plasma $\Delta t \sim D/v_T$, where $D \sim v_T/\omega_{\rm L}$ is the Debye radius (see (1.4)). Thus, the obtained formulas for ν_{ei} are only valid in the region $\omega_{\rm L} \gg \omega$ which is insignificant for the propagation of waves in isotropic plasma ($n^2_{\rm em}$ and $n^2_{\rm pl}$ are negative). The same validity range holds for absorption coefficients and decrements (6.6)-(6.9) if we use $\nu_{\rm eff} = \nu_{ei}$ there. On the other hand, the formulas (6.1), (6.2) which are valid in the region $\omega_{\rm L} \ll \omega$, can be presented in the form (6.8) with $\nu_{\rm eff} = \nu_{ei}$, where ν_{ei} is given by (6.12). However, the value A in this case should be put equal to 2/3 of the logarithmic factors (6.13), (6.14). Extrapolation of this new coefficient $A' \equiv \frac{2}{3}A$ to the frequencies $\omega \sim \omega_{\rm L}$ yields the values close to (6.13), (6.14). As a result, a viable approximation of the $\nu_{\rm eff}$ in the region $\omega \sim \omega_{\rm L}$ is given by (6.12) with logarithmic factors (6.13), (6.14) extrapolated from the region $\omega_{\rm L} \ll \omega$.

Since the duration of a collision with a neutral atom $\Delta t \sim a/v_T$ (a is atomic radius) is much less than that for an electron-ion collision, the above constraints on the formulas (6.6)-(6.9) are removed if $\nu_{\rm eff} \approx \nu_{ea}$. The frequency of collisions of an electron with a neutron atom (molecule) equals

$$\nu_{ea} = \frac{4\pi}{3} a^2 \bar{v} N_a = 8.3 \cdot 10^5 \, \pi a^2 T^{1/2} N_a, \tag{6.15}$$

where N_a is the density of neutral particles, πa^2 is the effective cross-section of collision. For example, $\pi a^2 \sim 10^{-16}\,{\rm cm}^2$ for atomic hydrogen. Collision frequency with neutral particles grows with the electron temperature as $T^{1/2}$. In most cases the collisions with neutrals are of minor importance in astrophysical plasmas: $\nu_{ea} \ll \nu_{ei}$, though one should take them into account in cold parts of the solar chromosphere and in planet ionospheres (for example in the D-layer of the Earth's ionosphere).

When calculating bremsstrahlung emission and absorption at higher frequencies in low-density plasma ($v \ll 1$) in which the wavelength is small compared to the Debye radius and the Debye screening is negligible the effect of particle collisions is described very well by the account of pair collisions only. The absorption coefficient in this case can be found by the Kirchhoff law (4.36) from the expression (5.119) for the bremsstrahlung emissivity:

$$\mu_{\rm em} = \frac{\pi^3}{\sqrt{6\pi}} \frac{32}{3} \frac{e^6 Z^2 N N_+}{m_{\rm e}^2 c^2 \kappa T \omega^2} \left(\frac{m_{\rm e}c^2}{\kappa T}\right)^{1/2} g(\omega, T) \frac{1 - e^{-\hbar\omega/\kappa T}}{\hbar\omega/\kappa T}, \tag{6.16}$$

[3] Treating the absorption due to collisions as the inverse process with respect to bremsstrahlung, one can get from the Kirchhoff law the formulas (6.8) with $\nu_{\rm eff} = \nu_{ei}$, where ν_{ei} is given by (6.12)-(6.14).

where the Gaunt factor $g(\omega, T)$ is given by the formula (5.120) (see also [85]). In the Rayleigh–Jeans limit $\hbar\omega \ll \kappa T$ this expression is reduced to

$$\mu_{em} = \frac{32\pi^2}{3\sqrt{2\pi}} \frac{e^6 Z^2 N N_+}{(\kappa T m_e)^{3/2} c\omega^2} \left(\ln \frac{4\kappa T}{\delta \hbar\omega} \right) \tag{6.17}$$

which, in its turn, becomes identical to (6.2) at $Z = 1$ and $N = N_+$. In the opposite Wien limit $\hbar\omega \gg \kappa T$ the bremsstrahlung absorption coefficient is given by

$$\mu_{em} = \frac{32\pi^2}{3\sqrt{2\pi}} \frac{e^6 Z^2 N N_+}{(\hbar\omega m_e)^{3/2} c\omega^2}. \tag{6.18}$$

Note that the expressions (6.16)-(6.18) are only valid for plasmas in which the refractive indices are close to unity. If $n_j \neq 1$, the above absorption coefficients should be divided by n_j.

To judge on the form of the frequency spectrum of bremsstrahlung radiation from a plasma source, consider the simplest case of an isothermal layer of rarefied plasma. The brightness temperature of thermal radiation escaping a layer with temperature T is equal to $T_b = T(1 - e^{-\tau})$ where the optical depth $\tau = \int \mu_{em}\, dl$ (see section 4.1). The bremsstrahlung absorption coefficient μ_{em} in rarefied plasma is determined by the formulas (6.1), (6.2), so that the corresponding optical depth can be written in the form

$$\tau = \frac{\int N^2 dl}{\omega^2 T^{3/2}} B, \tag{6.19}$$

where

$$B = \frac{16\pi^2}{\sqrt{2\pi}} \frac{e^6}{(\kappa m_e)^{3/2} c} A', \tag{6.20}$$

and the factor A' is 2/3 of the logarithms (6.1), (6.2). This factor is the slowly varying function of frequency. It is possible to approximate this by the constant if we make the frequency ω in the logarithm argument equal to the critical one,

$$\omega_{cr} = \Big(\frac{B}{T^{3/2}} \int N^2\, dl\Big)^{1/2}, \tag{6.21}$$

at which $\tau = 1$.

The brightness temperature T_b of thermal bremsstrahlung from a layer of rarefied plasma as a function of frequency is plotted in Fig. 6.1. At $\omega \ll \omega_{cr}$ the optical depth $\tau \gg 1$, so $T_b \approx T$. In the region $\omega \gg \omega_{cr}$, otherwise, $\tau \ll 1$ and $T_{eff} \approx T\tau \propto \omega^{-2}$. The frequency spectrum of radiation intensity may be derived from $T_b(\omega)$ using the relation $I_\omega = \omega^2 \kappa T_{eff}/(2\pi)^3 c^2$ (see (1.38); note that in a low-density plasma $n = 1$ and $\vartheta = 0$). The function $I_\omega(\omega)$ is also plotted in Fig. 6.1. The intensity grows monotonously with frequency

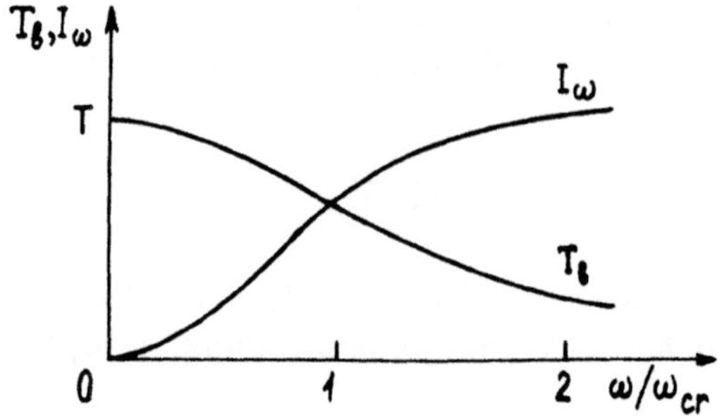

Figure 6.1. The frequency spectrum of thermal bremsstrahlung by an isothermal layer of rarefied plasma ($\omega_L \ll \omega_{cr}$)

reaching the constant value in the region $\omega \gg \omega_{cr}$; at lower frequencies $\omega \ll \omega_{cr}$ the intensity is proportional to ω^2.

Now we consider the role played by bremsstrahlung emission and absorption in cosmic plasmas. The estimates based upon (6.1) show that the absorption of radio waves in interstellar H I regions is weak because the optical depth of the Galaxy $\tau \ll 1$. In more dense H II regions, on the contrary, the absorption becomes significant at lower frequencies. Due to this fact, for example, the H II regions shadowing a discrete source with a power-law spectrum $I_\omega = \tilde{K}\omega^{-\alpha}$ distort it as follows:

$$I_\omega = \tilde{K}\omega^{-\alpha} e^{-\tau(\omega)}. \tag{6.22}$$

This effect leads to the depression in the observed spectra of some discrete sources at frequencies $\omega \lesssim \omega_{cr}$.

Note that H II regions themselves emit radiation with a spectrum similar to the one shown in Fig. 6.1. For example, the spectrum of the well-known Orion Nebula, the region of ionized hydrogen, corresponds to $T_b \approx 10^4$ K at frequencies far below the critical one which is estimated as $\omega_{cr} \approx 3 \cdot 10^9\,s^{-1}$ ($\lambda_{cr} \approx 60$ cm). This means that the kinetic temperature of the hydrogen in the nebula is close to 10^4 K which is consistent with values revealed by observations of optical emission lines of oxygen. Such a coincidence of the derived temperatures and the shape of the spectrum gives evidence that the radio emission of the Orion Nebula (as well as other H II regions) may be explained by the thermal bremsstrahlung mechanism. This allows one to find the so-called emission measure $EM = \int N^2 dl$ using the observed value of ω_{cr}. The estimate for the kinetic temperature T can be derived at least from the value of brightness temperature of radiation at frequencies $\omega \ll \omega_{cr}$. For $\omega_{cr} = 3 \cdot 10^9\,s^{-1}$ and $T_b \sim 10^4$ K we get $EM \sim 2 \cdot 10^{24}\,cm^{-5}$. Assuming that the distribution of ionized hydrogen over a source with the size $L \sim 10^2$ pc $\sim 3 \cdot 10^{20}$ cm is homogeneous, we find that $EM \sim N^2 L$ and, therefore, the mean electron density in the source $N \sim 10^2\,cm^{-3}$. More detailed data on emission measure over the source may be obtained using the angular distribution of radio emission at frequencies $\omega \gg \omega_{cr}$ where $I_\omega \propto T^{-1/2} \int N^2 dl$.

In the solar chromosphere and corona the difference of the refraction index from unity can no longer be neglected. Here the bremsstrahlung absorption coefficient μ_{em} and the corresponding optical depth $\tau = \int \mu_{em}\, dl$ should be calculated using (6.8), (6.12) with the logarithmic factor A in the form (6.13) for chromosphere (where $T \sim 10^4$ K) and in the form (6.14) for the corona ($T \sim 10^6$ K). The results of the calculations of ν_{eff} and τ are given in [3,71].

Effective collision frequency $\nu_{eff} = \nu_{ei}$ in the corona is about 1–10 s^{-1} at distances $r \leq 1.5 R_\odot$ from the solar center ($R_\odot$ is the optical radius of the Sun). The function $\tau(R) = \int_R^\infty \mu_{em} dR$ for $T = 10^6$ K and density distribution determined by the Baumbach-Allen formula (1.16) is shown in Fig. 6.2. The graphs are drawn up to $R = R^*$ at which the refractive index $n_{em} = 0$. According to this figure, the optical depth of the corona measured from the turnover point R^* decreases with increasing wavelength.

As for the free-free absorption of plasma waves in coronal plasma, this can be estimated using the formula (6.8). At coronal regions the kinetic temperature $T \approx 10^6$ K, so that $\nu_{eff} \sim 5\, s^{-1}$ and $v_T \simeq 4 \cdot 10^8$ cm/s. Then the absorption coefficient μ_{pl} for plasma waves with phase velocity $v_{ph} \approx 10 v_T$ is about $10^{-7}\, cm^{-1}$. This means that the e-folding length for such waves is about 10^7 cm which is much less than the typical size of active regions and sources of solar radio emission. One thus may conclude that plasma waves are absorbed within the region of their generation. The characteristic damping time equals $1/2\gamma_{pl}$ which is only 0.2 s for the adopted value of $\nu_{eff} \sim 5\, s^{-1}$ (see (6.9)).

Given the optical depth and temperature of the solar chromosphere and corona, one may judge the characteristics of the thermal bremsstrahlung of these regions. According to the ideas put forward in the early years of radio astronomy [160,276–278], this type of radiation is identified with the radio emission of "quiet" Sun. The main features of radio emission of this type are (i) the increase of brightness temperature of radiation T_b with wavelength from 10^4 K in a centimeter band up to 10^6 K in meter band, and (ii) the solar limb "brightening", i.e., the existence of regions with higher brightness temperature at the limb of the Sun. Analysis of the thermal bremsstrahlung mechanism of radio emission [279] allows the interpretation of these features of radio emission of the "quiet" Sun even within the simplest model of the solar atmosphere.

Assume that the electron density distribution in the solar atmosphere is spherically symmetric and can be described by the Baumbach-Allen formula (1.16), while the corona and the chromosphere are isothermal, the kinetic

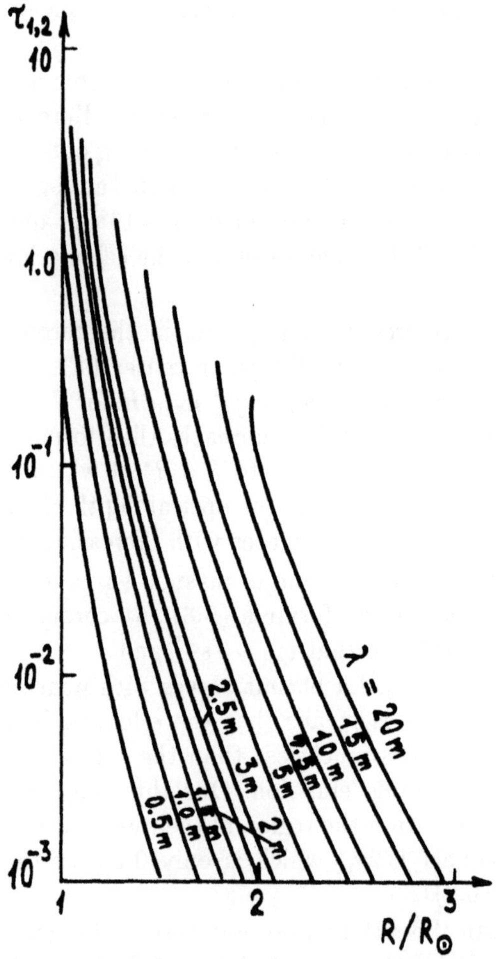

Figure 6.2. **Bremsstrahlung optical depth of the solar corona at different wavelengths for radial propagation of radio waves.**

temperatures being equal to $T_{\rm ch} = 3 \cdot 10^4$ K and $T_{\rm c} = 10^6$ K, respectively. The shape of a ray in such an atmosphere is shown in Fig. 3.5. The optical depth along such a ray from the distance R to infinity is determined by the expression $\tau = \int_R^\infty \mu(1 - r^2/R^2n^2)^{-1/2}dR$ in which the relation (3.26) between the line elements dl and dR is taken into account. Approximation of the isothermal corona and chromosphere allows us to apply the formulas (4.42)-(4.44) to analyze the brightness temperature of radiation. If the turnover point of a ray lies in the corona, the brightness temperature

$$T_{\rm b} = T_{\rm c}(1 - e^{-2\tau_{\rm c}^*}), \tag{6.23}$$

where

$$\tau_c^* = \int_{R^*}^{\infty} \mu(1 - \frac{r^2}{R^2 n^2})^{-1/2}\, dR \qquad (6.24)$$

is the optical depth along the ray from the turnover point R^* to infinity. But if the turnover point lies in the chromosphere, the brightness temperature of radiation escaping the corona equals

$$T_b = T_c(1 - e^{-\tau_c}) + T_{ch} e^{-\tau_c}. \qquad (6.25)$$

Here

$$\tau_c = \int_{R_{ch}}^{\infty} \mu(1 - \frac{r^2}{R^2 n^2})^{-1/2}\, dR \qquad (6.26)$$

is the optical depth of the coronal part of the ray and R_{ch} is the radius of the upper boundary of the chromosphere. The calculations show that, in the framework of the adopted model, the bremsstrahlung optical depth of the chromosphere exceeds unity if the turnover point R^* lies deeper than 100 km below R_{ch} (note than the total thickness of the of the chromosphere is hundred times larger). These are the rays which the formula (6.25) is written for. One may also account in this formula that bremsstrahlung optical depth of the corona $\tau_c \ll 1$ for rays entering the chromosphere ($R^* < R_{ch}$). Then [4]

$$T_b \approx T_c \tau_c + T_{ch}. \qquad (6.27)$$

Contribution of the corona to the brightness temperature of the observed radiation will be small for $\tau \ll T_{ch}/T_c \sim 3 \cdot 10^{-2}$. The bremsstrahlung optical depth of the corona $\tau_c \propto \omega^{-2} \propto \lambda^2$ (see, for example, (6.6)), so that $\tau_c \ll 3 \cdot 10^{-2}$ at wavelengths $\lambda \ll 10 - 20$ cm.

As a result, the radio emission of the "quiet" Sun in the centimeter band is generated mainly in the chromosphere, and $T_{eff} \approx T_{ch} \sim 3 \cdot 10^4$ K. At longer wavelengths the contribution of the corona becomes more significant, and T_b determined by (6.27) increases with increasing wavelength. At meter waves $R^* > R_{ch}$, $2\tau_c^* > 1$, and the brightness temperature (6.23) rises to the coronal temperature $T_c \sim 10^6$ K. That is how the frequency spectrum of radio emission of the "quiet" Sun at centimeter and meter waves is explained. However, the above isothermal model of the chromosphere fails to explain the millimeter band spectrum.

Limb "brightening" can not occur if τ_c in the center of the solar disk (for $r = 0$) exceeds unity: the brightness temperature T_b in this case reaches the maximum at $r = 0$ and can only decrease with increasing r. This case is realized at meter waves. In the decimeter band the quantity $\tau_c < 1$, rays enter chromosphere, so T_b is determined by the formulas (6.27), (6.26). Due to the increase of the length of the coronal part of the ray the optical

[4] The formula (6.27) accounts for the fact that $T_{ch} \ll T_c$ (cf. the formula (4.44).

depth of the corona τ_c now increases with distance from the disk center, and the brightness temperature T_b increases with τ_c. Such an enhancement of the brightness temperature of radio emission occurs if $r \lesssim R_{ch}$. At $r > R_{ch}$ the turnover point R^* rises from the chromosphere up to the corona where T_b is given by (6.23), (6.24). According to these formulas, τ_c^* and T_b rapidly decrease with increasing R^* due to the decreasing of the interval of integration in the expression for τ_c^*. Therefore, the relative importance of chromosphere and corona in the formation of the observed radio emission depends on the distance from the solar disk center. It is this effect that the limb brightening is related to. At shorter wavelengths, where the role of the corona is negligible, the limb brightening is caused by the temperature inhomogeneity of the chromosphere. This problem, as well as other issues concerning the origin of radio emission of "quiet" Sun, is discussed in detail in [3].

6.1.2. ČERENKOV ABSORPTION OF PLASMA WAVES. LANDAU DAMPING

As has been already been mentioned at the beginning of this section, Čerenkov absorption of plasma waves occurs in addition to the bremsstrahlung one. This kind of absorption is directly related to Čerenkov emission by distinct plasma particles and, in this sense, is nothing but the inverse Vavilov-Čerenkov effect.

The Čerenkov absorption coefficient $\mu^{\text{Čer}}$ can be found easily from the Čerenkov emissivity (5.126) using the Kirchhoff law. The absorption coefficient in an equilibrium plasma is given by [217]

$$\mu^{\text{Čer}} = \frac{a_\omega}{I_\omega^{(0)}} = \frac{1}{3}\sqrt{\frac{\pi}{2}}\frac{\omega_L^2\omega^3}{k^4 v_T^5}\exp\left(-\frac{\omega^2}{2v_T^2k^2}\right), \tag{6.28}$$

and the corresponding decrement equals

$$\gamma^{\text{Čer}} = \mu^{\text{Čer}}\frac{v_{gr}}{2} = \sqrt{\frac{\pi}{8}}\frac{\omega_L^2\omega^2}{k^3 v_T^3}\exp\left(-\frac{\omega^2}{2v_T^2k^2}\right). \tag{6.29}$$

Note that the relation between the decrement and the absorption coeffificent introduced in the beginning of this chapter was used to yield this formula. It was also taken into account here that the equilibrium intensity of plasma waves $I_\omega^{(0)} = n_{pl}^2\omega^2\kappa T/(2\pi)^3c^2$, where $n_{pl}^2 = \epsilon/3\beta_T^2$, and their group velocity $v_{gr} = 3v_T^2k/\omega$ (see (2.117)). The expression (6.29) is identical to the formula by Landau [238] which characterizes damping, not related to collisions (the so-called Landau damping), of longitudinal waves in an equilibrium isotropic plasma. It has been derived by the solution of the dispersion equation for plasma waves (2.47) with $\epsilon_{\|}$ in the form (2.88), i.e. in the framework of the kinetic approach. The above derivation, however, is much more evident and emphasizes the close relation of Landau damping with Vavilov-Čerenkov effect, this fact having been clarified in [280–282]. Moreover, it has been definitely pointed out in [283] that Landau damping and Čerenkov damping are the same physical phenomenon described in two different ways.[5]

The above formulas for $\mu^{\check{C}er}$ and $\gamma^{\check{C}er}$ are valid for weak absorption when $\mu^{\check{C}er} \ll k$ and $\gamma^{\check{C}er} \ll \omega$. It follows at least from the fact that the expressions for $I_\omega^{(0)}$ and a_ω used in the above calculations hold only for transparent (weakly absorbing) medium. From the expressions for $\mu^{\check{C}er}$ and $\gamma^{\check{C}er}$ it is clear that the criterion of weak absorption is met if $\omega^2/2k^2v_T^2 = v_{\rm ph}^2/2v_T^2 \gg 1$. It is this condition that has been used in the work by Landau to obtain the formula (6.29). Note that in the considered region of weak damping of plasma waves their frequency is close to $\omega_{\rm L}$. But if $v_{\rm ph} \sim v_T$, the above method of deriving the coefficients $\mu^{\check{C}er}$ and $\gamma^{\check{C}er}$ is no longer valid. By contrast, the method of kinetic equation can still be used in this region. The corresponding analysis of the dispersion relation (2.47), (2.88) by this method [83] have shown that here $\gamma \sim \omega$, i.e., waves are damped in a time $t \sim 1/\gamma \sim 1/\omega$ or, equivalently, at a distance $l \sim \lambda/2\pi$. The latter means that plasma waves with $v_{\rm ph} \sim v_T$ (i.e., $kD \sim 1$) cannot propagate due to strong Landau damping in plasma. The obvious physical considerations leading to this result are given in section 2.2.

It is Landau damping that prevents a plasma wave from escaping the region $\omega \approx \omega_{\rm L}$ into rarefied plasma (for example, from the solar corona to interplanetary medium). Indeed, in the course of propagation of a wave with frequency ω through inhomogeneous plasma with $\epsilon = 1 - \omega_{\rm L}^2/\omega^2$ its phase velocity $v_{\rm ph} = \sqrt{3}v_T/\sqrt{\epsilon}$ (see (2.117)) decreases with decreasing $\omega_{\rm L}$. Therefore, the plasma wave is completely absorbed in the regions where $v_{\rm ph}$ falls to the values comparable to v_T.

6.2. Absorption in magnetoactive plasma

The formulas describing the free-free (caused by collisions) and Čerenkov absorption become complicated for magnetoactive plasma. In addition, another kind of absorption, the cyclotron one, appears in the non-

[5] We do not consider here the classical treatment of Landau damping which takes into account that the electric field of a plasma wave accelerates electrons with velocities $v < v_{\rm ph}$ and decelerate ones with $v > v_{\rm ph}$. Prevailing of the first particle species in Maxwell plasma leads to the transfer of wave energy to particles, i.e., to the damping of the wave (see [280, 281] and any monograph on plasma physics).

relativistic case, $\kappa T \ll mc^2$.

6.2.1. ABSORPTION OF ELECTROMAGNETIC WAVES DUE TO PARTICLE COLLISIONS

In the framework of the elementary theory this kind of absorption is taken into account by introducing the term $-m\nu_{\text{eff}}\boldsymbol{v}$ into the right-hand side of the equation of motion of an electron (2.62). As a result, the expression (2.121) for the refractive indices squared, $n_{1,2}^2$, of ordinary and extraordinary modes is replaced by the following formula [71]:

$$(n + i\eta)_j^2 = 1 - \frac{2v(1 - v + i\nu)}{2(1 + i\nu)(1 - v + i\nu) - u\sin^2\alpha \mp G}, \tag{6.30}$$

where

$$v = \omega_L^2/\omega^2, \qquad u = \omega_B^2/\omega^2, \qquad \nu = \nu_{\text{eff}}/\omega,$$

$$G = \sqrt{u^2\sin^4\alpha + 4u(1 - v + i\nu)^2\cos^2\alpha},$$

and α is the angle between the wave vector $\boldsymbol{k}_j$ and the magnetic field $\boldsymbol{B}_0$. The minus and plus signs correspond to extraordinary ($j=1$) and ordinary ($j=2$) waves, respectively. The condition $\nu \ll 1$ is usually very well satisfied in cosmic plasma at radio frequencies. So, the expansion of (6.30) in powers of ν yields the refractive indices, $n_{1,2}$, which are nearly the same as (2.121), and also the bremsstrahlung (free-free) absorption indices given by

$$\eta_{1,2} \simeq \frac{v\nu}{n_{1,2}} \times \left\{\mp\sqrt{u^2\sin^4\alpha + 4u(1-v)^2\cos^2\alpha}\,[u\sin^2\alpha + 2(1-v)^2] - u^2\sin^4\alpha\right\} \times \frac{\left[2(1-v) - u\sin^2\alpha \mp \sqrt{u^2\sin^4\alpha + 4u(1-v)^2\cos^2\alpha}\right]^{-2}}{\pm\sqrt{u^2\sin^4\alpha + 4u(1-v)^2\cos^2\alpha}}. \tag{6.31}$$

The inequality

$$|1 - u - v + uv\cos^2\alpha| \gg \nu|3 - 2v - u| \tag{6.32}$$

is the validity criterion of this expression at small $\eta_{1,2}$. According to (6.31), the bremsstrahlung absorption in a magnetoactive plasma peaks sharply near the resonant frequencies (2.125) at which $n_{1,2} \to \infty$ in the absence of collisions.

The expression for $\eta_{1,2}$ is simplified in the cases of quasi-longitudinal and quasi-transverse propagation of waves with respect to the magnetic field. These cases are quite important for applications. In the quasi-longitudinal approximation the absorption indices are

$$\eta_{1,2} = \frac{v\nu}{2n_{1,2}(1 \mp |\sqrt{u}\cos\alpha|)^2}, \tag{6.33}$$

and the refractive indices $n_{1,2}$ are described by the formula (2.128). Respectively, in the quasi-transverse approximation the values $n_{1,2}$ are given by the expression (2.129), the absorption index of the extraordinary mode is determined by the formula

$$\eta_1 = \frac{v\nu[(1-v)^2 + u\sin^2\alpha]}{2n_1(1 - v - u\sin^2\alpha)}, \tag{6.34}$$

and the one of ordinary waves, η_2, does not differ from (6.6). Note that for $\nu \ll 1$ the validity criteria of the quasi-longitudinal and quasi-transverse approximations are practically the same as (2.126), (2.127).

If a plasma layer is rarefied enough (so that $v \ll 1$) and the magnetic field is weak ($|\sqrt{u}\cos\alpha| \ll 1$), its optical depth is equal to

$$\tau_{1,2} \approx \frac{\omega_{\text{cr}}^2}{\omega^2}\left(1 \pm 2\frac{\omega_B}{\omega}|\cos\alpha|\right) \tag{6.35}$$

within the wide interval of angles in which the quasi-longitudinal propagation is realized. Here $\omega_{\text{cr}}^2 = BT^{-3/2}N^2L$ and L is the geometrical thickness of the layer (see (6.21)). To write the formula (6.35), one should take into account that $\tau_{1,2} = \mu_{1,2}L \simeq 2(\omega/c)\eta_{1,2}L$. The quantities $\eta_{1,2}$ should be obtained by expanding the expression (6.33) in powers of $|\sqrt{u}\cos\alpha|$. The brightness temperature of thermal bremsstrahlung from the layer is $T_{\text{b}\,1,2} = T(1 - e^{-\tau_{1,2}})$. It is equal to T for both modes at frequencies $\omega^2 \ll \omega_{\text{cr}}^2$. The degree of polarization of radiation at these frequencies is equal to zero. In the region $\omega^2 \gg \omega_{\text{cr}}^2$ the layer becomes optically thin, so $T_{\text{eff}\,1,2} \simeq T\tau_{1,2}$. Since $\tau_1 > \tau_2$ (see (6.35)), the bremsstrahlung emission of the extraordinary waves is increased due to the magnetic field, whereas the corresponding emission of the ordinary ones is decreased. The total intensity, $I_\omega \propto T_{\text{b}1} + T_{\text{b}2}$, does not change (to be more correct, its increase is proportional to the higher powers of $|\sqrt{u}\cos\alpha|$). Thus, the frequency spectrum of the radiation, $I_\omega(\omega)$, is nearly the same as the one shown in Fig. 6.1. The degree of circular polarization of the radiation escaping the source is equal to [3]:

$$\zeta_c = \frac{T_{\text{b}1} - T_{\text{b}2}}{T_{\text{b}1} + T_{\text{b}2}} \approx \frac{\tau_1 - \tau_2}{\tau_1 + \tau_2} \approx 2\frac{\omega_B}{\omega}|\cos\alpha| \tag{6.36}$$

This formula allows us to estimate the longitudinal component of the magnetic field, $\boldsymbol{B}_0|\cos\alpha|$, in a source of thermal bremsstrahlung using the degree of circular polarization ζ_c. The measurements of ζ_c should be made in the frequency range $\omega^2 \gg \omega_{\rm cr}^2$ in which the source is optically thin. [6]

So, strong magnetic field can significantly change the characteristics of thermal bremsstrahlung radiation escaping the plasma. However, numerical calculations are necessary to study the optical depth of a source, as well as the intensity and polarization of radiation. Such calculations have been done in [284,285] for coronal and chromospheric conditions above sunspots.

6.2.2. CYCLOTRON ABSORPTION. MECHANISM OF THERMAL CYCLOTRON EMISSION

Čerenkov absorption (Landau damping) in a magnetoactive plasma has been studied in [286] by the method of kinetic equation. This problem is not discussed here; note only that the corresponding criteria of weak Čerenkov absorption based on the simple physical considerations are given in section 2.2. The possibility to neglect the Landau damping of plasma waves have been also discussed in that section. The same criteria hold for other types of waves in a magnetoactive plasma.

Cyclotron absorption has been analyzed by the method of kinetic equation in the papers [94, 287] in which the dispersion equation (2.41) with the tensor (2.99) has been solved. However, for the majority of cases this kind of absorption, which is closely related to the cyclotron emission by charged particles, can be calculated using the simpler approach based on the Kirchhoff law. The following resonant absorption coefficient at harmonics $s \geq 2$ is found with this method [230]:

$$\mu_{js} \approx \frac{a_{\omega j}}{I_\omega^{(0)}} = \mathcal{B}_{js} \frac{s^{2s}}{2^s s!} \frac{\omega}{c} \beta_T^{2s-3} \exp(-Z_{js}^2). \tag{6.37}$$

Here

$$\mathcal{B}_{js} = \sqrt{\frac{\pi}{2}}\, v \left[n_j^{2s-4} T_j^2 (1+g_j)^2 \cos\vartheta_j \right]_{\omega = s\omega_B} \frac{\sin^{2s-2}\alpha}{|\cos\alpha|},$$

$$Z_{js} = \frac{\omega - s\omega_B}{\sqrt{2}\omega n_j \beta_T \cos\alpha}, \tag{6.38}$$

$$\beta_T = v_T/c = \sqrt{\kappa T/m_e c^2},$$

[6] Thermal radiation escaping from an optically thick source into vacuum will be unpolarized: $T_{b\,1} = T_{b\,2}$, and $I_{\omega 1}^{(0)} = I_{\omega 2}^{(0)}$. However, degree of polarization of radiation inside the source may be non-zero due to the difference of the equilibrium intensities of ordinary and extraordinary waves in a magnetoactive plasma: $T_{b\,1} = T_{b\,2}$ but $I_{\omega 1}^{(0)} \neq I_{\omega 2}^{(0)}$ (see (1.37)).

α is the angle between the wave vector $\boldsymbol{k}_j$ and the magnetic field $\boldsymbol{B}_0$; ϑ_j is the angle between $\boldsymbol{k}_j$ and the group velocity $\boldsymbol{v}_{\text{gr}} = d\omega/d\boldsymbol{k}_j$. The parameters g_j, T_j and the refractive indices n_j are given by the formulas (2.142), (5.41) and (2.121). The formulas (5.132) and (1.37) have been also used to obtain (6.37). The inequalities (5.134) are the validity criteria of the obtained expressions for μ_{js} at harmonics $s \geq 2$.

Due to the exponential factor the coefficient $\mu_{js}(\omega)$ becomes quite small at frequencies not very close to $s\omega_B$ (i.e., in the wings of the absorption line, where $Z_{js}^2 \gg 1$). Therefore, the absorption spectrum looks like the series of discrete lines centered at frequencies $\omega = s\omega_B$.

The expression for μ_{js} is simplified when plasma is rarefied enough ($v \ll 1$). In this case the emissivity $a_{\omega j}$ is described by the formula (5.136). The cyclotron absorption coefficient has the form (6.37), but now [230]

$$\mathcal{B}_{js} = \sqrt{\frac{\pi}{2}}\, v \left[\frac{(1 + K_j \cos\alpha)^2}{1 + K_j^2}\right]_{\omega = s\omega_B} \frac{\sin^{2s-2}\alpha}{|\cos\alpha|}, \tag{6.39}$$

$$Z_{js} \equiv Z_s = \frac{\omega - s\omega_B}{\sqrt{2}\omega\beta_T \cos\alpha}.$$

The parameter K_j is given by the formula (5.55); $K_j = \pm 1$ for quasi-parallel propagation (2.126). The formulas (6.37), (6.39) are only valid at harmonics $s \geq 2$. The case of the first harmonic, $s = 1$ is more complicated. Cyclotron absorption coefficient at $\omega \simeq \omega_B$ can be found from the Kirchhoff law using the expression (5.133) for the cyclotron emissivity. This coefficient, however, is valid only in the wings of the absorption line (for $Z_{js}^2 \gg 1$). The entire profile of the line can only be obtained using the method of the kinetic equation. This method leads to the following expression for μ_{j1} [288]:

$$\mu_{j1} = \Lambda_{j1} \frac{\omega\beta_T}{c} |w(Z_{j1})|^{-2} \exp(-Z_{j1}^2) \tag{6.40}$$

where $w(Z_{j1})$ is the Kramp function (2.91) with the argument $Z_{j1} = (\omega - \omega_B)/\sqrt{2}\omega\beta_T n_j \cos\alpha$, and the factor Λ_{j1} is given by:

$$\begin{aligned}\Lambda_{j1} &= \sqrt{\frac{8}{\pi}}\frac{n_j}{v}(2n_j^2 \sin^2\alpha + 2v - 2 - \sin^2\alpha)^{-1} \\ &\times \left\{ n_j^4 (1 - v + \tfrac{7}{4} v \sin^2\alpha) + n_j^2 \left[(1 + v)(\frac{v}{2} - 1)\sin^2\alpha \right.\right. \\ &\left. + \frac{v^2}{4}\tan^2\alpha(1 + \cos^2\alpha) - (1 - v)(1 - \frac{v}{4})(1 + \cos^2\alpha)\right] \\ &\left. + (1 - v)(1 - \frac{v}{2}) + \frac{v^2}{4}(v - 2)\tan^2\alpha \right\} \cos\alpha \cos\vartheta_j.\end{aligned} \tag{6.41}$$

Being persistent enough one can check that the kinetic expression for μ_{j1} and the one obtained with the Kirchhoff law are identical in the region $Z_{js}^2 \gg 1$.

The above expression for Λ_{j1} is simplified significantly for the ordinary wave ($j=2$) in sufficiently rarefied plasma.[7] In this case the coefficient Λ_{21} has the form

$$\Lambda_{21} = \frac{v}{\sqrt{2\pi}} \frac{\sin^4 \alpha (1 + 2\cos^2 \alpha)^2}{(1 + \cos^2 \alpha)^3 |\cos \alpha|}, \tag{6.42}$$

and $Z_{12} = (\omega - \omega_B)/\sqrt{2}\beta_T \cos \alpha$. The expression for the absorption coefficient of extraordinary radiation at first harmonic in plasma with $v \ll 1$ is more complicated, since in this case $n_1^2 \to \infty$ unless thermal motion is taken into account. Characteristics of absorption of the ordinary waves at gyrofrequency have been investigated in [289] for the quasi-transverse and strictly transverse propagation in a dense plasma.

The above consideration has dealt with cyclotron absorption of electromagnetic waves, ordinary and extraordinary. However, if the frequencies of plasma resonance, ω_∞ (see (2.125)), coincide with one of the cyclotron harmonics $\omega = s\omega_B$ (i.e., the double plasma resonance is realized), then the cyclotron absorption of longitudinal plasma waves existing in the region $\omega \approx \omega_\infty$ becomes possible, too. The refractive index of these waves, n_3, is determined by the equation (2.153), and the cyclotron absorption coefficient is given by the formula (6.37) where [72]

$$\mathcal{B}_{js} = \mathcal{B}_{3s} =$$
$$\sqrt{\frac{\pi}{2}} v(1-u) n_3^{2s-2} \frac{\sin^{2s} \alpha}{|\cos \alpha|} (1 - u - v + uv \cos^2 \alpha)^{-1}, \tag{6.43}$$
$$Z_{js} = Z_{3s} = (\omega - s\omega_B)/\sqrt{2}\omega \beta_T n_3 \cos \alpha.$$

For all harmonics except the lowest ones the condition $s^2 \gg 1$ is well satisfied, so the expression for $\mathcal{B}_{3s}$ becomes simpler:

$$\mathcal{B}_{3s} = \sqrt{\frac{\pi}{18}} \frac{n_{\rm pl}^{2s-4}}{\beta_T^2} \frac{\sin^{2s} \alpha}{|\cos \alpha|}. \tag{6.44}$$

It was taken into account here that the parameter $u \ll 1$ for $s^2 \gg 1$, while the refractive index n_3 (see (2.153)) is reduced to $n_{\rm pl} \approx \sqrt{1-v}/\sqrt{3}\beta_T$ which describes plasma waves in an isotropic medium. The strong Čerenkov absorption (Landau damping) of these waves can be neglected if $1 - v \ll 1$. This inequality has been also taken into account when getting (6.44) from (6.43).

[7] It is supposed that $v \ll 1$ while the criterion (5.60) is still valid.

All the above formulas in this section are valid for the limit of dense plasma (5.60) which can be realized even at $v \ll 1$. If plasma is extremely rarefied, so that the criterion (5.61) is satisfied together with the inequality $v \ll 1$, the refractive indices $n_j \simeq 1$ for both modes. Cyclotron absorption of extraordinary waves in this case is described by the coefficient

$$\mu_{11} = \sqrt{\frac{\pi}{8}} \frac{\omega}{c} \frac{\omega_{\rm L}^2(1+\cos^2\alpha)}{\omega^2 \beta_T |\cos\alpha|} e^{-Z_1^2}, \tag{6.45}$$
$$Z_1 = (\omega - \omega_B)/\sqrt{2}\omega\beta_T \cos\alpha,$$

which can be obtained from (5.53) using the Kirchhoff law and taking into account (5.62). To get the corresponding absorption coefficient for the ordinary waves at the gyrofrequency, it is necessary to account for the corrections to the polarization coefficients (5.63) due to the thermal motion of plasma particles. Performing a procedure and using again the Kirchhoff law, we get

$$\mu_{21} = \frac{1}{\sqrt{2\pi}} \frac{\omega}{c} \frac{\omega_{\rm L}^2}{\omega^2} \beta_T \frac{\sin^4\alpha \cos^4\alpha (3+\tan^2\alpha)^2}{|W(Z_1)|^2 |\cos\alpha| (1+\cos^2\alpha)^3} e^{-Z_1^2}, \tag{6.46}$$

where $W(Z_1)$ is the Kramp function (2.91). The formulas (6.45) and (6.46) can also be derived by the method of kinetic equation if one takes into account the terms of the order of $b\beta_T^2$ in the tensor $\epsilon_{\alpha\beta}^{\rm pl}$ (2.145). The details of the calculation can be found in [104, 290]. At the same time, the formulas (6.37), (6.39) with the polarization coefficients K_j in the form (5.55) properly describe cyclotron absorption in rarefied plasma at $s \geq 2$.

Next, in plasma with temperature anisotropy $T_\perp \neq T_\parallel$ in which the electron distribution is determined by the function (5.142) the cyclotron absorption coefficients have the form:

$$\mu_{11} = \sqrt{\frac{\pi}{8}} \frac{\omega_{\rm L}^2}{\omega_B c} \beta_{T_\parallel}^{-1} \frac{1+\cos^2\alpha}{|\cos\alpha|} e^{-Z_1^2}, \tag{6.47}$$

$$\mu_{21} = \frac{\omega_{\rm L}^2}{\sqrt{2\pi}\omega_B c} \beta_{T_\parallel} \frac{\sin^4\alpha \cos^4\alpha (3+\tan^2\alpha)^2}{|W(Z_1)|^2 |\cos\alpha| (1+\cos^2\alpha)^3} e^{-Z_1^2} \tag{6.48}$$

$$\mu_{js}(s \geq 2) = \sqrt{\frac{\pi}{2}} \frac{\omega_{\rm L}^2}{\omega_B c} \frac{\beta_{T_\perp}^{2s-2}}{\beta_{T_\parallel}} \frac{s^{2s-1}}{2^s s!} \frac{\sin^{2s-2}\alpha (1+K_j\cos\alpha)^2}{|\cos\alpha|(1+K_j^2)} e^{-Z_s^2}. \tag{6.49}$$

They can be derived from the expressions (5.143)-(5.145). However, the Kirchhoff law in the form (4.36) cannot be used for thermally anisotropic plasma. It should be replaced by the equivalent relation

$$a_{\omega j} = \mu_{js} \frac{\omega^2 \kappa T_\perp}{(2\pi)^3 c^2}$$

which will be derived in the next section 7.1 for the case $T_\perp \neq T_\parallel$ (see (7.51)). Note that the above relation, as well as (7.51) takes into account that $n_j \simeq 1$ and $\vartheta_j = 0$ in rarefied plasma. If plasma is thermally isotropic, $T_\perp = T_\parallel = T$, the formulas (6.47)-(6.49) are reduced to (6.45), (6.46) and (6.39), respectively.

Comparison of the formulas (6.45) and (6.46) shows that cyclotron absorption of ordinary waves at gyrofrequency (in the region $Z^2 \lesssim 1$) in the limit of rarefied plasma is considerably weaker than that of extraordinary ones: $\mu_{21} \sim \mu_{11}\beta_T^2$. This fact has been already noted in the section 5.1. Weak absorption of the ordinary waves as compared to extraordinary ones is explained by the character of mode polarization. Polarization of extraordinary waves in the approximation (5.62), (5.63) is identical to that of cyclotron emission by a circularly moving electron in vacuum, whereas ordinary wave polarization is opposite. Evidently, such an electron emits and absorbs only extraordinary waves. Weak emission and absorption of the ordinary waves comparable to the corresponding effects at the second harmonic $s = 2$ appear due to the influence of electron thermal motion which makes polarization of the ordinary mode somewhat different from (5.63).

In the limit of dense plasma (5.60) the mode polarization is radically different. Electric field $\boldsymbol{E}$ of both modes has the component along the wave vector $\boldsymbol{k}$, and its projection onto the plane perpendicular to the magnetic field $\boldsymbol{B}_0$ rotates in the direction opposite to electron gyration. This leads to the depression of the cyclotron emission at gyrofrequency (see section 5.1) and the corresponding decrease of absorption coefficients of both modes at the first harmonic to the level typical for the second one [230].

If rarefied plasma is permeated by such a strong magnetic field that mode polarization is dominated by the effect of vacuum birefringence,[8] the formulas for cyclotron absorption coefficients are considerably simplified [104]:

$$\mu_{1s} = \sqrt{\frac{\pi}{2}} \frac{s^{2s}}{2^s s!} \frac{\omega}{c} \frac{\omega_{\rm L}^2}{\omega^2} \beta_T^{2s-3} \sin^{2s-2}\alpha |\cos\alpha| e^{-Z_s^2}, \tag{6.50}$$

$$\mu_{2s} = \sqrt{\frac{\pi}{2}} \frac{s^{2s}}{2^s s!} \frac{\omega}{c} \frac{\omega_{\rm L}^2}{\omega^2} \beta_T^{2s-3} \frac{\sin^{2s-2}\alpha}{|\cos\alpha|} e^{-Z_s^2}, \tag{6.51}$$

where $Z_s = (\omega - s\omega_B)/\sqrt{2}\omega\beta_T\cos\alpha$. These formulas are valid for an equilibrium, classical plasma ($\hbar\omega_B \ll \kappa T$) at any harmonic including the first one. The corresponding absorption coefficient at arbitrary ratio $\hbar\omega_B/\kappa T$ has been derived in [223]:

$$\mu_{js} = \sqrt{\frac{\pi}{2}} \frac{\omega_{\rm L}^2 \omega_B a_j}{\omega^2 \beta_T c |\cos\alpha|} \frac{s^2}{s!} \left(\frac{\hbar\omega^2 \sin^2\alpha}{2m_e c^2 \omega_B}\right)^{s-1} \frac{\sinh(\hbar\omega/2\kappa T)}{[\exp(\hbar\omega_B/\kappa T) - 1]^{-s}}$$

[8] This case is realized under the conditions (2.187), (2.189).

$$\times \exp\left[\frac{s\hbar\omega_B}{2\kappa T} - \frac{(\omega - s\omega_B)^2 - (\omega - s\omega_B)\hbar\omega^2/m_e c^2}{2\beta_T^2\omega^2\cos^2\alpha}\right]. \tag{6.52}$$

Here $a_1 = \cos^2\alpha$ and $a_2 = 1$ are polarization coefficients of extraordinary and ordinary modes in birefringent vacuum, and the parameter $\beta_T = \sqrt{\kappa T/m_e c^2}$ characterizes electron distribution over longitudinal velocities with respect to the magnetic field $\boldsymbol{B}_0$ and also over transverse velocities in the quasi-classical limit $\hbar\omega_B \ll \kappa T$. In the latter case the formula (6.49) is transformed into (6.47), (6.48). On the contrary, the absorption coefficients in quantized plasma $\hbar\omega_B \gg \kappa T$ have the following form:

$$\mu_{js} = \sqrt{\frac{\pi}{2}}\frac{\omega_L^2\omega_B a_j}{\omega^2\beta_T c|\cos\alpha|}\frac{s^2}{s!}\left(\frac{s\hbar\omega^2\sin^2\alpha}{2m_e c^2\omega_B}\right)^{s-1}$$
$$\exp\left[-\frac{(\omega - s\omega_B)^2 - (\omega - s\omega_B)(1+\cos^2\alpha)\hbar\omega^2/m_e c^2}{2\beta_T^2\omega^2\cos^2\alpha}\right]. \tag{6.53}$$

The second term in the exponent describes the effect of the recoil on the cyclotron line profile. The analogous formula for $s = 1$ has been obtained in [292].

In a system of plasma with $T_\perp \neq T_\parallel$ and birefringent vacuum the coefficient of resonance cyclotron absorption is also determined by the formula (6.49) in which, however, the parameter $\beta_T = \sqrt{\kappa T_\parallel/m_e c^2}$ describes the dispersion of electrons over longitudinal velocities, and the value T stands for the transverse electron temperature $T_\perp$ (see also [54]).

Note that the above coefficients of cyclotron absorption in plasma with temperature anisotropy play the role of coefficients of resonance cyclotron scattering if cyclotron radiation transfer is dominated by scattering rather than pure emission and absorption (see the section 8.2.1).

Compare cyclotron and bremsstrahlung absorption coefficients in astrophysical plasmas. The former one for ordinary (at harmonics $s \geq 1$) and extraordinary (at $s \geq 2$) waves propagating at angles $\alpha \sim 1$ in plasma with $v \ll 1$ can be estimated as

$$\mu_{2,s=1} \sim \frac{\omega}{c} v\,\beta_T, \qquad \mu_{j,s\geq 2} \sim \frac{s^{2s}}{2^s s!}\frac{\omega}{c} v\beta_T^{2s-3} \tag{6.54}$$

(see formulas (6.37), (6.39) and (6.40), (6.42)). Note that $\mu_{2,s=1} \sim \mu_{j,s=2}$ since the emission at gyrofrequency by electrons in magnetoactive plasma is weak (see section 5.1). The above estimates of μ_{js} correspond to the frequencies inside the absorption lines $Z_{js}^2 \leq 1$; cyclotron absorption becomes exponentially small in the wings of the lines where $Z_{js}^2 \gg 1$. Then, for $v \ll 1$ and $|\sqrt{u}\cos\alpha| \ll 1$, the bremsstrahlung absorption coefficient is equal to:

$$\mu_j^{\rm b} = 2\frac{\omega}{c}\eta_j \sim \frac{v\nu_{\rm eff}}{c} \tag{6.55}$$

(see (6.33)). The obtained rough estimates are enough to illustrate the importance of the effects of cyclotron absorption in astrophysical conditions. For example, at the first and the second harmonics $\mu_{js} \sim \omega v\beta_T/c$, so

$$\frac{\mu_{j,s=1,2}}{\mu_j^{\rm b}} \sim \frac{\omega\beta_T}{\nu_{\rm eff}} \tag{6.56}$$

In the solar corona $\beta_T \sim 10^{-2}, \nu_{\rm eff} \sim 10\,{\rm s}^{-1}$, so this ratio is about $6 \cdot 10^6$ at frequencies $\omega \sim 2\pi \cdot 10^9\,{\rm s}^{-1} (\lambda \sim 30\,{\rm cm})$. Thus, in coronal plasma at frequencies near $\omega \approx \omega_B$, $2\omega_B$ the cyclotron absorption coefficient is a few million times greater than that caused by particle collisions [116, 256].

The cyclotron absorption of plasma waves is even stronger. The corresponding absorption coefficient is approximately β_T^{-2s+2} times greater than that for electromagnetic waves. This can be easily deduced from the formulas (6.37), (6.38) and (6.44). These formulas at $u \ll 1$ and $1 - v \ll 1$ yield the following estimates of the cyclotron absorption coefficients for electromagnetic waves:

$$\mu_{1,2} \sim (s^{2s}/2^s s!)(\omega/c) n_{1,2}^{2s-4} \beta_T^{2s-3},$$

and for plasma waves:

$$\mu_3 \sim (s^{2s}/2^s s!)(\omega/c)(\beta_T n_{\rm pl})^{2s-4} \beta_T^{-1}.$$

However, in plasma with $u \ll 1$ the quantity $\beta_T n_{\rm pl} \approx n_{1,2} = 1 - v$, hence the ratio $\mu_3/\mu_{1,2} \sim \beta_T^{-2s+2} \gg 1$. This fact is explained by the resonance properties of plasma, which make the emission of plasma waves by particles more effective than that of electromagnetic waves. Cyclotron damping of plasma waves have been applied in [291] to explain some features of type III fast-drifting bursts, namely, the absorption bands in their dynamic spectra.

To judge more definitely on the influence of cyclotron absorption on the electromagnetic waves propagation in cosmic plasma, one should account for the inhomogeneity of the magnetic field. Whereas the spectrum of cyclotron absorption (i.e., the function $\mu_{js}(\omega)$) in a homogeneous magnetic field is the sequence of distinct "lines" at frequencies $\omega = s\omega_B$, it is smoothed out in a sufficiently inhomogeneous magnetic field. In the latter case the cyclotron absorption (and emission) at a given frequency ω takes place within the localized layers where the magnetic field strength matches the condition $\omega \approx s\omega_B = seB_0/m_e c$. The optical depth of such a gyroresonant layer is $\tau_{js} = \int \mu_{js} dl$. Using linear approximation for $\omega_B(l)$ within this layer, one can write the difference $\omega - s\omega_B(l)$ as

$$\omega - s\omega_B(l) = \frac{\omega l}{L_B}, \tag{6.57}$$

where $L_B = \omega_B |dl/d\omega_B|$ is the characteristic scale of the magnetic field inhomogeneity along a ray. The optical depth of a gyroresonant layer is found by integrating over l the cyclotron absorption coefficient μ_{js} given by (6.37) and taking into account (6.57), [230]:

$$\tau_{js} = \sqrt{2\pi} \frac{s^{2s}}{2^s s!} \frac{\omega}{c} \beta_T^{2s-2} n_j \mathcal{B}_{js} L_B |\cos\alpha|. \tag{6.58}$$

Here, as before, the factor $\mathcal{B}_{js}$ at harmonics $s \geq 2$ has the form (6.38). In a low-density plasma with $v \ll 1$ this factor is reduced to (6.39). Note, that (6.39) is also applicable at the first harmonic in a "vacuum" approximation, when the inequality (5.61) is satisfied. But if the opposite criterion (5.60) is valid, then the optical depth of the first gyroresonant layer should be obtained by integrating the expression (6.40). With (6.57) it leads to:

$$\tau_{js} = \sqrt{2}\Lambda_{j1} \frac{\omega L_B}{c} \beta_T^2 n_j |\cos\alpha| \int_{-\infty}^{\infty} |w(Z)|^{-2} \exp(-Z^2) dZ, \tag{6.59}$$

where the integral equals $\pi^{3/2}/2$.

The size of a gyroresonant layer along l can be estimated as follows. The absorption coefficient, μ_{js} at a given frequency ω drops exponentially in the regions where $Z_{js}^2 \gg 1$ (see (6.37), (6.40)). Thus, the main contribution to the τ_{js} comes from the thin layer located near the surfaces at which $\omega = s\omega_B$. In such a layer $Z_{js}^2 = [(\omega - \omega_B)/\sqrt{2}\omega\beta_T n_j \cos\alpha]^2 \leq 1$. From (6.57) it is clear that the latter is valid if $l \leq \sqrt{2} L_B \beta_T n_j |\cos\alpha|$. Therefore, the linear size of the layer is

$$L_{js} \approx 2\sqrt{2} L_B \beta_T n_j |\cos\alpha|. \tag{6.60}$$

Note that L_{js} does not depend on the number of the gyroresonant layer (on the harmonic number) s, and the inequality $L_{js} \ll L_B$ is satisfied due to the small value of $\beta_T n_j$ (see (5.134)). According to (6.57), the adjacent gyroresonant layers with numbers s and $s+1$ are spaced approximately L_B/s apart, so they do not overlap if $L_{js} \ll L_B/s$, i.e., under the condition

$$2\sqrt{2} s \beta_T n_j |\cos\alpha| \ll 1. \tag{6.61}$$

The frequency spectrum of cyclotron radiation from plasma with linear dimensions of L situated in an inhomogeneous magnetic field depends on the relationship between L, L_B and L_{js}. If $L \ll L_{js}$, the absorption spectrum exhibits a line structure which is typical for plasma in a homogeneous magnetic field. If $L_{js} \ll L \ll L_B/s$, then the spectrum, $\tau_j(\omega)$, remains "fragmented" though the shape of individual lines will strongly depend on the magnetic field distribution over the source. Finally, under the condition $L \gg L_B/s$ there exist several gyroresonant layers in plasma at any given frequency, so the function $\tau_j(\omega)$ becomes smooth. Of course, the line spectrum vanishes at any L,L_B and L_{js} under the condition opposite to (6.61) when adjacent harmonics are overlapped.

Discussions of astrophysical aspects of cyclotron radiation mechanism were started in the beginning of the sixtieth in [293] in connection with the discovery of the Jovian decimetric radio emission and also in [69]. The hypothesis on the cyclotron nature of the Jovian decimetric radio emission required the existence of the magnetic fields at the planet. The value of the Jovian magnetic field was unknown at that time. Later on it was found that this value is of the order of tens gauss, and thus the assumption on the cyclotron mechanism of the Jovian decimetric radio emission was ruled out. However, it is quite probable that a coherent cyclotron mechanism gives rise to the radio emission at lower frequencies, in the decametric band, though the detailed picture of the generation of this type of Jovian radio emission is far from clear. Investigations of this problem have been extensively carried out up to now [294, 295]. As for the theory of solar radio emission, the application of the cyclotron mechanism proved to be more fruitful. First, essential role of the thermal cyclotron mechanism in the generation of microwave radiation from the local sources on the Sun has been noted [116]. Then, the possible contribution of this mechanism to the radio emission from Ap-stars (which has not yet been detected) and flaring stars of UV Ceti type has been estimated [37, 296, 297] Finally, further investigations have clarified the role of the cyclotron effects in the formation of frequency spectra of astrophysical objects with strong magnetic fields: magnetic white dwarfs, X-ray pulsars and gamma-ray bursts. Since these effects are closely related to the process of cyclotron scattering, their analysis is postponed to chapter 8.

Turning to the more detailed analysis of the thermal cyclotron mechanism of solar radio emission, we note, first of all, that the condition $L \gg L_B/s$, where L is the size of the plasma region, L_B is the characteristic scale of the magnetic field inhomogeneity, and s is the harmonic number, can be most probably realized in active regions of the chromosphere and corona above sunspots. According to the above, there are several gyroresonant layers at any fixed frequency ω, and the cyclotron radiation spectrum is smoothed out. The gyroresonant layers $\omega = \omega_B$, $2\omega_B$, $3\omega_B \ldots$ in the magnetic field of a sunspot are sketched in Fig. 6.3. From this figure it

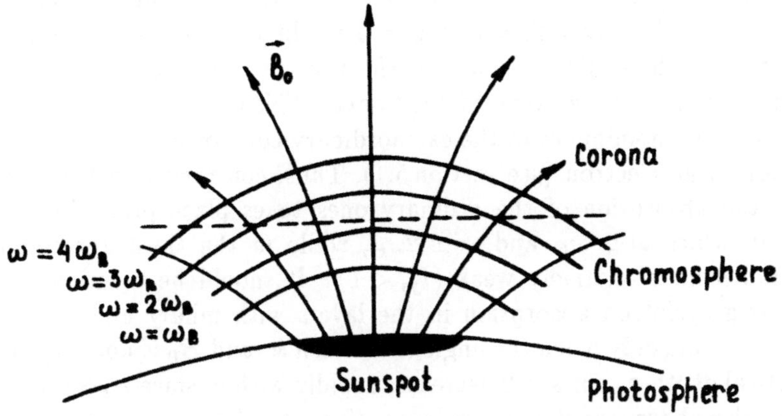

Figure 6.3. Location of gyroresonant layers in the unipolar magnetic field.

is clear that for a given frequency ω the layers corresponding to lower harmonics are situated closer to the sunspot, in a region at which the magnetic field is stronger. As frequency ω decreases, the entire system of gyroresonant layers shifts upward to the region in which the magnetic field is weaker.

The optical depth of the gyroresonant layers localized near the above mentioned levels can be easily estimated if one multiplies the cyclotron absorption coefficient μ_{js}, (6.54), by the effective thickness of the layer $L_{js} \sim L_B \beta_T$ (see (6.60)):

$$\tau_{2,s=1} \sim \frac{\omega}{c} v \beta_T^2 L_B, \qquad \tau_{j,s\geq 2} \sim \frac{s^{2s}}{2^s s!} \frac{\omega}{c} v \beta_T^{2s-2} L_B. \tag{6.62}$$

In the inner corona $N \sim 10^9\,\mathrm{cm}^{-3}$, $L_B \sim 10^9\,\mathrm{cm}$, $\beta_T \sim 10^{-2}$, so the following estimates of the optical depth of gyroresonant layers at frequencies $\omega \sim 2 \cdot 10^{10}\,\mathrm{s}^{-1} (\lambda \sim 10\,\mathrm{cm})$ can be given: $\tau_{2,s=1} \sim 5 \cdot 10^2$, $\tau_{j,s=2} \sim 10^3$, $\tau_{j,s=3} \sim 1$, $\tau_{j,s=4} \sim 10^{-3}$, etc. Although very rough, these estimates leave no doubt that the gyroresonant absorption in the solar corona leads to a substantial attenuation of radiation passing through the layers $\omega = \omega_B$, $\omega = 2\omega_B$, and $\omega = 3\omega_B$, whereas the layers $s \geq 4$ remain transparent for radio waves. Of course, this occurs only in the regions with strong magnetic fields. Indeed, at frequencies $\omega \sim 2 \cdot 10^{10}\,\mathrm{s}^{-1}$ the harmonics $s = 1, 2, 3$ correspond to the magnetic fields $B_0 \sim 3 \cdot 10^2 - 10^3\,\mathrm{G}$.

The conclusion on the role of the first three gyroresonant layers in the solar corona conditions has been generally confirmed by the rigorous calculations of the optical depth τ_{js} of the corona and chromosphere based on the formulas (6.58),(6.59) (see [284,285,298]). These calculations have shown that in a very wide wavelength range including centimeter, millimeter and meter bands the cyclotron absorption is caused mainly by the layers in which $\omega = 2\omega_B$ and $\omega = 3\omega_B$. In these layers $\tau_{js} \gtrsim 1$. The absorption of the ordinary waves in the regions $\omega = s\omega_B$ is less effective compared to that of the extraordinary waves. This can finally be attributed to a general predominance of the extraordinary component in the cyclotron radiation of an electron (see section 5.1). Thus, contrary to extraordinary waves, the absorption of the ordinary ones takes place primarily in the layers in which $\omega \simeq \omega_B$ and $\omega \simeq 2\omega_B$, while in the third gyroresonant layer it becomes relatively weak $(\tau_{js} < 1)$. It should be mentioned that the strong cyclotron absorption in the layers with mildly high harmonic numbers occurs only when the angle α between $\boldsymbol{k}_j$ and $\boldsymbol{B}_0$ is not very small: the optical depth τ_{js} for $s \geq 2$ decreases rapidly with α, since $\tau_{js} \propto \sin^{2s-2} \alpha$. Due to the above angular dependence of τ_{js} the layer $\omega \approx 3\omega_B$ becomes optically thick for the ordinary waves when α is sufficiently large. One should bear this fact in mind when performing the detailed analysis of the cyclotron absorption (and emission) in the corona and chromosphere (see [3]). For the sake of simplicity the dependence of τ_{js} on α will not be taken into account in the subsequent considerations.

The assumption on the existence of the optically thick gyroresonant layers in active regions above the sunspots with a strong magnetic field lies in the basis of the theory [298–300] which explains the main spectral and polarization characteristics of the slowly varying component (S-component) of the solar radio emission in terms of the thermal cyclotron mechanism.

The brightness temperature T_b of radiation from gyroresonant layers can be found easily owing to their small size along a ray, L_{js}, as compared to the characteristic scale of the kinetic temperature T. The value of T_b is determined by the simple formulas (4.42)-(4.44). They indicate that the brightness temperature of an optically thick gyroresonant layer $(\tau_{js} > 1)$ equals the kinetic temperature of plasma in the layer. For optically thin layers at higher harmonics $(\tau_{js} \ll 1)$ the brightness radiation temperature is much lower: $T_b \approx T\tau_{js} \ll T$.

As follows from Fig. 6.3 and the above analysis, the influence of the layers $\omega \approx s\omega_B$ for $s = 2$ and $s > 4$ on the extraordinary radiation from a coronal or chromospheric region above a sunspot is insignificant. The extraordinary waves from the layer $\omega \approx 2\omega_B$ cannot escape the corona since they are absorbed by the optically thick layer $\omega \approx 3\omega_B$ which is situated higher above the spot, in the region in which the magnetic field is weaker. The brightness temperature of the extraordinary radiation from the optically thin layers $\omega \approx 4\omega_B$, $5\omega_B, \ldots$ is very low: $T_b \ll T$. So, when analyzing the cyclotron mechanism of radiation by the local sources above sunspots, one can assume that the extraordinary component is emitted mainly by the third gyroresonant layer. The corresponding value of T_b is close to the kinetic temperature in this layer. Cyclotron emission of the ordinary component by the layers $\omega \approx \omega_B$, $\omega \approx 3\omega_B$, $4\omega_B, \ldots$ is negligible due to the same reasons as in the case of extraordinary waves. Thus, the role of the effectively emitting region is played here by the layer in which $\omega \approx 2\omega_B$. The resulting brightness temperature of the ordinary component is close to the kinetic temperature in the second gyroresonant layer.

The frequency spectrum of thermal cyclotron radiation is determined by the distribution of the temperature T and the magnetic field $\boldsymbol{B}_0$ above sunspots. For our analysis it is important that the magnetic field decreases gradually with height, whereas the kinetic temperature rises sharply in the transition region from the chromosphere to the corona. For large spots with a strong magnetic field the layers $\omega \approx 2\omega_B$, $3\omega_B$, which effectively emit in the decimeter band, lie in the corona. The kinetic temperature $T_c \simeq (1-3)\cdot 10^6$ K is nearly the same in both layers. So, in the decimeter band the brightness temperature of a local region above the spot $T_b \approx T_c$, and the specific intensity $I_\omega(\omega) \propto \omega^2 T_b$. In the centimeter band the intensity I_ω becomes a decreasing function of ω. This is caused by the shift of the layers which effectively emit at higher frequencies into the colder chromosphere. As a result, the brightness temperature of radiation T_b decreases rapidly (steeper than ω^{-2}), so the intensity $I_\omega(\omega) \propto \omega^2 T_b$ decreases with increasing ω. Such a frequency spectrum with maximum of the intensity at wavelengths $\lambda \sim 5-10$ cm and the nearly constant brightness temperature $T_b \sim (1-3)\cdot 10^6$ K at longer wavelengths fits well the observed spectrum of radio emission from typical local regions above sunspots, the sources of the S-component.

Rather strong polarization of the S-component in the centimeter domain (with the extraordinary radiation being dominant in the spectrum), as well as its absence in the decimeter band, are also well explained in terms of the thermal cyclotron mechanism. Indeed, at frequencies in the centimeter band the effectively emitting layers, $\omega \approx 2\omega_B$ and $\omega \approx 3\omega_B$, are located in the transition region from the chromosphere to the corona, where the kinetic temperature gradient is large. Since the layer $\omega \approx 3\omega_B$ is situated higher than the one $\omega \approx 2\omega_B$, the brightness temperature of the former one will be larger than that of the latter. The extraordinary component is emitted by the "hotter" layer $\omega \approx 3\omega_B$, and the ordinary one is related to the "colder" layer $\omega \approx 2\omega_B$. Thus, the extraordinary component prevails in the observed radiation, and the S-component becomes polarized at centimeter wavelengths. As wavelength increases, the brightness temperatures of the layers $\omega \approx 2\omega_B$ and $\omega \approx 3\omega_B$ become equal because both layers are shifted into the isothermal corona. This results in a strong decrease of the degree of polarization of the radiation by a local source in the decimeter band. [9]

[9] See [3, 284, 285] for more details on the origin of the S-component in solar radio emission.

It is however clear that frequency spectra of plasma sources in complex magnetic field structures can be much more complicated and exhibit fine structure, for example, distinct line features at cyclotron harmonics or high-frequency steps (the sharp decrease of the intensity with increasing frequency) at these harmonics [301–303]. Discovery and investigation of such spectral features can be an effective tool for plasma diagnostics in solar active regions, first and foremost, the magnetic field strength and the kinetic temperature of the coronal plasma.

X-ray observations of solar flares reveal that in some events an intensive X-ray emission comes from small cores which exist during a short time of about ten seconds at the tops of flaring loops filled by coronal plasma. According to modern concepts, such cores can be the locations of energy release regions which govern the whole scenario of the subsequent solar flare. Small cores of hot plasma of a size much smaller than the size of the gyroresonant layer is the simplest object which may exhibit distinct lines at cyclotron harmonics. Radiation in optically thick lines $\tau_s > 1$ has the blackbody intensity corresponding to the kinetic temperature of plasma in the core. Therefore, analysis of the radio emission spectrum may yield the values of the magnetic field, kinetic temperature and plasma density in such cores. At the same time, it is quite improbable that line features can be seen in the integral emission from the entire flaring loop, since any fine structure in the spectrum is smoothed out due to magnetic field inhomogeneity over the source. Cyclotron lines can only be registered using special methods of observation: they should be searched with narrow-beam antenna receiving radio emission from the small part of the loop in which the magnetic field inhomogeneity is insignificant.

According to one hypothesis on the origin of solar flares, the powerful energy release at the initial phase of a flare results from the process of reconnection of the magnetic field lines in neutral current sheets located in the flare region. The problem of detection of such layers is thus quite important for solar physics. In this respect, analysis of microwave spectra of hot current sheets and, in particular, cyclotron lines from these sources can be very useful for their studies. Note that the conclusion on the possibility to observe any fine structure in the current sheet spectrum seems to be quite non-trivial since the magnetic field strongly varies from the maximal value $B_0^{\max}$ at one edge to zero in the center and the again to the maximal field of the opposite direction at another edge. Nevertheless, the detailed analysis made in [301] has shown that the line spectrum can appear due to cyclotron radiation from the periphery of the current sheet where the magnetic field is relatively homogeneous. As for cyclotron emission from central parts of the sheet in which the magnetic field $|B_0| < B_0^{\max}$ is strongly inhomogeneous, it cannot escape the source since the layer $\omega = \omega_L$ restricting the propagation and escape of radiation at frequencies $\omega > \omega_L$ is located near the current sheet periphery.

Discovery and investigation of cyclotron lines in solar radio emission requires both excellent spectrographic equipment and antenna systems of extremely high directivity. The observations reported in [304] gave some evidence for the existence of cyclotron lines in solar radio emission. The analysis of these observations and their possible interpretation has been given in [305]. However, the reliable detection of cyclotron features and using them for the diagnostic of coronal plasma in solar active regions is a matter for future investigations.

In conclusion, we remark on the gyrosynchrotron absorption in thermal plasmas. The corresponding absorption coefficients can be found using the Kirchhoff law (4.36) and the gyrosynchrotron emissivity into each mode. However, the formulas (5.163) and (5.164) in the previous chapter characterize the total emissivity into both modes. Using these expression, one can only get some "total" absorption coefficient which, in fact, can only be used for the rough estimates of the role of synchrotron reabsorption in an extended source. Correct expressions for the coefficients of gyrosynchrotron absorption μ_j in thermal plasmas can only be found from the Kirchhoff law if one uses the gyrosynchrotron emissivities $a_{\omega j}$ of equilibrium plasma into each mode which has been derived in [267].

Reabsorption and amplification of radiation in non-equilibrium plasma

Variation of radiation intensity in plasma with non-equilibrium particle distribution is usually described by the radiation transfer equation (4.3). Intensity of radiation escaping a homogeneous layer of geometrical thickness L is equal to

$$I_\omega = \frac{a_\omega}{\mu_j}(1 - e^{-\mu_j L}) + I_{\omega 0} e^{-\mu_j L}. \tag{7.1}$$

In contrast to the case of an equilibrium plasma, the ratio a_ω / μ_j is not equal to the Plank intensity $I_\omega^{(0)}$ (the Kirchhoff law is violated). Moreover, this ratio can be negative for negative absorption coefficient μ_j. Radiation which propagates through a non-equilibrium system with $\mu_j < 0$ is amplified: $I_\omega \propto e^{|\mu_j L|}$, and the outgoing intensity emitted by optically thick plasma layer ($|\mu_j L| \gg 1$) rises to the value

$$I_\omega \approx a_\omega L \frac{e^{|\mu_j L|}}{|\mu_j L|}.$$

This intensity is much larger than $a_\omega L$, the total power of radiation by particles along the ray. Thus, according to the definitions of section 1.3, a coherent radiation mechanism is realized in plasma. Since the above enhancement of radiation occurs due to its amplification (negative reabsorption[1]) in the source, this coherent mechanism should be referred to maser type. Effective maser mechanism can only be realized in a system which satisfies both the condition $\mu_j < 0$ and the criterion $|\mu_j L| \gg 1$.

Appearance of the negative reabsorption and the value of the increment $(-\mu_j)$ essentially depend on the particle distribution function $f_0(\boldsymbol{p})$ in the non-equilibrium system. Investigation of maser emission mechanisms which this chapter is devoted to is reduced, to a great extent, to the calculation of the reabsorption coefficient (decrement) in a non-equilibrium plasma and deriving criteria for amplification of radiation. The next stage of the analysis is the study of the influence of the amplified radiation on the system of emitting particles. In what follows this is done in the framework of the so-called quasi-linear theory of wave processes.

[1] Extinction of radiation is called reabsorption if the radiation is absorbed by the same particles which emit it.

7.1. Method of kinetic equation and method of Einstein coefficients

Maser generation mechanisms in plasma physics are investigated mainly by the following two methods: the classical method of kinetic equation and the elementary quantum method of Einstein coefficients. This section gives general description of both methods, while their specific applications to the maser mechanisms in cosmic plasma are given in sections 7.2-7.4.

7.1.1. METHOD OF KINETIC EQUATION. AMPLIFICATION AND CONVECTIVE INSTABILITY

This method is based upon the solution and analysis of the system of electromagnetic field equations (2.19)-(2.22) and the kinetic equation (2.76), or, in other words, on the solution of the kinetic equation with a self-consistent field. The capabilities of this method have been demonstrated in section 2.1 which is devoted to the linear theory of waves in stationary homogeneous plasma. It is assumed in this theory (which is valid, strictly speaking, for waves of vanishing amplitude) that the unperturbed part of the distribution function, $f_0(\boldsymbol{p})$, which does not depend on the high-frequency fields, is constant in space and time, whereas the rapidly oscillating part $f_0'(\boldsymbol{p})$ is small enough to approve the linearization of the kinetic equation. Then the calculation of the absorption coefficient μ_j and the decrement γ_j is reduced to the solution of the dispersion equation (2.41), the condition at which the linear system of the field equations (2.25), (2.26) has a nontrivial solution. Note that this system contains the dielectric tensor $\epsilon_{\alpha\beta}(\omega, \boldsymbol{k})$ in the form (2.84) derived with the kinetic equation (2.79).

The dispersion equation relates two complex values, ω and k_j which correspond to a wave propagating in the direction $\boldsymbol{k}/k$. However, in some cases it is more convenient to assume k_j or ω to be real. In the first case the complex frequency

$$\omega \equiv \Omega - i\gamma = \varphi_1(k_j) \tag{7.2}$$

is found using real k_j, whereas in the second one the complex wave number

$$k_j \equiv \frac{\omega}{c}(n_j + i\eta_j) = \varphi_2(\omega) \tag{7.3}$$

is obtained for real ω (see (2.55) and (2.56)).

Dealing with the problem of stability of electromagnetic disturbances, i.e., the problem of temporal behavior of the energy of these disturbances $W(t)$, one should use real values of k_j and put $\operatorname{Im} k_j = 0$. Suppose that a local disturbance (for example, an electromagnetic fluctuation) appears in plasma at the initial moment $t = 0$. Fourier expansion of this disturbance over real $\boldsymbol{k}$ is

$$\boldsymbol{E}(\boldsymbol{r}, t=0) = \int \boldsymbol{g}_1(\boldsymbol{k}) \exp(i\boldsymbol{k}\boldsymbol{r}) d^3k. \qquad (7.4)$$

Any Fourier component gives rise to a normal wave at $t > 0$. The frequency ω of this wave is related to $\boldsymbol{k}$ according to (7.2). Then

$$\boldsymbol{E}(\boldsymbol{r}, t) = \int \boldsymbol{g}_1(\boldsymbol{k}_j) \exp(i\boldsymbol{k}_j \boldsymbol{r} - i\omega t) d^3k, \qquad \omega = \varphi_1(\boldsymbol{k}_j), \qquad (7.5)$$

where

$$\int |\boldsymbol{E}|^2 \, dV = \int |\boldsymbol{g}_1|^2 e^{-2\gamma_j t} \, d^3k, \qquad (7.6)$$

as follows from the theory of Fourier integrals [306].

Note that the integral in the left-hand side is proportional to the energy of disturbance W. Thus, the behavior of the function $W(t)$ is completely determined by the value and sign of the decrement γ_j. Specifically, if the decrement is positive, $\gamma_j > 0$ for any $\boldsymbol{k}_j$ for which $\boldsymbol{g}_1(\boldsymbol{k}_j) \neq 0$, then $W(t) \to 0$ as $t \to \infty$. This means that the plasma is stable with respect to disturbances containing the modes with the above values of the wave vector $\boldsymbol{k}_j$. On the contrary, if $\gamma_j < 0$ in a region of $\boldsymbol{k}_j$ for which $\boldsymbol{g}_1(\boldsymbol{k}_j) \neq 0$, then $W(t)$ will infinitely grow as $t \to \infty$. The infinite growth of $W(t)$ appears only within the linear approximation. In fact, the increase of $W(t)$ will be limited by various non-linear effects (see sections 7.2,8.1).

According to [307], instabilities can be divided into two types: convective and absolute. Let $\boldsymbol{g}_1(\boldsymbol{k}_j) \exp(i\varphi_1(\boldsymbol{k}_j)t)$ be bounded at any $\boldsymbol{k}_j$. Then the field (7.5) describes a spatial "wave packet" in which $\boldsymbol{E}$ goes to zero as $\boldsymbol{r}$ becomes infinite.[2] If such a space-like packet is also a time-like one, i.e., $\boldsymbol{E}(\boldsymbol{r}, t) \to 0$ at $\boldsymbol{r} = \text{const}$ and $t \to \pm\infty$, then, by definition, the instability is convective. Otherwise, an instability is absolute. If convective instability is realized in a system, then small disturbances, while amplifying, are swept out from the point of their appearance (Fig. 7.1,a). Thus, the field at any point is bounded and goes to zero as $t \to \infty$. In case of absolute instability, on the contrary, the energy of any disturbance grows exponentially at the region of its appearance. It is also possible that finally the disturbance will extend over the whole system (see Fig. 7.1,b).

The formulation of the spatial problem of amplification and absorption is different. In this case $\operatorname{Im}\omega = 0$, so that the absorption coefficient (see (7.3)) $\mu_j = 2(\omega/c)\eta_j \cos\vartheta$ is found here instead of the decrement. Here, again, the tempo-spatial behavior of a disturbance comprising the following

[2] This statement on the behavior of the function $\mathbf{E}(\mathbf{r}, t)$ at infinity is the matter of the Riemann lemma [306].

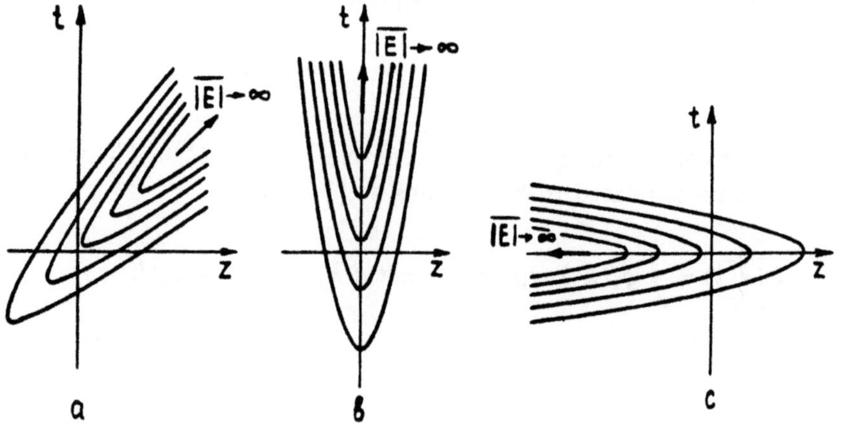

Figure 7.1. Isolines of electric field intensity averaged over the high-frequency oscillations ("envelope" of the packet $\overline{|\mathbf{E}(\mathbf{r},t)|}$) for *a*, convective instability and amplification, *b*, absolute instability, and *c*, "nonpropagation"

set of normal waves,

$$\boldsymbol{E}(\boldsymbol{r},t) = \int \boldsymbol{g}_2(\omega)\exp(i\boldsymbol{k}_j\boldsymbol{r} - i\omega t)d\omega, \qquad k_j = \varphi_2(\omega), \qquad (7.7)$$

differs from that of a single mode. Let $\boldsymbol{g}_2(\omega)$ and $\varphi_2(\omega)$ are the bounded functions at any ω. Then, according to the Riemann lemma, the field (7.7) is a time-like packet ($\boldsymbol{E}(\boldsymbol{r},t)\to 0$ at $\boldsymbol{r}=$const and $t\to\pm\infty$). If $\boldsymbol{E}(\boldsymbol{r},t)$ is also a space-like one (i.e., if $\boldsymbol{E}(\boldsymbol{r},t)$ tends to zero at infinity for any $t=$const), the radiation is called amplified or absorbed. The choice is made depending on the increase or decrease of $|\boldsymbol{E}|^2$ along the direction of the group velocity $\boldsymbol{v}_{\rm gr}=d\omega/d\boldsymbol{k}_j$ at which a tempo-spatial packet moves (see Fig. 7.1,a). But if the time-like packet is not space-like, the "nonpropagation" of waves occurs in the system (Fig. 7.1,c). A time-like packet in the frequency range where the plasma refractive index square is negative, $n_j^2(\omega)$, is an example of such a packet.

It is clear from the above definitions and Fig. 7.1,a that amplification is closely related to convective instability (and respectively, absorption, to convective stability). Both effects describe the behavior of the same tempo-spatial wave packets for which the group velocity $\boldsymbol{v}_{\rm gr}=d\omega/d\boldsymbol{k}_j$ and absorption coefficient μ_j can only be defined. Moreover, only radiation in amplifying/absorbing media can be described by the transfer equations for spectral intensity written in section 4.1.

Classification of the modes in non-equilibrium plasmas with respect to the type of instability demands special analysis of the dispersion equation. The criteria classifying an instability according to the shape of dispersion curves have been initially proposed in [308]. We will not present these results here, referring the reader to this paper and also to the monograph [72]. Here we merely note that in a homogeneous, infinite plasma the choice between convective and absolute instability depends on the reference frame of an observer. From Fig. 7.1 it is easy to understand that in the frame moving with the velocity $\boldsymbol{v}_{gr}$ a convective instability appears as an absolute one. In this case variation of energy density in space cannot be treated as amplification. Thus, the equation of radiation transfer cannot be used in this frame. However, all the instabilities below are considered in the frames ensuring their convective character.

Separation of convective and absolute instabilities in a finite plasma region becomes more definite. Here the specific reference frame appears, namely, the one at which plasma boundaries are stationary.[3] If an instability is absolute with respect to this specific frame, the energy density of high-frequency oscillations grows infinitely (as long as non-linear effects are not accounted for) at any point of the plasma. In terms relevant to electronic devices such a system is called a generator. But if convective instability arises in a finite plasma layer, the system behaves as an amplifier. In such a system the decrement is negative, so plasma amplifies both incoming radiation and the "background" radiation by its individual particles. Energy density at any point of the plasma remains finite even in the linear approximation. Concerning local sources of cosmic radiation, convective instability is usually realized there providing the amplification of radiation by individual particles in the source. Such sources can therefore be considered as noisy amplifiers. According to the terminology generally accepted in theoretical radio astronomy (see section 1.3), a coherent mechanism of maser type is realized in these objects. Of course, this does not mean that systems with absolute instability are of no interest for cosmic conditions.

The above discussion has been related to the linear theory of amplification and instabilities in plasma which is valid only at sufficiently small amplitudes of high-frequency field. It is this condition that allows us to neglect the effect of the finite amplitude of the field on the unperturbed part of the particle distribution $f_0(\boldsymbol{p})$ as well as the interaction of fields at different frequencies. The reversed influence of the high-frequency field on the non-oscillating part of the distribution function, $\overline{f}(\boldsymbol{p})$, is taken into account in the framework of the quasi-linear theory. According to this theory, the variation of $\overline{f}(\boldsymbol{p}, t)$ in the momentum space is a diffusion process, the "diffusion rate" being set by the energy density of the high-frequency oscillations. Increment and gain are, in turn, determined by the formulas of the linear theory in which the function $\overline{f}(\boldsymbol{p}, t)$ is substituted instead of $\overline{f}(\boldsymbol{p})$.

[3]Such a frame in a smoothly inhomogeneous plasma is the one in which the density distribution is time-independent.

As well as in the linear theory, non-linear wave interactions are neglected in quasi-linear approximation.

Quasi-linear equations have been derived by averaging the kinetic equations and field equations over the high-frequency oscillations. Such a procedure is quite complicated (see, for instance, [72, 309]). So, these equations will be obtained below by the elementary quantum method using Einstein coefficients. The desired quasi-linear equations appear in this method as population balance equations for particle quantum states.

7.1.2. METHOD OF EINSTEIN COEFFICIENTS

Starting from the papers [283,310], the elementary quantum method based on the Einstein coefficient scheme has been used together with the classical method of kinetic equation to treat the amplification and instabilities of electromagnetic and Langmuir waves in cosmic plasmas. This method has proved to be fruitful even if the problem considered is essentially classical and does not require the quantum description. Though the method of Einstein coefficients is not as general as the method of kinetic equation, it allowed (owing to its simplicity and clarity) discovery of new types of instabilities of waves in non-equilibrium plasma, for example, kinetic instability of electromagnetic waves at harmonics of gyrofrequency [310], beam instability of electromagnetic waves in the region of the anomalous Doppler effect [79], and synchrotron instability [311,312].

The method of Einstein coefficients is based on the quantum description of radiation in a media. A quantum of radiation (called photon in the case of electromagnetic waves, $j=1,2$, and plasmon for plasma waves, $j=3$) has energy $\hbar\omega$ and momentum $\hbar\boldsymbol{k}=\hbar(\omega/c)n_j\boldsymbol{s}$ where n_j is the refractive index of the j-th mode propagating along the wave vector $\boldsymbol{k}_j = k_j\boldsymbol{s}$. According to the quantum theory of radiation by Einstein, a system of two-level, incoherently-emitting centers with possible quantum states (m) and (n) interacts with a radiation field via three elementary processes: spontaneous emission, stimulated emission and true absorption. Spontaneous emission of a quantum $\hbar\omega$ results from a spontaneous transition of a two-level system from the "upper" state (m) to the "lower" one (n) which occur independently of the external radiation field. The classical analog of the spontaneous emission is the proper radiation by a particle. Spontaneous emission is characterized by the probability A_m^n. By definition, the value $N_m A_m^n d\omega d\Omega$, where N_m is the number density of particles at the "higher" state (m), equals the number of quanta with frequencies in the interval $d\omega$ and wave vectors in the solid angle element $d\Omega$ emitted by a unit volume during unit time. According to the correspondence principle [211], quantum probability of spontaneous emission is related to the classical spectral power of radiation by a quasi-classical particle (see section 5.1) as follows:

$$P_{\omega\Omega} = \hbar\omega A_m^n. \tag{7.8}$$

Under the influence of an external radiation field the additional probability appears for a system in the "upper" state to make the radiative transition $(m) \to (n)$ accompanied by the emission of a quantum $\hbar\omega$. This probability characterizes the process of stimulated (induced) emission. But if the system occupies the "lower" state (n), it can transit up to the state (m) by absorbing the quantum $\hbar\omega$. This process is called true absorption. Stimulated emission and true absorption are characterized by the coefficients B_m^n and B_n^m, respectively. These are determined from the condition that the numbers of quanta emitted or absorbed during the unit time under the action of radiation with the intensity I_ω equal $N_m B_m^n I_\omega d\omega d\Omega$ and $N_n B_n^m I_\omega d\omega d\Omega$, respectively.

An important relation between the Einstein coefficients A_m^n, B_m^n and B_n^m holds in thermodynamic equilibrium. This can be found by demanding the detailed balancing of these processes. A system of radiating particles is ensured to be stationary in this case if the mean rates of "upward", $(n)\to(m)$, and "downward", $(n)\to(m)$, transitions are the same:

$$N_n B_n^m I_\omega = N_m A_m^n + N_m B_m^n I_\omega. \tag{7.9}$$

In thermodynamic equilibrium, however, the radiation intensity $I_\omega = I_\omega^{(0)}$, where the equilibrium (or Planck, or blackbody) intensity $I_\omega^{(0)}$ is determined by the formula (1.37) in the Rayleigh-Jeans region $\hbar\omega \ll \kappa T$, and the populations are set by the Boltzmann formula

$$N_n/N_m = \exp[(\mathcal{E}_m - \mathcal{E}_n)/\kappa T], \tag{7.10}$$

where $\mathcal{E}_m$ and $\mathcal{E}_n$ are particle energies at the states (m) and (n), respectively. For the states considered the difference $\mathcal{E}_n - \mathcal{E}_m = \hbar\omega$, so the population ratio in the Rayleigh-Jeans region

$$N_n/N_m = 1 + \hbar\omega/\kappa T. \tag{7.11}$$

The equilibrium intensity infinitely grows as $T \to \infty$, so the equation (7.9) is reduced to $N_n B_n^m = N_m B_m^n$. On the other hand, the populations should become equal as $T \to \infty$: $N_n = N_m$. Therefore,

$$B_n^m = B_m^n. \tag{7.12}$$

Taking into account (1.37), (7.11), (7.12), we find from (7.9) that

$$A_m^n = \frac{n_j^2 \hbar\omega^3}{8\pi^3 c^2 |\cos\vartheta_j|} B_m^n. \tag{7.13}$$

The Einstein relations (7.12), (7.13) are quite universal. Of course, for the sake of simplicity, the above invoked thermodynamic equilibrium between the particle system and radiation field, as well as the condition $\hbar\omega \ll \kappa T$. However, since Einstein coefficients depend only on the properties of a radiating particle, these relations hold in the general case when both particle system and radiation are non-equilibrium. The Einstein relations for any specific radiating system can, in principle, be found by the appropriate quantum-mechanical calculations (see [211]), but this approach is quite complicated and in many cases cannot be followed up to the final result.

When radiation passes a unit length along the ray in a medium, the intensity is decreased by the value $-\Delta I_\omega = \hbar\omega(N_n B_n^m - N_m B_m^n) I_\omega$ due to the radiative transitions $(m) \leftrightarrow (n)$. The absorption coefficient μ_j is defined as the relative loss of radiation at a unit ray length:

$$\mu_j = -\frac{\Delta I_\omega}{I_\omega} = \hbar\omega(N_n B_n^m - N_m B_m^n). \tag{7.14}$$

From this definition it is clear that μ_j depends on the difference between the rates of true absorption and simulated emission. With the Einstein relations (7.12), (7.13) taken into account, we find from (7.14) that

$$\mu_j = \frac{8\pi^3 c^2 |\cos\vartheta_j|}{n_j^2 \omega^2} A_m^n N_m \left(\frac{N_n}{N_m} - 1 \right). \tag{7.15}$$

The absorption coefficient in a non-equilibrium system is thus proportional to the intensity of spontaneous emission (7.8) and the factor $(N_n/N_m - 1)$ which depends on the ratio of particle densities (populations) at higher (m) and lower (n) quantum states. The difference between particle energies in these states, $\mathcal{E}_m$ and $\mathcal{E}_n$, is equal to the photon energy $\hbar\omega$. Being applied to the equilibrium system where population ratio N_n/N_m is determined by (7.11), the formula (7.15) yields the following expression for the absorption coefficient:

$$\mu_j = \frac{8\pi^3 c^2 |\cos\vartheta_j|}{n_j^2 \omega^2 \kappa T} \hbar\omega A_m^n N_m. \tag{7.16}$$

Note, however, that the right-hand side of this formula is nothing but the ratio of the emissivity $a_\omega = \hbar\omega A_m^n N_m$ to the equilibrium intensity $I_\omega^{(0)}$. Thus, the absorption coefficient (7.15) of an equilibrium system with Boltzmann distribution of particles over the energy levels is reduced to the one based on the Kirchhoff law: $\mu_j = a_\omega / I_\omega^{(0)}$.

The formula (7.15) accounts for the transitions between two energy states of radiating particles. If absorption or emission of radiation at a frequency ω in direction $\boldsymbol{k}_j$ may also result from other transitions, one should sum the right-hand side of (7.15) over all such transitions to find the absorption coefficient:

$$\mu_j = \frac{8\pi^3 c^2 |\cos\vartheta_j|}{n_j^2 \omega^2} \sum_{(m)\leftrightarrow(n)} A_m^n N_m \left(\frac{N_n}{N_m} - 1\right). \qquad (7.17)$$

The corresponding emissivity of the system equals

$$a_\omega = \hbar\omega \sum_{(m)\leftrightarrow(n)} A_m^n N_m. \qquad (7.18)$$

If

$$\sum_{(m)\leftrightarrow(n)} A_m^n N_m \left(\frac{N_n}{N_m} - 1\right) > 0, \qquad (7.19)$$

the absorption coefficient is positive, so radiation is re-absorbed in the system. If, on the contrary,

$$\sum_{(m)\leftrightarrow(n)} A_m^n N_m \left(\frac{N_n}{N_m} - 1\right) < 0, \qquad (7.20)$$

the negative re-absorption occurs in the system.

So, the case of (7.20) corresponds to an amplifying medium: the intensity of radiation increases along a ray). Recalling, at the same time, that the absorption coefficient μ_j is related to the decrement $\gamma_j = \mu_j v_{gr}/2$ (see the introduction to chapter 6), it is easy to understand that in case of (2.19) initial electromagnetic disturbances in the system are convectively unstable, and their energy increases with time.

In terms of elementary quantum theory the process of amplification can be treated in the following way. The coefficient μ_j is proportional to the difference between the rate of true absorption and that of stimulated emission. For $\mu_j < 0$ the latter process dominates over the former one. As a result, the system adds energy to the radiation field, instead of removing it. Such a regime is realized because of the peculiar particle distribution ensuring the criterion (7.20): population N_m of the level *from which* particles make radiative transitions resulting in the emission of photons $\hbar\omega$ is greater than N_n of the level *at which* a particle appears as a result of photon emission.

The formulas (7.8), (7.17) mean that any elementary emission process in a non-equilibrium system which is characterized by some value A_m^n gives rise to a definite type of absorption (or re-absorption). In particular, Čerenkov, cyclotron and synchrotron absorptions are accompanied by the corresponding emissions. These absorptions may become negative under some conditions ensuring the validity of the criterion (7.20). Since $\gamma_j = \mu_j v_{gr}/2$, specific instability with an increment $-\gamma_j$ is related to any type of negative re-absorption. Therefore, Čerenkov, cyclotron and synchrotron instabilities may exist in a non-equilibrium plasma. Such a physically meaningful classification of instabilities according to the elementary emission process by single particles will be used in the following discussion of the problem of instabilities and amplification in plasma. Such a general classification does not depend on the form of the distribution function providing an instability, nor on the type of growing modes. These two characteristics nevertheless appear in the title of an instability. This is well justified by the vast variety of instabilities and their criteria. As an example, we mention here "beam" and "loss-cone" instabilities of plasma waves. The first title describes a process of excitation of such waves in a particle beam travelling through a background equilibrium plasma. The second one corresponds to the instability arising in plasma with the distribution function typical for magnetic traps (no particles with small pitch-angles within the loss-cone).

The above expressions for the absorption coefficient μ_j, (7.17), and emissivity a_ω, (7.18), allows one to find the intensity I_ω along a ray in a non-equilibrium plasma with definite "population" distribution of emitting particles over the quantum states using the equation of radiation transfer (4.30). Of course, it is possible to use instead the transfer equation for the energy density (4.28):

$$\frac{\partial W_{\mathbf{k}}}{\partial t} + v_{gr}\frac{dW_{\mathbf{k}}}{dl} = \alpha_{\mathbf{k}} - \beta_j W_{\mathbf{k}}, \tag{7.21}$$

in which

$$\alpha_{\mathbf{k}} = a_\omega v_{gr}|\cos\vartheta_j|k_j^{-2}, \qquad \beta_j = \mu_j v_{gr} = 2\gamma_j. \tag{7.22}$$

With (7.17), (7.18) taken into account, the expressions for the emissivity $\alpha_{\mathbf{k}}$ related to the unit volume in wave vector space and the decrement γ_j take the following form:

$$\alpha_{\mathbf{k}} = \hbar\omega\frac{v_{gr}|\cos\vartheta_j|}{k_j^2}\sum_{(m)\leftrightarrow(n)} A_m^n N_m, \tag{7.23}$$

$$\beta_j = \frac{8\pi^3 c^2|\cos\vartheta_j|v_{gr}}{n_j^2\omega^2}\sum_{(m)\leftrightarrow(n)} A_m^n N_m\left(\frac{N_n}{N_m} - 1\right). \tag{7.24}$$

In these formulas $v_{gr}|\cos\vartheta_j| = |v_{gr,\mathbf{k}}| = c|\partial(\omega n_j)/\partial\omega|^{-1}$.

7.1.3. POPULATION BALANCE EQUATION

The transfer equation (7.21) with $\alpha_{\mathbf{k}}$ and μ_j in the form (7.23), (7.24) can also be used to find the spectral density of radiation $W_{\mathbf{k}}(l, t)$ in the case when populations vary slowly on the wave period $2\pi/\omega$ and their spatial scale is much larger than the wavelength $2\pi/k_j$. Generally, the population distribution in plasma may keep stationary or change due to some factors not related directly to the radiation considered, for example, via injection of particles with definite velocity distribution or the influence of quasi-static electromagnetic fields. However, radiation with spectral density $W_{\mathbf{k}}$ may influence particle distribution, as well, due to the processes of true absorption and stimulated emission whose probabilities are proportional to $W_{\mathbf{k}}$ (or I_ω – see above). Such a modification of particle distribution can be described in the following way [309].

Consider the variation of a population N_m of a state (m) under the action of radiation at frequency ω. This occurs due to: (i) "upward" transitions of particles to the state (q) by absorbing quanta $\hbar\omega$ with the momenta $\hbar\boldsymbol{k}_j$, and (ii) "downward" transitions to the state (n) because of stimulated emission. The rates of these transitions are given by $N_m B_m^q I_\omega$ and $N_m B_m^n I_\omega$, respectively. In addition, the population considered increases in cases of induced emission from the higher level (q) (the corresponding rates are $N_q B_q^m I_\omega$ and $N_n B_n^m I_\omega$). As a result,

$$\frac{\partial N_m}{\partial t} = A_q^m N_q + B_q^m (N_q - N_m) I_\omega - [A_m^n N_m + B_m^n (N_m - N_n) I_\omega]. \quad (7.25)$$

Here the Einstein relation (7.12) is taken into account and the right-hand side is supplemented with the terms $A_q^m N_q$ and $-A_m^n N_m$ implying the net variation of the population of the state (m) due to spontaneous transitions $(q) \to (m)$ and $(m) \to (n)$. Since, evidently, N_m changes due to all the transitions from the level (m) and to (n) resulting in emission/absorption of quanta with different energies $\hbar\omega$ and momenta $\hbar\boldsymbol{k}_j$, the right-hand side of the balance equation (7.25) should finally be integrated over the frequencies and solid angles:

$$\begin{aligned}\frac{\partial N_m}{\partial t} &= \int \left[A_q^m N_q + B_q^m (N_q - N_m) I_\omega \right] d\omega d\Omega \\ &\quad - \int \left[A_m^n N_m + B_m^n (N_m - N_n) I_\omega \right] d\omega d\Omega. \qquad (7.26)\end{aligned}$$

For spatially inhomogeneous particle distributions the operator $\partial N_m/\partial t + \boldsymbol{v}\partial N_m/\partial \boldsymbol{r}$ should be used instead of $\partial N_m/\partial t$ in the left-hand side of (7.26) (cf. the kinetic equation (2.85) for particle distribution function). Taking this fact into account and using the Einstein relation (7.13) we finally arrive at the equation:

$$\frac{\partial N_m}{\partial t} + \boldsymbol{v}\frac{\partial N_m}{\partial \boldsymbol{r}} = \int (A_q^m N_q - A_m^n N_m)\, d\omega d\Omega$$

$$+\int[A_q^m(N_q-N_m)-A_m^n(N_m-N_n)]\frac{8\pi^3W_{\mathbf{k}}}{\hbar\omega}\,d\omega d\Omega, \qquad (7.27)$$

or, in somewhat different form,

$$\frac{\partial N_m}{\partial t}+\boldsymbol{v}\frac{\partial N_m}{\partial \boldsymbol{r}}=\int\{A_q^m[N_q(1+N_{\mathbf{k}})-N_mN_{\mathbf{k}}]$$
$$-A_m^n[N_m(1+N_{\mathbf{k}})-N_nN_{\mathbf{k}}]\}\,d\omega d\Omega. \qquad (7.28)$$

According to (1.34), the intensity $I_\omega = W_{\mathbf{k}}k_j^2|\cos\vartheta_j|^{-1}$ is used in (7.27). Moreover, the number of radiation quanta per normal wave (or mode),

$$N_{\mathbf{k}}=\frac{8\pi^3W_{\mathbf{k}}}{\hbar\omega}, \qquad (7.29)$$

is introduced in the equation (7.28). The last equation follows from the fact that $W_{\mathbf{k}}$ and $1/(2\pi)^3$ are energy and the number of modes per unit phase space volume (i.e., coordinate space and wave vector space; see (1.23) and (1.35)), respectively, so that their ratio, $W_{\mathbf{k}}(2\pi)^3$, equals the energy per mode. Equation (7.28) clearly shows that in many cases the effect of the spontaneous processes on population distribution is negligible if $N_{\mathbf{k}} \gg 1$.

In case of local thermodynamic equilibrium the detailed balancing principle in realized in the form (cf. (7.9))

$$\begin{aligned} A_n^m N_m + B_m^n N_m I_\omega &= B_n^m N_n I_\omega, \\ A_q^m N_q + B_q^m N_q I_\omega &= B_q^m N_m I_\omega. \end{aligned} \qquad (7.30)$$

These equalities nullify the right-hand side of the equation (7.26) and thus, ensure the steady-state $\partial N_m/\partial t = 0$ of a system "plasma + radiation" in a thermodynamic equilibrium.

If radiation is non-equilibrium, i.e., its angular pattern in an isotropic medium with $n_j \simeq 1$ is anisotropic and the intensity differs from the Planck one, the detailed balancing (7.30) is violated, and the system becomes non-stationary. Populations of energy levels and radiation intensity in such a system should vary in time. However, under stationary external conditions the system evolves to a steady state at which the following integral relations

$$\int\left[A_q^m N_q + B_q^m(N_q-N_m)I_\omega\right]d\omega d\Omega = 0 \qquad (7.31)$$

hold for any pair of quantum states for which radiative transitions $(q)\leftrightarrow(m)$ are possible. The analogous relation should thus be valid for the transitions $(m)\leftrightarrow(n)$ which has also been taken into account in (7.26). As a result, a steady state $\partial N_m/\partial t = 0$ is realized in the system.

In a medium with $n_j \simeq 1$ and $\cos\vartheta_j \simeq 1$ the Einstein relation (7.13) takes the form

$$B_q^m = A_q^m \frac{8\pi^3 c^2}{\hbar\omega_{qm}^3}, \qquad B_m^n = A_m^n \frac{8\pi^3 c^2}{\hbar\omega_{mn}^3}.$$

Hence, the sufficient condition (7.31) can be written as

$$\frac{N_m}{N_q} = 1 + \left[\frac{8\pi^3 c^2}{\hbar\omega_{qm}} \int A_q^m I_\omega \, d\omega d\Omega\right]^{-1} \int A_q^m \, d\omega d\Omega. \tag{7.32}$$

The obtained ratio of the populations of the states (m) and (q) can be described in terms of the brightness temperature T_{eff} of radiating particles introduced by the relation (7.11):

$$\frac{N_m}{N_q} = 1 + \frac{\hbar\omega_{qm}}{\kappa T_{\text{eff}}(q,m)}. \tag{7.33}$$

Then (7.32) yields

$$T_{\text{eff}}(q,m) = \frac{8\pi^3 c^2}{\omega_{qm}^2 \kappa} \frac{\int A_q^m I_\omega \, d\omega d\Omega}{\int A_q^m \, d\omega d\Omega}, \tag{7.34}$$

or, equivalently,

$$T_{\text{eff}}(q,m) = \frac{1}{\omega_{qm}^2} \frac{\int A_q^m T_{\text{b}} \omega^2 \, d\omega d\Omega}{\int A_q^m \, d\omega d\Omega}. \tag{7.35}$$

The formula (1.38) for the brightness temperature T_{b} of radiation at $n_j \simeq 1$ and $\cos\vartheta_j \simeq 1$ has been taken into account when deriving the last expression. Finally, since for any given direction $\mathbf{\Omega}$ the probability A_q^m is non-zero at the frequency $\omega = \omega_{qm}$, it can be written using the delta-function: $A_q^m \propto \delta(\omega - \omega_{qm})$. Thus, the expression for the effective temperature $T_{\text{eff}}(q,m)$ which describes the ratio of populations of energy states (q) and (m) can be presented in the form [183]

$$T_{\text{eff}}(q,m) = \frac{\oint_{4\pi} P_\Omega T_{\text{b}} \, d\Omega}{\oint_{4\pi} P_\Omega \, d\Omega} \tag{7.36}$$

where $P_\Omega = \int \hbar\omega A_q^m \, d\omega$ is the power of spontaneous emission.

The expression (7.36) has been derived in the quasi-classical limit (7.33) when $\hbar\omega_{qm} \ll \kappa T_{\text{eff}}$. For the arbitrary ratio between $\hbar\omega_{qm}$ and κT_{eff} the effective temperature should be introduced using the initial Boltzmann distribution (7.10) rather than the first terms of its power series expansion (7.11):

$$\frac{N_m}{N_q} = \exp(\hbar\omega_{qm}/\kappa T_{\text{eff}}). \tag{7.37}$$

The following expression should replace (7.35) in this case [41]:

$$T_{\text{eff}} = \frac{\hbar\omega_{qm}}{\kappa} \ln \frac{x+1}{x}, \tag{7.38}$$

where

$$x = \frac{\oint_{4\pi} N_{\mathbf{k}} \Psi(\boldsymbol{\Omega})\, d\Omega}{\oint_{4\pi} \Psi(\boldsymbol{\Omega})\, d\Omega} \tag{7.39}$$

and $N_{\mathbf{k}}$ is determined by the formula (7.29) at $\omega = \omega_{qm}$. The obtained formula for T_{eff} is reduced to (7.35) at $x \ll 1$.

Let the radiating system have an equidistant energy spectrum with a constant energy difference $\mathcal{E} = \hbar\omega_{qm}$ between adjacent levels. Thus, radiative transitions between any adjacent levels influence radiation at the same frequency $\omega = \mathcal{E}/\hbar$. Assume that spontaneous radiation power at this frequency can be presented as

$$P_{\Omega} = \varphi_{qm} \Psi(\boldsymbol{\Omega}), \tag{7.40}$$

where $\Psi(\boldsymbol{\Omega})$ is the angular pattern of spontaneous emission, the same for the transition between any pair of adjacent levels. Then particle distribution over energy states is described by the Boltzmann formula (7.10) with the effective temperature

$$T_{\text{eff}} = \frac{\oint_{4\pi} T_{\text{b}}(\omega, \boldsymbol{\Omega}) \Psi(\boldsymbol{\Omega})\, d\Omega}{\oint_{4\pi} \Psi(\boldsymbol{\Omega})\, d\Omega}, \tag{7.41}$$

the same for the entire equidistant spectrum. In other words, distribution of particles which interact with radiation at frequency ω over an equidistant energy level is quasi-equilibrium. The effective temperature of this quasi-equilibrium distribution is determined by the brightness temperature of radiation $T_{\text{b}}(\omega, \boldsymbol{\Omega})$ and the angular pattern of emission by these particles $\Psi(\boldsymbol{\Omega})$. In particular, in case of isotropic angular pattern $\Psi(\boldsymbol{\Omega}) = \text{const}$ the formula (7.41) is simplified in the following way:

$$T_{\text{eff}} = \frac{1}{4\pi} \oint_{4\pi} T_{\text{b}}(\boldsymbol{\Omega})\, d\Omega. \tag{7.42}$$

For example, if radiation from the photosphere of a star illuminates an optically thin plasma layer nearby, the intensity of radiation at any point of the layer equals the photospheric one. Therefore,

$$\oint_{4\pi} T_{\text{b}}(\boldsymbol{\Omega})\, d\Omega = \int_{2\pi} T_{\text{ph}}\, d\Omega = 2\pi T_{\text{ph}},$$

and the effective temperature $T_{\text{eff}} = T_{\text{ph}}/2$, where T_{ph} is the brightness temperature of the photospheric radiation.

To apply the obtained relations, consider now the example of interaction between radiation and rarefied, non-relativistic, magnetized plasma with $n_j \simeq 1$. Assume that the effect of particle collisions is negligible. Such an interaction appears as cyclotron emission and absorption at frequencies (see (11.16))

$$\omega = \frac{s\omega_B}{1 - \beta_{\parallel} \cos\alpha} \simeq s\omega_B, \tag{7.43}$$

where $s = 1, 2, \ldots$ is the harmonic number, $\beta_{\parallel} = v_{\parallel}/c$, $v_{\parallel}$ is the component of an electron velocity along the magnetic field $\boldsymbol{B}_0$, and α is the angle between $\boldsymbol{\Omega}$ and $\boldsymbol{B}_0$. As has been shown in section 5.2, cyclotron radiation in this case is contributed mainly by emission of extraordinary waves at the first harmonic $s = 1$. The corresponding radiation power equals

$$P_{\Omega 1}(s=1) = \frac{e^2 \omega_B^2 \beta_{\perp}^2}{8\pi c}(1 + \cos^2\alpha) \tag{7.44}$$

(see the formula (5.58) and the remarks thereafter). Since radiation power for ordinary mode at the same harmonic, $P_{\Omega 2}(s=1) \simeq \beta_{\perp}^2 P_{\Omega 1}(s=1)$, is small compared to $P_{\Omega 1}(s=1)$, the effect of this component of radiation on electron distribution function is negligible. The same is true for higher harmonic emission, since the corresponding radiation powers are also much less than (7.44) and rapidly decrease with increasing s. As for cyclotron radiation in the extraordinary mode, the angular pattern of emission,

$$\Psi(\boldsymbol{\Omega}) = 1 + \cos^2\alpha \tag{7.45}$$

is the same for all electrons. Hence, electron distribution over the energy states which are coupled by the radiative transitions involving the extraordinary mode at the first harmonic (the corresponding frequency given by (7.43) at $s = 1$) is quasi-equilibrium (Boltzmann) with the effective temperature

$$T_{\text{eff}} = \frac{\oint_{4\pi} T_{\text{b}}(\boldsymbol{\Omega})(1 + \cos^2\alpha)\, d\Omega}{\oint_{4\pi}(1 + \cos^2\alpha)\, d\Omega}. \tag{7.46}$$

Note that, according to (5.25), (5.26), cyclotron radiative transitions are accompanied by the variations of both transverse $(\Delta p_{\perp})$ and longitudinal $(\Delta p_{\parallel})$ momenta of an electron. However, in the classical limit which is realized at higher Landau levels the ratio of these values for non-relativistic particles is small:

$$\frac{\Delta p_{\parallel}}{\Delta p_{\perp}} = \frac{p_{\perp}}{m_e c} \cos\alpha \ll 1. \tag{7.47}$$

If electron distribution over longitudinal momenta is wide enough $(T_{\parallel} \gtrsim T_{\perp})$, the influence of $\Delta p_{\parallel}$ on cyclotron interaction between plasma and radiation can be neglected [313]. According to (7.47), cyclotron transitions affect only the transverse distribution of plasma electrons. This distribution becomes Maxwell with the temperature determined by (7.46):

$$T_\perp = \int_0^\pi T_b(\omega, \alpha)(1 + \cos^2\alpha)\sin\alpha\, d\alpha. \tag{7.48}$$

The effective temperature in the above expression should correspond to radiation which interacts with electrons having the momentum $p_\parallel$. Thus, according to the resonance condition (7.43) at $s = 1$ and $\beta_\parallel = \text{const}$, the frequency in $T_b(\omega, \alpha)$ is a function of α.

The formula (7.47) is valid at $\hbar\omega_B \ll \kappa T_\perp$. Using (7.38), one may get the general formula valid at arbitrary $\hbar\omega_B$ and $\kappa T_\perp$:

$$\kappa T_\perp = \frac{\hbar\omega_B}{\ln(1 + x_1^{-1})}, \tag{7.49}$$

where

$$x_1 = \int_0^\pi \frac{3}{8}(1 + \cos^2\alpha) N_{\mathbf{k}}^{(1)} \sin\alpha\, d\alpha \tag{7.50}$$

equals photon number in extraordinary mode, $N_{\mathbf{k}}^{(1)} = (2\pi)^3 W_{\mathbf{k}}/\hbar\omega$, averaged over the angular pattern of extraordinary cyclotron radiation; energy density of extraordinary radiation $W_{\mathbf{k}}^{(1)}$ is introduced according to (7.29).

Evidently, the Kirchhoff law in the form (4.36) is violated if $T_\parallel \neq T_\perp$. The expressions (7.17) and (7.18) can however be used to relate emissivity and absorption coefficient. Since the transverse electron distribution (7.33) in plasma with $T_\parallel \neq T_\perp$ is characterized by the effective temperature $T_{\text{eff}} = T_\perp$ determined by radiation (see (7.46)), the desired relation is given by

$$a_\omega = \mu_j \frac{\omega^2 \kappa T_\perp}{(2\pi)^3 c^2}. \tag{7.51}$$

In plasma with $n_j \simeq 1$ and $\cos\vartheta_j \simeq 1$ this formula is analogous to (4.36) in which T is substituted with $T_\perp$.

The above analysis has been made for a rarefied plasma under the conditions when the vacuum birefringence is negligible. On the contrary, if the criterion (2.187) is met, this effect dominates mode polarizations and refractive indices. In this case populations of Landau levels are set by both modes at the first harmonic $s = 1$. The influence of higher harmonics is negligible if their intensities are comparable to that of the first one. The effective transverse temperature of plasma electrons is now determined by the formula

$$\kappa T_\perp = \frac{\hbar\omega_B}{\ln\left[1 + (x_1 + x_2)^{-1}\right]} \tag{7.52}$$

where

$$x_1(p_{\|}) = \int_0^{\pi} \frac{3}{8} N_{\mathbf{k}}^{(1)}(p_{\|}, \alpha) \sin \alpha d\alpha,$$
$$x_2(p_{\|}) = \int_0^{\pi} \frac{3}{8} N_{\mathbf{k}}^{(2)}(p_{\|}, \alpha) \cos^2 \alpha \sin \alpha d\alpha \tag{7.53}$$

are photon numbers of extraordinary and ordinary modes at the first harmonic averaged over the corresponding angular patterns [54,314].

The obtained Boltzmann distributions of electrons over Landau levels (transverse velocities) does not account for the effect of particle collisions. This effect can be described the simplest way by the so-called "relaxation time approximation" in which the rate of collisional transition is given by

$$\left(\frac{\partial N_m}{\partial t}\right)_{\text{st}} = -\nu_{\text{eff}}(N_m - N_m^{(0)}). \tag{7.54}$$

Here the effective collision frequency ν_{eff} is determined by the formulas of the section 5.2,

$$N_m^{(0)} = N\left(1 - e^{-\hbar\omega_B/\kappa T_{\|}}\right) e^{-m\hbar\omega_B/\kappa T_{\|}} \tag{7.55}$$

are the populations corresponding to the Boltzmann distribution with the temperature $T_{\perp} = T_{\|}$,[4] and $N = \sum_{m=0}^{\infty} N_m$ is electron density in plasma. Introducing the term (7.54) into the right-hand side of the equations (7.26) and of its steady-state analogue (7.31), one can obtain the following non-equilibrium, steady-state distribution of electrons over Landau levels which is formed under the combined influence of collisions and extraordinary radiation at the first cyclotron harmonic [41]:

$$N_m = \frac{N}{1+x_1}\left\{1 - \epsilon \ln \frac{1 + \kappa T_{\|}/\hbar\omega_B}{1+x_1}\right\}\left(\frac{x_1}{x_1+1}\right)^m$$
$$+ \frac{\epsilon}{x_1+1}\sum_{q=1}^{m}\left(\frac{x_1}{x_1+1}\right)^{m-q}\frac{1}{q}\left[e^{-q\hbar\omega_B/\kappa T_{\|}} - \left(\frac{x_1}{x_1+1}\right)^q\right]. \tag{7.56}$$

Here m and q denote Landau level numbers, the frequency of cyclotron radiative transition at the first harmonic equals ω_B, and the ratio of characteristic time of cyclotron transition to the average time between electron collisions equals

$$\epsilon = \nu_{\text{eff}}/A, \tag{7.57}$$

[4]The longitudinal temperature $T_{\|}$ is assumed to be constant. It may be fixed, for example, due to electron-ion collisions or longitudinal acceleration (heating) of electrons.

where

$$A = \frac{A_m^{m-1}}{m} = A_1^0 = \frac{4}{3}\frac{e^2\omega_B^2}{m_e c^3} \tag{7.58}$$

is the probability of spontaneous cyclotron transition between the lowest Landau levels. The numerical value of the parameter ϵ is given by the following expression:

$$\epsilon = 1.25 \cdot 10^{10} N B_0^{-2} T_{\|}^{-3/2}. \tag{7.59}$$

This has been obtained using the expression for $\nu_{\text{eff}} = \nu_{\text{ei}}$ in the form (6.12) with the Coulomb logarithm (6.14) valid at $T \gg 4 \cdot 10^5$ K. Moreover, this logarithm has been made equal to 10 which further restricts the validity of (7.59) by the ranges $T \sim 10^7$ K and $N \sim 10^9 - 10^{13}\text{cm}^{-3}$.

The solution (7.56) is obtained under the condition $\epsilon \ll 1$ (the limiting case of "collisionless plasma") when the influence of collisions on the transverse electron distribution is relatively weak. In the opposite "collisional" limit $\epsilon \gg 1$ the role of collisions is dominant. Electron distribution in this case is close to the equilibrium one with $T_\perp \simeq T_\| = T$, and radiation with $T_b \lesssim T$ is unable to disturb it. The case of arbitrary ϵ has been analyzed in [315]. The corresponding distribution is given in chapter 8.

In active regions of the solar corona $T_\| \simeq T \sim 10^6$ K, $N \simeq 10^9\text{cm}^{-3}$ and $B_0 \simeq 10^2$ G, so the parameter $\epsilon \simeq 10^6$ is much larger than unity. Therefore, cyclotron radiation at the first harmonic cannot lead to the anisotropy of distribution function of radiating electrons: $T_\perp \simeq T_\|$. The same is true for the coronas of magnetic Ap-stars at $B_0 \simeq 10^4$ G. However, for plasmas on magnetic white dwarfs and neutron stars the opposite case $\epsilon \ll 1$ is realized. For example, at $T_\| \sim 10^6$ K and $B_0 \sim 10^8$ G, the values typical for magnetic white dwarfs, the limit of collisionless plasma is valid if $N \ll 10^{15}\,\text{cm}^{-3}$. On neutron stars with $T_\| \sim 10^6$ K and $B_0 \sim 10^{12}$ G the condition $\epsilon \ll 1$ is met for electron densities $N \ll 10^{23}\,\text{cm}^{-3}$. Therefore, transverse electron distribution in plasma envelopes of these objects is formed by radiation at gyrofrequency, while the effect of particle collisions in negligible. As a result, electron distribution in such plasmas is generally characterized by temperature anisotropy: $T_\| \neq T_\perp$ (see the details in [41, 183].

The expressions (7.49), (7.52) for electron transverse temperature and the electron distribution over Landau levels (7.56), as well as its analogue for arbitrary ϵ, will be used in chapter 8 to investigate radiation transfer in plasma with the effects of resonance cyclotron scattering taken into account.

Note that the value $T_\perp$ has been found above neglecting the spatial transfer of radiating particles. This process has been neglected in the initial equation (7.26) which the expression (7.46) for $T_{\text{eff}} = T_\perp$ follows from.

Such an approximation can be applied if the characteristic scale of the temperature $T_\perp$ in plasma is relatively large:

$$l_\perp \gg v_{T_\parallel} t_c. \tag{7.60}$$

Here $v_{T_\parallel}$ is the longitudinal thermal velocity of electrons, and

$$t_c = A^{-1} \simeq 5.1 \cdot 10^8 B_0^{-2}\,\mathrm{s} \tag{7.61}$$

is the characteristic time of cyclotron radiation which is independent of $T_\perp$. With this expression taken into account, the condition (7.60) is transformed to

$$l_\perp \gg 2 \cdot 10^4 T_\parallel^{1/2} B_0^{-2}\,\mathrm{cm}. \tag{7.62}$$

The transfer equation with the absorption coefficient μ_j or, equivalently, decrement γ_j set by the formulas (7.17), (7.24) with given populations N_n, N_m describes the distribution of energy density or intensity of radiation in the framework of the linear theory of propagation of waves in plasma with given, stationary particle distribution function $f_0(\boldsymbol{p})$. The system (7.21),(7.27) which accounts for both variation of the energy density of radiation $W_{\mathbf{k}}$ and the modification of populations N_m under the action of this radiation lies in the basis of the so-called quasi-linear theory of wave processes [316–318].

When dealing with classical plasma it is more convenient to rewrite the equations which initially contain populations and probabilities of spontaneous emission in terms of the momentum distribution function $f_0(\boldsymbol{p})$ and the radiation power $P_{\omega\Omega}$. Such a procedure will be performed in the following sections of this chapter when discussing the various types of amplification and instabilities in non-equilibrium plasma.

Concluding the section, we make a few remarks on the validity range of the quasi-linear theory (i.e., the Einstein coefficient method and population balance equations) to describe the wave processes in classical plasma. The absorption coefficient μ_j and the decrement γ_j are related to the system in which an amplification (re-absorption) is realized together with the corresponding instability (decay) of the convective type which is closely related to the former processes. Moreover, the above formulae for A_m^n, μ_j and γ_j are related to a single type of waves in the system. To find $\partial N_m/\partial t + \boldsymbol{v}\partial N_m/\partial \boldsymbol{r}$, one should therefore integrate the left-hand side of the population balance equation (7.28) over all the types of modes which interact with particles at a given state (m).

The method of Einstein coefficients, moreover, assumes that all absorption/emission processes occur in plasma with definite, fixed dispersion characteristics, including refraction index n_j, group velocity $v_{\mathrm{gr}} = d\omega/d\boldsymbol{k}_j$, and mode polarization vector e_j. This, in turn, is realized if the amount of "resonant" particles which undergo radiative transitions under the action of radiation considered is small compared to the amount of other, "non-resonant", ones which determine the dispersion characteristics of plasma as a whole. The "resonant" particles can thus be treated as a small perturbation leading to the absorption (amplification) of radiation.[5] One of the necessary conditions determining the possible "abundance" of the resonant particles is that μ_j, γ_j are small compared to k_j and ω. However, this criterion is obviously insufficient, since the amount of "resonant" particles may become too large, say, if the spectrum of radiation in the system is wide enough. Redistribution of particles over energy states may lead to the remarkable modification of dispersion properties of the system. This constrains the width of the radiation spectrum. Finally, for the dispersion characteristics to be independent on the radiation, the energy density $W = \int W_{\mathbf{k}}\, d^3k$ should be small compared to the kinetic energy density of plasma particles. This condition allows one to neglect the oscillatory motion of particles in the field of the wave when calculating the dispersion properties of the plasma.

An important feature of population balance equations (quasi-linear theory) in particular and the method of Einstein coefficients in general is the summation of the radiation powers over the transitions of different particles. "Incoherency" of the initial state of a particle system and radiation appearing as random phases of modes at different frequencies and wave vectors, as well as random phases of the spontaneous emission (the latter is ensured by the proper choice of the initial unperturbed distribution function in particle phase space), is the necessary condition justifying this procedure. The phasing of spontaneous emission remains random in the course of the further evolution of the system if the width of the radiation spectrum is large enough, so that any particle interacts with many incoherent modes comprising this spectrum. In other words, the conditions mentioned prevent the development of two processes driving the system beyond the limits of the quasi-linear theory: phasing of particles in the phase space and formation of a narrow packet of normal waves in the wave vector space.

Note, finally, that the scheme considered accounted for the quasi-linear interaction of waves with particles only, whereas the wave–wave interaction has not been taken into account. We mean here the coalescence and decay of waves considered in section 8.4. The probabilities of these processes are proportional to the wave energy density squared (or to higher powers of $W_{\mathbf{k}}$). Rather, the effects of wave–particle interaction are proportional to

[5]In this respect the search for the coefficients μ_j and γ_j using the Einstein coefficient scheme is analogous to the perturbative solution of the dispersion equation with real ω and $\mathbf{k}_j$ used as a zero-order approximation.

$W_{\mathbf{k}}$. Therefore, the influence of the wave–wave interactions on the quasi-linear evolution of the system is negligible for sufficiently small $W_{\mathbf{k}}$, though the rigorous constraints on the value of $W_{\mathbf{k}}$ depend on the nature of the system.

7.2. Čerenkov instability

The Čerenkov mechanism of amplification related to Čerenkov instability plays an important role among various maser mechanisms which work (or may work) in cosmic plasmas. This instability effectively excites plasma waves which is very important for the so=called "plasma mechanism" of the generation of extra-terrestrial radio emission. Čerenkov instability is realized under the conditions depending on the type of waves and the form of particle distribution function in a non-equilibrium system. We start the analysis of this instability with the simplest case of the excitation of plasma waves in a system comprising a beam of charged particles penetrating isotropic equilibrium plasma.

7.2.1. INSTABILITY AND AMPLIFICATION IN A SYSTEM "BEAM–PLASMA"

If particle density in the beam N_{S} is sufficiently small compared to that of the ambient plasma N, the dispersion properties of waves propagating in the system are determined by the latter. According to section 2.2, transverse electromagnetic and longitudinal plasma waves can propagate in such a plasma. The refractive index for the transverse waves $n_{\mathrm{el}} < 1$, so the Vavilov-Čerenkov effect (see section 5.1) and, thus, Čerenkov amplification and instability cannot occur for these waves.[6] It may however be realized for plasma waves in the parameter region where $n_{\mathrm{pl}}(\omega) > 1$.

To find explicitly the gain $(-\mu_j)$ and the increment $(-\gamma_j)$ as well as amplification and instability criteria, we apply the method of Einstein coefficients outlined in section 7.1.

According to (5.21), plasma waves in the direction $\boldsymbol{k}$ are emitted by the particles whose velocity projection onto this direction equal to the phase velocity of the waves: $v_{\mathbf{k}} = v_{\mathrm{ph}} = \omega/k$. Let $v_{\mathbf{k}}^{(m)}$ be the initial $\boldsymbol{k}$-component of particle velocity. Then, after the emission of a plasmon with energy $\hbar\omega$ and momentum $\hbar\boldsymbol{k} = \hbar(\omega n_{\mathrm{pl}}/c)(\boldsymbol{k}/k)$, it becomes equal to $v_{\mathbf{k}}^{(n)} = v_{\mathbf{k}}^{(m)} - \hbar k/m$. Note that the transverse velocity component $\boldsymbol{v}_{\perp}$ with respect to the wave vector $\boldsymbol{k}$ does not change in such a radiative transition. If $F_{\mathrm{S}}(\boldsymbol{v}_{\perp}, v_{\mathbf{k}})$ is the distribution function of particles in the beam, populations of the

[6]The case of magnetoactive plasma is different: the corresponding refractive index may exceed unity, so Čerenkov instability becomes possible (for example, for whistlers).

corresponding states (m) and (n) are equal to $N_m = F_S(\boldsymbol{v}_\perp, v_{\mathbf{k}}^{(m)})$ and $N_n = F_S(\boldsymbol{v}_\perp, v_{\mathbf{k}}^{(n)})$, respectively. If the characteristic scale of F_S as a function of $v_{\mathbf{k}}$ is large compared to $\hbar k/m_e$, the population difference $N_m - N_n$ can be expressed as follows:

$$N_m - N_n \approx \frac{\partial F_S}{\partial v_{\mathbf{k}}} \frac{\hbar k}{m_e}. \tag{7.63}$$

Substituting (7.63) and (7.8) into the expression (7.7) for μ_j and integrating over velocity to account for all the transitions, we find:

$$\mu = -\frac{8\pi^3}{\omega k m_e} \int P_{\omega\Omega} \frac{\partial F_S}{\partial v_{\mathbf{k}}} \, d^3v. \tag{7.64}$$

With the expression for the power of Čerenkov radiation (5.22) the absorption coefficient becomes

$$\mu = -\frac{4\pi^2 e^2}{m_e \epsilon(\omega)} \int \frac{\partial F_S}{\partial v_{\mathbf{k}}} \delta(\omega - \boldsymbol{k}\boldsymbol{v}) \, d^3v. \tag{7.65}$$

Since the integral over transverse velocities $\int (\partial F_S/\partial v_{\mathbf{k}}) \, d^2v_\perp = dF(v_{\mathbf{k}})/dv_{\mathbf{k}}$, where $F(v_{\mathbf{k}}) = \int F_S(\boldsymbol{v}_\perp, v_{\mathbf{k}}) \, d^2v_\perp$ is the distribution function over the velocity components $v_{\mathbf{k}}$, the Čerenkov absorption coefficient for a beam penetrating equilibrium plasma looks like [283]:

$$\mu = -\frac{4\pi^2 e^2 v_{\text{ph}}^3}{3\omega m_e v_T^2} \left. \frac{dF(v_{\mathbf{k}})}{dv_{\mathbf{k}}} \right|_{v_{\mathbf{k}} = v_{\text{ph}}}. \tag{7.66}$$

The corresponding decrement is given by

$$\gamma = \frac{1}{2}\mu v_{\text{gr}} = -2\pi^2 \frac{e^2 v_{\text{ph}}^2}{\omega m_e} \left. \frac{dF(v_{\mathbf{k}})}{dv_{\mathbf{k}}} \right|_{v_{\mathbf{k}} = v_{\text{ph}}}. \tag{7.67}$$

To come to (7.67) from (7.66) we recalled the relation between μ and γ pointed out in the introduction to chapter 6 and the expression for the group velocity of plasma waves: $v_{\text{gr}} = 3v_T^2/v_{\text{ph}}$.

According to (7.66), a beam of charged particles absorbs ($\mu > 0$) longitudinal waves with phase velocities for which

$$v_{\text{ph}} \left. \frac{dF(v_{\mathbf{k}})}{dv_{\mathbf{k}}} \right|_{v_{\mathbf{k}} = v_{\text{ph}}} < 0 \tag{7.68}$$

and amplifies ($\mu < 0$) the ones from the interval of phase velocities where

$$v_{\text{ph}} \left. \frac{dF(v_{\mathbf{k}})}{dv_{\mathbf{k}}} \right|_{v_{\mathbf{k}} = v_{\text{ph}}} > 0. \tag{7.69}$$

At the same time, the system considered is stable (ie.e, $\gamma 0$) with respect to plasma wave disturbances in the region (7.68), whereas it becomes unstable ($\gamma 0$) for the ones in the region (7.69).

To be definite, let the distribution function $F_S(\boldsymbol{v})$ be determined by (5.124). Then

$$F(v_{\mathbf{k}}) = N_S(\frac{m_e}{2\pi\kappa T_S})^{1/2} \exp\left[\frac{-m_e(v_{\mathbf{k}} - v_S\cos\theta)^2}{2\kappa T_S}\right]. \tag{7.70}$$

Under the condition $v_S \cos\theta \gg v_T$ ensuring the absence of Čerenkov damping in the background plasma the absorption coefficient (7.66) is reduced to [217]

$$\mu = \frac{1}{3}\sqrt{\frac{\pi}{2}\frac{\omega_{LS}v_{ph}^3}{\omega v_T^2 v_{T_S}^3}}(v_{ph} - v_S\cos\theta)\exp\left[\frac{-(v_{ph} - v_S\cos\theta)^2}{2v_{T_S}^2}\right]. \tag{7.71}$$

Here

$$\omega_{LS}^2 = 4\pi e^2 N_S/m_e, \qquad v_{T_S}^2 = \kappa T_S/m_e,$$

and θ denotes the angle between $\boldsymbol{k}$ and the bulk velocity of the beam $\boldsymbol{v}_S$. The relevant Čerenkov emissivity is described by the formula (5.125). The function $F(v_{\mathbf{k}})$ for the beam and the background plasma is sketched in Fig. 7.2. Beam particles from the interval I ($v_{ph} > v_S\cos\theta$) absorb plasma waves whereas those in the interval II ($v_{ph} < v_S\cos\theta$) amplify them. The latter can be naturally explained by the pre-eminence of stimulated emission over true absorption. Indeed, population inversion is realized in the region II owing to the positive slope of the function $F(v_{\mathbf{k}})$: the number of electrons in the initial state is larger than that in the state where particle appear by emitting a quantum $\hbar\omega$.

Examine now the system beam–plasma by the classical method based on the dispersion equation for plasma waves (2.47): $\epsilon_{\|}(\omega, \boldsymbol{k}) = 0$. Different approaches to the description of particle motion in the system lead to different forms of the longitudinal dielectric permittivity $\epsilon_{\|}(\omega, \boldsymbol{k})$. For example, the quasi-hydrodynamic approach (see section 2.1) yields the following expression:

$$\epsilon_{\|} = 1 - \frac{\omega_L^2}{\omega^2 - k^2 v_T^2} - \frac{\omega_{LS}^2}{(\omega - kv_S)^2 - k^2 v_{T_S}^2}. \tag{7.72}$$

The first two terms here describe the longitudinal permittivity of a stationary plasma, and the last one stands for the contribution of the beam moving with bulk velocity $\boldsymbol{v}_S \| \boldsymbol{k}$. Thermal velocities v_T and v_{T_S} in (7.72) account for the electron pressure in plasma and beam, respectively. Assuming $(\omega - kv_S)^2 \gg k^2 V_{T_S}^2$, we neglect the pressure by the beam particles and yield the

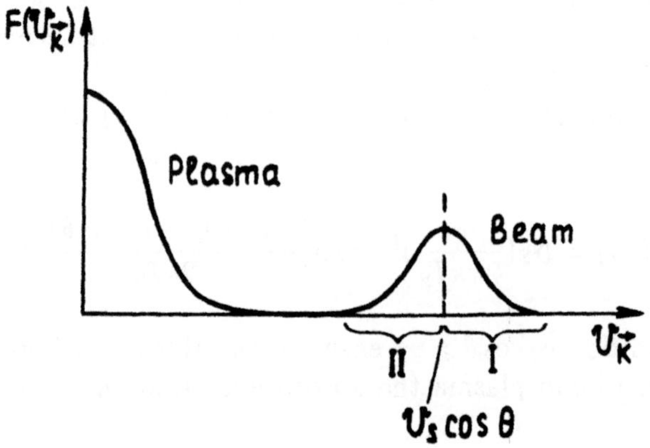

Figure 7.2. Velocity distribution function $F(v_{\mathbf{k}})$ in the system "equilibrium plasma + particle beam": I – absorbing region, II – region which amplifies plasma waves

dispersion equation

$$\frac{\omega_{\rm L}^2}{\omega^2 - k^2 v_T^2} + \frac{\omega_{\rm LS}^2}{(\omega - k v_{\rm S})^2} = 1 \tag{7.73}$$

in which $\omega_{\rm L}$ and $\omega_{\rm LS}$ are Langmuir frequencies of plasma and beam, respectively. To perform the stability analysis of the system, we derive the explicit expression for the frequency ω as a function of real k by the perturbative solution of the equation (7.73). Substituting $\omega = \Omega + \delta$, where Ω matches both the Čerenkov condition $\Omega = k v_{\rm S}$ and the "unperturbed" dispersion relation $\omega^2 = \omega_{\rm L}^2 + k^2 v_T^2$ for the background plasma (see (2.111)), into the equation (7.73) we get the relation

$$2\Omega\delta^3 + \delta^4 = \omega_{\rm LS}^2[\Omega^2 + 2\delta\Omega + \delta^2 - k^2 v_T^2].$$

Supposing that $\omega_{\rm LS}$ and the frequency correction δ are small and holding the lowest power of δ in this equation, we obtain that $\delta^3 \approx \omega_{\rm LS}^2 \omega_{\rm L}^2 / 2\Omega$. If the beam velocity $v_{\rm S} \gg v_T$, then $k^2 v_T^2 \ll k^2 v_{\rm S}^2 = \Omega^2$, so Ω is close to $\omega_{\rm L}$. The frequency of plasma waves is therefore

$$\omega = k v_{\rm S} + \sqrt[3]{1}(\omega_{\rm LS}^2 \omega_{\rm L}/2)^{1/3}, \tag{7.74}$$

where $\sqrt[3]{1} = 1;\ -1/2 \pm i\sqrt{3}/2$. One of these three solutions for plasma waves at frequency $\omega \approx k v_{\rm S}$ corresponds to a wave which increases with time, the increment being equal to[7]

$$-\gamma = \frac{\sqrt{3}}{2}\left(\frac{\omega_{\rm LS}^2 \omega_{\rm L}}{2}\right)^{1/3} = \frac{\sqrt{3}}{2^{4/3}}\omega_{\rm L}\left(\frac{N_{\rm S}}{N}\right)^{1/3}. \tag{7.75}$$

[7] Beam instability with the increment (7.75) has been discovered in [280,281,319,320].

Note that (7.75) is not the only unstable solution of the equation (7.73). In fact, there are other ones which do not obey the equation $\Omega^2 = \omega_L^2 + v_T^2 k^2$. Such solutions exist for small ω_{LS} if the term $\omega_{LS}^2/(\omega^2 - kv_S)^2 = \omega_{LS}^2/\delta^2$ in (7.73) remains finite as $\delta \to 0$. Bearing in mind that the beam velocity $v_S \gg v_T$, it is easy to find from the equation (7.73) that $\delta^2 = \omega_{LS}^2 k^2 v_S^2/(k^2 v_S^2 - \omega_L^2)$. One of two solutions for δ is unstable with the increment which, for $\Omega = kv_S < \omega_L$, is determined by

$$-\gamma = \frac{\omega_{LS} k v_S}{(\omega_L^2 - k^2 v_S^2)^{1/2}} = \frac{\omega_L \Omega}{(\omega_L^2 - \Omega^2)^{1/2}} \left(\frac{N_S}{N}\right)^{1/2}. \tag{7.76}$$

Thus, longitudinal waves with phase velocity $v_{ph} = \Omega/k$ equal to the beam velocity v_S are growing in a system beam–plasma, the increments being proportional to $(N_S/N)^{1/3}$ or $(N_S/N)^{1/2}$. The transition between these two regimes, (7.75) and (7.76), appears at the parameters for which the corresponding increments are similar:

$$1 - \frac{\Omega}{\omega_L} \approx \left(\frac{N_S}{N}\right)^{1/3}. \tag{7.77}$$

At frequencies near ω_L the difference $1 - \Omega/\omega_L \ll (N_S/N)^{1/3}$, so the expression (7.75) is valid, while the formula (7.77) holds for the frequencies far from ω_L where $1 - \Omega/\omega_L \gg (N_S/N)^{1/3}$. Since the instabilities considered have been found using the quasi-hydrodynamic approach, they are called hydrodynamic, in contrast to the kinetic instability considered below.

The longitudinal dielectric permittivity of the system beam–plasma obtained by the method of kinetic equation has the following form:

$$\epsilon_{\parallel}(\omega, \boldsymbol{k}) = 1 + \frac{\omega_L^2}{k^2 v_T^2}[1 + i\sqrt{\pi} Z w(Z)] + \frac{\omega_{LS}^2}{k^2 v_{T_S}^2}[1 + i\sqrt{\pi} Z_S w(Z_S)]. \tag{7.78}$$

In this formula $w(Z)$ and $w(Z_S)$ are the Kramp functions with arguments $Z = \omega/\sqrt{2} k v_T$ and $Z_S = (\omega - kv_S)/\sqrt{2} k v_{T_S}$, respectively. The first two terms in (7.78) characterize $\epsilon_{\parallel}$ of equilibrium plasma, whereas the last term is the contribution of a beam of charged particles with the bulk velocity $\boldsymbol{v}_S \| \boldsymbol{k}$ and the velocity distribution (5.124) (see in this respect (2.88) and the comments thereafter). Assuming the beam to be tenuous, i.e. ω_{LS} sufficiently small, we use the solution of the dispersion equation for the equilibrium plasma, $\epsilon_{\parallel}^{(0)}(\Omega, \boldsymbol{k}) = 0$ (see (2.112)), as the zero-order approximation to the desired solution $\omega = \Omega - i\gamma$ of the dispersion equation $\epsilon_{\parallel}(\omega, \boldsymbol{k}) = 0$. For $|Z| \gg 1$ the longitudinal permittivity $\epsilon_{\parallel}^{(0)}$ is given by (2.114); in what follows the exponential term which gives rise to the Landau damping in the background plasma is neglected. Denoting by $\epsilon_{\parallel}^{(1)}(\omega, \boldsymbol{k})$ the first-order correction to $\epsilon_{\parallel}(\omega, \boldsymbol{k})$ produced by the beam and applying the perturbation method, we find the following expression for the decrement of plasma waves:

$$\gamma = \frac{\operatorname{Im} \epsilon_{\parallel}^{(1)}(\Omega, \boldsymbol{k})}{(\partial \epsilon_{\parallel}^{(0)}/\partial \omega)_{\omega=\Omega}}. \tag{7.79}$$

As follows from (7.78), the imaginary part of $\epsilon_{\parallel}^{(1)}$ for real Ω and $\boldsymbol{k}$ (i.e., for the real argument Z_{S}) is determined by $\operatorname{Re} w(Z_{\mathrm{S}})$. However, $\operatorname{Re} w(Z_{\mathrm{S}}) = e^{-Z_{\mathrm{S}}^2}$ for $\operatorname{Im} Z = 0$, so

$$\operatorname{Im} \epsilon_{\parallel}^{(1)}(\Omega, \boldsymbol{k}) = \sqrt{\frac{\pi}{2}} \frac{\omega_{\mathrm{LS}}^2(\Omega - k v_{\mathrm{S}})}{k^3 v_{T_{\mathrm{S}}}^3} e^{-(\Omega - k v_{\mathrm{S}})^2/2k^2 v_{T_{\mathrm{S}}}^2}.$$

At the same time, $(\partial \epsilon_{\parallel}^{(0)}/\partial \omega)_{\omega=\Omega} \approx 2\omega_{\mathrm{L}}^2/\Omega^3$. As a result, we obtain from (7.79) that

$$\gamma = \sqrt{\frac{\pi}{8}} \frac{\omega_{\mathrm{LS}}^2 \Omega^3 (\Omega - k v_{\mathrm{S}})}{\omega_{\mathrm{L}}^2 k^3 v_{T_{\mathrm{S}}}^3} e^{-(\Omega - k v_{\mathrm{S}})^2/2k^2 v_{T_{\mathrm{S}}}^2}. \tag{7.80}$$

The decrement becomes negative for $\Omega < k v_{\mathrm{S}}$. Kinetic instability arises in this case in the system beam–plasma. Being transformed to the absorption coefficient, the above kinetic expression (7.80) is reduced to the one (7.71) derived by the method of Einstein coefficients if we put $\theta = 0$ and take into account that $\Omega/k = v_{\mathrm{ph}}$ and $\omega \approx \Omega \approx \omega_{\mathrm{L}}$. This fact reveals again that the classical kinetic approach and the elementary quantum approaches are two different ways to treat the same phenomenon, both leading to the same result in the parameter region where both methods are valid [283,321].

Note, furthermore, that the dispersion equation (7.78) is reduced to the hydrodynamic one (7.73) at $|Z|, |Z_{\mathrm{S}}| \gg 1$, the only slight discrepancy being the term $3k^2 v_T^2$ appearing instead of $k^2 v_T^2$ in (7.73). As was shown earlier, the latter describes hydrodynamic instabilities of plasma waves with $\Omega = k v_{\mathrm{S}}$. Therefore, the instability criteria for this case are reduced to the conditions $v_{\mathrm{S}} \gg v_T$, $|\delta| \gg \sqrt{2} k v_{T_{\mathrm{S}}}$. The latter inequality for the increment (7.75) takes the form:

$$\left(\frac{N_{\mathrm{S}}}{N} \right)^{1/3} \gg \frac{v_{T_{\mathrm{S}}}}{v_{\mathrm{ph}}}. \tag{7.81}$$

Under the condition opposite to (7.81) the kinetic instability with the increment $(-\gamma) \propto N_{\mathrm{S}}/N$ determined by (7.80) only exists in the system beam–plasma. But if the ratio N_{S}/N is large enough so that the inequality (7.81) is valid, the strong hydrodynamic instability with $(-\gamma) \propto (N_{\mathrm{S}}/N)^{1/3}$ appears at frequencies $\omega \approx \omega_{\mathrm{L}}$. The width of the frequency interval in the region $\omega < \omega_{\mathrm{L}}$ where this instability is realized increases with N_{S}/N.

7.2.2. BUNEMAN INSTABILITY

A particular case of beam instability is the instability in plasma in which electrons drift relative to ions thus generating an electric current. The dispersion equation for longitudinal waves in the reference frame of electrons

$$\frac{\omega_{\rm L}^2}{\omega^2 - k^2 v_T^2} + \frac{\omega_{\rm L i}^2}{(\omega - k v_{\rm S})^2} = 1 \tag{7.82}$$

obtained in the quasi-hydrodynamic approximation is analogous to (7.73) in which $\omega_{\rm LS}^2 = 4\pi e^2 N_{\rm S}/m_{\rm e}$ is substituted by $\omega_{\rm Li}^2 = 4\pi e^2 N_i/m_{\rm i}$ (N_i is the ion density equal to the electron density N, $v_{\rm S}$ is ion velocity). It is clear from (7.82) and (7.73) that a dense ($N_i = N$) ion beam is equivalent to a tenuous electron beam with $N_{\rm S} = N_i m_{\rm e}/m_{\rm i} = N m_{\rm e}/m_{\rm i}$. This allows us to find immediately the hydrodynamic increments from (7.75), (7.76) by putting $N_{\rm S} = N_e m_{\rm e}/m_{\rm i}$:

$$-\gamma = \frac{\sqrt{3}}{2^{4/3}}\omega_{\rm L}\left(\frac{m_{\rm e}}{m_{\rm i}}\right)^{1/3}, \qquad -\gamma = \frac{\omega_{\rm L}\Omega}{(\omega_{\rm L}^2 - \Omega^2)^{1/2}}\left(\frac{m_{\rm e}}{m_{\rm i}}\right)^{1/2}. \tag{7.83}$$

Phase velocity of the unstable plasma waves $v_{\rm ph} = \Omega/k \simeq v_{\rm S}$. The frequency Ω is close to $\omega_{\rm L}$ for the first increment in (7.82) and less than $\omega_{\rm L}$ for the second one. Note that all the above values correspond to the electron frame. Conversely, in the rest frame of ions the phase velocity and the frequency of the excited waves are nearly zero.

The instability characterized by the increments (7.82) is called Buneman [322]. It is realized under the condition

$$\frac{v_{T_{\rm i}}}{v_{\rm ph}} \ll \left(\frac{m_{\rm e}}{m_{\rm i}}\right)^{1/3} \tag{7.84}$$

which may be found from (7.81) by the ansatz $N_{\rm S} = N_e m_{\rm e}/m_{\rm i}$. If electron and ion temperatures are equal, the mean thermal velocities are related as $v_{T_{\rm i}} = v_T (m_{\rm e}/m_{\rm i})^{1/2}$. Then the condition (7.84) is met for ion beams with velocity $v_{\rm S} = v_{\rm ph} \gtrsim v_T$. However, Landau damping becomes strong at ion velocities comparable to v_T, so that the Buneman instability occurs only if the ratio $v_{\rm S}/v_T$ is sufficiently large. For example, in electron-proton plasma under the condition

$$v_{\rm S}/v_T > 1.27 \tag{7.85}$$

For such beams the Landau decrement is less than the Buneman increment determined by the first expression in (7.83). Kinetic instability is unimportant in this case since it is always obscured by the stronger hydrodynamic one. If electron and ion temperatures are different ($T_{\rm e} > T_{\rm i}$), a plasma with

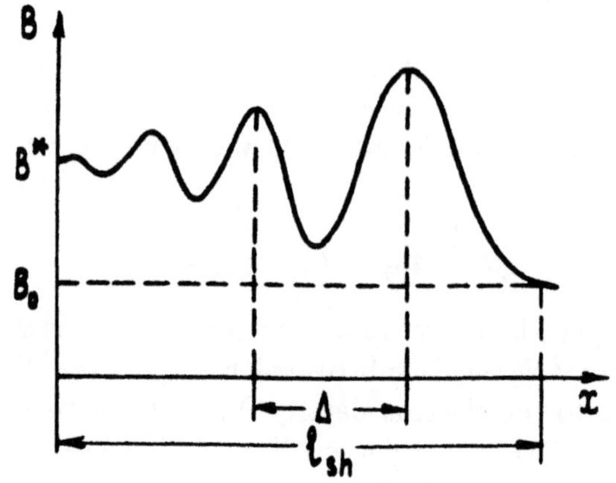

Figure 7.3. Profile of a shock wave in plasma

current is unstable with respect to ion-sound and ion-Langmuir waves even at $v_S < v_T$ (see [73]).

An example of the system favoring the Buneman instability is a collisionless shock wave [323]. The profile of the shock travelling across the magnetic field $\boldsymbol{B}_0$ in a rarefied plasma ($N\kappa T \ll B_0^2/8\pi$) is shown in Fig 7.3. The spatial period of the magnetic field equals

$$\Delta \approx (c/\omega_L\sqrt{M-1})\ln(U\omega_L\sqrt{M-1}/\nu_{eff}c)$$

at the beginning of the front and $\Delta \approx c/\omega_L\sqrt{M-1}$ at its end, and the total width of the shock front is $l_{sh} = U/\nu_{eff}$ (here $M > 1$ is the "magnetic" Mach number equal to the ratio of the velocity of the shock U to the Alfven velocity $v_A = B_0/\sqrt{4\pi m_i N_i}$). The magnetic field behind the shock rises to $B^* = \frac{1}{2}B_0(\sqrt{8M^2+1}-1)$. The profile shown can appear if $l_{sh} \gg \Delta$, i.e., $\nu_{eff} \ll \omega_L U/c$. Such a shock is called collisionless.

The magnetic field at the shock front changes due to the relative drift of electrons and ions resulting in the current $\boldsymbol{j}$ flowing along the plane of the shock. At Mach numbers

$$M > 1 + \frac{3}{4}\left(\frac{8\pi N\kappa T}{B_0^2}\right)^{1/3} \tag{7.86}$$

the drift velocity v_S exceeds v_T [323,324]. According to the criterion (7.85), Buneman instability with the increments (7.83) is developed in the shock front. Note that the influence of the static magnetic field B_0 has not been taken into account in the above analysis of the Buneman instability. So, the results obtained can be used for the shock considered if $\omega_L \gg \omega_B$ and $\omega \gg \omega_B$ in the electron rest frame.

7.2.3. LOSS-CONE INSTABILITY

Čerenkov instability may also arise in a system with particle distribution $F_S(\boldsymbol{v})$ typical for magnetic traps. Absence of particles with pitch angles within the loss cone $\vartheta < \vartheta_{cr}$ is the characteristic feature of such a distribution. The loss cone is formed due to the free escape of particles with small ϑ from the trap. Čerenkov instability in a non-equilibrium plasma with the above distribution has been discovered in [325]. A similar instability occurs in systems in which there is a deficiency of particles with small transverse (with respect to the magnetic field $\boldsymbol{B}_0$) velocities $v_\perp$. As an example of a system providing the loss-cone instability, we consider a dense, homogeneous, equilibrium plasma of density N in which there exists a rarefied energetic component ($N_S \ll N$), both threaded by a static magnetic field $\boldsymbol{B}_0$ [326, 327]. Our choice is somehow justified by the fact that, typically, the "cold" component of cosmic plasmas rapidly relaxes to the Maxwell distribution due to particle collisions, whereas the energetic particles with $v \gg v_T$ remain non-equilibrium during the longer time. Of course, in cosmic conditions there may also exist non-equilibrium particle systems without any equilibrium component.

To investigate the stability of plasma in a magnetic field with respect to plasma wave disturbances we use the dispersion equation $\epsilon_{||}(\omega, \boldsymbol{k}) = 0$ in which the longitudinal dielectric permittivity $\epsilon_{||}$ is described by the formula (2.96). For a non-equilibrium system of helically moving particles in a static magnetic field $\boldsymbol{B}_0$ it has the form (2.97). The last formula takes into account that unperturbed, stationary distribution $f_0(\boldsymbol{p})$ of charged particles should depend only on the longitudinal momentum and the transverse momentum squared: $f_0(\boldsymbol{p}) = f_0(p_{||}, p_\perp^2)$. Absorption of plasma waves caused by the energetic electrons with $N_S \ll N$ is described by the coefficient (7.79), where $\epsilon_{||}^{(0)}$ is the longitudinal permittivity of the equilibrium plasma, and $\epsilon_{||}^{(1)}$ is the contribution by the energetic component. Restricting ourselves to the analysis of the non-relativistic case, we assume the longitudinal ($k_{||}$) and transverse ($k_\perp$) components of the wave vector to meet the following conditions

$$k_{||}\Delta v_{||} \gg \omega_B, \qquad k_\perp v_\perp \gg \omega_B, \tag{7.87}$$

in which $v_\perp$ and $\Delta v_{||}$ are the characteristic transverse velocity and the spread of the longitudinal velocities of the energetic particles, respectively. Since different harmonics contribute $\epsilon_{||}^{(1)}(\omega, \boldsymbol{k})$ in this case, this function can be found by the complicated procedure of summing up over the harmonic number l (see (2.97)). Nevertheless, there is a simpler way to get it. Indeed, under the conditions (7.87) the wavelength of a plasma wave is much smaller than the curvature radius of the unperturbed particle trajectory. It is thus a good approximation to treat the unperturbed motion as uniform and use $\boldsymbol{v}(t)=\boldsymbol{v}_0=\text{const}$ in the expression (2.84) for the tensor $\epsilon_{ik}(\omega,\boldsymbol{k})$. Damping or growing of waves interacting with the doping energetic electrons can now be attributed to the Čerenkov radiation by uniformly moving particles. Substitution of (2.84) into (2.96) and subsequent integration over the directions of the transverse velocity lead to the following expression [73,325]:

$$\epsilon_{\parallel}^{(1)}(\omega,\boldsymbol{k}) = \tag{7.88}$$

$$-\frac{4\pi e^2}{m_e k^2}\int\left\{\frac{\partial F_S}{\partial v_\perp^2} - \frac{(\omega - k_\parallel v_\parallel)\partial F_S/\partial v_\perp^2 + k_\parallel \partial F_S/\partial v_\parallel}{[(\omega-k_\parallel v_\parallel)^2-(k_\perp v_\perp)^2]^{1/2}}\right\} v_\perp\, dv_\perp\, dv_\parallel .$$

For the sake of simplicity we further assume that, in addition to (7.87), the inequality $k_\parallel v_T \gg \omega_B$ is satisfied (see (2.101)). Then, according to section 2.1, $\epsilon_\parallel^{(0)}$ equals that of isotropic plasma, (2.61). The longitudinal permittivity for plasma waves with $v_{\text{ph}} \gg v_T$ equals $\epsilon_\parallel^{(0)} \approx 1-\omega_{\text{L}}^2/\omega^2$. Note that for $k_\parallel v_T \gg \omega_B$ the condition $v_{\text{ph}} \gg v_T$, which allows one to neglect Landau damping in the equilibrium component, can only be satisfied if $\omega_{\text{L}}^2 \gg \omega_B^2$.

Suppose, finally, that the distribution of energetic particles is described by the formula

$$F_S = \frac{N_S}{4\sqrt{2}\pi^{3/2}}\left(\frac{m_e}{\kappa T_S}\right)^{5/2} v_\perp^2 \exp\left[-\frac{m_e(v_\perp^2+v_\parallel^2)}{2\kappa T_S}\right]. \tag{7.89}$$

Substitution of (7.89) into (7.89) and then (2.128), (2.72) into (7.79) yields the following increment of the loss-cone Čerenkov instability:

$$-\gamma = \frac{1}{4}\sqrt{\frac{\pi}{2}}\frac{\omega_{\text{L}}^4}{N k^3 v_{T_S}^3}\,\frac{1-\omega^2 k^{-2} v_{T_S}^{-2} - 2\cot^2\alpha}{1+\cot^2\alpha}\, e^{-\omega^2/2k^2 v_{T_S}^2}, \tag{7.90}$$

where $v_{T_S} = \sqrt{\kappa T_S/m_e}$ and α is the angle between $\boldsymbol{k}$ and $\boldsymbol{B}_0$. The system considered is unstable ($\gamma < 0$) with respect to plasma waves with phase velocities

$$v_{\text{ph}}^2 < \left(\frac{\omega}{k}\right)^2_{\text{cr}} = v_{T_S}^2(1-2\cot^2\alpha). \tag{7.91}$$

According to (7.91), the disturbances propagating at small angles $\alpha < \alpha_{\text{cr}} = 1/\arctan(\sqrt{2})$ remain stable in this system.

It is interesting to note that the above increment (7.90) can be obtained trivially with the method of Einstein coefficients: one should just insert into (7.67) the distribution (7.89) integrated over the velocity component orthogonal to $\boldsymbol{k}$. Then the instability criterion $\alpha > \alpha_{\text{cr}}$ points out the range of phase velocities $v_{\mathrm{ph}} = v_{\mathbf{k}}$ corresponding to the positive derivative $dF(v_{\mathbf{k}})/dv_{\mathbf{k}}$ (see (7.69)).

7.2.4. QUASI-LINEAR THEORY OF BEAM INSTABILITY

The previous section has outlined the linear theory of Čerenkov instability. Consider now the quasi-linear regime of Čerenkov beam instability.[8]

According to section 7.1, one of the equations of the quasi-linear theory is the transfer equation (7.21) for the energy density $W_{\mathbf{k}}$. In a spatially homogeneous system it has the form

$$\frac{\partial W_{\mathbf{k}}}{\partial t} = -2\gamma_j W_{\mathbf{k}}. \tag{7.92}$$

The decrement γ_j is determined by the expression (7.67) for a slowly varying particle distribution function $F(v_{\mathbf{k}}, t)$. The temporal behavior of this distribution is described by the second equation of quasi-linear theory, (7.27). Omitting, for the sake of simplicity, the terms corresponding to spontaneous emission (as has already been done in (7.92))[9] and taking into account that $N_m = N_m(t)$, we write the second quasi-linear equation as

$$\frac{\partial N_m}{\partial t} = \int [A_q^m(N_q - N_m) - A_m^n(N_m - N_n)] \frac{8\pi^3 W_{\mathbf{k}}}{\hbar\omega} d\omega d\Omega. \tag{7.93}$$

Since $N_m = F_{\mathrm{S}}(\boldsymbol{v}_\perp, v_{\mathbf{k}})$ and population differences can be expressed in terms of F_{S} by the relations similar to (7.63), the above equation becomes

$$\frac{\partial F_{\mathrm{S}}}{\partial t} = \int \left\{ P_{\omega\Omega}(\boldsymbol{v}^{(q)}) \frac{\partial F_{\mathrm{S}}}{\partial \boldsymbol{v}}\Big|_{\boldsymbol{v}=\boldsymbol{v}^{(q)}} - P_{\omega\Omega}(\boldsymbol{v}^{(m)}) \frac{\partial F_{\mathrm{S}}}{\partial \boldsymbol{v}}\Big|_{\boldsymbol{v}=\boldsymbol{v}^{(m)}} \right\} \frac{8\pi^3 \boldsymbol{k} W_{\mathbf{k}}}{m_{\mathrm{e}}\hbar\omega^2} d\omega d\Omega, \tag{7.94}$$

where (7.7) has also been used. The spectral power of Čerenkov radiation $P_{\omega\Omega}$ (see (5.22)) is shown to depend explicitly on particle velocity $\boldsymbol{v}$. Next, since the difference $\boldsymbol{v}^{(q)} - \boldsymbol{v}^{(m)} = \hbar\boldsymbol{k}/m_{\mathrm{e}}$ is small compared to $\boldsymbol{v}^{(m)}$ and to the characteristic scale of the distribution F_{S}, the expression in the braces can be linearized with respect to the small parameter $\hbar\boldsymbol{k}/m_{\mathrm{e}}$. Taking into account (5.22), we finally obtain

$$\frac{\partial F_{\mathrm{S}}}{\partial t} = \int \boldsymbol{k} \frac{\partial}{\partial \boldsymbol{v}} \left[\frac{4\pi^2 e^2}{m_{\mathrm{e}}^2 v_{\mathrm{gr}}} \delta(\omega - \boldsymbol{k}\boldsymbol{v}) W_{\mathbf{k}} \frac{\partial F_{\mathrm{S}}}{\partial \boldsymbol{v}} \right] d\omega d\Omega. \tag{7.95}$$

[8]The quasi-linear theory of the loss-cone instability is given in [328].

[9]This is possible at sufficiently high $W_{\mathbf{k}}$ at which $N_{\mathbf{k}} \gg 1$ (see (7.29) and the remarks thereafter).

If the integration is made over the phase velocity space: $d\omega d\Omega = k^{-2} v_{\rm gr} d^3k$, this equation can be written in the following standard form:

$$\frac{\partial F_{\rm S}}{\partial t} = \frac{\partial}{\partial v_\alpha} D_{\alpha\beta} \frac{\partial F_{\rm S}}{\partial v_\beta}, \tag{7.96}$$

where the notation

$$D_{\alpha\beta} = 4\pi^2 \frac{e^2}{m_{\rm e}^2} \int W_{\mathbf{k}} \delta(\omega - \boldsymbol{kv}) \frac{k_\alpha k_\beta}{k^2}\, d^3k \tag{7.97}$$

has been introduced.

The equation (7.96) with the coefficient $D_{\alpha\beta}$ in the form (7.97) and the equations (7.92), (7.67) with $F(v_{\mathbf{k}}) = \int F_{\rm S}(\boldsymbol{v})\, d^2v_\perp$ comprise the closed system of quasi-linear theory equations. This theory has been developed in [316–318]. Note that the equation (7.96) describes particle diffusion in the velocity space, the diffusion coefficient depending on the energy density of plasma waves $W_{\mathbf{k}}$.

The complicated system (7.92), (7.96) is simplified if plasma wave spectrum in one-dimensional, i.e., all the wave vectors are collinear. The equations (7.96), (7.97) are then transformed to

$$\frac{\partial F}{\partial t} = \frac{\partial}{\partial v_{\mathbf{k}}} \left(D \frac{\partial F}{\partial v_{\mathbf{k}}} \right), \qquad D = \frac{4\pi^2 e^2}{m_{\rm e}^2 v_{\mathbf{k}}} W(k). \tag{7.98}$$

This equation, in combination with

$$\frac{\partial W}{\partial t} = -2\gamma W, \qquad \gamma = -2\pi^2 \frac{e^2 v_{\mathbf{k}}^2}{m_{\rm e} \omega_{\rm L}} \frac{\partial F}{\partial v_{\mathbf{k}}} \tag{7.99}$$

which follows from (7.92) and (7.67), describes quasi-linear evolution of the system with particle distribution over velocity projections $F(v_{\mathbf{k}})$ and the one-dimensional spectrum of plasma waves with energy $W(k)$ per unit interval of wave vectors $k = \omega / v_{\rm ph} \approx \omega_{\rm L} / v_{\mathbf{k}}$.

Using equations (7.98), (7.99), we consider below the development of the Čerenkov instability (quasi-linear relaxation) in a system beam–plasma. If a one-dimensional packet of plasma waves with the set of phase velocities $\omega / k = v_{\mathbf{k}}$ in the region of positive $dF/dv_{\mathbf{k}}$ exists at the initial moment $t = 0$, it will grow as described by the equation (7.99). The increase of $W(k)$ leads, according to (7.98), to the increase of the diffusion coefficient D and flattening of the distribution $F(v_{\mathbf{k}})$ in the region of instability. The latter, in its turn, causes the gradual decrease of the increment. The relaxation process finally results in the formation of the "plateau" with $dF/dv_{\mathbf{k}} = 0$ and the steady-state spectrum $W(\omega_{\rm L}/v_{\mathbf{k}})$ (Fig. 7.4).

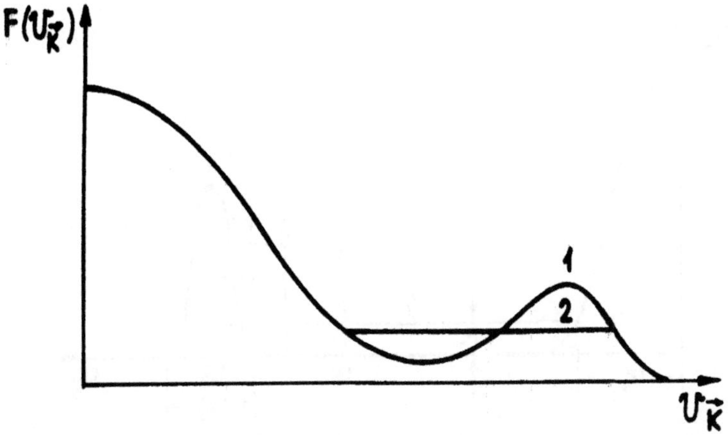

Figure 7.4. Quasi-linear relaxation of the system "beam–plasma": 1 – initial state; 2 – the plateau

The formation of the plateau is physically clear and may be explained as follows. In case of population inversion $dF(v_{\mathbf{k}})/dv_{\mathbf{k}} > 0$ the rate of transitions, which result in stimulated emission of a plasmon and the corresponding decrease of the velocity $v_{\mathbf{k}}$ of a radiating particle, exceeds that leading to absorption and the increase of $v_{\mathbf{k}}$. Thus, the net result of particle diffusion is their shift to smaller values of $v_{\mathbf{k}}$ and the flattening of the distribution $F(v_{\mathbf{k}})$. If, otherwise the spectrum of plasma waves corresponds to phase velocities $v_{\rm ph} = v_{\mathbf{k}}$ in the region of the negative slope of the function $F(v_{\mathbf{k}})$, the net particle flux is directed to the region of greater $v_{\mathbf{k}}$. Again, this results in the flattening of the distribution $F(v_{\mathbf{k}})$.

Since the total energy of the system "plasmons + electrons" is conserved during the quasi-linear relaxation, plasma wave energy in the plateau region can be related to the variation of the distribution function $F(v_{\mathbf{k}})$. Substituting $W\partial F/\partial v_{\mathbf{k}}$ from (7.99) into (7.98) we find that

$$\frac{\partial}{\partial t}\left[F - \frac{\partial}{\partial v_{\mathbf{k}}}\left(\frac{\omega_{\rm L}}{m_{\rm e} v_{\mathbf{k}}^2} W\right)\right] = 0. \tag{7.100}$$

If at the initial moment the spectrum of plasma waves $W_0(k)$ occupies the region from v_1 to v_2 and the distribution function is $F_0(v_{\mathbf{k}})$, the plateau stage is described by the relation

$$F_{\infty} - \frac{\partial}{\partial v_{\mathbf{k}}}\left(\frac{\omega_{\rm L}}{m_{\rm e} v_{\mathbf{k}}^3} W_{\infty}\right) = F_0 - \frac{\partial}{\partial v_{\mathbf{k}}}\left(\frac{\omega_{\rm L}}{m_{\rm e} v_{\mathbf{k}}^3} W_0\right),$$

where F_{∞} and W_{∞} are particle distribution and the plasma wave spectrum after the relaxation. Then it is clear that

$$W_{\infty}(k) = W_0(k) - m_{\rm e}\frac{\omega_{\rm L}^2}{k^3}\int_{v_1}^{\omega_{\rm L}/k} (F_0 - F_{\infty})\, dv_{\mathbf{k}}. \tag{7.101}$$

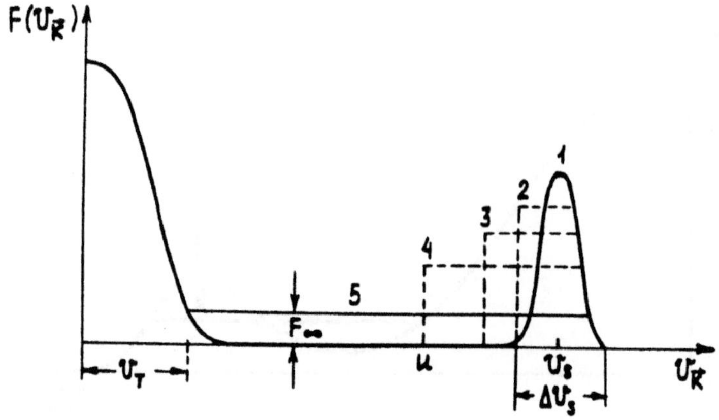

Figure 7.5. Relaxation of an electron beam with small velocity dispersion: 1 – the initial distribution; 2,3,4 – intermediate stages; 5 – the final stage.

The "height" of the final plateau is set by the requirement of particle conservation in the region $v_1 < v_{\mathbf{k}} < v_2$:

$$\int_{v_1}^{v_2} (F_0 - F_\infty)\, dv_{\mathbf{k}} = 0, \qquad F_\infty = (v_2 - v_1)^{-1} \int_{v_1}^{v_2} F_0\, dv_{\mathbf{k}} \tag{7.102}$$

Since the energy density $W(k)$ and the diffusion coefficient D are equal to zero outside the interval $v_1 < v_{\mathbf{k}} < v_2$, particles cannot leave this region and may only be redistributed within it. As a result, the energy density of plasma waves within the plateau region is uniquely related to initial distribution $F_0(v_{\mathbf{k}})$ and spectrum $W_0(k)$ by the formula (7.101). If plasma waves are excited effectively enough so that $W_\infty \gg W_0$, the final value W_∞ does not depend on the radiation energy density at the beginning of the relaxation process.

Consider in more detail the relaxation of a fast ($v_S > v_T$) beam of electrons with a small initial spread of velocities, $\Delta v_S \ll v_S$, penetrating an equilibrium plasma (see Fig. 7.5).[10] It is seen from the figure that the particle beam with the density N_S relaxes to the plateau with the height F_∞ and the width approximately equal to v_S. The particle conservation requirement (7.102) yields $F_\infty \approx N_S/v_S$ for $v_{\mathbf{k}} < v_S$ and $F_\infty = 0$ if $v_{\mathbf{k}} > v_S$. The beam energy at the plateau stage equals 1/3 of its initial value, so the remaining two thirds are transformed into plasma waves during the relaxation. The wave spectrum can be found easily from (7.101). The initial energy $W_0(k)$ is typically about the thermal fluctuation level and thus can be neglected compared to the final value. Taking into account the form of

[10] The value of Δv_S should nevertheless be large enough to satisfy the condition (7.81) and prevent the development of hydrodynamic instability.

F_∞, we find

$$W_\infty(k) = \begin{cases} m_e N_S \omega_L^3 / 2 v_S k^4 & k < \omega_L / v_S, \\ 0 & k < \omega_L / v_S. \end{cases} \tag{7.103}$$

Spectral energy density of plasma waves is proportional to $v_{ph}^4 = (\omega_L/k)^4$ and reaches maximum at $v_{ph} = v_S$.

The evolution of the beam particle distribution can be easily followed for $W_0/W_\infty \ll 1$ [329, 330]. Let u be some velocity far below v_S so that $F(u) \approx 0$. Let also $W_0 = 0$ for $v_{ph} < u$ and $W_0 \neq 0$ in the region $v_{ph} > u$. Then, according to (7.99), plasma waves with $v_{ph} < u$ cannot appear in the system, so that the diffusion coefficient $D \propto W(k)$ in this region is zero (see (7.98)). The relaxation process thus cannot change the distribution $F(v_{\mathbf{k}}) = 0$ in the interval $v_{ph} < u$. This means that the relaxation finishes with the establishing of the plateau $F = N_S(v_S - u)^{-1}$ in the region $u < v_{\mathbf{k}} < v_S$, whereas F remains zero outside this interval (see Fig. 7.5). In fact, the energy density of plasma waves $W(k) \neq 0$ for $v_{ph} < u$, though this value is rather small. This causes the slow shift of the lower edge of the plateau to the left, in the direction of small velocities. The intense diffusion of particles due to the excited plasma waves will constantly maintain the plateau in the region $v_{\mathbf{k}} > u$. The latter is illustrated by the plots 2, 3 and 4 in Fig. 7.5.

The velocity $u(t)$ of the lower edge of the plateau can be found by integrating the equation (7.99) over the small interval $v_1 < v_{\mathbf{k}} < v_2$ containing the value $v_{\mathbf{k}} = u$:

$$\frac{\partial \ln W}{\partial t} = \frac{\pi \omega_L}{N} v_{\mathbf{k}}^2 \frac{\partial F}{\partial v_{\mathbf{k}}}.$$

Since

$$\begin{aligned} \int_{v_1}^{v_2} \frac{\partial \ln W}{\partial t} d\, v_{\mathbf{k}} &\approx -\frac{du}{dt}(\ln W_2 - \ln W_1), \\ \int_{v_1}^{v_2} v_{\mathbf{k}} \frac{\partial F}{\partial v_{\mathbf{k}}} d\, v_{\mathbf{k}} &= v_2^2 F(v_2) \approx u^2 N_S (v_S - u)^{-1}, \end{aligned} \tag{7.104}$$

where W_2 and W_1, respectively, correspond to the plateau region and the initial fluctuation level, the velocity u obeys the following equation:

$$\frac{du}{dt} = -\frac{\pi \omega_L N_S}{\Lambda N} \frac{u^2}{v_S - u} \tag{7.105}$$

in which $\Lambda = \ln(W_2/W_1)$. Since $W_2/W_1 \gg 1$, the parameter Λ is weakly dependent on this ratio. As a good approximation, one may adopt $\Lambda \approx \ln(W_\infty/W_0)$ or $\Lambda \approx \ln(N_S m_e v_S^2 / 3 W_T)$. The latter is the logarithm of the ratio of plasma wave energy after the relaxation, $N_S m_e v_S^2/3$, to the energy of thermal fluctuations in an equilibrium plasma $W_T \sim N \kappa T / N_D$ (N_D is the number of particles in the Debye sphere). The equation (7.105) at $\Lambda = \text{const}$ has the following solution:

$$\frac{v_{\rm S}}{u} + \ln\frac{u}{v_{\rm S}} = 1 + \frac{\pi\omega_{\rm L}}{\Lambda}\frac{N_{\rm S}}{N}t. \tag{7.106}$$

At small $\Delta v = v_{\rm S} - u \ll v_{\rm S}$ this solution looks like $(\Delta v/v_{\rm S})^2 = 2\pi\omega_{\rm L} N_{\rm S} t/N\Lambda$. Hence, a narrow electron beam relaxes to the plateau of the width Δv in a time

$$t = \frac{1}{2\gamma}\Lambda, \qquad \gamma = \pi\omega_{\rm L}\frac{N_{\rm S}}{N}\left(\frac{\Delta v}{v_{\rm S}}\right)^2. \tag{7.107}$$

The entire process of the quasi-linear relaxation is characterized by the time

$$t_{\rm q-l} \sim \frac{\Lambda}{2\pi}\frac{N}{\omega_{\rm L} N_{\rm S}}. \tag{7.108}$$

The assumption of the one-dimensional form of plasma wave spectrum which has been used to get the quasi-linear equations (7.98), (7.99) significantly restricts the generality of the above conclusions. In fact, the plateau distribution F_∞ is unstable with respect to plasma waves with different orientation of the wave vector $\boldsymbol{k}$. The excitation of such waves leads to the formation of plateau distributions along other $\boldsymbol{k}$. A "global" plateau $F(\boldsymbol{v}) =$ const should form as a result of this cascade. But the energy density of particles with such a distribution is infinitely large: the integral $\int v^2 F(v)\, d^3v$ is nonconvergent. The process of the relaxation should therefore result in the formation of either purely one-dimensional plasma wave spectrum and the plateau distribution or a stable particle distribution with the Landau damping of the excited plasma waves. More details on the quasi-linear relaxation in case plasma wave spectrum is not one-dimensional may be found in [331,332].

The results mentioned concern the relaxation of a non-relativistic beam in plasma. The behavior of a relativistic beam differs in several features. The most important one is that the relativistic beam undergoes quasi-one-dimensional relaxation even if the initial spectrum of plasma waves drastically differs from the one-dimensional [333,334]. This can be explained by the strong angular dependence of the Čerenkov instability increment and diffusion coefficient which favors mainly the excitation of plasma waves with wave vector $\boldsymbol{k}$ along the bulk velocity of the beam $\boldsymbol{v}_{\rm S}$. At the same time, diffusion over directions of particle momenta $\boldsymbol{p}$ is much slower than over their absolute values.

The above discussion has been restricted to the quasi-linear relaxation of a homogeneous system in which $F_{\rm S}(\boldsymbol{v}, t)$ and $W_{\bf k}(\boldsymbol{k}, t)$ vary only in time but do not depend on spatial coordinates. But the "spatial", or boundary-value, problem of quasi-linear relaxation when $F_{\rm S}(\boldsymbol{v}, \boldsymbol{r})$ and $W_{\bf k}(\boldsymbol{k}, \boldsymbol{r})$ are time-independent though vary in space is also of interest for cosmic plasmas. Quasi-linear regimes in such systems are considered in [309,330,335].

7.2.5. PLASMA MECHANISM OF RADIO WAVE GENERATION

The term "plasma mechanism" used in the theory of generation of cosmic radio waves means the excitation of plasma waves and their subsequent conversion into electromagnetic radiation. Plasma mechanisms are usually applied to explain the radiation of cosmic plasmas if the Langmuir frequency falls into the radio wave band and plasma waves are excited more easily than electromagnetic ones.[11]

A classical example of plasma mechanism is the generation of fast-drifting bursts (or type III bursts) of solar radio emission [3]. This radiation is generated as follows. A beam of electrons with the velocity $v_S \sim c/3$ ejected from a solar flare region penetrates the coronal plasma. The criteria of beam–plasma Čerenkov instability are satisfied along the beam path. The Čerenkov absorption coefficient in the system becomes negative, and plasma waves emitted via the Vavilov–Čerenkov effect by individual particles of the beam are amplified (maser mechanism is realized). Intense plasma waves then partially transform into electromagnetic ones due to the scattering on particles or plasmons (see sections 8.1,8.2). Thus, observable radio emission occurs which is detected as type III bursts. The characteristic frequency drift observed in such bursts is caused by the penetration of the particle beam into the more rarefied coronal region which results in a gradual decrease of the frequency $\omega \approx \omega_L = (4\pi e^2 N/m_e)^{1/2}$ of the excited plasma waves and, respectively, of radio wave frequency.

Temporal or spatial growth (amplification) of plasma waves in a system beam–plasma may only occur if the Čerenkov increment γ^{Cer} in the beam exceeds the decrement due to the dissipation in the background plasma. As was shown in section 6.1, the latter is caused by absorption due to particle collisions and Landau damping. In the case considered the main source of dissipation is the bremsstrahlung absorption.[12] For the maser mechanism to be realized it is thus necessary that $\gamma^{Cer} + \gamma^b < 0$. Substituting $v_{ph} \sim v_S \sim 10 v_T$, $v_{ph} - v_S \sim v_{T_S} \sim v_T$, and $\nu_{eff} \simeq 5\,s^{-1}$ into (6.9) and (7.80) we find that this condition can be met if $N_S > 10^{-10} N$. So, even a beam with incredibly small density $N_S > 10^{-2}\,cm^{-3}$ penetrating coronal plasma with

[11] However, there is a kind of plasma mechanism for which the plasma frequency ω_L is much lower than the frequency of electromagnetic radiation. In this case the frequency ω is increased up to the observed values due to Compton scattering of plasma waves on relativistic particles (see section 6.2).

[12] At typical coronal conditions ($N \sim 10^8\,cm^{-3}$, $v_T \sim 4 \cdot 10^8$ cm/s, $\nu_{eff} \sim 5\,s^{-1}$) bremsstrahlung absorption dominates over Landau damping for plasma waves with phase velocities $v_{ph} > 6 v_T$.

$N \sim 10^8\,\mathrm{cm}^{-3}$ gives rise to the kinetic Čerenkov instability (amplification). This instability exists in a wide range of N_S up to the limit $N_S \sim 10^{-3} N \sim 10^5\,\mathrm{cm}^{-3}$. If N_S exceeds this limiting value, the hydrodynamic instability with the increments (7.75), (7.76) appears.

The process of quasi-linear relaxation in a system with the one-dimensional plasma wave spectrum results in the formation of the plateau particle distribution in a time of about t_{q-l} determined by (7.108). A beam with density $N_S \sim 10\,\mathrm{cm}^{-3}$ propagating through the coronal plasma with $N \sim 10^8\,\mathrm{cm}^{-3}$ relaxes in a time $t_{q-l} \sim 3 \cdot 10^{-3}\,\mathrm{s}$ (this estimate was made using (7.108) at $\ln \Lambda \sim 1$). The generation of plasma waves stops at the plateau stage, and the waves excited should dissipate in a time $t \sim 1/2\nu_{\mathrm{eff}} \sim 0.1\,\mathrm{s}$. So, the problem arises to provide the non-zero slope of the energetic particle distribution in the region of plateau formation during the longer-lasting motion of the beam through the corona.

There are two factors capable to support a distribution with $dF/dv_{\mathbf{k}} > 0$ in the region of plateau formation during a long time [336,337]: (i) collisions of electrons comprising the beam with coronal plasma particles, and (ii) inhomogeneity of the spatial distribution of beam particles. The first factor is connected with the strong dependence of collision frequency on particle velocity: $\nu_{\mathrm{eff}} \propto v^{-3}$ (see section 6.1). The relative deficit of particles with small velocities occurs due to this reason which, in other words, means the positive slope of the plateau $dF/dv_{\mathbf{k}} > 0$. The second factor works in a similar way: in the course of the beam motion the particles with relatively small velocities are delayed with respect to the faster ones, so they become deficient at the beam front. A problem concerning the dynamics of an electron cloud injected into the corona from a solar flare region becomes very important in this respect [338–340]. Note that one-dimensional expansion proceeds self-similarly, i.e., energy density of plasma waves W and electron distribution F are the functions of z/t, z being the coordinate along the direction of expansion). As for one-dimensional particle diffusion in the velocity space and formation of a quasi-one-dimensional spectrum of plasma waves in the system beam–plasma, we refer the reader to [341].

Along with the beam instability, an important role may be played by Buneman instability in astrophysical plasmas. For example, collisionless shocks may be the source of intense radio emission. This is generated by the excitation (due to Buneman instability) of plasma waves in the shock front and their subsequent conversion into electromagnetic radiation. The shocks moving in the solar corona from the flare region excite radio emission with slow frequency drift called type II bursts [3,342,343].

Finally, the development of the loss-cone Čerenkov instability is quite probable in solar corona conditions. Excitation of plasma waves with $\omega \approx \omega_L$

may occur in magnetic traps formed above the bipolar groups of sunspots and, generally, in solar active regions with a complicated configuration of magnetic field lines. Here the above instability, as well as the loss-cone cyclotron instability considered in section 7.3, ensures the effective action of plasma mechanism which is applied to explain some types of type IV radio emission [327, 344].

We also mention the attempts to use plasma mechanism to explain pulsar radio emission and refer to the reviews [63, 345, 346] for discussion of these items.

7.3. Cyclotron instability

This section deals with the instability related to the emission by electrons at distinct harmonics of the gyrofrequency. This instability occurs in a system of particles permeated by a static magnetic field $\boldsymbol{B}_0$. The corresponding increments are determined by the solutions of the dispersion equation (2.41) for the normal waves in a magnetoactive plasma. The dielectric permittivity tensor $\epsilon_{\alpha\beta}(\omega, k)$ in the form (2.94) enters this equation. The dispersion equation for longitudinal waves is reduced to $\epsilon_{||}(\omega, \boldsymbol{k}) = 0$ where $\epsilon_{||}(\omega, \boldsymbol{k})$ is determined by (2.97). Both $\epsilon_{\alpha\beta}$ and $\epsilon_{||}$ are infinite sums over the harmonic number l. If the spread of particle velocities is small enough, the main contribution to the dielectric tensor comes from one or two neighboring harmonics, while the influence of the others can be neglected. For non-relativistic particles this is possible if

$$k_{||}\Delta v_{||} \ll \omega_B. \tag{7.109}$$

The above condition is opposite to the first one in (7.87) which determines the parameter region of Čerenkov instability/absorption in plasma.

7.3.1. CYCLOTRON INSTABILITY OF LONGITUDINAL WAVES

Concerning longitudinal waves, we restrict ourselves to the discussion of the cyclotron loss-cone instability in a non-equilibrium particle system characterized by the distribution $f_S(\boldsymbol{p}) = f_S(p_{||}, p_\perp^2)$ with zero mean velocity along the magnetic field $\boldsymbol{B}_0$ and the deficit of particles with small $p_\perp$ ($p_{||}$ and $p_\perp$ are the longitudinal and the transverse components of the electron momentum $\boldsymbol{p}$ with respect to the magnetic field). Such a system with $f_S \propto \delta(p_\perp - p_\perp^{(0)})$ has been proved to be unstable with respect to longitudinal waves [347]. Then it was shown in [348] than this instability is also realized in the case when such non-equilibrium particles are submerged in a dense, equilibrium plasma. In what follows an example of the cyclotron loss-cone instability is considered in the two-component system comprising the equilibrium plasma and the energetic particle component with the distribution

$$f_S(p_{\|}, p_{\perp}^2) = Ap_{\perp}^2 e^{-(p_{\|}^2+p_{\perp}^2)/2a^2} \tag{7.110}$$

where $A = N_S/4\sqrt{2}\pi^{3/2}a^5$.

Suppose that velocities of energetic particles are large compared to the thermal velocity of electrons in the equilibrium plasma but still much less than the speed of light:

$$m_e^2 v_T^2 \ll a^2 \ll m_e^2 c^2. \tag{7.111}$$

On the one hand, this makes it possible to neglect the Landau damping in the equilibrium plasma. On the other hand, the non-equilibrium component can be treated as weakly relativistic. Then, the conditions (7.109) can only be met for energetic particles with $\Delta v_{\|} \approx a/m_e$ and a from the interval (7.111) if $k_{\|} v_T \ll \omega_B$. In this case the properties of plasma waves in the equilibrium plasma are drastically changed. Cyclotron absorption (see section 6.2) leads to the strong damping of these waves in the interval $|\omega - l\omega_B| \lesssim k_{\|} v_T \ll \omega_B$ which is of no interest to the present analysis. However, for $|\omega - l\omega_B| \gg k_{\|} v_T$ the absorption becomes weak, and the dispersion properties of the longitudinal waves are determined by the equation (2.157) which describes Bernstein modes and waves in the hybrid band (see section 2.2 and Fig. 2.9,2.10 therein which shows the dispersion curves of these waves).

If the density of energetic particles N_S is much less than N, the electron density of the equilibrium plasma, then $\mathrm{Re}\,\omega$ and k are related by the equation (2.157), while the decrement γ may be found by the perturbation method. As a result, we arrive again at the formula (7.79) for the decrement in which the contribution of the equilibrium plasma, $\epsilon_{\|}^{(0)}$, is now given by (2.157) and that of the non-equilibrium component, $\epsilon_{\|}^{(1)}$, by (2.97).

The pole of the integrand in (2.97) is determined by the resonance condition

$$m(p)\omega - lm_e\omega_B - k_{\|}p_{\|} = 0 \tag{7.112}$$

which is just the Doppler condition (5.34) for the magnetic bremsstrahlung. The mass of a weakly relativistic particle equals $m(p) \approx m_e(1 + p^2/2m_e^2c^2)$. The relativistic dependence $m(p)$ can be neglected in (7.112) if the ratio $\omega p^2/2m_e c^2 \ll k_{\|} p_{\|}$. According to (7.110), $p \sim a$ and $p_{\|} \sim a$, hence the latter means that

$$|k_{\|}| \gg \omega a/2m_e c^2. \tag{7.113}$$

So, the Doppler condition (7.112) becomes non-relativistic for an arbitrary orientation of the vector $\boldsymbol{k}$ except in the small angular interval close to the direction perpendicular to $\boldsymbol{B}_0$. That is why the increments of the cyclotron instability in the region (7.113) is called non-relativistic, in contrast to the relativistic increments in the case when the term $k_{\|}v_{\|}$ is negligible and the condition (7.112) becomes essentially relativistic describing the so-called transverse Doppler effect.

7.3.2. KINETIC INSTABILITY OF BERNSTEIN MODES

Let us find $\mathrm{Im}\,\epsilon_{\|}^{(1)}$ from (2.97) under the condition (7.113). This value is determined by the residue of the integral over $p_{\|}$ in (2.97) at the point $m_e\omega - lm_e\omega_B - k_{\|}p_{\|} = 0$ [96]:

$$\mathrm{Im}\epsilon_{\|}^{(1)} = \sqrt{\frac{\pi}{2}}\frac{\omega_{\mathrm{LS}}^2\omega_B}{k_{\|}k_{\perp}^2v_{\mathrm{e}}^3}\sum_{l=-\infty}^{+\infty} le^{-Z_l^2}[\delta_l\varphi_l + (\delta_l+1)\chi_{\mathrm{e}}\varphi_l']. \tag{7.114}$$

The following notations are used here:

$$\varphi_l = e^{-\chi_{\mathrm{e}}}I_l(\chi_{\mathrm{e}}), \qquad Z_l = \frac{\omega - l\omega_B}{\sqrt{2}k_{\|}v_{\mathrm{e}}}, \qquad \delta_l = \frac{\omega - l\omega_B}{l\omega_B}$$
$$\chi_{\mathrm{e}} = \frac{k_{\perp}^2v_{\mathrm{e}}^2}{\omega_B^2}, \qquad v_{\mathrm{e}} = \frac{a}{m_{\mathrm{e}}}, \qquad \omega_{\mathrm{LS}}^2 = 4\pi e^2N_{\mathrm{S}}/m_{\mathrm{e}}, \tag{7.115}$$

and $I_l(\chi_{\mathrm{e}})$ is the modified Bessel function of the order l. If

$$2k_{\|}^2v_{\mathrm{e}}^2 \ll \omega_B^2, \tag{7.116}$$

all the terms in the sum over l except $l = s-1$ and $l = s$ are exponentially small in the interval $(s-1)\omega_B \leq \omega \leq s\omega_B$. Keeping only these two terms in the sum, we get

$$\mathrm{Im}\,\epsilon_{\|}^{(1)} \approx \sqrt{\frac{\pi}{2}}\frac{\omega_{\mathrm{LS}}^2\omega_B}{k_{\|}k_{\perp}^2v_{\mathrm{e}}^3}\Big\{e^{-Z_{s-1}^2}(s-1)[\delta_{s-1}(\varphi_{s-1} + \chi_{\mathrm{e}}\varphi_{s-1}') + \chi_{\mathrm{e}}\varphi_{s-1}]$$
$$+e^{-Z_s^2}s[\delta_s(\varphi_s + \chi_{\mathrm{e}}\varphi_s') + \chi_{\mathrm{e}}\varphi_s']\Big\}. \tag{7.117}$$

If the frequency ω is very close to the harmonic $s\omega_B$, so that $|\delta_s| \ll 1$ and $|Z_{s-1}^2| \gg 1$, the expression (7.117) is simplified as follows:

$$\mathrm{Im}\epsilon_{\|}^{(1)} \approx \frac{\sqrt{\pi}s\omega_{\mathrm{LS}}^2}{\omega_B(\omega - s\omega_B)}\varphi_s'(\chi_{\mathrm{e}})Z_se^{-Z_s^2}. \tag{7.118}$$

The formulas (7.117) and (7.118) correspond to the cases when cyclotron radiation at two neighboring cyclotron harmonics or at a single one is taken into account.

The derivative $(\partial\epsilon_{\parallel}^{(0)}/\partial\omega)_{\omega=\Omega}$, where Ω is the frequency satisfying the dispersion equation $\epsilon_{\parallel}^{(0)}(\omega,\boldsymbol{k})=0$, is positive in an equilibrium plasma at frequencies below the hybrid frequency $\omega_{\mathrm{h}}=\sqrt{\omega_{\mathrm{L}}^2+\omega_B^2}$. Therefore, the increment of the cyclotron instability $-\gamma$, where γ is given by (7.79), becomes positive if $\operatorname{Im}\epsilon_{\parallel}^{(1)}<0$. The left-hand side in the last inequality is determined by (7.114), (7.117), (7.118). For example, if the increment for the Bernstein modes is determined by a single cyclotron harmonic, the value $\operatorname{Im}\epsilon_{\parallel}^{(1)}$ is given by (7.118), and the system is unstable in the region $\varphi_s'(\chi_{\mathrm{e}})<0$. The limiting value $\chi_{\mathrm{e}}^{\mathrm{cr}}$ corresponding to the instability threshold is determined by the equation $\varphi_s'(\chi_{\mathrm{e}}^{\mathrm{cr}})=0$. The system is instable provided

$$\chi_{\mathrm{e}}\equiv k_{\perp}^2 v_{\mathrm{e}}^2/\omega_B^2>\chi_{\mathrm{e}}^{\mathrm{cr}}\approx s^2. \tag{7.119}$$

In terms of the parameter $\chi\equiv k_{\perp}^2 v_T^2/\omega_B^2$, where v_T is the thermal velocity of the equilibrium plasma electrons, the instability criterion looks like

$$\chi>\chi_{\mathrm{cr}}\approx\left(\frac{s v_T}{v_{\mathrm{e}}}\right)^2. \tag{7.120}$$

This criterion and the dispersion curve profiles $\omega(\chi)$ drawn in Fig. 2.9 reveal that cyclotron instability of the Bernstein modes occurs at the lower-frequency part of the frequency range between the cyclotron harmonics $(s-1)\omega_B$ and $s\omega_B$, whereas absorption appears at higher frequencies. The frequency dependence of the non-relativistic cyclotron increment for the Bernstein modes is thus different from that of the Čerenkov one (7.80) for Langmuir waves in isotropic plasma. In the latter case the absorption band occupies the frequencies below the region of instability. The difference is caused by the different dispersion properties of Bernstein modes and plasma waves. The former are characterized by the anomalous dispersion (i.e., ω is decreasing with k; see Fig. 2.9) at frequencies below ω_{h} whereas the latter are of normal dispersion (ω is increasing with k; see (2.115)).

The explicit expressions for the non-relativistic increments of Bernstein modes follow from (7.79), (2.160), (2.163) and (7.118). In the limiting cases of small and large χ they look like [96]

$$\begin{aligned}
-\gamma &\approx \sqrt{\pi}\frac{\omega_{\mathrm{LS}}^2}{\omega_{\mathrm{L}}^2}\omega_B s(s^2-1)Z_s e^{-Z_s^2}\varphi_s'(\chi_{\mathrm{e}}), \quad \chi\ll 1,\ Z_s<0\\
-\gamma &\approx -\sqrt{\pi}\frac{\omega_{\mathrm{LS}}^2}{\omega_B}sZ_s e^{-Z_s^2}\varphi_s'(\chi_{\mathrm{e}}), \qquad \chi\to\infty,\ Z_s>0.
\end{aligned} \tag{7.121}$$

The behavior of γ as a function of $k_{\parallel}$ is determined by the expression $Z_s e^{-Z_s^2}$. The maximal increment corresponds to $Z_s=1/\sqrt{2}$ or

$$k_{\parallel}^{\max}=\frac{\omega-s\omega_B}{v_{\mathrm{e}}}. \tag{7.122}$$

In combination with the criterion (7.113) of the non-relativistic instability this inequality yields

$$|\delta_s| \equiv \left| \frac{\omega - s\omega_B}{s\omega_B} \right| \gg \frac{v_e^2}{2c^2}. \tag{7.123}$$

The violation of this inequality drives $k_{\parallel}^{max}$ out of the region where the cyclotron instability is possible without accounting for the relativistic correction of the electron mass.

To investigate the dependence of $-\gamma$ on ω and $k_{\parallel}$ in a wide range of χ, Z_s, Z_{s-1}, one should turn from the formulas (7.121) to the expression (7.79) in which the tensors $\epsilon_{\parallel}^{(0)}$ and $\epsilon_{\parallel}^{(1)}$ are given directly by the formulas (2.157) and (7.117). The results here can only be obtained numerically. Fig. 7.6 shows the plots of $-\gamma(\omega, k_{\parallel}^{max})$ versus ω. It is clear that, for example, in the range $\omega_B < \omega < 2\omega_B$ the region of cyclotron absorption spans over a relatively small interval of about $0.2\omega_B$. The rest corresponds to the cyclotron instability with the maximal increment $-\gamma_{max} \approx 10^{-2}\omega_B N_S/N$. The width of the instability band is weakly dependent on the harmonic number s. This means that the frequency spectrum of the excited Bernstein modes comprises a series of quasi-equidistant bands separated by intervals of about ω_B.

At small $k_{\parallel}$ for which the condition (7.123) is violated the position of the pole (7.112) on the plane $p_{\parallel}, p_{\perp}$ is significantly affected by the relativistic effects. In case of strictly perpendicular propagation $k_{\parallel} = 0$ the pole position is entirely determined by the relativistic dependence of the mass of energetic electrons on their velocity. The width of the frequency band corresponding to the relativistic cyclotron instability can be estimated by the following quantum considerations.

Cyclotron radiation by an electron in the direction perpendicular to $\boldsymbol{B}_0$ results in the decrease of $p_{\perp}^2$ at constant $p_{\parallel}$ (the normal Doppler effect, see section 5.1). So, population inversion for the particle distribution (7.110) is realized if $\partial f_S / \partial p_{\perp}^2 > 0$, i.e. at $p_{\perp} < \sqrt{2}a$. On the other hand, radiation at frequency ω can be absorbed/emitted by particles with the momentum

$$p = \tilde{p} \equiv m_e c \sqrt{\frac{s^2 \omega_B^2}{\omega^2} - 1} \tag{7.124}$$

which may be found by solving the Doppler equation (7.112) for the harmonic $l = s$ at $k_{\parallel} = 0$. Therefore, the transitions in the region $\partial f_S / \partial p_{\perp}^2 > 0$ at $p_{\parallel} = 0$ correspond to the frequencies

$$0 < m_e c \sqrt{\frac{s^2 \omega_B^2}{\omega^2} - 1} < \sqrt{2}a. \tag{7.125}$$

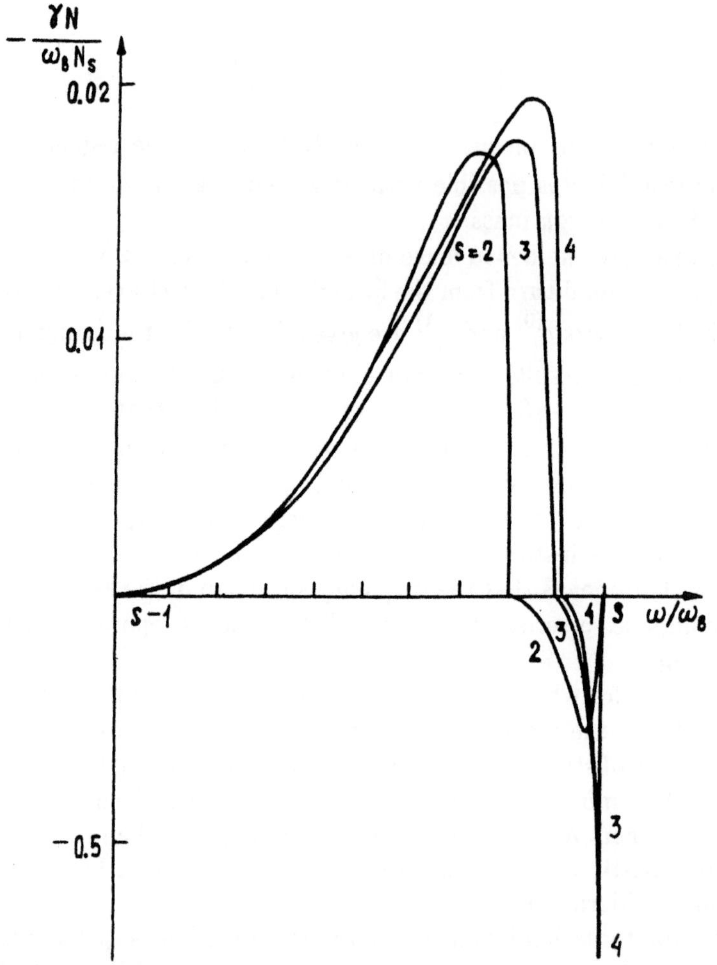

Figure 7.6. The non-relativistic cyclotron increment of Bernstein modes plotted as a function of frequency for $\omega_L/\omega_B=5$, $v_e/v_T=4$ and $s=2,3,4$ (note the different scales at $\gamma>0$ and $\gamma<0$).

The increment $-\gamma < 0$ in this frequency range, i.e., instability develops in the system. Owing to the small value of the parameter $a^2/m_e^2c^2$, the instability band of the width

$$\Delta\omega \sim \frac{s\omega_B v_e^2}{c^2} \tag{7.126}$$

adjoins the harmonic $\omega = s\omega_B$ from below. At higher s the width of the instability band may become comparable to ω_B even at $v_e^2/c^2 \ll 1$. More rigorous kinetic analysis [96] leads to the value

$$\Delta\omega = s\omega_B \frac{v_e^2}{c^2}\frac{B_1}{B_2}, \tag{7.127}$$

where

$$B_1 = \int_0^{\pi/2} J_s^2(\xi)\sin\vartheta\, d\vartheta \qquad B_2 = \int_0^{\pi/2} J_s^2(\xi)\sin^3\vartheta\, d\vartheta,$$

the argument $\xi = k_\perp \tilde{p}\sin\vartheta/m_e\omega_B$, and ϑ is the electron pitch angle. The values B_1 and B_2 are of the same order of magnitude, though $B_1 > B_2$: for example, $B_1/B_2 = 7/6$ at $\xi \ll 1$. Some increase of the $\Delta\omega$ as given by (7.127) compared to (7.126) seems natural. Indeed, the instability takes place even if part of the radiative transitions occurs in the region $\partial f_S/\partial p_\perp^2 < 0$: the increment changes sign only when the rate of such transitions rises high enough compared to the ones in the region $\partial f_S/\partial p_\perp^2 > 0$.

According to [96], relativistic increment of the cyclotron instability is given by

$$-\gamma = \sqrt{2\pi}\frac{\omega_{LS}^2}{\omega^2}\left(\frac{\partial\epsilon_{\parallel}^{(0)}}{\partial\omega}\right)^{-1}\frac{v_T^2 c^2}{v_e^4}\frac{sy}{\chi}\left(B_1 - \frac{1}{2}y^2 B_2\right)e^{-y^2/2}, \tag{7.128}$$

where $y = \tilde{p}/a$ (see (7.124)). For Bernstein modes at frequencies near $s\omega_B$ [13]

$$y^2 = \frac{2(s-\omega/\omega_B)c^2}{sv_e^2} \simeq \frac{c^2}{v_e^2}\frac{s^2-1}{s!}\left(\frac{\chi}{2}\right)^{s-1}, \tag{7.129}$$

$$\frac{\partial\epsilon_{\parallel}^{(0)}}{\partial\omega} \simeq \frac{2\omega_L^2}{\omega_B^3}\frac{(s-2)!}{(s^2-1)(s+1)}\left(\frac{\chi}{2}\right)^{-s+1}. \tag{7.130}$$

The attention to the generation of Bernstein modes in astrophysical plasmas has arisen due to the interpretation of some short-time events in solar radio emission characterized by peculiar frequency spectrum. This spectrum and its temporal behavior [14] are given in Fig. 7.7 [349]. It comprises the band in emission at lower frequencies and the adjoining strong band in absorption at higher frequencies. This absorption feature, in its turn, is ended by the bright, narrow "eye". It is seen that the frequency dependence of the non-relativistic cyclotron increment of Bernstein modes (Fig. 7.6) conforms very well to the observed spectrum of the event in both the emission band

[13] The formulas (7.129) and (7.130) are valid for plasmas with $\omega_B \ll \omega_L$ at harmonics $s \ll \omega_L/\omega_B$ and under the condition $\chi \ll 1$ (see (2.160), (2.161)).

[14] In radio astronomy the two-dimensional plot of radiation intensity as a function of frequency and time is called dynamical spectrum.

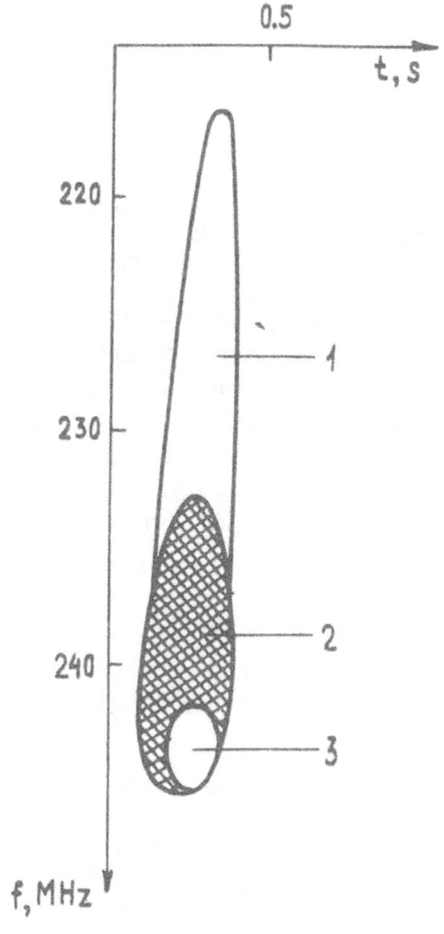

Figure 7.7. Dynamical spectrum of a distinct element in the solar radio emission on March 2, 1970: 1 – the band in emission; 2 – the region in absorption; 3 – the bright "eye".

and the absorption interval, the former corresponding to the cyclotron instability and the latter to strong cyclotron absorption. The bright "eye" occurs due to the positive relativistic increment in the narrow frequency band near $s\omega_B$. Parameters of non-equilibrium component, the values of ω_L and ω_B in the coronal plasma, and constraints on the homogeneity of the magnetic field in the source can be found in [97]. We do not touch on these items here and only note that the frequency spectrum of electromagnetic waves escaping the solar corona retains the main features of that for the Bernstein modes if the radio emission appears due to Raman scattering (coalescence) of the Bernstein modes (see section 8.4).

7.3.3. KINETIC INSTABILITY IN THE HYBRID BAND

The cyclotron instability in this frequency range has some features originating from the differences of the dispersion curves from that of Bernstein modes at frequencies below the hybrid one (see Fig. 2.10). This difference appears at $\chi = k_\perp^2 v_T^2/\omega_B^2 < s$. The anomalous dispersion is removed by the normal one; the dispersion relation has the form (2.162). The cyclotron increment of the longitudinal waves in the hybrid band (the kinematic increment) is given again by the expression (7.79) with $\mathrm{Im}\,\epsilon_{\parallel}^{(1)}$ determined by (7.117) and by the formula (7.128). The derivative $\partial\epsilon_{\parallel}^{(0)}/\partial\omega$ which enters the denominators in (7.79) and (7.128) is however much less in the normal dispersion region of the hybrid band than in the anomalous dispersion region of this band (in which $\chi > s$) or in case of Bernstein modes.[15] Thus, one may expect the increment (7.79) at frequencies near the hybrid one to be greater than for Bernstein modes.

In case of the cyclotron instability in the region of non-relativistic Doppler effect (7.113) rather high increments are realized at sufficiently large ratios v_e/v_T at which the instability threshold χ_{cr} is shifted into the region of normal dispersion ($\chi_{cr} < s$). If, however, v_e/v_T is not so large, the parameter χ_{cr} exceeds s and the instability develops solely in the region of anomalous dispersion where it is characterized by the increments similar to that for Bernstein modes.

The plots of the cyclotron increment (maximal over $k_\parallel$) in the hybrid band as a function of the parameter χ are given in Fig. 7.8. The solid curves mark the region $|k_\parallel^{max}| \gg \omega v_e/2c^2$ (see (7.113)), the dashed ones correspond to the case when the maximal increment at a given frequency is reached at $|k_\parallel^{max}| \sim \omega v_e/2c^2$, i.e., at the border between the non-relativistic and the relativistic ("transverse") Doppler effect.

It is clear from the figure that under the condition $\chi_{cr} < s$ which is satisfied for all the plots the increment in the hybrid band depends quite sensitively on the ratio ω_h/ω_B. Namely, it increases as soon as the double plasma resonance occurs in the system, i.e., the hybrid frequency ω_h approaches one of the harmonics $s\omega_B$. For the selected values of v_e/v_T and s the regions of high increments span over the interval $\Delta\omega \lesssim (0.1 - 0.2)\omega_B$ near $s\omega_B$. The peak value of the increment is several times larger than the corresponding value of $(-\gamma)$ in the hybrid band far from the double resonance and one-two orders of magnitude greater than the increment for Bernstein modes.

The expression for the relativistic increment describing the case of trans-

[15] At $\chi \ll 1$, for example, the derivative at the hybrid frequency $\partial\epsilon_{\parallel}^{(0)}/\partial\omega \approx 2/\omega_h$ which is much less than the corresponding value (7.130) for Bernstein modes.

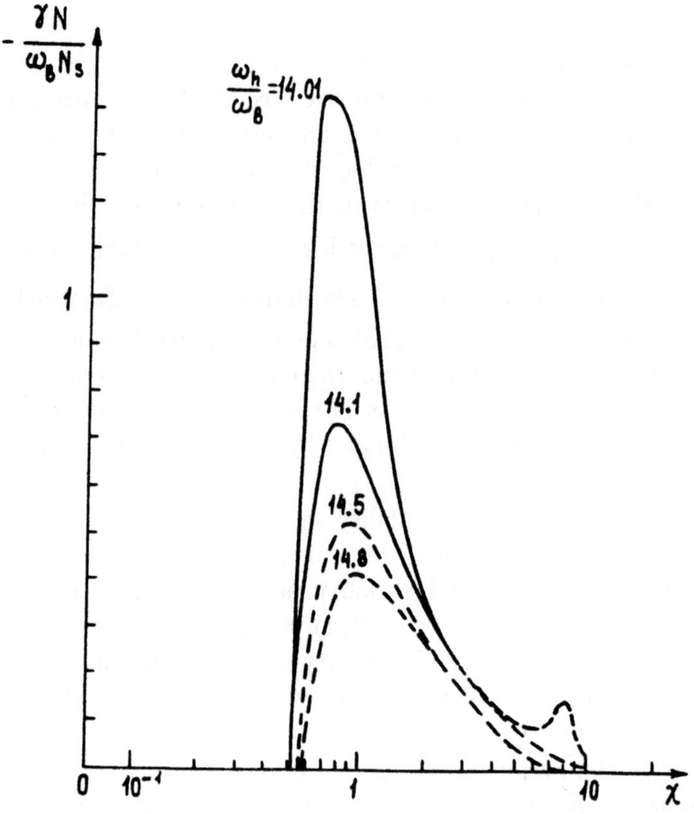

Figure 7.8. The increment of the non-relativistic cyclotron instability in the hybrid band $14\omega_B < \omega < 15\omega_B$ as a function of the parameter $\chi = k_\perp^2 v_T^2/\omega_B^2$ at various ratios ω_h/ω_B.

verse propagation of the longitudinal waves is determined again by the formula (7.128). But now one should take into account the different behavior of $\partial\epsilon_\parallel^{(0)}/\partial\omega$ in the hybrid band. In particular, $\partial\epsilon_\parallel^{(0)}/\partial\omega|_{\epsilon_\parallel^{(0)}=0} \approx 2/\omega_h$ in the region (2.162) of normal dispersion. If $sv_e^2/c^2 \ll 1$, the relativistic instability can appear only in the narrow interval (7.127) adjoining the value $s\omega_B$ from below. However, at sufficiently high ratio ω_h/ω_B (or, in other words, at sufficiently high s) the parameter $sv_e^2/c^2 \sim 1$. Here the interval of the relativistic cyclotron instability $\Delta\omega$ occupies the larger part of the hybrid band. Note that for $\Delta\omega \gtrsim \omega_B$ more than one harmonic $l = s$ must be accounted in the formula for the increment, in contrast to (7.128).

Enhanced generation of longitudinal waves in case of double plasma resonance brings the possibility to interpret the "zebra" structures seen occasionally in the spectra of solar radio emission. Such a structure looks like the system of parallel, drifting bands in emission which are separated by approximately equal frequency intervals. This can probably be formed in the following way [97, 350]. Let a narrow magnetic flux tube in the corona be filled by energetic electrons with the distribution of the type (7.110) providing the development of the loss-cone cyclotron instability (Fig. 7.9). Possible variations of the hybrid frequency $\omega_{\rm h} = \sqrt{\omega_{\rm L}^2 + \omega_B^2}$, the gyrofrequency ω_B and its harmonics $s\omega_B$ along the flux tube are shown in Fig. 7.10. The regions of double plasma resonance $\omega_{\rm h} = s\omega_B$ corresponding to the selected coordinates $l = l_s$ along the tube are determined by the intersection of the curves in this figure. As was already pointed out, longitudinal waves at frequencies $\omega_s \approx \omega_{\rm h}$ are effectively generated in these regions.

The observed radio emission can result from either coalescence of the excited plasma waves to the electromagnetic one at the frequency $\omega = 2\omega_s$ or scattering of plasma waves at particles (the frequency of radio emission will then be close to ω_s; see sections 8.1,8.4). In both cases the system of bands parallel to the time axis will appear on the dynamical spectrum of the event. As follows from the above discussion, distinct bands are generated in the different, localized parts within the source. Looking at the Fig. 7.10, it is easy to understand that the distance between the neighboring layers of double plasma resonance l_{s-1} and l_s equals

$$\Delta l \simeq \frac{L_B L_N}{(s-1)L_B - sL_N}, \tag{7.131}$$

where $L_B = \omega_B (d\omega_B/dl)^{-1}$ and $L_N = \omega_{\rm L}(d\omega_{\rm L}/dl)^{-1}$ are characteristic scales of magnetic field and electron density in the corona along the flux tube. The formula (7.131) has been obtained under the assumptions that $\omega_{\rm L} \gg \omega_B$, hence $\omega_{\rm h} \approx \omega_{\rm L}$. Then, the frequency interval between two nearest emission bands in the radio spectrum, $\Delta\omega = 2|\omega_s - \omega_{s-1}|$, can be written in the form [98]

$$\Delta\omega \approx \frac{2(s-1)\omega_B L_B}{|(s-1)L_B - sL_N|}. \tag{7.132}$$

Depending on the ratio L_N/L_B, the interval $\Delta\omega$ varies from $2\omega_B$ (for $L_N/L_B \ll 1$) to $2\omega_B L_B/L_N$ (for $L_N/L_B \gg 1$). Note that these values correspond to the case when radio emission originates at double plasma frequency. But if the observed radio emission occurs mainly at frequencies $\omega \approx \omega_{\rm L}$, the intervals $\Delta\omega$ will be two times less. Note, finally, that temporal variation of the magnetic field and coronal plasma along the flux tube change the intervals between the adjacent bands and lead to the drift of the system of the emission bands on the dynamical spectrum, the feature observed in "zebra" structures.

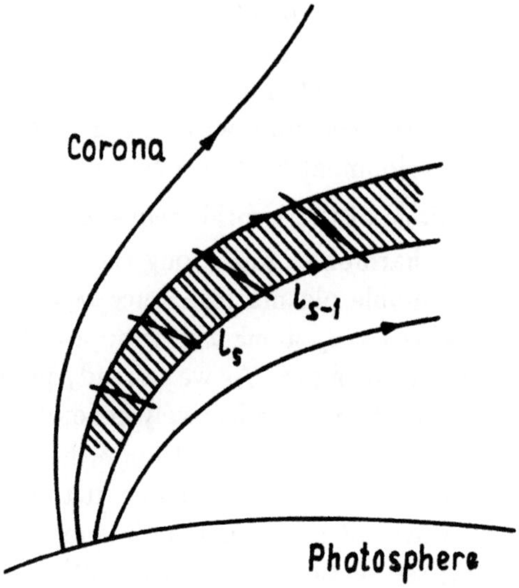

Figure 7.9. A model for an extended source of solar radio emission with "zebra" structure. The hatched region is filled by energetic electrons, l is the coordinate along the magnetic flux tube.

Note also that loss-cone cyclotron instability in solar corona magnetic loops is considered as one of the possible reasons of generation of plasma waves giving rise to some kinds of type IV solar radio emission [344].

7.3.4. HYDRODYNAMIC INSTABILITY AT DOUBLE PLASMA RESONANCE

The above analysis dealt with the kinetic cyclotron instability which is related to the residue in the integral expression (2.97) for $\mathrm{Im}\epsilon_{\|}^{(1)}$. Under the classical treatment of the problem this instability may be found only in the framework of the kinetic equation, which this instability is named after. But if the density of energetic particles is sufficiently high or their velocity spread is small, stronger instability connected with $\mathrm{Re}\,\epsilon_{\|}^{(1)}$ may exist, while the contribution of the energetic component to the increment $\mathrm{Im}\,\epsilon_{\|}^{(1)}$ can be neglected.[16] Similar to the instability in a system comprising plasma and mono-velocity beam which is also connected with $\mathrm{Re}\,\epsilon_{\|}^{(1)}$ and may be described in terms of quasi-hydrodynamic equations (see section 7.2), the above regime is called hydrodynamic.

In case of transverse propagation ($k_{\|}=0$, $k_{\perp}=k$) the resonant condition

[16]The separation of the real and imaginary part of $\epsilon_{\|}^{(1)}$ is done here assuming ω and $\mathbf{k}$ to be real.

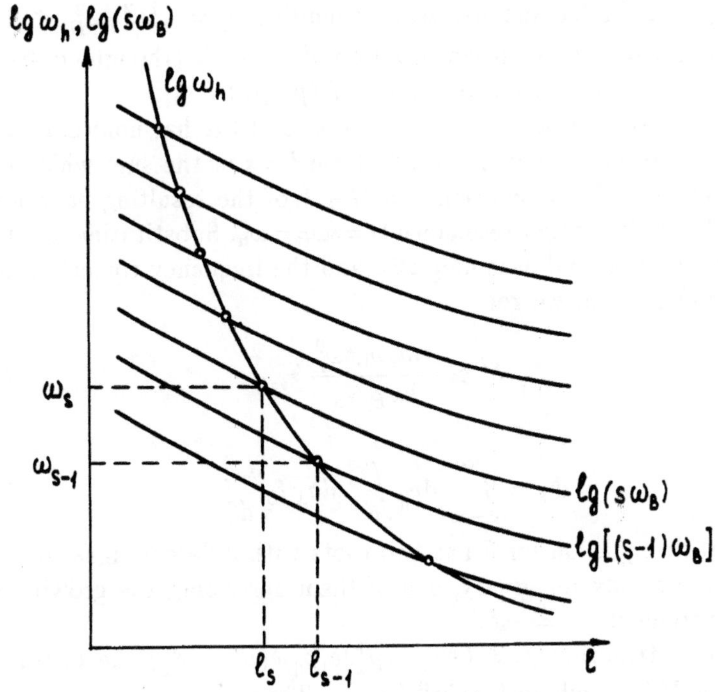

Figure 7.10. Distributions $\omega_{\rm h}(l)$, $\omega_B(l)$ and $s\omega_B(l)$ along the magnetic fluxtube.

(7.112) for weakly relativistic particles takes the form

$$(1 + p^2/2m_{\rm e}^2c^2)\omega - l\omega_B = 0. \tag{7.133}$$

If the spread of particle velocities is small enough, one can neglect the relativistic dependence of electron mass on the velocity in the dispersion equation $\epsilon_{||}(\omega, \boldsymbol{k}) = 0$, as well as the contribution of the residue to the integral (2.97). For distribution (7.110) this may be done for

$$|\omega - l\omega_B| \gg \omega^2 a^2/2m_{\rm e}^2c^2 = \omega v_{\rm e}^2/2c^2. \tag{7.134}$$

The equation $\epsilon_{||}(\omega, \boldsymbol{k}) = 0$ may then be written in the following form:

$$1 - \frac{\omega_{\rm L}^2}{\omega^2 - \omega_B^2} + \frac{8\pi^2 e^2 m_{\rm e}}{k^2} \times \sum_{l=-\infty}^{+\infty} \frac{l\omega_B}{\omega - l\omega_B} \int_{-\infty}^{\infty} dp_{||} \int_0^{\infty} dp_{\perp} J_l^2(\xi) \frac{\partial f_{\rm S}}{\partial p_{\perp}} = 0, \tag{7.135}$$

where $\xi = kp_{\perp}/m_{\rm e}\omega_B$. The first two terms here characterize the longitudinal dielectric permittivity of the background plasma at frequencies near $\omega_{\rm h} =$

$\sqrt{\omega_{\rm L}^2+\omega_B^2}$, i.e., under the necessary condition $\chi = k^2v_\perp^2/\omega_B^2 \ll 1$ (see (2.160)). The last term is the contribution of energetic (though non-relativistic) electrons with distribution $f_{\rm S}(\boldsymbol{p}) = f_{\rm S}(p_\|, p_\perp)$.

If the frequency of a wave is close to one of the harmonics, $\omega \approx s\omega_B$, it is possible to retain only the single term $l = s$ in the sum which enters (7.135). Let us find the solution $\omega = \Omega + \delta$ of the resulting equation for the case of double plasma resonance $\Omega = s\omega_B = \omega_{\rm h}$. Substituting ω into the dispersion equation and assuming $\omega_{\rm LS}$ and the frequency correction factor δ to be small enough, we get

$$\delta^2 \simeq \frac{\pi m_e \omega_{\rm LS}^2 \omega_{\rm L}^2}{k\omega_B N_{\rm S}} \Phi_s, \tag{7.136}$$

where

$$\Phi_s = \int_{-\infty}^{\infty} dp_\| \int_0^\infty dp_\perp f_{\rm S} \frac{dJ_s^2}{d\xi}. \tag{7.137}$$

The quadratic equation for δ has two roots with different signs. If $\Phi_s < 0$, the roots are purely imaginary, one of them describing the growing wave with the increment $-\gamma = -i\delta$.

For the distribution $f_{\rm S} = (N_{\rm S}/2\pi p_\perp^0)\delta(p_\|)\delta(p_\perp - p_\perp^0)$ the factor $\Phi_s = (N_{\rm S}/2\pi p_\perp^0) dJ_s^2/d\xi_0$, where $\xi_0 = kp_\perp^0/m_e\omega_B$. Thus,

$$\delta^2 \simeq \frac{\omega_{\rm LS}^2 \omega_{\rm L}^2}{2\omega_B^2} \frac{1}{\xi_0} \frac{dJ_s^2}{d\xi_0} \tag{7.138}$$

and the hydrodynamic cyclotron instability appears in the system of wave vector bands in which $dJ_s^2/d\xi_0 < 0$.

In a system comprising energetic particles with the distribution (7.110) the hydrodynamic instability may occur only at sufficiently large density $N_{\rm S}$. Indeed, the function Φ_s in this case looks like the following:

$$\Phi_s = \frac{N_{\rm S}}{2\pi m_e v_{\rm e}} \sqrt{\chi_{\rm e}} \varphi_s'(\chi_{\rm e}), \tag{7.139}$$

where $\varphi_s(\chi_{\rm e})$ and $\chi_{\rm e}$ are determined by (7.115). According to (7.136), the correction δ to the zero-order approximation for the frequency $\Omega = s\omega_B = \omega_{\rm h}$ is determined by the equation

$$\delta^2 \simeq \frac{\omega_{\rm LS}^2 \omega_{\rm L}^2}{2\omega_B^2} \varphi_s'(\chi_{\rm e}). \tag{7.140}$$

The function $\varphi_s'(\chi_{\rm e})$ changes the sign at $\chi_{\rm e} = \chi_{\rm e}^{\rm cr} \approx s^2$, so that $\varphi_s' > 0$ for $\chi_{\rm e} < \chi_{\rm e}^{\rm cr}$ and negative otherwise. Thus, the hydrodynamic instability with the increment

$$-\gamma \approx \frac{\omega_{\rm LS}\omega_{\rm L}}{\omega_B} \sqrt{-\frac{1}{2}\varphi_s'(\chi_{\rm e})} \tag{7.141}$$

occurs in this region.[17] The instability considered is possible only under the condition

$$|\gamma| \gg \omega_{\rm L} v_{\rm e}^2/2c^2, \tag{7.142}$$

which can be obtained from (7.134) by putting $l = s$, $\omega = \Omega - i\gamma$, and $\Omega = s\omega_B = \omega_{\rm h}$ there. Taking into account the expression (7.141) for the hydrodynamic increment, we find the following criterion from (7.142):

$$\frac{\omega_{\rm LS}}{\omega_B}\sqrt{-2\varphi_s'(\chi_{\rm e})} \gg \frac{v_{\rm e}^2}{c^2}. \tag{7.143}$$

The obtained criterion of the hydrodynamic regime of cyclotron instability can only be satisfied at sufficiently large $\omega_{\rm LS}$, i.e., densities of energetic electrons $N_{\rm S}$. For example, the maximum of the function $-\varphi_s'(\chi_{\rm e})$ is about $2\cdot 10^{-4}$ for $s = 5$, so that the hydrodynamic instability is realized for $\omega_{\rm LS}/\omega_B \gg 50 v_{\rm e}^2/c^2$. In solar corona conditions this criterion demands rather high densities of non-equilibrium electrons: say, for $v_{\rm e} \sim 10 v_T$ and $B_0 \sim 20$ G ($\omega_B \sim 4{\cdot}10^8\,{\rm s}^{-1}$) the density $N_{\rm S}$ should be in excess of $10^7\,{\rm cm}^{-3}$. The criteria of the kinetic cyclotron instability considered in the previous section are less restrictive.

More details on the hydrodynamic cyclotron instability may be found in [95,348]. Note that, under certain conditions, this instability arises in the absence of the dense equilibrium plasma, i.e., in one-component, non-equilibrium plasmas [347,351]. Kinetic and hydrodynamic instabilities in such systems have been studied in [95,352,353].

In this section we considered only the cyclotron instability of longitudinal waves. Of course, this does not diminish the role of the cyclotron instability of electromagnetic waves. It is quite possible that this instability is realized in solar corona plasma and in Jovian magnetosphere and gives rise to some components in the radio emission of these objects [354].

The possibility of the cyclotron instability of electromagnetic waves at distinct cyclotron harmonics has been pointed out in [310] for the case of a system with a deficit of particles at small transverse momenta. Then it was studied in [84,355–366] (see also [72,73,367] and references therein) for different distributions of energetic particles. Cyclotron instability of electromagnetic disturbances has also been found in a system with temperature anisotropy when the distribution function has the form

$$F(\boldsymbol{v}) \propto \exp(-mv_\perp^2/2\kappa T_\perp - mv_\parallel^2/2\kappa T_\parallel)$$

with $T_\perp \neq T_\parallel$ [358].

[17] Note that the region of the parameter $\chi_{\rm e}$ where the hydrodynamic instability at double plasma resonance may appear is the same as the one for the non-relativistic kinetic instability of the Bernstein modes (see (7.119)).

Cyclotron instability can play a role of the "beam" one in a system "tenuous particle beam – equilibrium plasma" at frequencies in the region of anomalous Doppler effect [79]. This instability can be explained as follows. As was clarified in section 5.1, the emission of a quantum $\hbar\omega$ by a particle with pitch angle α in the region of anomalous Doppler effect (5.36) leads to the increase of $p_\perp^2$. Therefore, if the distribution $f_S(\boldsymbol{p}) = f_S(p_\parallel, p_\perp^2)$ decreases with $p_\perp^2$, the number of electrons in the state before the emission exceeds that in the final state ($N_m/N_n > 1$). Such a population inversion leads to amplification and instability (see section 7.1). On the contrary, this distribution is stable in the region of normal Doppler effect. Here the emission of a photon leads to the decrease of $p_\perp^2$. The latter means no population inversion in the system ($N_m/N_n < 1$). The instability criterion for a particle beam with the distribution (5.124) has the following form [79]:

$$n_j \frac{v_S}{c} \cos\alpha > 1. \tag{7.144}$$

Since both $p_\perp^2$ and $p_\parallel$ vary due to emission, this condition is slightly different from that of the anomalous Doppler effect (5.36). Note that such a variation of the longitudinal component of particle momentum has not been taken into account in the above considerations.

7.4. Synchrotron reabsorption and instability

Contrary to the previous sections dealing mainly with the instability of longitudinal (plasma) waves, we turn here to the instability of plasma with respect to electromagnetic disturbances. We illustrate instabilities of this kind by the example of synchrotron instability. It is closely related to the process of synchrotron emission by relativistic particles in a magnetic field. This instability occurs if the synchrotron reabsorption coefficient μ_j becomes negative. According to the adopted terminology, coherent synchrotron mechanism is realized in this case. But if $\mu_j > 0$, incoherent synchrotron mechanism works in the source. In the latter case the reabsorption can change the form of the synchrotron frequency spectrum.

7.4.1. SYNCHROTRON REABSORPTION

To find the reabsorption coefficient of the system of relativistic electrons in a magnetic field, we apply the method of Einstein coefficients [311]. According to this method, μ_j is determined by the general formula (7.17). To obtain the coefficient of synchrotron reabsorption with this formula, one should take into account the essential feature of synchrotron radiation: it is emitted into the small solid angle along the electron velocity $\boldsymbol{v}$, while the frequency spectrum of this radiation is practically continuous (see section 5.1). For the sake of simplicity, we also assume that relativistic electrons have isotropic distribution: $f_S(\boldsymbol{p}) = f_S(p)$.

Since a synchrotron quantum $\hbar\omega$ is emitted along the instant velocity of an electron, the momentum conservation leads to the following relation: $p^{(m)} = p^{(n)} + \hbar k$, where $p^{(m)}$ and $p^{(n)}$ are electron momenta before and after the emission, respectively. The corresponding populations N_m, N_n of the states (m) and (n) are $f_S(p^{(m)})$ and $f_S(p^{(n)})$. If the electron distribution is changed weakly at an interval $\Delta p \sim \hbar k$, the latter may be written as

$$f_S(p^{(n)}) = f_S(p^{(m)} - \hbar k) \approx f_S(p^{(m)}) - \frac{df_S}{dp}\hbar k. \tag{7.145}$$

According to (7.145), the population difference entering the formula (7.17) for μ_j takes the following form:

$$N_n/N_m - 1 \approx -\frac{\hbar k}{f_S(p^{(m)})}\frac{df_S}{dp}.$$

Hence, the synchrotron reabsorption coefficient is given by

$$\mu_j = -\frac{8\pi^3 c^2}{n_j^2\omega^2}|\cos\vartheta_j| \int A_m^n\, \hbar k\, \frac{\partial f_S}{\partial p}\, d^3p. \tag{7.146}$$

Here the integral over the momentum space is used instead of the sum over the states $(m) \leftrightarrow (n)$. Taking into account that $k = (\omega/c)n_j$, $d^3p = p^2 dp d\Omega$ and using the relation (7.8) between the probability of spontaneous emission A_m^n and spectral power of radiation $P_{\omega\Omega}$, we find

$$\mu_j = -\frac{8\pi^3 c}{n_j\omega^2}|\cos\vartheta_j| \int P_{\omega\Omega}\, \frac{df_S}{dp}\, p^2 dp d\Omega, \tag{7.147}$$

where

$$P_{\omega\Omega} = P_{\omega j}\delta(\boldsymbol{\Omega}_{\mathbf{k}} - \boldsymbol{\Omega}), \tag{7.148}$$

$P_{\omega j}$ is the spectral power of synchrotron radiation in unit frequency range into j-th normal wave, $\delta(\boldsymbol{\Omega}_{\mathbf{k}} - \boldsymbol{\Omega})$ is the delta-function, $\boldsymbol{\Omega}_{\mathbf{k}}$ and $\boldsymbol{\Omega}$ are unit vectors along $\boldsymbol{k}$ and $\boldsymbol{p}$, respectively. Here we have taken into account again that the synchrotron radiation is emitted into a small solid angle along the electron velocity (momentum).

In what follows it is convenient to use the particle energy spectrum $N(\mathcal{E})$ instead of the momentum distribution $f_S(p)$. Using the formulas (7.148) and (5.177), we obtain from (7.147) that [30, 247]

$$\mu_j = -\frac{8\pi^3 c}{n_j\omega^2}|\cos\vartheta_j| \int_0^\infty P_{\omega j}(\omega, \mathcal{E})\frac{d}{d\mathcal{E}}\left[\frac{N(\mathcal{E})}{\mathcal{E}^2}\right]\mathcal{E}^2 d\mathcal{E}. \tag{7.149}$$

In a rarefied plasma ($\omega_{\rm L}^2/\omega^2 \ll 1$) with weak magnetic field ($\omega_B/\omega \ll 1$) the refractive index is given by $n_j = 1 - \omega_{\rm L}^2/\omega^2$; it is close to unity, as well as $\cos\vartheta_j$. The above formula is then reduced to the following one [247,311]:

$$\mu_j \approx -\frac{8\pi^3 c^2}{\omega^2}\int_0^\infty P_{\omega j}(\omega,\mathcal{E})\frac{d}{d\mathcal{E}}\left[\frac{N(\mathcal{E})}{\mathcal{E}^2}\right]\mathcal{E}^2 d\mathcal{E}. \tag{7.150}$$

The value $P_{\omega j}$ in (7.150) characterizes the spectral power of synchrotron radiation by an electron with $\boldsymbol{p} \parallel \boldsymbol{k}$. The pitch angle ϑ of such an electron is, obviously, equal to the angle α between the wave vector $\boldsymbol{k}$ and the magnetic field $\boldsymbol{B}_0$.

For $\omega_{\rm L}^2/\omega^2 \ll 1$ and $\omega_B/\omega \ll 1$ the quasi-longitudinal propagation is realized in the wide angular interval $\omega_B^2 \sin^4\alpha/4\omega^2\cos^2\alpha \ll 1$ (see (2.126)). Ordinary and extraordinary modes are circularly polarized in this case, and the spectral power of synchrotron radiation per mode is [18]

$$P_{\omega j}(\omega,\mathcal{E}) = \frac{1}{2}P_\omega(\omega,\mathcal{E}), \tag{7.151}$$

where $P_\omega(\omega,\mathcal{E})$ is the total power of synchrotron radiation (5.94) into both modes. (Note that the gyrofrequency ω_B should be replaced by $\omega_B\sin\vartheta$ in this formula for $\vartheta \neq \pi/2$.) Thus, the synchrotron reabsorption coefficient in a rarefied plasma in the case of quasi-longitudinal propagation is described by the formula (7.150) in which

$$P_{\omega j}(\omega,\mathcal{E}) = \frac{1}{2\pi\sqrt{3}}\frac{e^2\omega}{c}\left[\left(\frac{m_e c^2}{\mathcal{E}}\right)^2 + \frac{\omega_{\rm L}^2}{\omega^2}\right]\int_{\omega/\omega_c'}^\infty K_{5/3}(x)\,dx. \tag{7.152}$$

Here

$$\omega_c' = \frac{3}{2}\omega_B\sin\alpha\left(\frac{\mathcal{E}}{m_e c^2}\right)^2\left[1 + \frac{\omega_{\rm L}^2}{\omega^2}\left(\frac{\mathcal{E}}{m_e c^2}\right)^2\right]^{-3/2}. \tag{7.153}$$

At arbitrary α the expression for $P_{\omega j}(\omega,\mathcal{E})$ has the following form [368]:

$$P_{\omega j}(\omega,\mathcal{E}) = \frac{1}{2\pi\sqrt{3}}\frac{e^2\omega}{c}\left[\left(\frac{m_e c^2}{\mathcal{E}}\right)^2 + \frac{\omega_{\rm L}^2}{\omega^2}\right] \times\left[\int_{\omega/\omega_c'}^\infty {\rm K}_{5/3}(x)\,dx + \frac{1-K_j^2}{1+K_j^2}{\rm K}_{2/3}(\omega/\omega_c')\right], \tag{7.154}$$

where ${\rm K}_{5/3}$ and ${\rm K}_{2/3}$ are McDonald functions, K_j is the polarization coefficient for j-th normal wave (see (2.136)). For quasi-longitudinal propagation $K_j = \pm 1$, so (7.154) is reduced to (7.152).

[18] The angle-averaged polarization of synchrotron radiation is linear with accuracy up to the terms of about $m_e c^2/\mathcal{E}$ and $\sqrt{1-n^2}$. That is why the total power of radiation P_ω can be written as the sum of two equal, circularly polarized components (7.151).

TABLE 7.1.

γ	1	2	3	4	5
$g(\gamma)$	6.03	4.39	4.08	4.33	5.21
$\overline{g}(\gamma)$	4.33	2.95	2.51	2.76	2.89

7.4.2. REABSORPTION COEFFICIENT FOR POWER-LAW PARTICLE ENERGY SPECTRUM

In the previous parts of the book (see, for example, section 5.2) we have already shown that synchrotron emission by electrons with a power-law energy spectrum $N(\mathcal{E}) = \mathcal{K}\mathcal{E}^{-\gamma}$ plays an important role in astrophysical conditions. It is clear from (7.152) that the influence of the media on $P_{\omega j}(\omega, \mathcal{E})$ can be neglected for frequencies and energies satisfying the condition (5.95). If one, however, takes into account that radiation at frequency ω is emitted and absorbed by particles with energy (5.82) $\mathcal{E} \sim \mathcal{E}_{\max}$, the criterion allowing one to neglect the influence of the refraction index of a medium on the synchrotron reabsorption coefficient is reduced to (5.102). At these frequencies the value μ_j determined by (7.150), (7.151) can be presented, for the power-law particle spectrum $N(\mathcal{E}) \propto \mathcal{E}^{-\gamma}$, in the following form [247]:

$$\mu_j = g(\gamma)\frac{e^3}{m_e}\left(\frac{3e}{m_e^3c^5}\right)^{\gamma/2}\mathcal{K}(B_0 \sin\alpha)^{\frac{\gamma+2}{2}}\omega^{-\frac{\gamma+4}{2}}, \tag{7.155}$$

where

$$g(\gamma) = \frac{\pi\sqrt{3}}{2}\Gamma\left(\frac{3\gamma+2}{12}\right)\Gamma\left(\frac{3\gamma+22}{12}\right). \tag{7.156}$$

The values of the factor $g(\gamma)$ for several γ are given in Table 7.4.2. Substituting numerical values of the constants into (7.155) we get [247]

$$\mu_j = 3\cdot 10^{-3} g(\gamma)(3.5\cdot 10^9)^\gamma \mathcal{K}(B_0 \sin\alpha)^{\frac{\gamma+2}{2}} f^{-\frac{\gamma+4}{2}}\ \text{cm}^{-1}. \tag{7.157}$$

The coefficients (7.155), (7.157) describe the absorption in a homogeneous magnetic field. They depend both on the field strength B_0 and on the field orientation α. Synchrotron absorption in a chaotic field can be described by the effective absorption coefficient $\overline{\mu}$ which is just the ray average of μ_j. Assuming for simplicity that the magnetic field changes orientation chaotically retaining the magnitude ($B_0 =$ const), we get [66]

$$\overline{\mu} = \int_0^{\pi/2}\mu_j(\alpha)\sin\alpha\, d\alpha = \overline{g}(\gamma)\frac{e^3}{m_e}\left(\frac{3e}{m_e^3c^5}\right)^{\gamma/2}\mathcal{K}\cdot B_0^{\frac{\gamma+2}{2}}\omega^{-\frac{\gamma+4}{2}}, \tag{7.158}$$

where

$$\overline{g}(\gamma) = \frac{\pi\sqrt{3\pi}}{4}\Gamma\left(\frac{\gamma+6}{4}\right)\Gamma\left(\frac{3\gamma+2}{12}\right)\Gamma\left(\frac{3\gamma+22}{12}\right)\Gamma^{-1}\left(\frac{\gamma+8}{4}\right). \quad (7.159)$$

The values of $\overline{g}(\gamma)$ for several γ are also given in Table 7.4.2. After substitution of numerical values of the constants into (7.158) this formula takes the form

$$\overline{\mu} = 3 \cdot 10^{-3}\,\overline{g}(\gamma)(3.5 \cdot 10^{9})^{\gamma} \mathcal{K}\, B_0^{\frac{\gamma+2}{2}} f^{-\frac{\gamma+4}{2}}\ \mathrm{cm}^{-1}. \quad (7.160)$$

If $\overline{\mu} l_0 \ll 1$, where l_0 is the characteristic scale of the magnetic field, the solution of the radiation transfer equation with averaged reabsorption coefficient (7.158) and emissivity $\overline{a}_\omega$, (5.189), approximates very well the true value of the radiation intensity I_j at any point of the ray. The above criterion demands that the region of the size l_0 with quasi-homogeneous magnetic field is optically thin. Since the intensity I_j is formed on the ray segment with the length $l \sim \mu^{-1} \gg l_0$, it can be described in terms of the ray-averaged transfer coefficients $\overline{a}_\omega$ and $\overline{\mu}$. Then the influence of a cloud of the size l_0 on the intensity at a given point of a ray is quite weak.

The effect of synchrotron reabsorption on the radiation intensity can be completely neglected if the optical depth of the emitting region $\tau_j = \mu_j L \ll 1$ (L is the linear size of the source). The outgoing intensity in this case is $I_j = a_\omega L$. For the power-law spectrum of relativistic electrons $I_j \propto \omega^{-\frac{\gamma-1}{2}}$ (see section 5.2 and formulas (5.185), (5.186) in particular). But the intensity escaping the optically thick source $\tau_j \gg 1$ is equal to $I_j = a_\omega/\mu_j$. Since the transfer coefficients for synchrotron radiation behave as $a_\omega \propto \omega^{-\frac{\gamma-1}{2}}$, $\mu_j \propto \omega^{-\frac{\gamma+4}{2}}$, an optically thick synchrotron source $\tau_j \gg 1$ emits radiation with the standard frequency spectrum

$$I_j \propto \omega^{5/2} \quad (7.161)$$

with the index $\alpha = -5/2$. Note that this index does not depend on the spectral index of particle distribution γ.

Owing to the growth of the synchrotron reabsorption coefficient μ_j (and also $\overline{\mu}$) with decreasing ω the emitting region becomes optically thick at sufficiently low frequencies. The above considerations mean that the intensity of synchrotron radiation emitted by relativistic electrons with the power-law spectrum depends on frequency as follows (see Fig. 7.11): I_j increases with decreasing ω at higher frequencies where $\tau_j \ll 1$ and decreases with decreasing ω at lower ones where the opposite inequality, $\tau_j \gg 1$, holds. The maximum of the radiation spectrum appears at $\omega \sim \omega_{\mathrm{cr}}$, where ω_{cr} is determined by the condition $\tau(\omega_{\mathrm{cr}}) = 1$. Since the optical depth equals $\mu_j L$ or $\overline{\mu} L$ with μ_j and $\overline{\mu}$ given by (7.155) and (7.158), respectively,

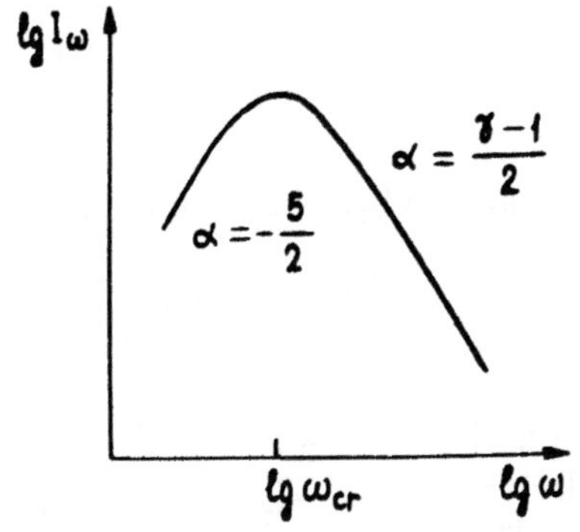

Figure 7.11. Frequency spectrum of synchrotron radiation by relativistic electrons with the power-law energy spectrum in vacuum. The effect of reabsorption is taken into account.

the last condition relates L, $\mathcal{K}$ and $B_0 \sin\alpha$ (or simply B_0) in a discrete source. If, in addition, the size of the source L is known, the position of the maximum in the synchrotron frequency spectrum, ω_{cr}, yields another missing relation, the one between $\mathcal{K}$ and B_0. It is thus possible to determine these values for a synchrotron source (see in this respect the notes at the end of section 5.2).

The effect of synchrotron reabsorption allows one to explain the observed falldown of the radiation spectrum of the Crab pulsar PSR 0532 in the IR range compared to the optical one and to estimate the source parameters. This becomes possible under the assumption that the entire spectrum of this pulsar from IR to X-rays is formed by a non-coherent radiation mechanism in the neutron star magnetosphere (for more details see [68, 369, 370]). At the same time, synchrotron reabsorption is not important for diffuse radio emission of the Galaxy, because here $f_{cr} = \omega_{cr}/2\pi \sim 0.1$ MHz (this estimate was obtained from (7.160) for $L \sim 10$ kpc $\approx 3 \cdot 10^{22}$ cm, $B_0 \sim 3 \cdot 10^{-6}$ G, $\mathcal{K} \sim 10 \cdot^{-14}$ erg cm^{-3}, and $\gamma = 2$).

7.4.3. SYNCHROTRON INSTABILITY

Let us now find the condition under which the synchrotron reabsorption becomes negative and the radiation coming into the system is amplified rather than absorbed in it. Such a system of relativistic particles amplifies its proper radiation as well, so that the intensity of the radiation escaping the source exceeds that of spontaneous emission by all the particles in it. In other words, the coherent synchrotron mechanism is realized in this case. Negative reabsorption ($\mu_j < 0$) corresponds to the positive increment $-\gamma_j$. For the initial value problem this means the appearance of synchrotron instability.

If rarefied plasma with a weak magnetic field ($\omega_B/\omega \ll 1$, $1 - n^2 = \omega_{\rm L}^2/\omega^2 \ll 1$) comprises a system of relativistic electrons with isotropic distribution, the synchrotron reabsorption coefficient is given by (7.150). Integrating by parts, one can modify this formula as follows:

$$\mu_j = \frac{8\pi^3 c^2}{\omega^2} \int_0^\infty \frac{N(\mathcal{E})}{\mathcal{E}^2} \frac{d}{d\mathcal{E}} [\mathcal{E}^2 P_{\omega j}(\omega, \mathcal{E})] d\mathcal{E}. \tag{7.162}$$

The sign of μ_j depends on the behavior of $\mathcal{E}^2 P_{\omega j}$. If $\mathcal{E}^2 P_{\omega j} = \text{const}$, then $\mu_j = 0$. The positive derivative $d(\mathcal{E}^2 P_{\omega j})/d\mathcal{E} > 0$ provides the positive synchrotron reabsorption. Finally, if $d(\mathcal{E}^2 P_{\omega j})/d\mathcal{E} < 0$ over some interval of energies $\mathcal{E}$, the synchrotron reabsorption coefficient μ_j can be made negative by the proper choice of the electron distribution $N(\mathcal{E})$.

Assume at first that the case of the quasi-longitudinal propagation is realized, and the normal waves are circularly polarized. Then the spectral power of synchrotron radiation $P_{\omega j}$ into mode j is determined by (7.152). Using this expression, one may obtain that the influence of the medium on $P_{\omega j}$ can be neglected in the energy interval

$$\frac{\omega_{\rm L}^2}{\omega^2} \left(\frac{\mathcal{E}}{m_{\rm e} c^2} \right)^2 \ll 1 \tag{7.163}$$

which we call region I. On the contrary, this influence is dominant in the region II where

$$\frac{\omega_{\rm L}^2}{\omega^2} \left(\frac{\mathcal{E}}{m_{\rm e} c^2} \right)^2 \gg 1. \tag{7.164}$$

Determine first the contribution to the reabsorption coefficient of the electrons from the region I. According to (7.152), in this region

$$P_{\omega j} = \frac{1}{2\pi\sqrt{3}} \frac{e^2 \omega}{c} \left(\frac{m_{\rm e} c^2}{\mathcal{E}} \right)^2 \int_z^\infty K_{5/3}(x)\, dx, \tag{7.165}$$

where

$$z \equiv \frac{\omega}{\omega_{\rm c}'} \approx \frac{2}{3} \frac{\omega}{\omega_B \sin\alpha} \left(\frac{m_{\rm e} c^2}{\mathcal{E}} \right)^2. \tag{7.166}$$

These formulas yield $d(\mathcal{E}^2 P_{\omega j})/d\mathcal{E} > 0$, i.e., the region I provides positive contribution to the reabsorption coefficient

$$\mu^{\rm I} = \frac{16\pi^2}{3\sqrt{3}} \frac{e^2 c}{\omega_B \sin\alpha} (m_{\rm e} c^2)^4 \int_0^{\mathcal{E} \ll \mathcal{E}_*} \frac{N(\mathcal{E})}{\mathcal{E}^5} K_{5/3}(z)\, d\mathcal{E}. \tag{7.167}$$

Here $\mathcal{E}_* = m_{\rm e} c^2 \omega/\omega_{\rm L}$, so that the upper limit of the integral is set to meet the criterion (7.163) of the weak influence of the medium.

Note that in vacuum ($\omega_L \to 0$) the inequality (7.163) is satisfied for arbitrary particle energy and, therefore, the upper limit in the integral entering (7.167) may be replaced by infinity. The resulting formula describes the total synchrotron reabsorption coefficient for circularly polarized waves in vacuum. It is positive for isotropic particle distribution with arbitrary energy spectrum $N(\mathcal{E})$ [30,371], so the synchrotron instability in this case is absent.

Consider now the region II, (7.164), where the role of the medium is quite significant. Here (7.152) is transformed in the following way:

$$P_{\omega j} = \frac{1}{2\pi\sqrt{3}} \frac{e^2}{c} \frac{\omega_L^2}{\omega} \int_z^\infty K_{5/3}(x)\,dx, \tag{7.168}$$

where now

$$z \equiv \frac{\omega}{\omega_c'} \approx \frac{2}{3} \frac{\omega_L^3}{\omega^2 \omega_B \sin\alpha} \frac{\mathcal{E}}{m_e c^2}. \tag{7.169}$$

Then the derivative

$$\frac{d}{d\mathcal{E}}(\mathcal{E}^2 P_{\omega j}) = \frac{\sqrt{3}}{4\pi} \frac{e^2}{c} \frac{\omega\omega_B \sin\alpha}{\omega_L} m_e c^2 \Phi(z), \tag{7.170}$$

where

$$\Phi(z) = 2z \int_z^\infty K_{5/3}(x)\,dx - z^2 K_{5/3}(z). \tag{7.171}$$

It is easy to see that $\int_0^\infty \Phi(z) dz = 0$, so $\Phi(z)$ is the alternating function. It is also seen on the plot in Fig. 7.12 that $\Phi(z) > 0$ if $0 < z < \tilde{z} \approx 1.35$ and $\Phi(z) < 0$ otherwise.

Taking into account (7.168) and (7.170), we obtain from (7.162) that the electrons from the region II (7.164) provide the following contribution to the reabsorption coefficient μ_j:

$$\mu^{\mathrm{II}} = 2\pi^2\sqrt{3}\, e^2 c \frac{\omega_B \sin\alpha}{\omega_L \omega} m_e c^2 \int_{\mathcal{E} \gg \mathcal{E}_*}^\infty \frac{N(\mathcal{E})}{\mathcal{E}^2} \Phi(z)\, d\mathcal{E}, \tag{7.172}$$

where $\mathcal{E}_* = m_e c^2 \omega / \omega_L$. The argument z is related to $\mathcal{E}$ via (7.169). The lower limit in the integral is chosen to satisfy the criterion (7.164) of significant influence of the medium.

The total coefficient of synchrotron reabsorption in a "cold" plasma comprises, evidently, the sum of (7.167), (7.172) and the integral over the intermediate region between I and II. Unfortunately, the general expression for $P_{\omega j}$ cannot be simplified within this latter region. Thus, to get μ_j in the general case for the wide energy spectrum of relativistic electrons one should use the exact formula (7.152) rather than (7.165) or (7.168). However, it is

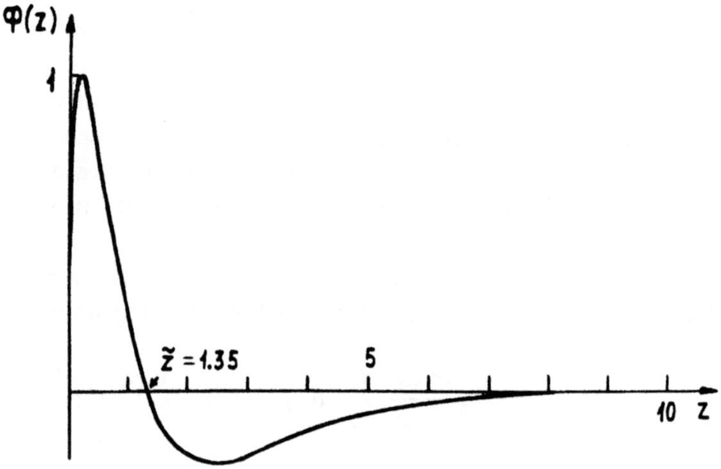

Figure 7.12. The plot of the function $\Phi(z)$.

clear that reabsorption may become negative for the special form of $N(\mathcal{E})$, namely, when the main contribution to the reabsorption coefficient comes from the energy region

$$\mathcal{E} > \tilde{\mathcal{E}} \approx 2m_e c^2 \omega^2 \omega_B \omega_L^{-3} \sin\alpha \gg \mathcal{E}_*, \qquad (7.173)$$

where $z > \tilde{z} \approx 1.35$ and $\Phi(z) < 0$. The synchrotron radiation will be amplified in such a system of emitting relativistic electrons [311].

As follows from the above, the effect of negative synchrotron reabsorption and the synchrotron instability occur due to the strong dependence of spectral power $P_{\omega j}(\omega, \mathcal{E})$ on particle energy $\mathcal{E}$ in plasma, as compared to the case of vacuum. Indeed, in the latter case the quantity $\mathcal{E}^2 P_{\omega j}$ increases monotonically with $\mathcal{E}$, whereas in the former one this function increases only in the region $\mathcal{E} < \tilde{\mathcal{E}}$ and decreases at $\mathcal{E} > \tilde{\mathcal{E}}$. Thus, $\mathcal{E}^2 P_{\omega j}$ has the maximum at $\mathcal{E} = \tilde{\mathcal{E}}$ Such a behavior of the quantity $\mathcal{E}^2 P_{\omega j}$ in the region of essential influence of the medium, $\mathcal{E} > \mathcal{E}_*$ corresponds to the condition $\tilde{\mathcal{E}} \gg \mathcal{E}_*$. If, on the contrary, $\tilde{\mathcal{E}} \lesssim \mathcal{E}_*$, the function $\mathcal{E}^2 P_{\omega j}$ decreases in the entire region $\mathcal{E} \gg \mathcal{E}_*$.

7.4.4. NEGATIVE REABSORPTION IN A SYSTEM OF RELATIVISTIC ELECTRONS WITH A "MONOENERGETIC" OR POWER-LAW ENERGY SPECTRUM

In this section we demonstrate the effect of negative synchrotron reabsorption with two specific examples of electron energy spectrum $N(\mathcal{E})$ [311].

Consider, first of all, the system of relativistic electrons with a narrow ("monoenergetic") spectrum which has the maximum at $\mathcal{E}_0$. The width of particle spectrum $\Delta\mathcal{E}$ is assumed to be small enough to take out of the integrals in (7.167), (7.172) the functions $\mathcal{E}^{-5}K_{5/3}(z)$ and $\mathcal{E}^{-2}\Phi(z)$ at $\mathcal{E}=\mathcal{E}_0$.[19] This procedure yields

$$\mu_j = \mu^{\rm I} = \frac{4\pi}{3\sqrt{3}}\frac{e^2c}{\omega_B \sin\alpha}\frac{(m_ec^2)^4}{\mathcal{E}_0^5}N_{\rm S}K_{5/3}(z_0),\qquad z_0 = \frac{2}{3}\frac{\omega}{\omega_B\sin\alpha}\left(\frac{m_ec^2}{\mathcal{E}_0}\right)^2 \tag{7.174}$$

for $\mathcal{E}_0 \ll \mathcal{E}_*$ and

$$\mu_j = \mu^{\rm II} = \frac{\pi\sqrt{3}}{2}e^2c\frac{\omega_B\sin\alpha}{\omega_{\rm L}\omega}\frac{m_ec^2}{\mathcal{E}_0^2}N_{\rm S}\Phi(z_0),\qquad z_0 = \frac{2}{3}\frac{\omega_{\rm L}^3}{\omega^2\omega_B\sin\alpha}\frac{\mathcal{E}_0}{m_ec^2} \tag{7.175}$$

in the region $\mathcal{E}_0 \gg \mathcal{E}_*$. We recall that $\mathcal{E}_* = m_ec^2\omega/\omega_{\rm L}$ and that $N_{\rm S} = 4\pi\int N(\mathcal{E})d\mathcal{E}$ denotes the density of relativistic electrons.

The reabsorption coefficient varies with frequency in the following way. At higher frequencies, in the region I, the function $\mu_j(\omega)$ is given by (7.174). The reabsorption is positive in this case. The reabsorption coefficient is exponentially small at high ω and rises quickly as frequency decreases. Below the intermediate region $\omega \sim \omega_{\rm L}\mathcal{E}_0/m_ec^2$ where $\mu_j(\omega)$ can not be described by either (7.174) or (7.175) we find the region II at which the expression (7.175) is valid. One can judge the behavior of $\mu_j(\omega)$ using Fig. 7.13 which shows the plot of $\sqrt{z_0}\Phi(z_0)\propto\mu_j$ versus $1/\sqrt{z_0}\propto\omega$.

Depending on the system parameters, two cases are possible in the region II. If the frequency

$$\omega \approx \left(0.5\frac{\omega_{\rm L}^3}{\omega_B\sin\alpha}\frac{\mathcal{E}_0}{m_ec^2}\right)^{1/2} \tag{7.176}$$

which corresponds to $z_0 = \tilde{z} \approx 1.35$ is much less than the characteristic frequency $\omega = \omega_{\rm L}\mathcal{E}_*/m_ec^2$ which can be found from the definition $\mathcal{E}_* = m_ec^2\omega/\omega_{\rm L}$, or, equivalently, if

$$\frac{\omega_{\rm L}}{2\omega_B\sin\alpha} \ll \frac{\mathcal{E}_0}{m_ec^2}, \tag{7.177}$$

[19] The value $\Delta\mathcal{E}$ should nevertheless be much bigger than the interval between the neighboring energies of particles which radiate at frequency ω. According to (5.83), this interval is equal to $m_ec^2\omega_B/\omega$, and, therefore, the "monoenergetic" spectrum considered should satisfy the condition $\Delta\mathcal{E} \gg m_ec^2\omega_B/\omega$. Only in this case one can use the "smoothed out" synchrotron spectrum (7.152), (7.154) instead of the exact discrete one and also use the formulas for reabsorption coefficient derived in this section [372, 373].

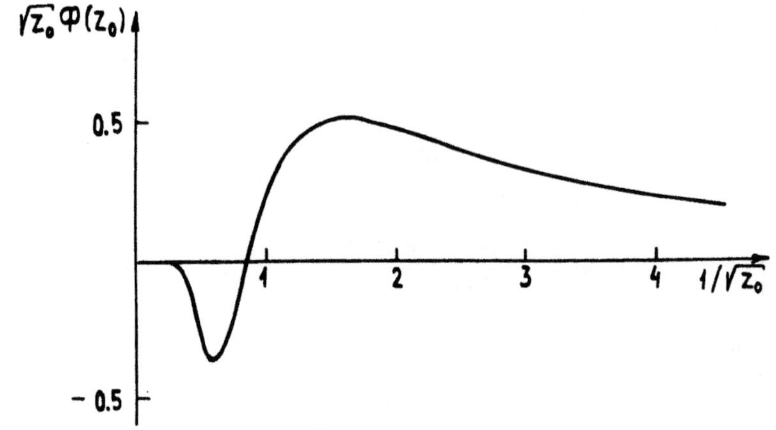

Figure 7.13. The plot of $\sqrt{z_0}\Phi(z_0)$ versus $1/\sqrt{z_0}$.

the function $\Phi(z_0)$ changes sign in the region II. The reabsorption coefficient is then positive in the high-frequency part of this region and may become negative only below the frequency (7.176). The absolute value of reabsorption coefficient in this region, $-\mu_j$, reaches maximum at the frequency

$$\omega_{\max} \approx \left(0.24\frac{\omega_L^3}{\omega_B \sin\alpha}\frac{\mathcal{E}_0}{m_e c^2}\right)^{1/2} \tag{7.178}$$

corresponding to $z_{\max} \approx 2.8$ and then falls off quickly as $\mu_j \propto \omega^{-4}e^{-a/\omega^2}$. It is seen from Fig. 7.13 that the frequency range corresponding to the synchrotron instability and amplification is quite narrow: $|\omega - \omega_{\max}| \lesssim 0.3\omega_{\max}$. In the other case when

$$\frac{\omega_L}{2\omega_B \sin\alpha} \gtrsim \frac{\mathcal{E}_0}{m_e c^2}, \tag{7.179}$$

the coefficient μ_j is negative in the entire range $\omega^2 \ll \omega_L^2(\mathcal{E}_0/m_e c^2)^2$ where the effect of the medium is quite strong. However, under the strong inequality $\omega_L/2\omega_B \sin\alpha \gg \mathcal{E}_0/m_e c^2$ the value $z_0 \gg 1$ over the entire region II, so the absolute value of the reabsorption coefficient becomes exponentially small. This occurs due to the rapid drop of the spectral power of the synchrotron radiation of an electron in this case (see (5.96)).

In the second example, we consider the system of relativistic electrons with power-law energy spectrum of the form

$$N(\mathcal{E}) = \begin{cases} \mathcal{K}\mathcal{E}^{-\gamma} & \text{if } \mathcal{E}_1 < \mathcal{E} < \mathcal{E}_2, \\ 0 & \text{if } \mathcal{E} < \mathcal{E}_1 \text{ or } \mathcal{E} > \mathcal{E}_2. \end{cases} \tag{7.180}$$

Let, for the sake of simplicity, $\mathcal{E}_1$ satisfy the condition (7.164). Then the reabsorption coefficient is determined only by the region of the strong effect of the medium:

$$\mu_j = \mu^{\rm II} = 2\pi^2\sqrt{3}\, e^2 c\, \frac{\omega_B \sin\alpha}{\omega_{\rm L}\omega}\, m_e c^2\, \mathcal{K} \int_{\mathcal{E}_1}^{\mathcal{E}_2} \mathcal{E}^{-\gamma-2}\Phi(z)d\mathcal{E} \tag{7.181}$$

The ansatz (7.169) transforms this formula into the following one:

$$\mu_j = 2\pi^2\sqrt{3}\, e^2 c\, \frac{\omega_B \sin\alpha}{\omega_{\rm L}\omega} \times \left(\frac{3\omega^2\omega_B \sin\alpha}{2\omega_{\rm L}^3}\right)^{-\gamma-1} (m_e c^2)^{-\gamma}\mathcal{K} \int_{z_1}^{z_2} z^{-\gamma-2}\Phi(z)\, dz. \tag{7.182}$$

The limits z_1 and z_2 are set by the values $\mathcal{E}_1$ and $\mathcal{E}_2$. The expression for μ_j is simplified if $z_1 \ll 1$ and $z_2 \gg 1$, i.e., in the frequency range

$$\frac{2}{3}\,\frac{\omega_{\rm L}^3}{\omega_B \sin\alpha}\,\frac{\mathcal{E}_1}{m_e c^2} \ll \omega^2 \ll \frac{2}{3}\,\frac{\omega_{\rm L}^3}{\omega_B \sin\alpha}\,\frac{\mathcal{E}_2}{m_e c^2}. \tag{7.183}$$

Since the function $\Phi(z) \propto z^{3/2}e^{-z}$ at $z \gg 1$, the integrand in (7.182) decreases rapidly at large z. Thus, the inequality $z_2 \gg 1$ allows the use of infinity as the upper limit in the above integral. At the same time, the inequality $z_1 \ll 1$ allows substitution of the lower limit in this integral by zero, provided the integral of $z^{-\gamma-2}\Phi(z)$ is converging. Since $z^{-\gamma-2}\Phi(z) \propto z^{-\gamma-5/3}$, the integral mentioned exists if $\gamma < -2/3$. Assuming this condition to be valid, we obtain the following increment at frequencies (7.183):

$$\mu_j \approx 2\pi^2\sqrt{3}\, e^2 c\, \frac{\omega_B \sin\alpha}{\omega_{\rm L}\omega}\left(\frac{3\omega^2\omega_B \sin\alpha}{2\omega_{\rm L}^3}\right)^{-\gamma-1} \times (m_e c^2)^{-\gamma}\mathcal{K} \int_0^\infty z^{-\gamma-2}\Phi(z)\, dz.$$

Substituting the expression (7.171) for $\Phi(z)$ and taking the first integral by parts, we transfrom the above expression to the following:

$$\mu_j \approx -2\pi^2\sqrt{3}\, e^2 c\, \frac{\omega_B \sin\alpha}{\omega_{\rm L}\omega}\left(\frac{3\omega^2\omega_B \sin\alpha}{2\omega_{\rm L}^3}\right)^{-\gamma-1} \times (m_e c^2)^{-\gamma}\mathcal{K}\left(\frac{2}{\gamma}+1\right) \int_0^\infty z^{-\gamma} K_{5/3}(z)\, dz. \tag{7.184}$$

The integral in the above expression is always positive. Thus, the reabsorption coefficient is negative for

$$\gamma < -2. \tag{7.185}$$

We have noticed earlier that the momentum distribution $f_S(p) \propto \mathcal{E}^{-2}N(\mathcal{E})$, i.e., the value $\mathcal{E}^{-\gamma-2}$ for the power-law spectrum. The criterion of the synchrotron instability (7.185) thus means that the distribution function $f_S(p)$ should increase with p.

7.4.5. CONCLUDING REMARKS

The possibility of the synchrotron instability in an isotropic system of relativistic electrons has been pointed out in the analysis of frequency dependence of synchrotron reabsorption with the effect of the medium taken into account [311] (see also [316]). This investigation was done by the method of Einstein coefficients in a way similar to that of the above considerations. These earlier results have then been confirmed by the kinetic theory of synchrotron instability developed in [195,196,368,373,374]. The kinetic approach has presented an opportunity to study the modification of the instability with the increasing width of the spectrum of relativistic electrons: from the hydrodynamic instability of the cyclotron type for a distribution $f_S(p) \propto \delta(p - p_0)$ to the kinetic instability at a single harmonic and, further, at sufficiently wide distribution function $f_S(p)$, synchrotron instability. The latter occurs if the emission and absorption are effectively contributed to by particles from a wide energy range (5.83) corresponding to different harmonic numbers s (see the footnote at the beginning of the previous section).

It should be emphasized that the validity of the above synchrotron reabsorption coefficients does not necessary demand the pure isotropy of particle momentum distribution. It is quite sufficient that the function $f_S(p)$ slowly varies in the pitch-angle range $\Delta\vartheta \sim [(m_e c^2/\mathcal{E})^2 + 1 - n^2]^{1/2}$ near the direction of wave propagation, i.e., within the characteristic width of the angular pattern of synchrotron radiation in a medium (see section 5.2). Moreover, if particle distribution in the momentum space is strongly anisotropic and $f_S(p)$ varies significantly over the above interval $\Delta\vartheta$, the synchrotron instability occurs even under the negligible effect of the medium. This circumstance has been pointed out in [197,375]; see also [376] and references therein on the synchrotron instability in systems with anisotropic electron distributions.

It is however clear that the formation of sharply anisotropic distributions of relativistic particles is quite problematic. Therefore, in astrophysical conditions one should expect the synchrotron instability and negative reabsorption to occur more probably in systems of relativistic particles with quasi-isotropic energy spectra which are imbedded in the "cold" plasma. There is no convincing evidence now concerning the existence of synchrotron instability in astrophysical conditions, though the relative softness of the corresponding criteria makes the appearance of synchrotron maser sources quite probable [372]. The objects with negative synchrotron reabsorption should be searched for at sufficiently low frequencies $\omega \lesssim \omega_L/2\omega_B$ at which the influence of the medium leads to the noticeable depression of the synchrotron radiation (see the inequality (5.102)).

In addition to the general arguments on possible non-stationary behavior and complexity of the frequency spectrum due to the exponential growth of amplified radiation, an important indication of negative reabsorption in a synchrotron source could be the strong circular polarization of the observed radiation. Indeed, synchrotron reabsorption is characterized by the coefficients μ_j ($j=1,2$) related to the normal waves of two types (see (7.150)). In the case of quasi-longitudinal propagation, however, the spectral power of synchrotron radiation $P_{\omega j}$ entering the expression for μ_j is described by the formula (7.151) which is valid up to the terms of the order $m_e c^2/\mathcal{E}$ and $\sqrt{1-n^2}$. Therefore, with the above accuracy $\mu_1 = \mu_2$. More rigorous kinetic analysis shows that $\mu_1 \neq \mu_2$, though μ_1 is close to μ_2 [368].

If the inequality $|\mu_1 - \mu_2| L \gtrsim 1$ is valid for a source with linear size L, the synchrotron radiation escaping the region of negative reabsorption will be circularly polarized, the degree of polarization ζ_c to be of about unity (see section 4.2).

Of course, the above does not mean that the coherent mechanism can only provide the strong circular polarization. If the mode polarization differs from circular, the radiation escaping a synchrotron maser with $|\mu_1 - \mu_2| L \gtrsim 1$ will have different polarization.

The above analysis of synchrotron instability has been made under the assumption that the refractive index and polarization of normal waves are determined by the "cold" plasma, whereas the relativistic electrons directly provide the emission or absorption of radiation. Such a separation of the components is possible when dealing with objects containing rather dense non-relativistic plasma in which relativistic particles exist (for, example, the interstellar medium, supernova nebulas, or solar corona). Conversely, in neutron star magnetosphere the bulk of plasma particles may be relativistic and, thus, such an approach becomes impossible.

According to [195, 196, 374], the "cold" plasma plays the leading role until $|n_1 - n_2| \gg (c/2\omega)\mu_j$, where $n_1 - n_2$ is the difference between refractive indices of ordinary and extraordinary waves in "cold" plasma, and μ_j – the reabsorption coefficient provided by the relativistic particles. The values of plasma density matching this condition can be easily obtained if one takes into account the expressions for the values entering the above inequality. For example, in the quasi-longitudinal approximation (at $\omega_L^2/\omega^2 \ll 1$ and

$\omega_B/\omega \ll 1$) the difference $n_1 - n_2$ is determined by (4.88) while the reabsorption coefficient for the power-law energy spectrum of relativistic particles in the approximation (5.102) is given by the formula (7.155). The mode polarization is formed by the "cold" plasma if its density

$$N \gg g(\gamma)\frac{m_e c^2}{8\pi}\mathcal{K}\left(\frac{3e}{m_e^3 c^5}\frac{B_0}{\omega}\right)^{\gamma/2}\frac{\sin^{\frac{\gamma+2}{2}}\alpha}{\cos\alpha}. \quad (7.186)$$

For monoenergetic, isotropic distribution of relativistic particles the following criterion should be used instead of (7.186):

$$N \gg N_S\sqrt{(m_e c^2/\mathcal{E})^2 + \omega_L^2/\omega^2}, \quad (7.187)$$

where N_S is the density of relativistic electrons.

When the inequalities opposite to (7.186) and (7.187) are satisfied, the mode polarization depends on frequency and is determined by the characteristics of relativistic plasma. For example, in a system with an isotropic monoenergetic spectrum of relativistic electrons the modes propagating at angles $\alpha \gg m_e c^2/\mathcal{E}$ with respect to the magnetic field are linearly polarized at frequencies $\omega \sim \omega_c \equiv (3/2)\omega_B \sin\alpha(\mathcal{E}/m_e c^2)^2$. In a system with a power-law distribution $N(\mathcal{E})$ the mode polarization could be both linear and circular (and also elliptical) depending on the values $\mathcal{E}_1$ and $\mathcal{E}_2$, the boundaries of the power-law spectrum (7.180) [195,196]. The expressions for synchrotron reabsorption coefficients in a system with a power-law energy spectrum and linearly polarized modes are given and discussed in [208].

If the distribution of relativistic electrons is strongly anisotropic, the synchrotron instability is possible in the absence of the "cold" plasma. Such an instability may occur even for isotropic particle distribution if the role of the media providing the refractive index different from unity is played by radiating electrons themselves [374]. As was clarified earlier, in a system comprising a "monoenergetic", isotropic spectrum of relativistic electrons in a "cold" plasma the synchrotron instability develops at frequencies $\omega \sim \omega_{max}$, (7.178), if $(m_e c^2/\mathcal{E})^2 \ll 1 - n^2(\omega) \ll 1$. In the absence of "cold" plasma $1 - n^2(\omega) \approx \Omega_L^2/\omega^2$, where $\Omega_L = \sqrt{(4\pi e^2 N_S/m_e)(m_e c^2/\mathcal{E}_0)}$ is the relativistic plasma frequency (see section 2.2). Taking into account that in the relativistic system considered the frequency of growing waves ω is obtained from (7.178) by the substitution of ω_L with Ω_L, the above criteria of synchrotron instability may then be represented in the form [377]

$$\left(\frac{m_e c^2}{\mathcal{E}_0}\right)^2 \ll \left(\frac{B_0^2\sin^2\alpha}{\mathcal{E}_0 N_S}\right)^{1/2} \ll 1. \quad (7.188)$$

For angles $\alpha \sim 1$ the right inequality in (7.188) demands that the energy of the particle system is much greater than that of the magnetic field. It is not an easy matter to realize such a strong requirement in astrophysical conditions. At angles $\alpha \ll 1$ this criterion gets softer.

Note, finally, that we have considered here the synchrotron instability in systems with refractive index $n < 1$, $1 - n \ll 1$. The characteristics of the instability in a medium with $n > 1$, $1 - n \ll 1$ are sharply different [378]. Evidently, this is connected, first of all, with the modification of emission by a relativistic particle in a media with $n > 1$ compared to the case of $n < 1$ (see section 5.1). The excitation of waves with both $n(\omega) < 1$ and $n(\omega) > 1$ has been studied in [377] for the case of synchrotron instability of plasma waves in the frequency range where the refractive index $n_{\rm pl}(\omega)$ is close to unity, and the difference $n_{\rm pl}(\omega) - 1$ changes the sign. The instability in the region $n_{\rm pl}(\omega) < 1$ is similar to the synchrotron instability of electromgnetic waves but with greater increments, whereas in the interval $n_{\rm pl}(\omega) > 1$ the function $\mu_{\rm pl}(\omega)$ is oscillating and enters the bands of amplification $\mu_{\rm pl}(\omega) < 0$ at sufficiently large amplitude of these oscillations. Kinetic cyclotron, gyro-synchrotron and synchrotron instabilitites are also discussed in the review [379] and in the paper [380].

Scattering of radiation in plasma

Until now, emission and absorption processes (including negative reabsorption in non-equilibrium plasma) have been considered as the main factors affecting transfer of radiation in astrophysical plasmas. Emission or absorption of photons or plasmons by charged plasma particles have been the basic processes of interaction of radiation with plasma in this case. There are however other processes which can significantly affect radiation transfer in plasmas. We refer to scattering by waves and particles which photon-photon or photon-particle collisions contribute to. In some cases these processes can radically transform angular and frequency spectra of radiation and also lead to the conversion (transformation) of radiation from one mode to another. The latter is especially important for plasma mechanisms of radio emission which can only act effectively if energy is transferred from the excited plasma waves into electromagnetic radiation capable of escaping the plasma (see section 3.2). Resonant cyclotron scattering leads to the formation of absorption features in spectra of magnetic degenerate stars. Inverse influence of radiation onto plasma leads to radiation pressure force. In some cases this force which originates due to strong cyclotron scattering exceeds the gravitational one and drives plasma ejection from magnetic degenerate stars. The ejected matter can accumulate in magnetospheres of these objects and form extended plasma envelopes around them. Note, finally that this chapter contains only the information necessary to pick up the initial knowledge on scattering processes, their influence on plasma and astrophysical applications of the relevant theory. To study these issues in more detail we refer the reader to monographs and reviews [10,72,89,309, 381–384] and to the original papers cited below.

8.1. Scattering of waves by particles in isotropic plasma

8.1.1. SPONTANEOUS AND STIMULATED SCATTERING

It has been already mentioned previously that, according to the quantum description, a scattering event is treated as an "elastic" collision of a photon (or plasmon) with a plasma particle in which the energy and momentum of the interacting particles is conserved:

$$\mathcal{E} + \hbar\omega \quad = \quad \mathcal{E}' + \hbar\omega', \tag{8.1}$$

$$\boldsymbol{p} + \hbar\boldsymbol{k} = \boldsymbol{p}' + \hbar\boldsymbol{k}', \tag{8.2}$$

where $\mathcal{E}$ and $\boldsymbol{p}$ are particle energy and momentum, $\hbar\omega$ and $\hbar\boldsymbol{k}$ are photon energy and momentum before the collision, whereas primes mark the corresponding values after collision. Assuming the relative change of particle energy in a scattering to be small enough, the energy $\mathcal{E}'$ can be presented as $\mathcal{E}' = \mathcal{E} + (d\mathcal{E}/d\boldsymbol{p})\Delta\boldsymbol{p}$, where $\Delta\boldsymbol{p} = \boldsymbol{p}' - \boldsymbol{p} = \hbar(\boldsymbol{k} - \boldsymbol{k}')$ and $d\mathcal{E}/d\boldsymbol{p} = \boldsymbol{v}$, the latter being valid both for relativistic and for non-relativistic particles. Using then the energy conservation law in the form (8.1) we conclude that the relation

$$\omega - \omega' = (\boldsymbol{k} - \boldsymbol{k}')\boldsymbol{v} \tag{8.3}$$

should hold for a scattering of a quantum by a particle with velocity $\boldsymbol{v}$.

Determine the probability of scattering of radiation $w(\boldsymbol{p}, \boldsymbol{k}, \boldsymbol{k}')$ so that the value $w(\boldsymbol{p}, \boldsymbol{k}, \boldsymbol{k}')(d^3k'/8\pi^3)$ is the probability per unit time of a transition of a photon with the momentum $\hbar\boldsymbol{k}$ into phase space volume d^3k' due to scattering by a particle with momentum $\boldsymbol{p}$.[1] The probability w is assumed to be independent of the radiation intensity. It therefore characterizes spontaneous scattering, in a similar way as the probability A_m^n which also does not depend on the intensity describes the process of spontaneous emission (see section 7.1). In addition to the spontaneous one, there also exists induced (or stimulated) scattering. This is described by the probability $\tilde{w}N_{\mathbf{k}'}$ proportional to $N_{\mathbf{k}'}$, the number of photons in a unit volume per mode with the wave vector $\boldsymbol{k}'$.[2] Taking into account the relation (7.29) between $N_{\mathbf{k}'}$ and $W_{\mathbf{k}'}$, we conclude that the induced scattering probability is proportional to the energy density of scattered radiation (i.e., the radiation into which an initial photon is scattered).

The probability of spontaneous scattering $w(\boldsymbol{p}, \boldsymbol{k}, \boldsymbol{k}')$ depends on the specific properties of a "scattering center" and those of the normal waves being scattered. There is, however, the general relation between $w(\boldsymbol{p}, \boldsymbol{k}, \boldsymbol{k}')$, $\tilde{w}(\boldsymbol{p}, \boldsymbol{k}, \boldsymbol{k}')$, $w(\boldsymbol{p}', \boldsymbol{k}', \boldsymbol{k})$, and $\tilde{w}(\boldsymbol{p}', \boldsymbol{k}', \boldsymbol{k})$, where the pair of latter quantities describe the inverse scattering from $\boldsymbol{k}'$ into $\boldsymbol{k}$. This may be found using the detailed balance conditions for scattering processes under thermodynamic equilibrium [386]. The procedure is analogous to that used in section 7.1 concerning the Einstein coefficients.

The rate of scattering transitions from the phase space volume d^3k near $\boldsymbol{k}$ into the one d^3k' near $\boldsymbol{k}'$ equals

$$[N_{\mathbf{k}}w(\boldsymbol{p}, \boldsymbol{k}, \boldsymbol{k}') + N_{\mathbf{k}}N_{\mathbf{k}'}\tilde{w}(\boldsymbol{p}, \boldsymbol{k}, \boldsymbol{k}')]\, N_{\mathbf{p}}\, \frac{d^3k d^3k'}{(2\pi)^6}.$$

[1] We repeat that $d^3k'(2\pi)^{-3}$ is the number of modes in the unit volume of the coordinate space and in the volume d^3k' of the wave vector space (see 1.35).

[2] The term "induced scattering" was introduced by Pauli [385].

The corresponding rate of the inverse transitions $\boldsymbol{k}' \to \boldsymbol{k}$ is

$$[N_{\mathbf{k}'} w(\boldsymbol{p}', \boldsymbol{k}', \boldsymbol{k}) + N_{\mathbf{k}'} N_{\mathbf{k}} \tilde{w}(\boldsymbol{p}', \boldsymbol{k}', \boldsymbol{k})] \, N_{\mathbf{p}'} \frac{d^3k d^3k'}{(2\pi)^6}.$$

Here $N_{\mathbf{p}}$ and $N_{\mathbf{p}'}$ are the densities of scattering particles with the momenta $\boldsymbol{p}$ and $\boldsymbol{p}'$, respectively. It is also taken into account in the above expressions that the number of quanta in the phase space volume d^3k is $N_{\mathbf{k}} d^3k (2\pi)^{-3}$.

For thermodynamic equilibrium to be established it is sufficient that the rates of direct and inverse transitions are equal for the equilibrium values of quantum numbers $N_{\mathbf{k}}$ and $N_{\mathbf{k}'}$ and "populations" $N_{\mathbf{p}}$ and $N_{\mathbf{p}'}$:

$$\begin{aligned} &[N_{\mathbf{k}} w(\boldsymbol{p}, \boldsymbol{k}, \boldsymbol{k}') + N_{\mathbf{k}} N_{\mathbf{k}'} \tilde{w}(\boldsymbol{p}, \boldsymbol{k}, \boldsymbol{k}')] \, N_{\mathbf{p}} = \\ &\qquad [N_{\mathbf{k}'} w(\boldsymbol{p}', \boldsymbol{k}', \boldsymbol{k}) + N_{\mathbf{k}'} N_{\mathbf{k}} \tilde{w}(\boldsymbol{p}', \boldsymbol{k}', \boldsymbol{k})] \, N_{\mathbf{p}'} \end{aligned} \tag{8.4}$$

The ratio $N_{\mathbf{p}'}/N_{\mathbf{p}}$ in this equation is determined by the Boltzmann formula (7.10) which, in accord with the conservation law (8.1), can be presented in the following form:

$$\frac{N_{\mathbf{p}'}}{N_{\mathbf{p}}} = \exp\left[\frac{\hbar(\omega' - \omega)}{\kappa T}\right]. \tag{8.5}$$

On the other hand, the equilibrium values of the photon numbers $N_{\mathbf{k}}$ and $N_{\mathbf{k}'}$ are determined by the Planck formula [55]:

$$N_{\mathbf{k}} = (e^{\hbar\omega/\kappa T} - 1)^{-1}, \qquad N_{\mathbf{k}'} = (e^{\hbar\omega'/\kappa T} - 1)^{-1}. \tag{8.6}$$

In the Wien limit $\hbar\omega \gg \kappa T$ the above formulas yield

$$N_{\mathbf{k}} = e^{-\hbar\omega/\kappa T}, \qquad N_{\mathbf{k}'} = e^{-\hbar\omega'/\kappa T}, \tag{8.7}$$

whereas in the opposite, Rayleigh-Jeans, limit ($\hbar\omega \ll \kappa T$)

$$N_{\mathbf{k}} = \frac{\kappa T}{\hbar\omega}, \qquad N_{\mathbf{k}'} = \frac{\kappa T}{\hbar\omega'}. \tag{8.8}$$

The last expressions also straightforwardly follow from the formulas (1.36) and (7.29). At low temperatures $T \to 0$ the equation (8.4) together with (8.5) and (8.6) leads to

$$w(\boldsymbol{p}, \boldsymbol{k}, \boldsymbol{k}') = w(\boldsymbol{p}', \boldsymbol{k}', \boldsymbol{k}). \tag{8.9}$$

At high temperatures ($T \to \infty$) it follows from (8.4), (8.5), and (8.8) that

$$\tilde{w}(\boldsymbol{p}, \boldsymbol{k}, \boldsymbol{k}') = \tilde{w}(\boldsymbol{p}', \boldsymbol{k}', \boldsymbol{k}). \tag{8.10}$$

Finally, taking into account (8.5), (8.6) and the derived above (8.9), (8.10) we find from (8.4) that

$$w(\boldsymbol{p}, \boldsymbol{k}, \boldsymbol{k}') = \tilde{w}(\boldsymbol{p}, \boldsymbol{k}, \boldsymbol{k}'). \tag{8.11}$$

The equality of the probabilities (8.9) and (8.10) of spontaneous scattering $\boldsymbol{k} \to \boldsymbol{k}'$ and $\boldsymbol{k}' \to \boldsymbol{k}$ and also the relation (8.11) between the spontaneous and induced scattering probabilities are valid in the absence of thermodynamic equilibrium as well. The reasons are the same as for the relations between Einstein coefficients A_m^n, B_m^n, and B_n^m (see section 7.1).

8.1.2. INTEGRAL KINETIC EQUATION FOR PHOTONS AND RADIATION TRANSFER EQUATION

Given the relations between the scattering probabilities, one may easily write the equations describing the temporal evolution of number of quanta. Such equations are called kinetic equations for photons (plasmons) taking scattering processes into account. The number of photons leaving, due to spontaneous and induced scattering, the unit phase space per unit time is equal

$$\int N_{\mathbf{k}}(N_{\mathbf{k}'} + 1) w(\boldsymbol{p}, \boldsymbol{k}, \boldsymbol{k}') f_0(\boldsymbol{p}) \frac{d^3p d^3k'}{(2\pi)^6}. \tag{8.12}$$

Similarly, the number of photons entering this volume per unit time is

$$\int N_{\mathbf{k}'}(N_{\mathbf{k}} + 1) w(\boldsymbol{p}', \boldsymbol{k}', \boldsymbol{k}) f_0(\boldsymbol{p}') \frac{d^3p' d^3k'}{(2\pi)^6}. \tag{8.13}$$

Here $f_0(\boldsymbol{p})$ and $f_0(\boldsymbol{p}')$ are distribution functions over particle momenta. However, $\boldsymbol{p}' = \boldsymbol{p} + \hbar(\boldsymbol{k} - \boldsymbol{k}')$ and $w(\boldsymbol{p}', \boldsymbol{k}', \boldsymbol{k}) = w(\boldsymbol{p}, \boldsymbol{k}, \boldsymbol{k}')$, so the last expression may be transformed to

$$\int N_{\mathbf{k}'}(N_{\mathbf{k}} + 1) w(\boldsymbol{p}, \boldsymbol{k}, \boldsymbol{k}') f_0(\boldsymbol{p} + \hbar\boldsymbol{k} - \hbar\boldsymbol{k}') \frac{d^3p d^3k'}{(2\pi)^6}.$$

Since the number of photons in unit phase space is $N_{\mathbf{k}}(2\pi)^{-3}$, it is clear from the above that

$$\begin{aligned} \frac{\partial N_{\mathbf{k}}}{\partial t} = \int w(\boldsymbol{p}, \boldsymbol{k}, \boldsymbol{k}') N_{\mathbf{k}'} f_0(\boldsymbol{p}) \frac{d^3p d^3k'}{(2\pi)^6} \\ - N_{\mathbf{k}} \int w(\boldsymbol{p}, \boldsymbol{k}, \boldsymbol{k}') f_0(\boldsymbol{p}) \frac{d^3p d^3k'}{(2\pi)^3} \\ + N_{\mathbf{k}} \int w(\boldsymbol{p}, \boldsymbol{k}, \boldsymbol{k}') \hbar(\boldsymbol{k} - \boldsymbol{k}') N_{\mathbf{k}'} \frac{\partial f_0}{\partial \boldsymbol{p}} \frac{d^3p d^3k'}{(2\pi)^3}. \end{aligned} \tag{8.14}$$

The obtained kinetic equation for $N_{\mathbf{k}}$ [3] accounts for the fact that in classical plasma the change of particle momentum due to scattering $\Delta \boldsymbol{p} = \hbar(\boldsymbol{k} - \boldsymbol{k}')$ is small compared to the characteristic width of the particle distribution $f_0(\boldsymbol{p})$. The latter therefore may be expanded in powers of $\hbar(\boldsymbol{k} - \boldsymbol{k}')$ keeping the two leading terms only. It is such a procedure that has led to the equation (8.14).

Since scattering does not provide sources or sinks of photons, it may only cause their redistribution in the momentum space $\hbar \boldsymbol{k}$. It is clear that the kinetic equation (8.14) describes the processes conserving the total number of photons in the phase space as well. From the form of (8.14) it is clear that spontaneous processes described by the first two terms in the right-hand side of this equation increase the number of photons in the processes $\boldsymbol{k}' \leftrightarrow \boldsymbol{k}$ if $N_{\mathbf{k}'} > N_{\mathbf{k}}$; on the contrary, this number decreases due to spontaneous scattering $\boldsymbol{k}' \leftrightarrow \boldsymbol{k}$ if $N_{\mathbf{k}'} < N_{\mathbf{k}}$. Spontaneous scattering thus leads to the redistribution of photons (plasmons) from the phase space regions with large $N_{\mathbf{k}}$ to those with smaller number of quanta. In other words, spontaneous scattering equilibrates the distribution $N_{\mathbf{k}}$ over wave vectors.

The effect of induced scattering on photon redistribution over $\boldsymbol{k}$-space is more complicated. It depends on the sign of the third term in the equation (8.14) which accounts for this process. The analysis is simplified if particle distribution function is isotropic: $f_0(\boldsymbol{p}) = f_0(p)$. In this case the derivative [4]

$$\frac{\partial f_0}{\partial \boldsymbol{p}} = \frac{\partial f_0}{\partial \mathcal{E}} \frac{d\mathcal{E}}{d\boldsymbol{p}} = \boldsymbol{v} \frac{\partial f_0}{\partial \mathcal{E}} = -\boldsymbol{v} \frac{f_0(p)}{\kappa T}, \tag{8.15}$$

where $\mathcal{E}$ is the energy of a particle with momentum $\boldsymbol{p}$. Taking into account this expression together with (8.3), we arrive at the following form for the third term in the kinetic equation (8.14):

$$N_{\mathbf{k}} \int w\hbar(\omega - \omega') N_{\mathbf{k}'} \frac{\partial f_0}{\partial \mathcal{E}} \frac{d^3p d^3k'}{(2\pi)^6}. \tag{8.16}$$

In plasma with $\partial f_0 / \partial \mathcal{E} < 0$ (for example, in equilibrium plasma; see (8.15)) the sign of the integrand in (8.16) is opposite to that of the difference ω–ω'. It is thus easy to conclude that the induced scattering which redistributes photons between two phase space elements d^3k and d^3k' increases the number of photons with the frequency ω if $\omega < \omega'$ and decreases it in

[3]Kinetic equation for photons being scattered by plasma particles has been derived in [387, 388]. Note also that the equation (8.14) as well as the condition (8.3) do not take into account the influence of a static magnetic field $\mathbf{B}_0$ on the state of scattering particles. Thus, the kinetic equation for magnetoactive plasma is different from (8.14) (see, e.g., [389, 390]).

[4]The latter equality in (8.15) is valid for the Maxwell distribution of particles.

the case $\omega > \omega'$. As a result, the induced scattering leads to the photon redistribution from the phase space region with a higher frequency to those with a lower one. Photon energy $\hbar\omega$ decreases and that of particles increases due to induced scattering events. Note that the above indicated direction of photons transfer over the spectrum corresponds to the case of plasma with isotropic $f_0(\boldsymbol{p})$ and $\partial f_0/\partial\mathcal{E} < 0$. Violation of these conditions may alternate this direction. This occurs, for example, in the case of scattering by charged particle beams in plasma [391].

The kinetic equation for photon number $N_{\mathbf{k}}$ may be transformed to the transfer equation for radiation energy density $W_{\mathbf{k}}$ using the relation (7.29):

$$\frac{\partial W_{\mathbf{k}}}{\partial t} + v_{gr}\frac{\partial W_{\mathbf{k}}}{\partial l} = \int w\,\frac{\omega}{\omega'}\,W_{\mathbf{k}'} f_0(\boldsymbol{p})\,\frac{d^3p d^3k'}{(2\pi)^3}$$
$$-W_{\mathbf{k}}\int w f_0(\boldsymbol{p})\,\frac{d^3pd^3k'}{(2\pi)^3} + W_{\mathbf{k}}\int w\,\frac{\boldsymbol{k}-\boldsymbol{k}'}{\omega'}\,W_{\mathbf{k}'}\,\frac{\partial f_0}{\partial \boldsymbol{p}} d^3pd^3k', \quad (8.17)$$

where $w = w(\boldsymbol{p}, \boldsymbol{k}, \boldsymbol{k}')$. The term $v_{gr}(\partial W_{\mathbf{k}}/\partial l)$ added to the left-hand side of this equation accounts for the possible inhomogeneity of the energy density $W_{\mathbf{k}}$ along a ray (cf. to the transfer equation (4.28)).

Further comparison of the equations (8.17) and (4.28) shows also that the role of the emissivity $\alpha_{\mathbf{k}}$ is played by the first term in the right-hand side of the equation (8.17) which characterizes the spontaneous scattering from any $\boldsymbol{k}'$ to a given $\boldsymbol{k}$:

$$\alpha_{\mathbf{k}} = \int w\,\frac{\omega}{\omega'}\,W_{\mathbf{k}'} f_0(\boldsymbol{p})\,\frac{d^3pd^3k'}{(2\pi)^3} \quad (8.18)$$

The sum of the second term describing spontaneous scattering from a given $\boldsymbol{k}$ to any $\boldsymbol{k}'$ and the third one accounting for direct $(\boldsymbol{k} \to \boldsymbol{k}')$ and inverse $(\boldsymbol{k}' \to \boldsymbol{k})$ induced scattering yield the value $-\beta_j W_{\mathbf{k}} = -2\gamma_j W_{\mathbf{k}}$, where γ_j means the decrement due to scattering:

$$\gamma_j = \int w\, f_0(\boldsymbol{p})\,\frac{d^3pd^3k'}{(2\pi)^3} - \int w\,\frac{\boldsymbol{k}-\boldsymbol{k}'}{\omega'}\,W_{\mathbf{k}'}\,\frac{\partial f_0}{\partial \boldsymbol{p}}\,d^3pd^3k'. \quad (8.19)$$

Furthermore, with (4.29) the equation (8.17) may be transformed to the transfer equation for radiation intensity (4.30) with the absorption coefficient $\mu_j = 2\gamma_j v_{gr}^{-1}$ being determined γ_j from (8.19) and the emissivity $a_\omega = k^2\alpha_{\mathbf{k}}(|\cos\vartheta_j|v_{gr})^{-1}$ by $\alpha_{\mathbf{k}}$ from (8.18).

The first term in the decrement (8.19) denoted in what follows as γ_j^{sp} appears due to spontaneous scattering of radiation with a wave vector $\boldsymbol{k}$.[5]

[5] In astrophysics the value $\mu_j^{sp} = 2\gamma_j^{sp} v_{gr}^{-1}$ is usually used which is called the extinction coefficient.

This process always yields the positive contribution to γ_j, thus effectively increasing the damping of radiation with energy density $W_{\mathbf{k}}$. The second term, $\gamma_j^{\rm ind}$, describes the induced scattering from $\boldsymbol{k}$ to any $\boldsymbol{k}'$ and vice versa. This term is non-zero for $W_{\mathbf{k}'} \neq 0$ only. Evidently, the sign of $\gamma_j^{\rm ind}$ depends on those of the difference $\boldsymbol{k} - \boldsymbol{k}'$ and the derivative $\partial f_0/\partial \boldsymbol{p}$ of the scattering particle distribution. The following expression holds for isotropic particle distribution:

$$\gamma_j^{\rm ind} = -\int w \frac{\omega - \omega'}{2\omega} W_{\mathbf{k}'} \frac{\partial f_0}{\partial \mathcal{E}} d^3p d^3k'. \qquad (8.20)$$

Since $\partial f_0/\partial \mathcal{E} < 0$ in equilibrium plasma, the frequency ranges $\omega' < \omega$ and $\omega' > \omega$ give positive and negative contributions to $\gamma_j^{\rm ind}$, respectively. The resulting sign of $\gamma_j^{\rm ind}$ depends on the distribution of the radiation energy density $W_{\mathbf{k}}$ over the spectrum. Namely, $\gamma_j^{\rm ind} > 0$ if the energy of waves interacting with radiation of the frequency ω is contributed to mainly by the higher frequency range.

The transfer equation (8.17) describes the variation of energy density $W_{\mathbf{k}}$ due to scattering by particles. The effect of true emission and absorption may be easily accounted for by the inclusion the corresponding term $\alpha_{\mathbf{k}}^{\rm em} - 2\gamma_j^{\rm abs} W_{\mathbf{k}}$, where $\alpha_{\mathbf{k}}^{\rm em}$ is for spontaneous emission and $\gamma_j^{\rm abs}$ characterizes absorption due to particle collisions or Landau damping (see in this respect sections 5.2,6.1), into the right-hand of the equation (8.17). Then the total decrement in the transfer equation takes the form

$$\gamma_j = \gamma_j^{\rm sp} + \gamma_j^{\rm ind} + \gamma_j^{\rm abs}. \qquad (8.21)$$

If the induced scattering coefficient is negative and its absolute value is large enough, $-\gamma_j^{\rm ind} > \gamma_j^{\rm sp} + \gamma_j^{\rm abs} > 0$, the total decrement becomes negative, too. As a result, the instability due to induced scattering appears at frequencies ω. Contrary to instabilities due to induced emission which occur in case of particle population inversion (section 7.1), such an instability may develop in equilibrium plasma. However, the spectral distribution of radiation energy density $W_{\mathbf{k}}$ should be strongly non-equilibrium for it to appear.

8.1.3. SCATTERING OF RADIATION WITH ONE-DIMENSIONAL SPECTRUM. DIFFERENTIAL AND INTEGRAL REDISTRIBUTION

The complexity of the integro-differential transfer equation (8.17) forces search for conditions allowing the analysis to be somehow simplified. For example, the effect of induced scattering and of absorption due to dissipation and/or spontaneous scattering on the energy density $W_{\mathbf{k}}$ may be neglected when dealing with time intervals $\Delta t \ll 1/\gamma_j$ or spatial scales $\Delta l \ll 1/\mu_j$ (here $\mu_j = 2\gamma_j v_{\rm gr}^{-1}$ is the absorption coefficient determined by the total decrement γ_j from (8.21)). This means the first term characterizing the emissivity $\alpha_{\mathbf{k}}$ due to the spontaneous scattering of waves with $\boldsymbol{k}'$ into ones with $\boldsymbol{k}$ may only be left in the equation (8.17)

Alternatively, the substantial simplification of the transfer equation (8.17) is possible if radiation density is sufficiently high. In this case one may neglect spontaneous scattering compared to the induced one and keep only the last term $2\gamma_j^{\text{ind}} W_{\mathbf{k}}$ in the equation (8.17). Greater simplification and illustrativity are achieved by the certain assumptions on the form of the angular spectrum of radiation, for example, on the isotropic or one-dimensional distribution of $W_{\mathbf{k}}$, $W_{\mathbf{k}'}$ in the wave vector space.

The transfer equation (8.17) accounting for the induced scattering in non-relativistic plasma has the form

$$\frac{\partial W_{\mathbf{k}}}{\partial t} + v_{\text{gr}} \frac{\partial W_{\mathbf{k}}}{\partial l} = -2\gamma_j^{\text{ind}} W_{\mathbf{k}}, \tag{8.22}$$

$$2\gamma_j^{\text{ind}} = \int w_\alpha(\boldsymbol{k}, \boldsymbol{k}') \frac{\omega - \omega'}{\omega' \kappa T_\alpha} W_{\mathbf{k}'} d^3 k', \tag{8.23}$$

where the relation (8.15) for Maxwell distribution is taken into account and the definition

$$w_\alpha(\boldsymbol{k}, \boldsymbol{k}') = \int w(\boldsymbol{p}, \boldsymbol{k}, \boldsymbol{k}') f_0(\boldsymbol{p}) d^3 p \tag{8.24}$$

is introduced, the index α marking scattering particle species. The scattering probability $w(\boldsymbol{p}, \boldsymbol{k}, \boldsymbol{k}')$ is non-zero under the condition (8.3) only. This means it is proportional to the delta-function $\delta(\omega_- - \boldsymbol{k}_- \boldsymbol{v})$, where $\omega_- = \omega - \omega'$ and $\boldsymbol{k}_- = \boldsymbol{k} - \boldsymbol{k}'$. Since $f_0(\boldsymbol{p}) \propto e^{-p^2/2m_\alpha \kappa T_\alpha}$ in equilibrium plasma, the integration of this delta-function in (8.24) leads to the formula

$$w_\alpha(\boldsymbol{k}, \boldsymbol{k}') = w_\alpha^*(\boldsymbol{k}, \boldsymbol{k}') \frac{\exp(-\omega_-^2 / 2\boldsymbol{k}_-^2 v_{T_\alpha}^2)}{|\boldsymbol{k}_-|}. \tag{8.25}$$

The form of $w_\alpha^*(\boldsymbol{k}, \boldsymbol{k}')$ depends on the specific properties of scattering species in plasma (see (8.53) and (8.54) below).

Bearing in mind the further applications, consider the transfer of radiation with one-dimensional spectra

$$\begin{aligned} W_{\mathbf{k}} d^3 k &= W(k) \delta(\boldsymbol{\Omega} - \boldsymbol{\Omega}_0) \, dk \, d\Omega \\ W_{\mathbf{k}'} d^3 k' &= W(k') \delta(\boldsymbol{\Omega}' - \boldsymbol{\Omega}_0') \, dk' \, d\Omega' \end{aligned} \tag{8.26}$$

in more detail. The values $W(k)$ and $W(k')$ in the above formula are one-dimensional distributions of radiation energy density, $d\Omega$ and $d\Omega'$ are the solid angle elements, $\boldsymbol{\Omega} = \boldsymbol{k}/k$ and $\boldsymbol{\Omega}' = \boldsymbol{k}'/k'$ the corresponding unit vectors. Now we substitute (8.23) and (8.26) into the equation (8.22), multiply it by

d^3k and integrate term-by-term over solid angles $d\Omega$ and $d\Omega'$. As a result, we get the transfer equation for one-dimensional spectra $W(k)$ and $W(k')$:

$$\frac{\partial W(k)}{\partial t} + v_{gr}\frac{\partial W(k)}{\partial l} = -\int w_\alpha(k,k')\,\frac{\omega_-}{\omega'\kappa T_\alpha}\,W(k)W(k')\,dk' \qquad (8.27)$$

Here $w_\alpha(k,k')$ is determined by the formula (8.25) for fixed $\boldsymbol{k}$ and $\boldsymbol{k}'$ directed along $\boldsymbol{\Omega}_0$ and $\boldsymbol{\Omega}_0'$, respectively. The exponential factor with the argument depending on the relative orientation of $\boldsymbol{k}$ and $\boldsymbol{k}'$ enters this probability. If θ_1 is the angle between $\boldsymbol{k}$ and $\boldsymbol{k}'$, then this exponent increases with increasing θ_1 reaching the maximum at $\theta_1 = \pi$. If $w_\alpha(k,k')|_{\theta_1=\pi} \gg w_\alpha(k,k')|_{\theta_1=0}$, it is a good approximation to assume that the scattering of waves having a one-dimensional spectrum with wave vectors $\boldsymbol{k}$ occurs mainly into waves in the opposite direction of $\boldsymbol{k}'$ (in other words, $\boldsymbol{\Omega}_0' = -\boldsymbol{\Omega}_0$ and $\boldsymbol{k}_-^2 = (k+k')^2$).

During the above discussion of the kinetic equation (8.14) it was noticed that induced scattering in homogeneous plasma leads to photon spectral redistribution to the region of smaller frequencies ω. Spectral evolution has a "differential character" if the frequency width of the spectrum $W(k')$ obeys the condition

$$\Delta\omega \gg |\boldsymbol{k}_\Gamma|v_{T_\alpha}. \qquad (8.28)$$

This inequality means that the frequency shift in a single scattering is relatively small: $|\omega_-| = |\omega - \omega'| \ll \Delta\omega$ (see (8.3)). For plasma waves occupying, according to the dispersion equation (2.115), the frequency range $\Delta\omega \lesssim 3k^2v_T^2/2\omega_L$ the differential redistribution is only possible for scattering by ions with $v_{T_\alpha} = v_{T_i} \ll v_T^2/v_{ph}$, where v_T is electron thermal velocity and v_{ph} the phase velocity of plasma waves.

If the condition (8.28) is realized, the integro-differential equation (8.27) can be approximated by the differential one [393]. As an example of such a procedure, consider plasma to plasma waves scattering by ions. Here we can put

$$w_\alpha^*(\boldsymbol{k},\boldsymbol{k}') = w_i^*(\boldsymbol{k},\boldsymbol{k}') = \frac{\sqrt{2\pi}}{16}\frac{\omega_L^2}{N v_{T_i}}\cos^2\theta_1, \qquad (8.29)$$

where θ_1 is the angle between $\boldsymbol{k}$ and $\boldsymbol{k}'$ (see (8.53)). The scattering of waves with opposite wave vectors means $\cos^2\theta_1 = 1$. It then follows from the condition (8.28) that the integrand in the equation (8.27) with the value w_α in the form (8.25), (8.29) is the product of the "sharp" function $\omega_-\exp(-\omega_-^2/2k_-^2v_{T_\alpha}^2)$ and the smooth ones. Thus, this sharp function in the integrand can be approximated by the delta-function derivative:

$$-\frac{\omega_-}{\sqrt{2\pi}|k_-|^3v_{T_\alpha}^3}\exp(-\omega_-^2/2k_-^2v_{T_\alpha}^2) \to \delta'(\omega_-), \qquad (8.30)$$

with an accuracy of the order of $|k_-|v_{T_\alpha}/\Delta\omega$. The factors preceding the exponent are chosen to ensure the usual normalizing condition for the delta-function: $\int \delta(\omega_-)d\omega_- = 1$. We repeat that in the above relation $v_{T_\alpha} = v_{T_i}$, $k_- = k + k'$ and $\omega_- = \omega - \omega'$. Integrating (8.27) over k' we finally find:

$$\frac{\partial W(k)}{\partial t} + v_{gr}\frac{\partial W(k)}{\partial l} = \frac{\pi}{18}\frac{\omega_L^3}{m_i N v_T^4} W(k)\frac{\partial W(-k)}{\partial k}. \tag{8.31}$$

Under the condition opposite to (8.28),

$$\Delta\omega \ll |\boldsymbol{k}_-|v_{T_\alpha}, \tag{8.32}$$

the "integral" redistribution takes place. Here the frequency changes with the value much larger than $\Delta\omega$ in a single scattering, so the scattered photons are removed from the band occupied by the radiation being scattered. If the spectra $W(k)$ and $W(k')$ are narrow enough,

$$\Delta k \ll k, \qquad \Delta k' \ll k', \tag{8.33}$$

to be well approximated by the following delta-functions:

$$W(k) = W\delta(k - k_0), \qquad W(k') = W'\delta(k' - k_0'), \tag{8.34}$$

where W and W' are the total energy densities in the spectra $W(k)$ and $W(k')$ which span the narrow bands near k_0 and k_0', the equation (8.27) can also be significantly simplified. Taking into account the above relations (8.34) and integrating (8.27) over k' and then the equation (8.27) over k, we obtain

$$\frac{\partial W}{\partial t} + v_{gr}\frac{\partial W}{\partial l} = -\eta W W', \tag{8.35}$$

where

$$\eta = w_\alpha(k_0, k_0')\frac{\omega(k_0) - \omega(k_0')}{\omega(k_0')\kappa T_\alpha}. \tag{8.36}$$

The similar equation holds for W', so the complete system comprises (8.35) and the equation

$$\frac{\partial W'}{\partial t} + v_{gr}\frac{\partial W'}{\partial l} = \eta W W' \tag{8.37}$$

with the same coefficient η. We thus arrive again at the transfer equation in differential form. This equation, however, is written now in terms of the integral characteristics of radiation, W and W'.

8.1.4. STABILIZATION OF BEAM INSTABILITY

As was shown in section 7.2, Čerenkov beam instability may occur in plasma, the corresponding increment to be proportional to $dF/dv_{\mathbf{k}}$, the derivative of electron distribution function in the beam with respect to the projection of electron velocity $\boldsymbol{v}$ onto the wave vector $\boldsymbol{k}$ of the excited plasma waves (see (7.67)). The growth of energy density of plasma waves in the resonance region (i.e., within the interval of phase velocities $v_{\rm ph}$ corresponding to beam particles with $v_{\mathbf{k}} = v_{\rm ph}$) and the corresponding variation of the distribution $F(v_{\mathbf{k}})$ under the action of plasma waves is described by the system of quasi-linear equations (7.98), (7.99). The analysis of this system shows that the evolution of $F(v_{\mathbf{k}})$ and $W(k)$ called quasi-linear relaxation proceeds to the formation of the plateau in the particle distribution $F(v_{\mathbf{k}})$ with a high intensity of plasma waves above it. The analysis of section 7.2 did not take into account the effect of scattering plasma waves. Under certain conditions, however, the scattering radically changes the whole process of quasi-linear relaxation and may lead to the non-linear stabilization of the beam instability [394].

Stabilization may occur due to the effective induced scattering of plasma waves in the region of high intensity which leads to the redistribution of photons to smaller frequencies (higher phase velocities). Such a redistribution removes plasma waves from the resonant region and breaks their interaction with the beam. If the redistribution of energy over the spectrum proceeds much faster than the quasi-linear relaxation, the latter will not be completed at all. Indeed, plasma wave energy is now concentrated in the non-resonant region, the intensity of plasma waves in the resonant region remains low and, therefore, particle distribution function is only slightly disturbed due to the resonant plasma waves.

The quasi-linear relaxation of a "narrow" particle beam with the initial velocity dispersion $\Delta v_{\rm S} \ll v_{\rm S}$ which leads to excitation of plasma waves with $\boldsymbol{v}_{\rm ph} \| \boldsymbol{v}_{\rm S}$ has been considered in detail in section 7.2. If quasi-linear relaxation is prevented by the effect of stabilization, the temporal evolution of plasma waves is described by the combined system (7.99), (8.35) and (8.37):

$$\frac{\partial W}{\partial t} = -2\overline{\gamma} W - \eta W W' - 2\gamma^{\rm abs} W, \tag{8.38}$$

$$\frac{\partial W'}{\partial t} = \eta W W' - 2\gamma^{\rm abs} W'. \tag{8.39}$$

Here η is determined by the formula (8.36), $2\gamma^{\rm abs} = \nu_{\rm eff}$ in the case of collisional damping (see (6.9)), and

$$\overline{\gamma} = -2\pi^2 \frac{e^2 v_{\rm S}^2}{m_e \omega_{\rm L}} \overline{\frac{\partial F_0}{\partial v_{\mathbf{k}}}} \approx -2\pi^2 \frac{e^2 v_{\rm S}^2}{m_e \omega_{\rm L}} \frac{N_{\rm S}}{(\Delta v_{\rm S})^2} \tag{8.40}$$

is a mean value of Čerenkov decrement in the instability region for a beam with the initial particle distribution $F_0(v_{\mathbf{k}})$. The formula (8.40) accounts for the fact that the particle distribution remains close to the initial one during the whole stabilization process: $\partial F/\partial v_{\mathbf{k}} \approx \partial F_0/\partial v_{\mathbf{k}}$ and $v_{\mathbf{k}} \approx v_S$. The equations (8.38), (8.39) describe a homogeneous system in which all the values and parameters are independent of spatial coordinates.

Consider, first, the stabilization under the absence of absorption of plasma waves ($\nu_{\text{eff}} = 0$) and find the energy density $W'(\infty)$ in the non-resonant region established at the end of this process due to excitation and subsequent induced scattering of plasma waves. Integrating (8.38) and (8.39) one can easily find that

$$W(t) + W'(t) = -\frac{2\overline{\gamma}}{\eta} \ln \frac{\eta W'(t)}{-2\overline{\gamma}} + C, \tag{8.41}$$

where C is the integration constant. Putting $t = 0$ and $t = \infty$, we get two relations from (8.41); their term-by-term subtraction yields the desired quantity

$$W'(\infty) \approx -\frac{2\overline{\gamma}}{\eta} \ln \frac{W'(\infty)}{W'(0)} \approx -\frac{2\overline{\gamma}}{\eta} \ln \frac{-2\overline{\gamma}}{\eta W'(0)}. \tag{8.42}$$

The initial values $W(0)$ and $W'(0)$ as well as the final energy density $W(\infty)$ in the resonance region are neglected in this relation compared to the final energy density in the non-resonant region.

The stabilization process which leads to the final state with $W'(\infty)$ determined by (8.42) proceeds as follows. First, energy density of plasma waves interacting with the particle beam grows due to the instability as $W(t) = W(0)e^{-2\overline{\gamma}t}$. The induced scattering is negligible at this stage because of the low intensity of plasma waves in the non-resonant region. Then, as W' increases, the induced scattering which removes plasma waves from the resonant region and breaks their interaction with the beam becomes more and more effective. The value of W reaches the maximum $W(t_{\max}) \sim W'(\infty)$ at the moment $t_{\max}$ when $\partial W/\partial t = 0$ and $W' = 2\overline{\gamma}/\eta$ (see (8.38)). Further, the energy density W starts decreasing while W' continues increasing and reaches the value $W'(\infty)$. The characteristic time of the decrease of W is also about $t_{\max}$. The interval from $t = 0$ to $t = t_{\max}$ can therefore be adopted as the characteristic time of the whole stabilization process. Assuming for the estimate that $W(t) \approx W(0)e^{-2\overline{\gamma}t}$ over this interval and that $W(t_{\max}) \sim W'(\infty)$, where the value $W'(\infty)$ is determined by (8.42), we find

$$t_{\text{st}} \sim -\frac{1}{2\overline{\gamma}} \ln \frac{-2\overline{\gamma}}{\eta W(0)}. \tag{8.43}$$

To find the stabilization criterion, i.e., the condition at which the variation of the particle distribution function in the beam can be neglected and the evolution of instability can be well described by the system (8.38), (8.39) with fixed $\overline{\gamma}$, one should refer to the equation (7.98). Being integrated over time, this equation becomes the following:

$$\delta F \equiv F_\infty - F_0 = \frac{\partial}{\partial v_{\mathbf{k}}}\left(\frac{a}{v_{\mathbf{k}}}\frac{\partial F_0}{\partial v_{\mathbf{k}}}\int_0^\infty W(k,t)\,dt\right). \tag{8.44}$$

Here $a = (2\pi)^2 e^2/m_e^2$ and $F_\infty = F(v_{\mathbf{k}})|_{t\to\infty}$. Assuming in advance the variation of particle distribution to be rather small, we put $\partial F/\partial v_{\mathbf{k}} \approx \partial F_0/\partial v_{\mathbf{k}}$ in the right-hand side. Then, according to (8.39),

$$\int_0^\infty W(t)\,dt = \frac{1}{\eta}\ln\frac{W'(\infty)}{W'(0)},$$

and, therefore,

$$\frac{\partial}{\partial v_{\mathbf{k}}}\delta F \sim \frac{a}{v_S(\Delta v_S)^2\eta\Delta k}\frac{\partial F_0}{\partial v_{\mathbf{k}}}\ln\frac{W'(\infty)}{W'(0)}. \tag{8.45}$$

To come to (8.45) from (8.45), we took into account that

$$\int_0^\infty W(k,t)\,dt \sim \frac{1}{\Delta k}\int_0^\infty W(t)\,dt,$$

where Δk is the characteristic width of the resonant spectrum. In addition, since the integrand consists of the functions with characteristic scales $\Delta v_{\mathbf{k}} \sim \Delta v_S$, taking the $v_{\mathbf{k}}$-derivative has been approximated by the division by Δv_S. The beam instability is stabilized if

$$\frac{\partial}{\partial v_{\mathbf{k}}}\delta F \ll \frac{\partial}{\partial v_{\mathbf{k}}}F.$$

Using (8.45) we find from the above inequality the criterion of beam instability stabilization:

$$\frac{a}{v_S(\Delta v_S)^2\eta\Delta k}\ln\frac{W'(\infty)}{W'(0)} \ll 1. \tag{8.46}$$

Taking into account the estimate (8.40) and putting $\Delta k \approx \omega_L\Delta v_S/v_S^2$, we write it in the form

$$-\frac{2\overline{\gamma}}{N_S m_e v_S \Delta v_S \eta}\ln\frac{-2\overline{\gamma}}{\eta W'(0)} \ll 1. \tag{8.47}$$

According to (8.25), (8.29), (8.36), the coefficient η for induced scattering by ions is determined by the formula

$$\eta = \frac{3}{16}\sqrt{\frac{\pi}{2}}\frac{\omega_L(k_0 - k_0')}{m_e N v_{T_i}}\exp\left[-\frac{(\omega_0 - \omega_0')^2}{2v_{T_i}^2(k_0 + k_0')^2}\right], \tag{8.48}$$

where $\omega_0 - \omega_0' = 3(k_0^2 - k_0'^2)v_T^2/2\omega_L$ (see (2.115)).

Since energy density of excited plasma waves is given by (8.42), the condition (8.47) means that this value is much less than the decrease of the beam energy due to formation of a "plateau" of the width Δv_S: $W'(\infty) \ll N_S m_e v_S \Delta v_S$. However, the process of quasi-linear relaxation of the "narrow" beam results in the formation of the plateau of the width $v_S \gg \Delta v_S$ with the transfer of about 2/3 of the beam energy to plasma waves (see section 7.2). Thus, the efficiency of plasma waves excitation is greatly reduced in case of stabilization.

In the above analysis of beam instability stabilization we did not account for collisional damping of plasma waves. This process, however, may drastically change the stabilization process making it oscillatory [395,396]. This is most clearly seen if the damping time of plasma waves, $\sim \nu_{\text{eff}}^{-1}$, is much larger than the characteristic time of stabilization $t_{\text{st}} \sim -1/\overline{\gamma}$ determined by (8.43). Damping of plasma waves in the non-resonant region over a time of about $1/\nu_{\text{eff}}$ restarts the generation process and ensures the new increase of W' due to the induced redistribution. Periodical repetition of this process with a period $\tau \sim 1/\nu_{\text{eff}}$ or, to be more correct,

$$\tau \approx \frac{1}{\nu_{\text{eff}}} \ln \frac{-2\overline{\gamma}}{\eta W'(0)}, \tag{8.49}$$

constitutes the pulsed regime of plasma wave generation by a stabilized beam [397,398].

Variations of particle distribution $F(v_{\mathbf{k}})$, though small in a single oscillation, are gradually accumulated and finally result in the non-linear relaxation of a stabilized beam to the plateau state. The relaxation of a stabilized beam occurs much slower, as compared to the case of a non-stabilized one. Since energy density $W'(\infty)$ is typically transferred from beam particles to plasma waves over a single period, the beam loses 2/3 of the initial energy $N_S m_e v_S^2/2$ and relaxes to the plateau in a time

$$t_{\text{nl}} \sim N_S m_e v_S^2 / 6\nu_{\text{eff}} W'(\infty). \tag{8.50}$$

This time of non-linear relaxation is much longer than the time of quasi-linear relaxation t_{ql} determined by (7.108). The dynamics of the pulsed regime for $t \ll t_{\text{nl}}$ is described by the equations (8.38), (8.39) with $\overline{\gamma} = \text{const}$. These equations are analyzed in [399].

The principal possibility for stabilization of an instability in astrophysical plasma (either a beam one or loss-cone, Čerenkov, synchrotron etc.) is no longer argued [10,396,400]. However, the detailed analysis is the only way to decide whether stabilization really does occur in a specific case. Such an analysis performed for type III bursts in the solar corona, the classical example of plasma generation mechanism, has shown that the criterion (8.47) is not satisfied for the majority of bursts generated by electron beams [336,337], see also [401–403]. Such beams, therefore, generate plasma waves in the regime of quasi-linear relaxation described in section 7.2. At the same time, the criterion (8.47) can be realized more easily for ion beams, since the corresponding Čerenkov increment is m_i/m_e times less than the one for electrons (see (7.67)). Some of the observed type III bursts exhibit quasi-periodic intensity variations with $\tau \sim 2$–3 s. This gives some evidence in favor of stabilization of the beam instability in the solar corona. The above quasi-periodic variations may be due to the pulsed regime of the instability. The corresponding period (8.49) is about seconds or a fraction of a second in solar corona conditions with $\nu_{\text{eff}} \sim 1 - 10\,\text{s}^{-1}$ (see section 6.1). Quasi-periodic profiles of radio emission $I_\omega(t)$ are more frequently observed among the type IV events and some specific types of solar radio bursts. These phenomena may also be connected with the pulsed regime of beam or loss-cone Čerenkov instability. Stabilization and pulsed regime of loss-cone instability for Earth magnetosphere conditions is discussed in [396].

8.1.5. RAYLEIGH SCATTERING IN NON-RELATIVISTIC PLASMA

When calculating a probability of scattering it is necessary to take into account the correlation between plasma particles due to the screening of the Coulomb potential of a charge at distances $r \sim D$, where D is the Debye radius (see (1.4), (1.5)). This correlation can drastically change the efficiency of scattering.

The above mentioned screening caused by the appearance of an opposite sign space charge depends of the motion of a particle. In particular, the region of increased electron density of the size D is formed around an ion with the charge $+e$ and velocity $v \ll v_{T_i} = \sqrt{\kappa T/m_i}$. The total charge of these electrons equals $-e/2$. Another half of the ion charge is screened due to the lack of ions in the Debye sphere. On the contrary, the positive space charge $+e$ which screens the field of an electron with velocity v in the limits $v_{T_i} \ll v \ll v_T$ is entirely provided by the lack of electrons in the Debye sphere. The ion density is not disturbed in this case, since slow and heavy ions with velocities of the order of v_{T_i} are not able to redistribute under the action of a particle with $v \gg v_{T_i}$. Finally, a charged particle with the velocity $v \gg v_T$ cannot be screened at all by the slow plasma particles.

Due to the absence of the screening envelope around fast particles with $v \gg v_T$ the scattering of radiation with arbitrary wavelength by such particles is similar to the scattering by free charges in vacuum.[6] The free

[6] Here we do not take into account the Cerenkov emission of plasma waves by fast particles in plasma which also contributes to the total scattering cross-section. Scattering of radiation by plasma waves is a kind of Raman scattering. It is considered in section 8.4.

particle approximation is also valid for the scattering of electromagnetic waves in plasma if $\lambda \ll D$ (i.e., $kD \gg 1$), since the contribution to the scattered radiation by the Debye region becomes small and the effective scattering is caused by the central particle. This case is typical for the scattering of optical and X-ray radiation.

On the contrary, the effects of Debye screening must be taken into account for the scattering of electromagnetic waves by non-relativistic plasma particles with $v \lesssim v_T$. The condition $kD \ll 1$ is usually realized for radio waves. In this case radiation is scattered not by a single particle but rather by the system of a particle and the surrounding Debye cloud. Scattering by a central particle is nothing but emission due to forced oscillations in the field of the infalling wave. If a particle is surrounded by the Debye cloud, the additional effect, emission by a current induced by an infalling wave in this cloud, contributes to the scattered radiation.

The efficiency of scattering in this case is strongly dependent on the nature of the scattering center. For example, in case of an ion ($v \ll v_{T_i}$) surrounded by an electron cloud with the total charge $-e/2$ the effect of the central particle is small compared to the contribution of the electron cloud due to the large ion mass. Therefore, the scattering probability in this case is mainly contributed by the screening electrons. But if scattering by an electron with the velocity v in the interval $v_{T_i} \ll v \ll v_T$ is considered, the effect of the central particle is largely compensated by the surrounding space charge $+e$ formed due to the lack of electrons in the Debye sphere. The full compensation does not occur because of the finite ratio of the wavelength to the size of the Debye region. As a result, the scattering probability by screened electrons is drastically reduced (by a factor of $(kD)^2$) compared to that by screened ions [387,388]. The compensation is somewhat reduced due to the influence of ions on the Debye region formation around an electron. The ions change the total space charge formed due to the lack of electrons, and the resulting scattering probability is increased [382,394].

It was assumed earlier that the scattering of radiation in plasma is nothing but the emission of radiation as a result of the forced oscillations of a particle and surrounding screening region in the field of the initial wave. Such an interpretation is valid in the case of scattering of a transverse electromagnetic wave. However, it becomes incomplete for longitudinal plasma waves. The reason is the "transition scattering" which appears together with emission due to forced oscillations. This term is used [392,404] to describe the process of radiation by a charged particle, stationary or moving, in the wave of dielectric permeability $\epsilon = \epsilon_0 + \epsilon_1 \cos(\boldsymbol{kr} - \omega t)$. Such a wave leads to the periodic variation of the polarization of the medium around a particle, and it gives rise to the scattered radiation. The needed variation of dielectric permeability occurs for longitudinal plasma waves connected with plasma density variations but absent for transverse electromagnetic waves. So, it is clear that transition scattering should be taken into account when calculating the probability of plasma wave scattering.

The procedure of calculating the probability is not fulfilled here (we refer the reader to [89, 382, 383]). Rather, we just give the resulting expressions characterizing the scattering probability by screened ions in isothermal plasma. This probability for the process of plasma wave to plasma waves scattering ($l \to l'$) has the following form:

$$w \equiv w_{ill'} = \frac{(2\pi)^3 e^4}{m_e^2 \omega_L^2} \cos^2\theta_1 |2 + i\sqrt{\pi} Z_i w(Z_i)|^{-2} \delta(\omega_- - \boldsymbol{k}_- \boldsymbol{v}). \tag{8.51}$$

The next expression holds for the case of plasma waves scattering into electromagnetic ones ($l \to t'$):

$$w \equiv w_{ilt'} = \frac{(2\pi)^3 e^4}{m_e^2 \omega_L^2} \sin^2\theta_1 |2 + i\sqrt{\pi} Z_i w(Z_i)|^{-2} \delta(\omega_- - \boldsymbol{k}_- \boldsymbol{v}). \tag{8.52}$$

In the above formulas $w(Z_i)$ is the Kramp function (2.91) with the argument $Z_i = \omega_- / \sqrt{2} |k_-| v_{T_i}$; ω_- and $\boldsymbol{k}_-$ denote the differences $\omega - \omega'$ and $\boldsymbol{k} - \boldsymbol{k}'$, respectively; and θ_1 is the angle between $\boldsymbol{k}$ and $\boldsymbol{k}'$. The delta function in the scattering probabilities accounts for the scattering condition (8.3) which follows from energy and momentum conservation.

The coefficients of spontaneous and induced scattering in equilibrium plasma γ_j^{sp}, γ_j^{ind} comprise the probability $w_\alpha(\boldsymbol{k}, \boldsymbol{k}')$, (8.24), of scattering of a wave $\boldsymbol{k}$ into the wave $\boldsymbol{k}'$ by particles from the unit volume of the medium (see (8.19) and (8.23)). For the case of scattering by the screened ions the corresponding probability $w_\alpha(\boldsymbol{k}, \boldsymbol{k}') = w_i(\boldsymbol{k}, \boldsymbol{k}')$ can be easily found by integrating (8.51) and (8.52) over the Maxwell distribution function $f_0(\boldsymbol{p})$. The probability $w_i(\boldsymbol{k}, \boldsymbol{k}')$ then takes the form of (8.25) with

$$w_\alpha^* \equiv w_{ill'}^* = \frac{\sqrt{2\pi}}{4} \frac{\omega_L^2}{N v_{T_i}} \cos^2\theta_1 |2 + i\sqrt{\pi} Z_i w(Z_i)|^{-2} \tag{8.53}$$

for the case of plasma wave to plasma wave scattering and

$$w_\alpha^* \equiv w_{ilt'}^* = \frac{\sqrt{2\pi}}{4} \frac{\omega_L^2}{N v_{T_i}} \sin^2\theta_1 |2 + i\sqrt{\pi} Z_i w(Z_i)|^{-2} \tag{8.54}$$

for the scattering of plasma waves into electromagnetic ones. If great accuracy is not needed, one may put $|2 + i\sqrt{\pi} Z_i w(Z_i)|^{-2} \sim 1/4$ in the expressions (8.53), (8.54). The error of about a factor of the order unity is compensated by the simplicity of the resulting expressions for w_α^*. Such an approximation is used quite frequently. In particular, the formula (8.29) is valid in this approximation.

Exact extinction coefficients $\mu_j^{\rm sp} = 2\gamma_j^{\rm sp} v_{\rm gr}^{-1}$ can be found by integrating $w_{\rm i}(\boldsymbol{k}, \boldsymbol{k}')$ over $\boldsymbol{k}'$ (see (8.19) and (8.24)). The resulting coefficient for the process $l \to l'$ looks like

$$\mu_j^{\rm sp} = 2\pi e^4 N / 27 m_{\rm e}^2 v_T^4, \tag{8.55}$$

and the one for the process $l \to t'$ has the form

$$\mu_j^{\rm sp} = 4\pi e^4 N / 3\sqrt{3} m_{\rm e}^2 c^3 v_T. \tag{8.56}$$

The ratio of these coefficients equals $c^3/6\sqrt{3} v_T^3$, much greater than unity in non-relativistic plasma.

If electron density perturbations in the Debye sphere of the radius D are treated as fluctuations near the equilibrium value which are caused by the thermal motion of plasma particles, an alternative description of scattering by screened ions in plasma may be proposed [283]. An electron density perturbation $\delta N(\boldsymbol{r}, t)$ may be presented in the following form: $\delta N = \delta N' + \delta N''$. The first term just follows the ion density fluctuations: $\delta N' = \delta N_{\rm i}$. This term is related to the quasi-neutral fluctuations of plasma density $\delta\rho \approx m_{\rm i} \delta N$ on scales larger than D. The second term accounts for electron density variations resulting in the space charge $\delta\rho_{\rm e} = -e\delta N''$. It is not difficult to relate the term $\delta N'$ to the space charge clouds which screen ions. Therefore, the scattering by these fluctuations is equivalent to the scattering by screened ions. On the other hand, the long-wavelength ($kD \ll 1$) part of the fluctuation spectra, $\delta N''$, is nothing but the ensemble of plasma waves of thermal origin.

Frequency spectrum of scattered radiation is determined by the form of temporal evolution of $\delta N'$ and $\delta N''$. Since density variations $\delta\rho$ are slow compared to the period $2\pi/\omega$ of an infalling wave with the frequency $\omega > \omega_{\rm L}$, the scattering of the latter by perturbations $\delta N'$ does not lead to the significant frequency variation: $|\omega - \omega'| \lesssim k v_{T_{\rm i}}$ (see (8.3)). Such a process is called Rayleigh scattering. The spectrum of $\delta N''$ has the sharp maximum at frequencies $\omega \approx \omega_{\rm L}$ typical for plasma waves. The resulting frequency variation in a scattering event may be quite large: $|\omega - \omega'| \approx \omega_{\rm L}$. This is the case of Raman scattering. The effects due to wave–wave scattering are postponed to the following section 8.4. Now we consider the Rayleigh scattering of plasma waves into electromagnetic radiation.

Since plasma dielectric constant ϵ depends on electron density N, the fluctuations $\delta N'$ caused by slow density variations $\delta\rho$ lead to the variations $\delta\epsilon$. Bearing in mind the slow time dependence of the latter, we neglect such a dependence and put $\delta\epsilon = \delta\epsilon(\boldsymbol{r})$. A wave $\boldsymbol{E} = \boldsymbol{E}_0 \cos(\omega t - \boldsymbol{k}\boldsymbol{r})$ propagating in the medium induces the additional polarization $\delta \boldsymbol{P} = \delta\epsilon \boldsymbol{E}/4\pi$ (we note that $\boldsymbol{P} = (\epsilon - 1)\boldsymbol{E}/(4\pi)$, see (2.5), (2.12)). The induced polarization plays the role of as an external source which emits both electromagnetic and plasma waves. Thus, the scattering of waves by the fluctuations $\delta N'$ is accompanied by wave conversion from one type to another — fact which is quite important for plasma generation mechanisms.

Let us find the radiation power scattered by a volume V of the linear size L in the range $D \ll L \ll \lambda/2\pi$, where λ is the wavelength of a propagating plasma wave. Due to the inequality $L \ll \lambda/2\pi$ the radiation generated by external polarization $\delta\boldsymbol{P} = \delta\epsilon\boldsymbol{E}/4\pi$ can be considered dipole. As is well known [241], the power of electromagnetic dipole radiation equals $\mathcal{P} = (n_{\text{el}}\omega^4/3c^3)\boldsymbol{d}_0^2$, where $\boldsymbol{d}_0$ is the amplitude of the dipole moment $\boldsymbol{d}$ varying in time with frequency ω. In our case ω is just the frequency of the propagating wave and $\boldsymbol{d} = \int_V \delta\boldsymbol{P}\, dV$. Thus, the mean radiation power is given by

$$\mathcal{P} = \frac{n_{\text{el}}(\omega)\omega^4 E_0^2 V^2}{48\pi^2 c^3}\overline{(\delta\epsilon)_V^2}. \tag{8.57}$$

Here $(\delta\epsilon)_V = V^{-1}\int_V \delta\epsilon\, dV$ and $\overline{(\delta\epsilon)_V^2}$ is the mean square value of dielectric constant fluctuations in the volume V. Since in plasma $\epsilon - 1 \propto N \propto \rho$, the following relations hold:

$$(\delta\epsilon)_V = \frac{\partial\epsilon}{\partial\rho}(\delta\rho)_V = \frac{\epsilon - 1}{\rho}(\delta\rho)_V = \frac{4\pi e^2}{m_e\omega^2}(\delta N_{\text{i}})_V.$$

Thus,

$$\mathcal{P} = \frac{n_{\text{el}}(\omega)e^4 E_0^2 V^2}{3m_e^2 c^3}\overline{(\delta N_{\text{i}})_V^2}. \tag{8.58}$$

The theory of fluctuations in equilibrium plasma (see, for example, [72]) yields $\overline{(\delta N_{\text{i}})_V^2} = N_{\text{i}}/2V$ for the mean square value of ion density fluctuations in the volume $V \gg D^3$. This value is different from the corresponding one, $\overline{(\delta N)_V^2} = N/V$, for a system of non-interacting ("free") particles. Such a difference seems to be quite natural, since plasma ions are correlated due to the effect of Debye screening.

Substituting $\overline{(\delta N_{\text{i}})_V^2} = N_{\text{i}}/2V$ into (8.58), we get [283]

$$\mathcal{P} = \frac{n_{\text{el}}(\omega)e^4 E_0^2 V N_{\text{i}}}{6m_e^2 c^3}. \tag{8.59}$$

Since fluctuations $\delta\epsilon$ and δN_{i} in different volumes $V \gg D^3$ are non-correlated, the powers of their dipole radiation should be summed. This means that the formula (8.59) is valid for any $V \gg D$ up to $V \gtrsim \lambda^3$.

Introduce now the total scattering cross-section $\sigma_{\rm tot}$. This is defined as the ratio of the power scattered by a single particle over all the frequencies and directions to the intensity of the infalling wave. The formula (8.59) and expressions (2.110), (2.119) for radiation intensity $I_{\mathbf{k}}$ then yield the total cross-sections

$$\sigma_{\rm tot} \equiv \sigma_{itt'} = \frac{1}{2}\sigma_{\rm T} \tag{8.60}$$

$$\sigma_{\rm tot} \equiv \sigma_{ilt'} = \frac{1}{2}\sigma_{\rm T}\frac{c}{\sqrt{3}v_T} \tag{8.61}$$

for scattering of electromagnetic and plasma wave, respectively, into an electromagnetic one by screened ions. The value

$$\sigma_{\rm T} = (8\pi/3)(e^2/m_{\rm e}c^2)^2 \tag{8.62}$$

denotes the so-called Thomson scattering which characterizes the scattering of electromagnetic waves by free electrons in vacuum in the absence of a constant magnetic field $\boldsymbol{B}_0$.

Note that dipole emission of plasma waves in plasma is $0.1(c^3/v_T^3)$ times more effective that of the electromagnetic waves [405]. Thus, the same factor describes the ratio of the cross-sections $\sigma_{itl'}$ and $\sigma_{ill'}$ of scattering processes $t \to l'$ and $l \to l'$ to the corresponding ones $\sigma_{itt'}$, (8.60), and $\sigma_{ilt'}$, (8.61) [406].

When dealing with plasma mechanisms of radio emission the quantity Q, the coefficient of the transformation of plasma waves into electromagnetic ones is frequently used [3]. By definition, Q equals the ratio of power of electromagnetic radiation scattered by a unit volume to plasma wave intensity $I_{\mathbf{k}}$. This definition and the one for the total cross-section $\sigma_{\rm tot}$ clearly reveal that $Q = \sigma_{ilt'}N_{\rm i}$, where $\sigma_{ilt'}$ is determined by the formula (18.59). The coefficient of transformation Q is the particular form of the coefficient of spontaneous scattering

$$\mu_j^{\rm sp} = \int \sigma_{\rm tot} f_0(\boldsymbol{p}) d^3p. \tag{8.63}$$

According to (8.60)-(8.62), this coefficient for electromagnetic to electromagnetic waves scattering is given by

$$\mu_j^{\rm sp} = \frac{4\pi e^4 N_{\rm i}}{3m_{\rm e}^2c^4} \tag{8.64}$$

in plasma with $kD \ll 1$ and by

$$\mu_j^{\rm sp} = \frac{8\pi e^4 N_{\rm i}}{3m_{\rm e}^2c^4} \tag{8.65}$$

at $kD \gg 1$. The corresponding coefficient for plasma to electromagnetic wave scattering equals

$$\mu_j^{\rm sp} = \frac{4\pi e^4 N_{\rm i}}{3\sqrt{3} m_{\rm e}^2 c^3 v_T}. \qquad (8.66)$$

The last expression is identical to the formula (8.56).[7] Of course, this is not surprising, since the coefficient of spontaneous scattering (8.63) has the same meaning as the above defined extinction coefficient

$$\mu_j^{\rm sp} = 2\gamma_j^{\rm sp} v_{\rm gr}^{-1} = v_{\rm gr}^{-1} \int w f_0(\boldsymbol{p}) \frac{d^3p d^3k'}{(2\pi)^3}. \qquad (8.67)$$

Note, finally, that the total cross-section $\sigma_{\rm tot}$ is related to the scattering probability w by the formula

$$\sigma_{\rm tot} = \int v_{\rm gr}^{-1} w(\boldsymbol{p}, \boldsymbol{k}, \boldsymbol{k}') \frac{d^3k'}{(2\pi)^3}. \qquad (8.68)$$

The integrand is called the differential scattering cross-section. This value characterizes the scattering into the unit volume of the phase space of the wave vectors $\boldsymbol{k}'$.

8.1.6. SCATTERING BY FREE PARTICLES. INVERSE COMPTON EFFECT

It has already been mentioned in the previous section that the Debye screening in plasma is negligible for scattering of short-wavelength radiation $\lambda \ll D$ or for scattering by fast particles with $v \gg v_T$. Under these conditions a scattering particle can be treated as "free": its motion is not correlated with the motion of other particles in plasma. The scattering by free particles is non-coherent, and the total radiation power $\mathcal{P}$ by a volume V is the sum of the corresponding powers of scattering by distinct particles.

The energy of photons scattered by free particles with arbitrary (including relativistic) energies in vacuum,

$$\mathcal{E}_\gamma \equiv \hbar\omega' = \frac{\epsilon(1 - \beta\cos\theta)}{1 - \beta\cos\theta_2 + \epsilon\mathcal{E}^{-1}(1 - \cos\theta_1)}, \qquad (8.69)$$

can be found from the conservation laws (8.1),(8.2). Here θ is the angle between $\boldsymbol{k}$ and $\boldsymbol{p}$; θ_1 is the angle between $\boldsymbol{k}$ and $\boldsymbol{k}'$; θ_2 is the one between $\boldsymbol{k}'$ and $\boldsymbol{p}$; $\mathcal{E}$, $\boldsymbol{p}$, and $v = \beta c$ denote the initial energy, momentum and velocity of the electron, respectively; and, finally, $\epsilon = \hbar\omega$ is the energy of the infalling photon.

[7] The plasma considered is assumed to be composed of "single-charged" ions.

If a photon with $\epsilon \ll mc^2$ is scattered by an electron at rest ($v=0$), the energy $\mathcal{E}_\gamma$ is close to ϵ. Then the scattering cross-section can be determined in the framework of classical theory. It is taken into account in this case that the electron undergoes the forced oscillations $\boldsymbol{r} = -e\boldsymbol{E}/m_e\omega^2$ in the field of the infalling wave $\boldsymbol{E} = \boldsymbol{E}_0 \cos(\omega t - \boldsymbol{kr})$ and emits as a dipole with the moment $\boldsymbol{d} = e\boldsymbol{r}$. The power of dipole radiation into the unit solid angle Ω' in vacuum is equal to [241]

$$\mathcal{P}_{\Omega'} = \frac{c}{8\pi} E_0^2 \left(\frac{e^2}{m_e c^2} \right)^2 \sin^2 \Theta, \tag{8.70}$$

where Θ is the angle between $\boldsymbol{k}'$ and $\boldsymbol{E}$. Therefore, the differential scattering cross-section per unit solid angle Ω' equals

$$\sigma_{\Omega'} \equiv \frac{\mathcal{P}_{\Omega'}}{(c/8\pi)E_0^2} = \left(\frac{e^2}{m_e c^2} \right)^2 \sin^2 \Theta. \tag{8.71}$$

If radiation infalling an electron is unpolarized (this assumption will be used thereafter), the cross-section is obtained by the averaging of (8.71) over all directions of $\boldsymbol{E} \perp \boldsymbol{k}$:

$$\sigma_{\Omega'} = \frac{1}{2} \left(\frac{e^2}{m_e c^2} \right)^2 (1 + \cos^2 \theta_1). \tag{8.72}$$

The total cross-section is equal to Thomson one (see (8.62)):

$$\sigma_{\text{tot}} = \int \sigma_{\Omega'}\, d\Omega' = \frac{8\pi}{3} \left(\frac{e^2}{m_e c^2} \right)^2 = \sigma_{\text{T}}. \tag{8.73}$$

(see (8.62))

The process of scattering of "soft" photons by energetic (relativistic) electrons ($\mathcal{E} \gg \epsilon = \hbar\omega$) is called "inverse" Compton effect, contrary to the usual Compton effect which corresponds to the scattering of "hard" photons by low-energetic electrons ($\mathcal{E} \ll \epsilon$). We consider the inverse Compton scattering[8] in the Thomson limit when the energy of infalling photon $\tilde{\epsilon} \ll m_e c^2$ in the proper frame of a relativistic electron. Since $\tilde{\epsilon} = \epsilon(1 - \beta\cos\theta)(1 - \beta^2)^{-1/2}$ (see (1.74)), the criterion of the Thomson limit is expressed in the following way through the value measured in a laboratory frame:

$$2\epsilon \frac{\mathcal{E}}{m_e c^2} \ll m_e c^2. \tag{8.74}$$

[8] This process is often called "Compton radiation"

As follows from the formula (8.69), the scattered photon energies range within the interval

$$\frac{1}{4}\epsilon\left(\frac{m_e c^2}{\mathcal{E}}\right)^2 \leq \mathcal{E}_\gamma \leq 4\epsilon\left(\frac{\mathcal{E}}{m_e c^2}\right)^2. \tag{8.75}$$

The maximal energy $\mathcal{E}_\gamma$ is achieved in case of counter collision of photon with the electron and subsequent scattering of the photon in the direction of initial electron momentum $\boldsymbol{p}$, i.e., for $\theta = \theta_1 = \pi$, $\theta_2 = 0$. The minimal energy of the scattered photon corresponds to $\theta = 0$, $\theta_1 = \theta_2 = \pi$. Since the scattering electron is relativistic, the scattered radiation occupies the narrow angular interval $\sim m_e c^2/\mathcal{E}$ along $\boldsymbol{p}$. This can be easily understood from the considerations similar to those concerning the width of synchrotron radiation pattern which has been given in section 5.1. Evidently, the maximal energy of scattered photons $\mathcal{E}_\gamma \gg \epsilon$ is gained due to the loss of electron energy. At the same time, (8.74), (8.75) mean that the electron loses just a small part of its energy in each scattering: $\mathcal{E}_\gamma \ll \mathcal{E}$.

Though the limit $\tilde{\epsilon} \gg m_e c^2$ can be realized as well in cosmic conditions (for example, the scattering of optical and X-ray photons by high energy electrons in the Crab nebula, quasars etc., for more detail, refer to [208]), the opposite limit, $\tilde{\epsilon} \ll m_e c^2$, is the most important for astrophysical applications. Note that the energy losses of an electron in the case $\tilde{\epsilon} \gg m_e c^2$ are catastrophic: the energy $\mathcal{E}_\gamma \sim \mathcal{E}$ is transferred to radiation in every scattering event.

To understand the following, one should bear in mind that, generally, there are three different physical values usually called scattering cross-sections: the ratio of scattered radiation power to the intensity of infalling radiation $\sigma^W_{\Omega'}$; the ratio of number of scattered photons to the infalling photon flux density $\sigma^N_{\Omega'}$; and, finally, the ratio of number of scattered photons to the number of photons infalling an electron $\sigma_{\Omega'}$. The subscript Ω' indicated that the scattering into the unit solid angle is considered in the above definitions. In other words, the above definitions introduce differential cross-sections which relate to each other in the following way:

$$\frac{\omega}{\omega'}\sigma^W_{\Omega'} = \sigma^N_{\Omega'} = (1 - \beta\cos\theta)\sigma_{\Omega'}. \tag{8.76}$$

The factor ω/ω' accounts for the change of frequency due to scattering, and $(1 - \beta\cos\theta)$ describes the difference between the photon flux density $cN_{\mathbf{k}}$ in a laboratory frame from the corresponding value $(c - v\cos\theta)N_{\mathbf{k}}$ in the electron co-moving frame.[9]

[9]The value $c - v\cos\theta$ is the projection of the relative velocity of the photon and electron in the laboratory frame onto the photon momentum $\hbar\mathbf{k}$.

In our above considerations (see (18.67), (8.72)) we have used only one of the defined cross-sections, $\sigma^W_{\Omega'}$. However, for Thomson scattering by a non-relativistic electron ($\omega \approx \omega'$, $\beta \ll 1$) all three definitions are identical. The cross-section $\sigma_{\Omega'}$ is the most convenient to study the inverse Compton scattering by relativistic electrons, since $\sigma_{\Omega'} d\Omega'$ is invariant with respect to the transitions from one inertial frame to another. This can be easily seen by recalling that the photon number in phase space does not depend on the frame and use the above definition for $\sigma_{\Omega'}$. The pointed out invariance and the identity of all three cross-section in the electron proper frame lead to the following relations between these values in laboratory frame:

$$\sigma^N_{\Omega'} d\Omega' = (1 - \beta \cos\theta)\sigma_{\Omega'} d\Omega' = (1 - \beta\cos\theta)\sigma_{\tilde{\Omega}'} d\tilde{\Omega}'. \tag{8.77}$$

Here $\sigma_{\tilde{\Omega}'}$ is the Thomson cross-section in the electron proper frame (see (8.72)). Taking into account also the transformation rule for solid angle element $d\tilde{\Omega}' \to d\Omega'$ which follows from the aberration law (1.73), one can find the cross-section

$$\sigma^N_{\Omega'} = \frac{1}{2}\left(\frac{e^2}{m_e c^2}\right)^2 \frac{(1 - \beta\cos\theta)(1 + \cos^2\tilde{\theta}_1)(1 + \beta\cos\tilde{\theta}_2)}{1 - \beta^2}. \tag{8.78}$$

Here $\tilde{\theta}_1$ is the angle between the wave vectors of infalling and scattered radiation $\tilde{\boldsymbol{k}}$ and $\tilde{\boldsymbol{k}}'$ in the electron co-moving frame, and $\tilde{\theta}_2$ is the angle between $\tilde{\boldsymbol{k}}'$ and $\boldsymbol{p}$, the electron momentum in the laboratory frame.

Assume now for the sake of simplicity that the angular distribution of infalling "soft" photons is isotropic, and their density in the energy interval $d\epsilon$ equals $n_{\rm ph}(\epsilon)d\epsilon$. Then the number of photons scattered by an electron into the energy interval $d\mathcal{E}_\gamma$ is equal to

$$\frac{d\mathcal{E}_\gamma}{4\pi}\int \sigma(\mathcal{E}, \epsilon, \mathcal{E}_\gamma) c n_{\rm ph}(\epsilon) d\epsilon, \tag{8.79}$$

where

$$\sigma(\mathcal{E}, \epsilon, \mathcal{E}_\gamma) = \int \sigma^N_{\Omega'} \delta\left[\mathcal{E}_\gamma - \Phi(\epsilon, \mathcal{E}, \theta, \theta_1, \theta_2)\right] d\Omega d\Omega'. \tag{8.80}$$

The delta-function of $\mathcal{E}_\gamma$ is introduced to account for the fact that the scattered photon with energy $\mathcal{E}_\gamma$ can only occur under certain combination of the parameters ϵ, $\mathcal{E}$, θ, θ_1, θ_2 determined by the conservation laws for a scattering event. The function Φ denotes the expression in the right-hand side of (8.69). Since the energy of scattered photon $\mathcal{E}_\gamma$ enters the argument of the delta-function, it is clear from (8.80) that $\sigma(\mathcal{E}, \epsilon, \mathcal{E}_\gamma)$ is differential cross-section averaged over directions of infalling photons and related to the unit interval of energies $\mathcal{E}_\gamma$. The formulas (8.69), (8.78) and (8.80) under the condition (8.74) and in the region $\epsilon \le \mathcal{E}_\gamma \le 4\epsilon(\mathcal{E}/m_e c^2)^2$ yield [407]

$$\sigma(\mathcal{E},\epsilon,\mathcal{E}_\gamma) = \tag{8.81}$$
$$\frac{3}{32}\sigma_T \frac{(m_e c^2)^4}{\epsilon^2 \mathcal{E}^3}\left\{2\frac{\mathcal{E}_\gamma}{\mathcal{E}} - \frac{(m_e c^2)^2 \mathcal{E}_\gamma^2}{\epsilon \mathcal{E}^3} + \frac{8\epsilon\mathcal{E}}{(m_e c^2)^2} + 4\frac{\mathcal{E}_\gamma}{\mathcal{E}}\ln\frac{(m_e c^2)^2 \mathcal{E}_\gamma}{4\epsilon\mathcal{E}^2}\right\}.$$

The total scattering cross-section is equal to the Thomson one:

$$\sigma_{tot} = \int \sigma(\mathcal{E},\epsilon,\mathcal{E}_\gamma)d\mathcal{E}_\gamma = \sigma_T. \tag{8.82}$$

Next, the average inverse Compton energy loss by an electron is given by

$$-\left(\frac{d\mathcal{E}}{dt}\right)_C = c\int \sigma(\mathcal{E},\epsilon,\mathcal{E}_\gamma) n_{ph}(\omega)\, d\mathcal{E}_\gamma d\epsilon = cN_{ph}\sigma_T \cdot \frac{4}{3}\bar{\epsilon}(\mathcal{E}/m_e c^2)^2. \tag{8.83}$$

Here $\bar{\epsilon}$ denotes the average photon energy and N_{ph} is the number of photons in the unit volume:

$$N_{ph} = \int n_{ph}(\epsilon)d\epsilon, \qquad \bar{\epsilon} = \frac{1}{N_{ph}}\int \epsilon n_{ph}(\epsilon)d\epsilon. \tag{8.84}$$

It is taken into account in (8.83) that an electron loses the energy $\Delta\mathcal{E} \approx \mathcal{E}_\gamma$ in every single scattering. On the other hand, electron energy losses per unit time may be expressed in terms of the mean energy of scattered quanta $\overline{\mathcal{E}}$ as $-(d\mathcal{E}/dt)_C = cN_{ph}\sigma_{tot}\overline{\mathcal{E}}_\gamma$. Taking into account (8.82) and (8.83) we find that

$$\overline{\mathcal{E}} = \frac{4}{3}\bar{\epsilon}\left(\frac{\mathcal{E}}{m_e c^2}\right)^2. \tag{8.85}$$

It is clear now that for arbitrary electron energy $\mathcal{E}$ the characteristic frequency of Compton scattering $\mathcal{E}_\gamma/\hbar$ is identical to the characteristic frequency of synchrotron radiation ω_{max} (see (5.80)) under the condition $8\bar{\epsilon}/3 = \hbar\omega_B$. At the same time, Compton energy loss rate (8.83) depends on $\mathcal{E}$ in a similar way as the synchrotron one (5.99). These values become equal for comparable energy densities of "soft" photons $W_{ph} = \bar{\epsilon}N_{ph}$ and static magnetic field $W_B = B_0^2/8\pi$ (to be more correct, at $2W_{ph}/3 = W_B$).

The Compton emissivity

$$a_\omega = \mathcal{E}_\gamma \int \sigma(\mathcal{E},\epsilon,\mathcal{E}_\gamma) c n_{ph}(\epsilon) N(\mathcal{E}) d\epsilon d\mathcal{E} \tag{8.86}$$

allows judgement on the frequency spectrum of Compton scattering by electrons. In particular, for the power-law energy spectrum of electrons

$N(\mathcal{E}) = K\mathcal{E}^{-\gamma}$, (5.184), the integral [407]

$$\int_{\mathcal{E}_{\min}}^{\infty} \sigma(\mathcal{E}, \epsilon, \mathcal{E}_\gamma) N(\mathcal{E}) d\mathcal{E} = \tag{8.87}$$

$$\frac{2^\gamma(\gamma^2+4\gamma+11)}{(\gamma+1)(\gamma+3)^2(\gamma+5)} 8\pi \left(\frac{e^2}{m_e c^2}\right)^2 \mathcal{K}(m_e c^2)^{1-\gamma} \epsilon^{\frac{\gamma-1}{2}} \mathcal{E}_\gamma^{-\frac{\gamma+1}{2}}.$$

The lower limit $\mathcal{E}_{\min} = m_e c^2 (\mathcal{E}_\gamma / 4\epsilon)^{1/2}$ is determined by the condition (8.75). The spectrum of equilibrium radiation is described by the Planck formula [55]

$$n_{\text{ph}}(\epsilon) = \frac{1}{\pi^2(\hbar c)^3} \frac{\epsilon^2}{e^{\epsilon/\kappa T} - 1}, \qquad \bar{\epsilon} \approx 2.7\,\kappa T. \tag{8.88}$$

The Compton emissivity in this case is determined by [407]

$$a_\omega = c(\gamma) \frac{2}{3} \sigma_{\text{T}} c N_{\text{ph}} \bar{\epsilon} (m_e c^2)^{1-\gamma} \left(\frac{4}{3}\bar{\epsilon}\right)^{\frac{\gamma-3}{2}} \mathcal{K} \mathcal{E}_\gamma^{-\frac{\gamma-1}{2}} \tag{8.89}$$

where

$$c(\gamma) = 4.74(1.05)^\gamma \frac{\gamma^2+4\gamma+11}{(\gamma+1)(\gamma+3)^2(\gamma+5)} \Gamma\left(\frac{\gamma+5}{2}\right) \zeta\left(\frac{\gamma+5}{2}\right), \tag{8.90}$$

$\Gamma(x)$ is gamma-function, $\zeta(x) = \sum_{n=1}^{\infty} n^{-x}$ is Riemann function. The factor (8.90) equals 0.84, 0.86, 0.99, 1.4 for $\gamma = 1, 2, 3, 4$, respectively. The formula (8.89) holds, with the accuracy of the factor of about unity, also for non-equilibrium distributions of "soft" photons with the same mean energy $\bar{\epsilon}$.

The formula (8.89) reveals that the spectral index $\alpha = (\gamma - 1)/2$ of Compton emission spectrum $a_\omega \propto \mathcal{E}_\gamma^{-\alpha} \propto \omega^{-\alpha}$ is identical to the spectral index of synchrotron radiation by electrons with the same spectral index γ (see for comparison the formulas (5.185), (5.186)). The emissivities at a given frequency ω are of the same order if $W_{\text{ph}} \sim W_B$.

Of course, the noted analogy of inverse Compton scattering and synchrotron radiation, both in spectral shapes and in energy losses by relativistic particles, is not accidental [208]. The radiation field of a classical charged particle is determined by its acceleration, i.e., the force $\boldsymbol{f} = e\boldsymbol{E} + (e/c)[\boldsymbol{vB}]$ acting on it (see, for example, [56]). The force by the field of an electromagnetic wave in vacuum $f \sim eE$, and the one by a static magnetic field $f \sim (e/c)vB_0$. These forces become equal for a relativistic particle $v \approx c$ at equal energy densities of electromagnetic field and static magnetic field. This ensures the "equivalence" of Compton scattering and synchrotron emission.

The inverse Compton scattering is usually applied to astrophysical plasmas when dealing with the problem of the origin of cosmic X-ray or gamma-ray radiation. It is quite possible that the major part of diffuse X-ray or gamma-ray radiation (the X-ray or gamma-ray background) is generated via scattering of the equilibrium relict radiation with temperature $T \approx 2.7\,\mathrm{K}$ by relativistic electrons in the Galaxy and intergalactic space [272]. One can not also exclude the possibility that the inverse Compton effect in the Crab pulsar PSR 0532, namely, the scattering of optical and soft X-ray photons by relativistic electrons, plays an important role in generating the observed gamma-radiation of this source [68]. More details on inverse Compton scattering and its astrophysical applications can be found in [208,250,407–409]. Note that Compton scattering of plasma waves with subsequent conversion into electromagnetic radiation is considered in [387,388], the induced Compton scattering is studied in [410–412] and many other publications.

It should be emphasized that the above derivation of the Compton cross-section has been essentially based on the possibility to find a frame where the electron motion is non-relativistic: $\tilde{v} \ll c$. In such a system which was called "electron proper frame" the velocity of the forced motion $\tilde{v} \sim e\tilde{E}/m_e\tilde{\omega}$. Then the low limit for the frequency in case of the inverse Compton effect appears: $\tilde{\omega} \gg e\tilde{E}/m_e c$. Since the ratio W/ω^2, where $W = E^2/8\pi$ is energy density in the electromagnetic wave, is invariant (see (2.109) and [56]), the Lorenz-invariant criterion for inverse Compton effect has the following form:

$$\delta = \frac{e\sqrt{8\pi W}}{m_e c\omega} \ll 1. \tag{8.91}$$

The emission of a particle moving in the wave field of arbitrary amplitude (or, in other words, at arbitrary value of the parameter δ) is called synchro-Compton. This radiation is reduced to the Compton one in the limiting case $\delta \ll 1$ and exhibits the features of synchrotron radiation in the opposite case $\delta \gg 1$. The latter is especially clear for a particle in the wave field $E \ll B$, for example, in an Alfven wave. In this case $W = B^2/8\pi$ and $\delta = eB/m_e c\omega = \omega/\omega_B$. Here the criterion $\delta \gg 1$ means that relativistic particles emit in the same way as in a constant magnetic field: the field of the wave does not vary much over the time $1/\omega_B$ when an observer is within the radiation pattern (see section 5.1).

Synchro-Compton emission by a system of electrons with a power-law energy spectrum is considered in [413].

The interest in synchro-Compton radiation in astrophysics has arisen in connection with the problem of pulsar radio emission. It has been assumed that such an emission occurs due to the motion of relativistic electrons in a low-frequency electric field generated by the oblique magnetic rotator (the neutron star with the magnetic moment oblique with respect to the rotation axis) [414]. Synchro-Compton radiation has been applied to interpret the radio emission by the Crab nebula and some extragalactic radio sources [415,416].

8.1.7. KOMPANEETS DIFFERENTIAL EQUATION AND "COMPTONIZATION" OF RADIATION

At the beginning of this section the integral kinetic equation (8.14) has been written which describes variations of frequency and angular spectrum of radiation due to its scattering, spontaneous or induced, by particles with a definite distribution function over velocities. The solution of this complex equation in only possible under specific assumptions on the form of angular spectrum of radiation. The analysis of this equation for a one-dimensional angular spectrum of radiation has been given above. However, the frequency shift due to Thomson (Compton) scattering in non-relativistic plasma is relatively small: $\omega - \omega' \ll \omega$, one may suspect that temporal variation of photon density at frequency ω, or the derivative $\partial N_{\mathbf{k}}/\partial t$ depends only on the values of $N_{\mathbf{k}}$ at neighboring frequencies. This means that the above derivative can be approximately related to $N_{\mathbf{k}}$ and the derivatives $\partial N_{\mathbf{k}}/\partial\omega$, $\partial^2 N_{\mathbf{k}}/\partial\omega^2$, etc. at the same frequency ω.

The initial integral equation

$$\frac{\partial N_{\mathbf{k}}}{\partial t} = \int N_{\mathbf{k}}(N_{\mathbf{k}}+1)w(\boldsymbol{p}',\boldsymbol{k}',\boldsymbol{k})f_0(\boldsymbol{p}')\frac{d^3p'd^3k'}{(2\pi)^6} - \int N_{\mathbf{k}}(N_{\mathbf{k}'}+1)w(\boldsymbol{p},\boldsymbol{k},\boldsymbol{k}')f_0(\boldsymbol{p})\frac{d^3pd^3k'}{(2\pi)^6} \tag{8.92}$$

which follows from (8.12) and (8.13) can be reduced to the differential one by expanding the integrands in the right-hand side in powers of small quantity $\omega - \omega'$ and keeping the small terms up to $(\omega-\omega')^2$. Let the distribution $f_0(\boldsymbol{p})$ of scattering electrons be equilibrium (Maxwellian), and radiation, isotropic and unpolarized. Performing the above described procedure, we arrive at the so-called Kompaneets equation [393]:[10]

$$\frac{\partial N_{\mathbf{k}}}{\partial t} = N\sigma_{\mathrm{T}}\frac{\hbar}{m_e c}\frac{1}{\omega^2}\frac{\partial}{\partial\omega}\left[\omega^4\left(\frac{\kappa T}{\hbar}\frac{\partial N_{\mathbf{k}}}{\partial\omega} + N_{\mathbf{k}} + N_{\mathbf{k}}^2\right)\right]. \tag{8.93}$$

Here $N_{\mathbf{k}}$ is photon number per mode, N and T are electron density and temperature, respectively. Stationary solution $\partial N_{\mathbf{k}}/\partial t = 0$ of the obtained equation corresponds to the equilibrium (Planck) frequency spectrum of radiation (8.6) and is realized under thermodynamic equilibrium between

[10] Later on this equation is derived in [417,418].

particles and radiation. Using the dimensionless variables

$$x = \frac{\hbar\omega}{\kappa T}, \qquad y = u\frac{\kappa T}{m_e c^2}, \qquad u = N c \sigma_T t \tag{8.94}$$

and denoting, for simplicity, $N_{\mathbf{k}} \equiv n$,[11] we get the Kompaneets equation in the following form:

$$\frac{\partial n}{\partial y} = \frac{1}{x^2}\frac{\partial}{\partial x}\left[x^4\left(\frac{\partial n}{\partial x} + n + n^2\right)\right]. \tag{8.95}$$

No parameters enter this equation.

The equation (8.95) accounts for the deformation of the frequency spectrum of radiation with characteristic photon energy $\hbar\omega \ll m_e c^2$ due to its scattering in non-relativistic plasma ($\kappa T \ll m_e c^2$). Photon energy is changed due to recoil and Doppler effect in every single scattering. If $\kappa T \gg \hbar\omega$, the average photon energy increases due to Doppler effect by the quantity $\hbar\omega\kappa T/m_e c^2$. If, on the contrary, the opposite inequality is satisfied: $\kappa T \ll \hbar\omega$, the average photon energy decreases due to the recoil effect by $(\hbar\omega)^2/m_e c^2$. Spectral evolution of non-equilibrium radiation due to multiple scattering in an equilibrium plasma is usually called "comptonization". This effect has been extensively discussed in [412,419–423,425] etc. on the basis of the Kompaneets equation (8.95).

The process of comptonization plays an important role in the interaction of astrophysical plasmas with radiation. The Kompaneets equation is especially convenient for the analysis of temporal problems relevant to the cases when parameters of homogeneous plasma and homogeneous, isotropic radiation are coordinate-independent. Comptonization also plays an essential role in the formation of the spectra of compact X-ray sources: accretion disks around black holes, neutron stars or white dwarfs, supernova envelopes. The use of the Kompaneets equation for bounded plasma configurations is approved by the large optical depth of these sources in Thomson scattering.

Turning to the analysis of the equation (8.95) we note, first of all, that the non-linear term comprising n^2 accounts for the induced scattering. This effect can be neglected in the Wien domain of an equilibrium spectrum or if radiation intensity is considerably less than the equilibrium one. Then the following equation can be considered instead of (8.95):

$$\frac{\partial n}{\partial y} = \frac{1}{x^2}\frac{\partial}{\partial x}\left[\left(x^4\frac{\partial n}{\partial x} + n\right)\right]. \tag{8.96}$$

With this equation it is not difficult to find e-folding time for the photon energy $\hbar\omega \ll \kappa T$ in the course of comptonization. Multiplying (8.96) by x^3

[11] This quantity is related to the radiation energy density $W_{\mathbf{k}}$ by the formula (7.29).

and integrating over x from 0 to ∞, we get

$$\frac{d}{dy}\int_0^\infty nx^3dx = 4\int_0^\infty nx^3dx - \int_0^\infty nx^4dx. \tag{8.97}$$

At $\hbar\omega \ll \kappa T$, or $x \ll 1$, the second integral in the right-hand side is negligible compared to the first one. Then the solution of the equation (8.97) looks like

$$\hbar\omega = \hbar\omega_0 \exp(t/t_C) \tag{8.98}$$

where

$$t_C = \frac{m_e c^2}{4\kappa T N c\sigma_T} \tag{8.99}$$

is the characteristic time of comptonization.

The term proportional to $\partial n/\partial x$ in the equation (8.96) describes the Doppler shift of the frequency of a scattered photon, and the one comprising n accounts for the frequency variation due to the recoil effect. If the resulting frequency shift is dominated by the Doppler effect, the solution of the corresponding differential equation has the form [426]

$$n(x,y) = \frac{1}{\sqrt{4\pi y}}\int_0^\infty \frac{n(z)}{z}\exp\left[-\frac{(\ln x + 3y - \ln z)^2}{4y}\right]dz. \tag{8.100}$$

It describes temporary evolution of an arbitrary initial spectrum $n(z) \equiv n(\omega, t=0)$. If the initial spectrum is monochromatic, $W_{\mathbf{k}} = A\delta(x - x_0)$, then

$$W_{\mathbf{k}}(x,y) = \frac{A}{\sqrt{4\pi y x_0}}\exp\left[-\frac{(\ln x_0 - \ln x + 3y)^2}{4y}\right]. \tag{8.101}$$

It follows from this solution that a spectral line is broadened with time, and its center is shifted to higher frequencies:

$$\frac{x_{\max}}{x_0} = e^{3y}, \tag{8.102}$$

where $x_{\max}$ corresponds to the maximum of the energy density $W_{\mathbf{k}}$.

If the first term comprising the derivative $\partial n/\partial x$ is neglected in the right-hand side of the equation (8.96), the remaining terms will describe the evolution of the radiation spectrum due to the recoil effect. According to [423,424], the quantity $n\omega^4$ is constant along the characteristics

$$\frac{d\omega}{du} = -\frac{\hbar\omega^2}{m_e c^2}, \qquad du = c\sigma_T N\, dt. \tag{8.103}$$

The equation (8.96) in its general form has been solved by A.S. Kompaneets in [393]. The solution is expressed in terms of Whittaker function of pure imaginary index. This solution implies, in particular, that average photon frequency $\bar{\omega}$ goes to $3\kappa T/\hbar$ regardless the initial value ω_0.

The solution of the Kompaneets equation for a homogeneous infinite medium is insufficient to describe comptonization of radiation in compact X-ray sources and to determine the spectra of outgoing radiation. Such a solution can nevertheless be useful for above mentioned problems if, in addition, the distribution of photon escape times is known.

Consider escape of photons injected by a central source at $r = 0$ into a surrounding homogeneous plasma cloud of the radius R as a result of their multiple scattering. The basic parameter which determines the desired escape time distribution is the optical depth of the cloud to Thomson scattering: $\tau_0 = N\sigma_T R$. The dimensionless time $u = \sigma_T N ct$ introduced by the formula (8.94) characterizes the number of scatterings of a photon before the escape from the cloud. Transfer of radiation in the case $u \gg 1$ is described by a diffusion-type problem, so the distribution of photons over the time of diffusion inside the cloud, $P(t)$, can be found more easily.

Photon diffusion in the cloud is described by the equation

$$\frac{\partial I}{\partial t} = \frac{1}{r^2}\frac{\partial}{\partial r}\left(r^2 D \frac{\partial I}{\partial r}\right), \tag{8.104}$$

in which the diffusion coefficient

$$D = \frac{c}{2\sigma_T N(r)} \tag{8.105}$$

does not depend on frequency ω or the coordinate r in the considered case of Thomson scattering in a homogeneous cloud. The following equality poses the boundary condition for the radiation intensity $I(r,t)$:

$$\left(\frac{\partial I}{\partial t} + \frac{3}{2}\frac{\partial I}{\partial r}\right)_{\tau=\tau_0} = 0. \tag{8.106}$$

It holds that no photons are incident on the cloud boundary $\tau = \tau_0 = N\sigma_T R$ from the outside. The initial condition to the problem considered is given by the relation

$$I\bigg|_{u=0} = \frac{1}{4\pi a^2}\delta(\tau), \tag{8.107}$$

which means that the photon source at the initial moment $u = 0$ is localized at the cloud center where $\tau = N\sigma_T r = 0$.

The solution of the equation (8.104) with the boundary conditions (8.106) and (8.107) has been given in [419].[12] Further analysis follows in

[12]This paper deals also with more complex cases of an inhomogeneous cloud and inhomogeneous distribution of photon sources.

the consideration of this paper. Mean time of photon diffusion in the cloud, or the escape time, equals

$$t_0 = \int_0^\infty tP(t)\,dt = R\tau_0/2c, \tag{8.108}$$

and the mean number of scatterings

$$\bar{u} = \tau_0^2/2. \tag{8.109}$$

The function $P(u)$ is plotted in Fig. 8.1. The maximum $P = P_{\max}$ is reached at $t = 0.3R\tau_0/c$ or $u = 0.3\tau_0^2$. The region $u \gg \bar{u}$ is well described by the asymptotic formula

$$P(u) \simeq \frac{2\pi^2}{3(\tau_0 + 2/3)^2} \exp\left(-\frac{\pi^2 u}{3(\tau_0 + 2/3)^2}\right), \tag{8.110}$$

whereas the approximation

$$P(u) \simeq \frac{3}{4}\sqrt{\frac{3}{\pi}}\frac{\tau_0^2}{u^{3/2}} \exp\left(\frac{-3\tau_0^2}{4u}\right) \tag{8.111}$$

is valid at $1 \ll u \ll \bar{u}$.

Note that $P(u)$ for a homogeneous distribution of radiation sources over a spherical plasma cloud is given by

$$P(u) \simeq \frac{2}{\tau_0^2} \exp\left(-\frac{\pi^2 u}{3\tau_0^2}\right) \tag{8.112}$$

at $u \gg \bar{u} = \tau_0^2/5$ and by

$$P(u) \simeq \frac{1}{\tau_0}\left(\frac{3}{\pi u}\right)^{1/2}\left[1 - \frac{2}{\tau_0}\left(\frac{\pi u}{3}\right)^{1/2}\right] \tag{8.113}$$

at $1 \ll u \ll \bar{u}$. If photon sources are distributed according to the function

$$\varphi(\tau) = \frac{\tau_0}{\pi\tau} \sin\frac{\pi\tau}{\tau_0}, \tag{8.114}$$

which describes an intermediate case between the central source and homogeneous source distribution, then at arbitrary u the function

$$P(u) = \beta e^{-\beta u}, \tag{8.115}$$

where

$$\beta = \frac{\pi^2}{3(\tau_0 + 2/3)^2} \simeq \frac{3}{\tau_0^2}. \tag{8.116}$$

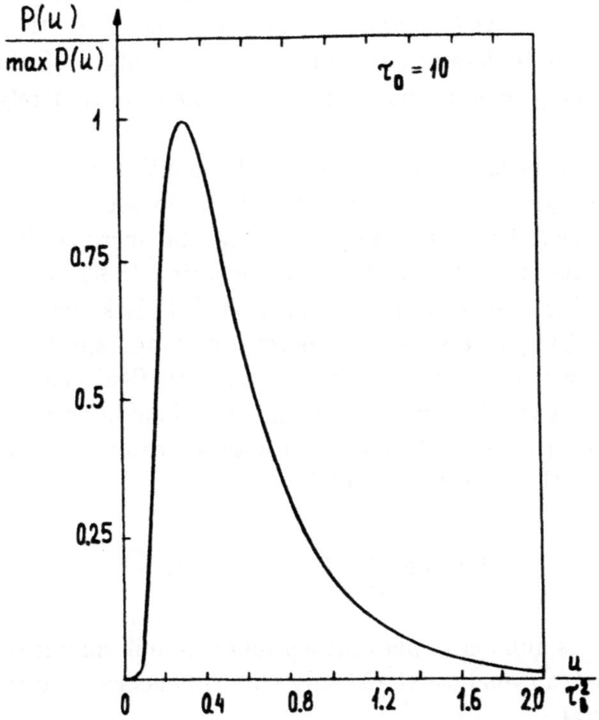

Figure 8.1. **Photon distribution over escape time for a spherical plasma cloud of the optical depth $\tau_0 = 10$. Photons are injected by the central source instantaneously at the initial moment $u = 0$. [419].**

The last equality in the above formula holds at $\tau_0 \gg 1$. The mean number of photon scattering in a cloud for which $P(u)$ is determined by (8.115) equals $\bar{u} = \beta^{-1}$. By comparison of the formulas (8.110), (8.112) and (8.115) it is clear that the behavior of the function $P(u)$ at $u \gg \bar{u}$ is nearly independent of the source distribution over the cloud.

The profile of the function $P(t)$ is of considerable interest for the theory of X-ray bursters. If the characteristic time of energy release in a burst source is small compared to the observed duration of the burst, then the latter can be determined by photon diffusion in a scattering cloud which delays their escape. As a result, the observed time history of a burst $I(t)$ is identical to the profile of $P(t)$. A typical X-ray burst has a sharp rising edge followed by an exponentially decaying profile $I \propto \exp(-t/t_0)$, so the entire burst profile is similar to the curve $P(u)$ in Fig. 8.1. Since

$$t_0 = \frac{3\tau_0^2}{\pi^2 \sigma_T N c} = \frac{3}{\pi^2} \frac{R\tau_0}{c}, \tag{8.117}$$

comparison of this curve with observed profiles of X-ray bursts yields an estimate of the Thomson optical depth τ_0 of the scattering plasma cloud. Thus, using the observed t_0 and the known radius R of the scattering cloud, it is possible to determine τ_0 and electron density N. Unfortunately, for unknown energy release time such an estimate of τ_0 is merely an upper limit.

It has already been pointed out that the solution of a homogeneous Kompaneets equation is insufficient to find the frequency spectrum of a steady-state object. However, the problem can be successfully treated if, in addition to the solution of the linear equation (8.96) which neglects the effect of induced scattering, the function $P(t)$ is known. Indeed, the solution of the Kompaneets equation describes photon spectrum after $u = \sigma_T N ct$ scattering. At the same time, the function $P(u)$ characterizes the number of photons which managed to escape the cloud in a time t. Then the observed spectrum by a scattering cloud with a steady-state source inside is determined by the convolution [427]

$$I_\omega(\omega) \propto \int_0^\infty I_\omega(\omega, t) P(t)\, dt. \tag{8.118}$$

If $N(x) \equiv N_{\mathbf{k}}(x)$ is number of photons per mode in unit volume of the phase space, then one can analyze the steady-state frequency spectrum (8.118) using the integral

$$N(x) = \int_0^\infty n(x, u) P(u)\, du. \tag{8.119}$$

This quantity is related to the radiation intensity by the formula

$$I_\omega(x) = \frac{\hbar\omega^3}{8\pi^3 c^2} N_{\mathbf{k}}(x), \tag{8.120}$$

which follows from (1.34), (7.29). The form of the spectrum $I_\omega(x)$, where $x = \hbar\omega/\kappa T$, is thus similar to the function $x^3 N(x)$:

$$I_\omega(x) = A x^3 N(x). \tag{8.121}$$

Determine the steady-state spectrum (8.119) which corresponds to the escape time distribution (8.115). Multiply the equation (8.96) by $\beta e^{-\beta u}$ and integrate over u from 0 to infinity under the initial condition $n(x, u = 0) = f(x) x^{-3}$ which is set by the spectrum $f(x)$ of a source with spatial distribution (8.114). This leads to the equation

$$\frac{1}{x^2} \frac{d}{dx} \left[x^4 \left(\frac{dN}{dx} + N \right) \right] = \gamma N - \frac{\gamma f(x)}{x^3}, \tag{8.122}$$

in which

$$\gamma = \frac{m_e c^2}{\kappa T}\beta = \frac{\pi^2 m_e c^2}{3\kappa T(\tau_0 + 2/3)^2}. \tag{8.123}$$

If soft photons are scattered in a hot plasma, $x \ll 1$, the second term in the parentheses in the left-hand side of the equation (8.122) can be neglected. This negligible term accounts for the recoil effects, so the resulting equation describes the influence of the Doppler shift of photon frequencies. If photons injected into the cloud have monochromatic spectrum $f(x) = x_0\delta(x - x_0)$, the spectral intensity of the outgoing radiation is given by

$$I_\omega(x) = Ax^3N(x) = A\frac{\alpha(\alpha+3)}{(2\alpha+3)}\left(\frac{x}{x_0}\right)^{\alpha+3}, \qquad 0 \le x \le x_0; \tag{8.124}$$

$$I_\omega(x) = A\frac{\alpha(\alpha+3)}{(2\alpha+3)}\left(\frac{x}{x_0}\right)^{-\alpha}, \qquad x_0 \le x \ll 1. \tag{8.125}$$

The spectral index α is described by the following formula [428]:

$$\alpha = \left(\frac{9}{4} + \gamma\right)^{1/2} - \frac{3}{2}. \tag{8.126}$$

If the injected photons have Planck spectrum with temperature $T_{rad} \ll T_e$, the spectrum of the outgoing radiation in the interval $\kappa T_{rad} \ll \hbar\omega \ll \kappa T_e$ is identical to (8.125) [419]. Thus, comptonization causes the transformation of high-frequency domain of the initial Planck spectrum into a power-law one.

In the opposite case of scattering of hard photons in relatively cold plasma ($x \gg 1$) the leading role is played by the recoil effect. Here one can drop the term dN/dx in the equation (8.122) and obtain the solution for an arbitrary spectrum of source photons:

$$I_\omega(z) = A\frac{\beta}{z}\exp\left(-\frac{\beta}{z}\right)\int_z^\infty f(\xi)\exp\left(\frac{\beta}{\xi}\right)\frac{d\xi}{\xi}. \tag{8.127}$$

Since the comptonized photon spectrum does not depend on the electron temperature in this case, it is convenient to use the variable $z = \hbar\omega/m_e c^2$ instead of x. Comptonization of a monochromatic spectrum $f(z) = z_0\delta(z - z_0)$ yields the following spectrum of the radiation escaping from the cloud:

$$I_\omega(z) = A\frac{\beta}{z}\exp\left(-\frac{\beta}{z} + \frac{\beta}{z_0}\right)\Theta(z_0 - z), \tag{8.128}$$

where

$$\Theta(z_0 - z) = \begin{cases} 0 & z > z_0, \\ 1 & z < z_0, \end{cases} \tag{8.129}$$

and the ratio

$$\frac{\beta}{z} = \frac{\pi^2 m_e c^2}{3(\tau_0 + 2/3)^2 \hbar\omega} \tag{8.130}$$

It is clear from the above formulas that the recoil effect can only lead to the decrease of photon energies.

Comptonization does not change the shape of a power-law spectrum. A source with $f(z) = Cz^{-\alpha}$ is comptonized into the power-law spectrum with different spectral index:

$$I_\omega(z) = \frac{A\beta C}{\alpha} z^{-(\alpha+1)}. \tag{8.131}$$

This formula results from (8.127) at $z \gg \beta$ when the exponential factors in the latter are close to unity. Since the total number of photon in a scattering medium is conserved, the decrease of the intensity $I_\omega(z)$ in the region $z \gg \beta$ is accompanied with its increase at $z \simeq \beta$. The spectrum at small $z \ll \beta$ is nearly unchanged.

The solution of the equation (8.122) for the entire spectrum has been obtained in [419]. In case of a monochromatic source with $f(x) = x_0\delta(x - x_0)$ and $x_0 \ll 1$ the high-frequency wing $x \gg x_0$ of the spectrum of radiation escaping from the plasma cloud is described by the formula

$$I_\omega(x) = A\frac{\alpha(\alpha+3)x_0^\alpha}{\Gamma(2\alpha+4)} x^3 e^{-x} \int_0^\infty \xi^{\alpha-1} e^{-\xi}\left(1 + \frac{\xi}{x}\right)^{\alpha+3} d\xi, \tag{8.132}$$

where Γ denotes the gamma-function and the spectral index α is determined by the formula (8.126). We emphasize that α does not depend on the photon source distribution and is set only by the electron temperature and Thomson optical depth of the cloud. The spectrum (8.132) is reduced to (8.125) at $x \ll 1$, whereas the Wien-type photon distribution

$$I_\omega(x) = A\frac{\alpha\Gamma(\alpha)(\alpha+3)x_0^\alpha}{\Gamma(2\alpha+4)} x^3 e^{-x} \tag{8.133}$$

corresponds to the region $x \gg 1$. The solutions for arbitrary source spectrum $f(x)$ may also be found in [419].

The role of comptonization in the formation of observed X-ray spectra has been demonstrated in [429] for the object Cyg X-1, the Galactic black hole candidate. As seen from the Fig. 8.2, the observed spectrum of Cyg X-1 is fit well by the formula (8.132) at $\alpha \simeq 0.56$ ($\gamma \simeq 2$) and $\kappa T \simeq 27$ keV which corresponds to the optical depth $\tau_0 \simeq 5$ (see (8.123)). Note that the comptonized spectrum formed by the plasma with the relatively low temperature "goes" up to 150 keV. Good agreement between the observed and theoretical spectra implies that comptonization plays the leading role

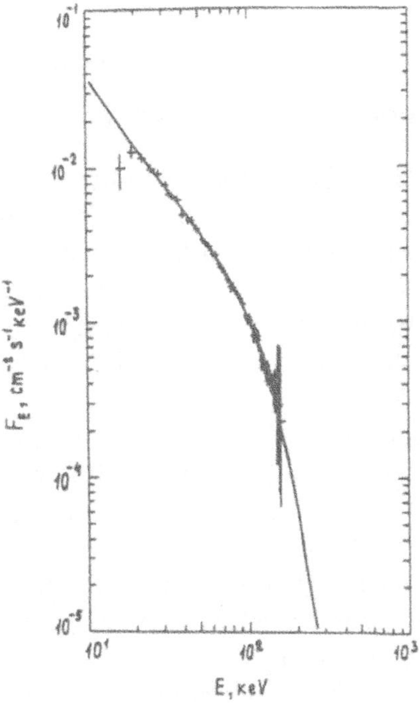

Figure 8.2. Comparison of the observed spectrum (spectral flux of photons F_E versus photon energy E) of Cyg X-1 with the spectrum formed by the comptonization of low-frequency photons in a plasma cloud with $\tau_0 = 5$ and $\kappa T = 27$ keV [429].

in the formation of the spectrum of Cyg X-1. The example considered is not unique. According to [420], the X-ray spectrum of the quasar 3C 273 is well described by the formula (8.132) at $\kappa T = 16\,\mathrm{keV}$ and $\tau_0 = 4$. The spectral fitting in this case has been done taking into account the influence of cosmological redshift and disk geometry of the source.[13] The latter effect influences the relation between γ and α which, instead of (8.126), is now given by

$$\gamma = \frac{\pi^2}{12} \frac{m_e c^2}{\kappa T (\tau_0 + 2/3)^2}. \tag{8.134}$$

All the analysis of this section has been made for non-relativistic plasma. Analytical treatment of the cases of mildly relativistic ($\kappa T \sim m_e c^2$) and relativistic ($\kappa T \gg m_e c^2$) plasma is not effective. These cases are usually

[13]The effect of disk geometry of the source is taken into account by assuming that the scattering plasma region is a plane-parallel layer with the scattering optical depth $2\tau_0$.

treated numerically (as a rule, by using the Monte-Carlo technique [430]). The results of such calculations are summarized in [420].

8.2. Transfer of radiation at cyclotron frequencies

8.2.1. RESONANT CYCLOTRON SCATTERING

Section 8.1 has dealt with the scattering in isotropic plasma. The magnetic field $\boldsymbol{B}_0$ not only changes scattering cross-sections but also gives rise to qualitatively new effects connected with resonant cyclotron scattering: anisotropy of particle distribution function, scattering in geometrically thin gyroresonant layers in an inhomogeneous magnetic field, appearance of strong radiation pressure force on plasma. These effects are considered in this and the next sections.

Bearing in mind the applications of the theory given in the sections 8.2, 8.3 to plasmas in strong magnetic fields of white dwarfs and neutron stars, we assume thereafter that the criterion of rarefied plasma (2.144) and the inequality $kD \ll 1$ are satisfied. Due to the first condition the refractive indices of normal waves are close to unity, while the second one allows us to neglect the Debye screening when calculating scattering cross-sections (see the remarks in section 1.1).

It is known [431] that the cross-section of scattering by an electron in a magnetic field is proportional to

$$\sigma_{\mathrm{T}}\left[\frac{1}{2}+\frac{1}{2}\frac{(\omega')^2}{(\omega'-\omega_B)^2+\Gamma^2/4}\right], \tag{8.135}$$

where σ_{T} is the Thomson cross-section (8.73), $\Gamma = 4e^2\omega_B^2/3m_ec^3$ is the natural line width (or inverse time of a spontaneous transition between adjacent Landau levels), ω_B is the electron gyrofrequency, and ω' is the frequency of radiation in the frame which moves along the magnetic field with the electron longitudinal velocity $v_{\|}$. The corresponding radiation frequency in the frame of the stationary plasma is related to ω' via the Doppler formula (1.74):

$$\omega' = \omega(1 - p_{\|}\cos\alpha/m_ec). \tag{8.136}$$

Here $p_{\|}$ is the component of the electron momentum along the magnetic field $\boldsymbol{B}_0$, α is the angle between the wave vector $\boldsymbol{k}$ and $\boldsymbol{B}_0$ (cf. with the formulas (1.74) and (5.34)). Relativistic effects are not taken into account in (8.136) and the further analysis, as well. The form of the expression (8.135) is similar to the one which describes the resonant fluorescence in a two-level systems (see [210]). Strictly speaking, the angular pattern the of scattering cross-section by an electron in a magnetic field is quite complex. It is more important for our purposes that the cross-section can be separated into two parts: the constant term and the resonant part. This fact is reflected by the formula (8.135) which exhibits a resonance at $\omega=\omega'$. If the characteristic width of the radiation spectrum is much larger than the width of the resonance $\Delta\omega\simeq\Gamma$, the scattering at frequencies $(\omega'-\omega_B)^2\lesssim\Gamma^2/4$ can be treated as a sequence of two processes: absorption due to the transition at the higher Landau level followed by emission due to inverse transition from this higher level to the initial lower one. Electron life time at the excited level is non-vanishing in this case. Note that, according to the analysis performed in the section 2.1, steady-state distribution of electrons over transverse momenta $p_\perp$ (or Landau levels) is determined by the radiation at the resonant frequency. On the other hand, scattering of photons at frequencies outside the interval $(\omega'-\omega_B)^2\lesssim\Gamma^2/4$ is non-resonant.

According to (8.136), resonant scattering at frequencies $|\omega'-\omega_B|\lesssim\Gamma/2$ is provided by particles from the following interval of longitudinal momenta:

$$m_e c\frac{\omega-\Gamma/2-\omega_B}{\omega}\lesssim p_\parallel\cos\alpha\lesssim m_e c\frac{\omega+\Gamma/2-\omega_B}{\omega}. \tag{8.137}$$

Hence, a wave with a frequency ω propagating at the angle α to the magnetic field $\boldsymbol{B}_0$ is scattered by electrons with longitudinal momenta from the vicinity of the resonant value

$$p_\parallel=m_e c\,\frac{\omega-\omega_B}{\omega\cos\alpha}. \tag{8.138}$$

The wave scattered by such particles has a different frequency $\tilde\omega$ and angle $\tilde\alpha$ which nevertheless are related by the Doppler condition (8.138) to the same value of the electron longitudinal momentum:[14]

$$p_\parallel=m_e c\frac{\tilde\omega-\omega_B}{\tilde\omega\cos\tilde\alpha}. \tag{8.139}$$

As a result, radiation at frequencies $\omega(\alpha)$ selected by the condition (8.138) for a given $p_\parallel$ is transferred independently of the radiation at frequencies corresponding to another value of the longitudinal momentum. In contrast, non-resonant scattering is provided by particles with arbitrary longitudinal momentum outside the interval (8.137). In what follows we assume that the Doppler width $\Delta\omega\sim\beta_{T_\parallel}\omega_B$ of the frequency profile of cyclotron scattering, where $\beta_{T_\parallel}=(\kappa T_\parallel/m_e c^2)^{1/2}$, is large compared to the natural width of the line $\Gamma=4e^2\omega_B^2/3m_e c^3$. This condition can be easily satisfied for astrophysical plasmas.

According to the above, nearly all the electrons in plasma contribute to the coefficient of non-resonant scattering. Therefore, this coefficient is

[14]This is true if the quantum recoil due to cyclotron transition is negligible (see the section 5.1 and the formula (5.33), in particular).

proportional to electron density N — the total number of particles in the unit volume:

$$\mu_{\mathrm{sc}} \simeq \frac{\sigma_{\mathrm{T}} N}{2}\left[1+\frac{\omega^2}{(\omega-\omega_B)^2}\right] \tag{8.140}$$

(cf. the corresponding expression (8.65) for isotropic plasma). Conversely, the coefficient of resonant scattering is proportional to the amount of particles with the "resonant" momentum (8.138) which interact with radiation from the unit frequency interval. This particle number, in its turn, is proportional to $f(p_{\|})|\partial p_{\|}/\partial\omega|$, where $f(p_{\|})$ is the electron distribution over longitudinal momenta. If the resonant momentum (8.138) corresponding to the parameters ω and α is sufficiently large, $p_{\|}/2m_e\kappa T_{\|} \gg 1$, the number of resonant particles set by the Maxwell distribution function becomes exponentially small, as well as the coefficient of resonant cyclotron scattering. This case corresponds to frequencies

$$|\omega-\omega_B| \gg \beta_{T_{\|}}\omega_B|\cos\alpha|. \tag{8.141}$$

Transfer of radiation at these frequencies is determined by non-resonant scattering and described by the transfer equation (4.12) with the scattering coefficient (8.140).

8.2.2. TRANSFER EQUATION FOR RADIATION AT CYCLOTRON FREQUENCIES IN COLLISIONLESS PLASMA

Transfer of radiation at frequencies

$$|\omega-s\omega_B| \lesssim \beta_{T_{\|}}|\cos\alpha| \tag{8.142}$$

is connected with resonant cyclotron transitions. The steady-state transfer equation for radiation at cyclotron harmonics can be derived from the general one, (4.2), which we write in the following form:

$$\frac{dI_\omega(\omega,\alpha)}{dl} = a_{js} - \mu_{js} I_\omega(\omega,\alpha). \tag{8.143}$$

Here $I_\omega(\omega,\alpha)$ is the specific intensity of radiation at frequency ω in the direction α, the functions a_{js} and μ_{js} are cyclotron emissivity and the absorption coefficient for j-th mode at s-th harmonic. Consider, first of all, extraordinary radiation $j=1$ at the first harmonic $s=1$. Generally, cyclotron transitions of electrons between the Landau levels under the action of radiation with intensity $I_\omega(\omega,\alpha)$ make their velocity distribution anisotropic. If collisions (collisional transitions between Landau levels) are neglected, electron distribution over transverse momenta $p_\perp$ is Boltzmann, (7.37), and the corresponding transverse temperature $T_\perp = T_{\mathrm{b}}$ is set by the radiation at the first cyclotron harmonic (see (7.35), (7.41)). On the contrary, the longitudinal distribution is independent on radiation if quantum recoil is neglected. Therefore, particle distribution over longitudinal momenta $p_{\|}$ may be assumed Maxwellian with a given temperature $T_{\|}$ which can be regarded as a free parameter of the problem. Such a plasma with temperature anisotropy is characterized by radiation transfer coefficients a_{js} and μ_{js} given in sections 5.2 and 6.2 (see the formulas (5.142)–(5.145) and (6.47)–(6.49)).

To obtain the transfer equation which describes cyclotron scattering, one should choose, in accord with the physical parameters of the plasma considered, the appropriate formulas for the transfer coefficients a_{js}, μ_{js} and substitute into these the expression (7.38) or (7.41) for the effective temperature. For example, if mode polarization is determined by rarefied plasma and vacuum birefringence is negligible, the transfer equation in the absence of collisions takes the form [183, 184]:

$$\frac{dI_\omega(\omega,\alpha)}{dl} = -\mu_{11}I_\omega(\omega,\alpha) + \frac{\mu_{11}}{4\pi}\oint_{4\pi}\frac{3}{4}(1+\cos^2\tilde{\alpha})I_\omega(\tilde{\omega},\tilde{\alpha})d\Omega. \qquad (8.144)$$

Here the integral is taken over the solid angle interval 4π, and the values $\tilde{\omega}$ and $\tilde{\alpha}$ in the integrand are related via the formula (8.139). The longitudinal momentum $p_{\|}$ in this formula should be determined using the "untilded" parameters ω and α. The equation (8.144) describes transfer of extraordinary radiation at the first harmonic of the electron gyrofrequency, so the factor $(3/4)(1+\cos^2\tilde{\alpha})$ is the angular pattern of this radiation, and μ_{11} is the corresponding absorption coefficient. According to (6.47) this coefficient equals

$$\mu_{11} = \sqrt{\frac{\pi}{8}}\frac{\omega_L^2}{\omega_B c\beta_{T_{\|}}}\frac{1+\cos^2\alpha}{|\cos\alpha|}e^{-Z_1^2}$$
$$= \frac{\sqrt{2}\pi^{3/2}e^2N}{(m_e\kappa T_{\|})^{1/2}\omega_B}\frac{1+\cos^2\alpha}{|\cos\alpha|}e^{-Z_1^2}, \qquad (8.145)$$

where

$$Z_1 = \frac{\omega-\omega_B}{\sqrt{2}\beta_{T_{\|}}\omega\cos\alpha} = \frac{p_{\|}}{\sqrt{2}m_e c\beta_{T_{\|}}}. \qquad (8.146)$$

The equation (8.144) describes transfer of the radiation which determines the electron transverse distribution which is established due to cyclotron radiative transitions. Thus, both $I_\omega(\omega,\alpha)$ and $I_\omega(\tilde{\omega},\tilde{\alpha})$ characterize angular and frequency spectrum of the extraordinary radiation at the first cyclotron harmonic. Comparison of (8.144) and (4.12) shows that the former is nothing but the transfer equation of a scattering type, and the cyclotron absorption coefficient plays the role of the coefficient of resonant cyclotron scattering. This principal fact should be specially emphasized since is has frequently led to misunderstanding. The above identity which has been noted in [183, 184] makes special calculation of cyclotron scattering coefficients unnecessary since their role is played by the cyclotron absorption coefficients.

It has been pointed out earlier that cyclotron interaction of radiation with rarefied plasma is most effective for the extraordinary mode at the first harmonic. Cyclotron transition probabilities for ordinary waves at the first harmonic and for both modes at higher harmonics are considerably less. As a result, emissivity and the absorption coefficient for $j=2$, $s=1$ are $\beta_{T_\parallel}^{-2}$ times less than a_{11} and μ_{11}, while the corresponding values for $j=1,2$ and $s \geq 2$ decrease as $\beta_{T_\perp}^{2s}$. This allows one to neglect the influence of higher harmonics and the ordinary mode at the first harmonic on the transverse temperature $T_\perp$ of electrons in plasma. The transfer equations for these radiation components have the form (8.143). However, the corresponding transfer coefficients a_{js} and μ_{js} comprise the quantity $T_\perp$, i.e., depend on the intensity of extraordinary radiation at the first harmonic which, in its turn, is determined by the equation (8.144). Thus, when dealing with the equation (8.143), it is possible to treat the transfer coefficients a_{js}, μ_{js} as the given functions independent of the desired intensity I_ω. The resulting transfer equation is not of the scattering type, and its solution is similar to (4.40)–(4.45). In particular, if the cyclotron optical depth τ_{js} exceeds unity, the brightness temperature of radiation escaping a plasma layer is close to the electron transverse temperature $T_\perp$.

Note that the quantity $p_\parallel$ enters the relation (8.138) between ω and α and the similar one (8.139) for $\tilde{\omega}$ and $\tilde{\alpha}$ as an independent parameter. It is thus possible to present the transfer equation (8.144) in a different form. Multiply this equation by the delta-function $\delta\{\omega(1 - p_\parallel \cos\alpha/m_e c) - \omega_B\}$ and integrate over frequency ω. As a result, the transfer equation becomes the following [433]:[15]

$$\frac{\partial I_\omega(p_\parallel,\alpha)}{\partial l} - \frac{d\omega_B}{dl}\frac{m_e c}{\omega_B \cos\alpha}\frac{\partial I_\omega(p_\parallel,\alpha)}{\partial p_\parallel} = \mu_{11}(p_\parallel,\alpha)[S - I_\omega(p_\parallel,\alpha)] \quad (8.147)$$

where

$$S = \int_0^\pi \frac{3}{8}(1+\cos^2\alpha) I_\omega(p_\parallel,\alpha)\sin\alpha\, d\alpha, \quad (8.148)$$

[15] A similar equation has been used to analyze radiation transfer due to resonant scattering in a two-level atomic gas moving with spatially varying velocity [432]. In both cases the transfer equation is written for the radiation which is scattered by particles with a given velocity.

is the source function [16] and

$$I_\omega(p_{\|}, \alpha) = \int I_\omega \delta\{\omega(1 - p_{\|} \cos\alpha/m_e c) - \omega_B\}\, d\omega \tag{8.149}$$

denotes the intensity of radiation in the direction α which is scattered by electrons with the longitudinal momentum $p_{\|}$. The derivative $\partial I_\omega/\partial l$ characterizes spatial variation of the intensity of radiation which is scattered by electrons with a given longitudinal momentum $p_{\|}$. The term comprising the derivative $\partial I_\omega/\partial p_{\|}$ accounts for variation of I_ω at a fixed spatial point l due to variation of the longitudinal momentum of resonant electrons, i.e., describes "transfer" of radiation over the resonance momenta $p_{\|}$.

When deriving the equation (8.147) we have neglected the terms on the left-hand side which comprise the derivative $d(\cos\alpha)/dl$. Analysis shows that these terms are small compared to the ones containing $d\omega_B/dl$. Indeed, in an inhomogeneous magnetic field the resonance condition (8.142) for radiation at a given frequency ω can only be met within the localized gyroresonant layer of finite size. If the magnetic field inhomogeneity appears as spatial variation of the field strength along a ray, the size of the gyroresonant region equals $\beta_{T_{\|}} L_B |\cos\alpha|$, where $L_B = \omega_B |d\omega_B/dl|^{-1}$. On the contrary, if the violation of the criterion (8.142) is caused by an inhomogeneity of the field direction, the resulting size of the gyroresonant layer is determined by the scale $L_\alpha = |\cos\alpha\, d(\cos\alpha)/dl|^{-1}$. If $L_B \sim L_\alpha$, the ratio of the above two sizes equals $\beta_{T_{\|}} |\cos\alpha| \ll 1$. Therefore, the process of cyclotron radiation transfer is much more sensitive to spatial variations of the strength of the magnetic field than to variations of the magnetic field direction.

8.2.3. SOLUTION OF RADIATION TRANSFER EQUATION IN A HOMOGENEOUS AND INHOMOGENEOUS MAGNETIC FIELD

Let us find the solution of the transfer equation (8.147) in the case of a weakly inhomogeneous magnetic field. This case, typical for white dwarf conditions, [17] is realized if the thickness of a plasma layer is small compared to the size of the gyroresonant layer: $H \ll \beta_{T_{\|}} L_B$. The inhomogeneity of the magnetic field as well as the transfer of radiation over resonant longitudinal momenta $p_{\|}$ can be neglected in this case. Hence, the term comprising the derivative $\partial I_\omega/\partial p_{\|}$ can be omitted in the transfer equation. Moreover, assume for the sake of simplicity that the magnetic field $\boldsymbol{B}_0$ is orientated along the z-axis which is perpendicular to the boundaries of the plasma

[16] For the sake of simplicity the tilde over the angle α has been dropped in this expression.

[17] If the density profile of a hot plasma with $T \sim 10^6$–10^7 K on a magnetic white dwarf is formed by the gravitational force only, the characteristic scale $H = 2\kappa T/m_p g \sim 10^6 - 10^7$ cm is equal to $10^{-3} - 10^{-2}$ of the star radius $R_* \sim 10^9$ cm. Hence, such a plasma layer is pressed to the star surface.

layer. As a result, the transfer equation takes the following form:

$$\cos\alpha\frac{\partial I_\omega}{\partial z} = \mu_{11}(p_{\|},\alpha)[S - I_\omega], \tag{8.150}$$

where $I_\omega = I_\omega(z, p_{\|}, \alpha)$ and $S = S(z, p_{\|})$. The boundary conditions

$$\begin{aligned} I_\omega\big|_{z=0,\,\cos\alpha>0} &= I_\omega^{\rm inc}(\omega,\alpha), \\ I_\omega\big|_{z=H,\,\cos\alpha<0} &= 0 \end{aligned} \tag{8.151}$$

correspond to the case when no radiation enters the plama layer from the outside at the "upper" boundary $z = H$, while it is illuminated by the radiation with the known specific intensity $I_\omega^{\rm inc}(\omega,\alpha)$ at the "lower" boundary $z=0$.

The solution of the transfer equation (8.150) is found here by the method of Schwarzschild–Shuster described in section 4.1. However, this method completely ignores angular dependence of the radiation intensity. Such a dependence can be approximately found if the above method is modified in the following way. Let us first apply the Schwarzschild–Shuster method in its initial form to the equation (8.150). First, we multiply this equation by $|\cos\alpha|\sin\alpha$ and integrate over α first from 0 to $\pi/2$ and then from $\pi/2$ to π. Then approximate the expression $|\cos\alpha|\cos\alpha$ entering the integrals in the left-hand side by the corresponding mean values, 1/3 and -1/3, at the intervals of integration. Similarly, the expression $(1+\cos^2\alpha)$ on the right-hand side is approximated by the corresponding mean value 4/3. The solution of the resulting system of equations

$$\begin{aligned} \frac{1}{3}\frac{\partial I_1}{\partial z} &= \frac{2}{3}\mu_{11}(\alpha=0)\left[-I_1 + \frac{1}{2}(I_1+I_2)\right] \\ -\frac{1}{3}\frac{\partial I_2}{\partial z} &= \frac{2}{3}\mu_{11}(\alpha=0)\left[-I_2 + \frac{1}{2}(I_1+I_2)\right] \end{aligned} \tag{8.152}$$

for the unknown angle-averaged intensities

$$I_1 = \int_0^{\pi/2} I_\omega \sin\alpha\, d\alpha, \qquad I_2 = \int_{\pi/2}^{\pi} I_\omega \sin\alpha\, d\alpha, \tag{8.153}$$

yields the following expression for the source function:

$$S = \frac{I_1+I_2}{2} = \frac{I^{\rm inc}(1+2\tau_{\rm c})}{2(1+\tau_{\rm c})} - \frac{I^{\rm inc}\tau_{\rm c}}{(1+\tau_{\rm c})}\frac{z}{H}, \tag{8.154}$$

which is valid in the Schwarzschild–Shuster approximation. Here

$$\tau_{\rm c} = \mu_{11}(\alpha=0) = \frac{(2\pi)^{3/2}e^2}{(m_e\kappa T_{\|})^{1/2}\omega_B} N H \exp\left[-\frac{p_{\|}^2}{2\kappa T_{\|}}\right] \tag{8.155}$$

is the optical depth of the plasma layer to cyclotron scattering of the extraordinary radiation at the first harmonic $s=1$ in the direction $\alpha=0$, and

$$I^{\rm inc} = \int_0^{\pi/2} I^{\rm inc}_\omega(\omega,\alpha)\sin\alpha\, d\alpha. \tag{8.156}$$

It is assumed in the last formula that the intensity of the incoming extraordinary radiation $I^{\rm inc}_\omega(\omega,\alpha)$ is independent on α and slowly varies with frequency ω over the cyclotron resonance band (8.142). The desired angle-dependent intensity of the outgoing radiation,

$$I^{\rm out}_\omega = \tag{8.157}$$
$$I^{\rm inc} e^{\Psi} + \frac{I^{\rm inc}}{1+\tau_{\rm c}}\frac{1+5\cos^2\alpha}{2(1+\cos^2\alpha)}\left(1-e^{\Psi}\right) - \frac{I^{\rm inc}\tau_{\rm c}}{(1+\tau_{\rm c})}e^{\Psi},$$

where

$$\Psi = -\frac{1+\cos^2\alpha}{2\cos^2\alpha}\tau_{\rm c}, \tag{8.158}$$

is found by solving the equation (8.150) with the obtained approximate source function (8.154). The formula (8.158) has been obtained in [433]; note that the last term has been missed in this paper. The intensity of radiation escaping from an optically thick layer ($\tau_{\rm c}\gg 1$) is described by the formula

$$I^{\rm out}_\omega = \frac{I^{\rm inc}}{1+\tau_{\rm c}}\frac{1+5\cos^2\alpha}{2(1+\cos^2\alpha)}. \tag{8.159}$$

According to this formula, the cyclotron line in absorption is formed in the spectrum of radiation propagating through a scattering plasma layer in a homogeneous magnetic field. The intensity of the extraordinary component of the outgoing radiation decreases to the value $I^{\rm out}\sim I^{\rm inc}\tau_{\rm c}^{-1}$. The line width $\Delta\omega\sim\beta_{T_\parallel}\omega_B$ is determined by the Doppler broadening.

The above described approach to the analysis of cyclotron radiation transfer in rarefied plasma is used for the interpretation of absorption bands in the spectra of some magnetic white dwarfs.

Consider now the case of a strongly inhomogeneous magnetic field in which the condition $H\gg\beta_{T_\parallel}L_B$ is satisfied. Here the effective cyclotron scattering can occur only within the gyroresonant layer of the size $L_{11}\simeq 2\sqrt{2}\beta_{T_\parallel}L_B|\cos\alpha|$ (cf. the formula (6.60) for the similar case of strong cyclotron absorption in an inhomogeneous magnetic field). This limiting case of cyclotron radiation transfer can be realized on magnetic white dwarfs either in an accreting column above the magnetic poles or in an extended plasma envelope formed and supported by strong radiation pressure at cyclotron frequencies which exceeds the gravity force near the star (see section 8.2.5 and also the next section 8.3). Scattering of the photospheric radiation in such an envelope can significantly affect the observed optical spectrum of the star.

In the limiting case of a strongly inhomogeneous magnetic field the spatial transfer operator in the equation (8.147) is negligible compared to the one describing the "transfer" over longitudinal momenta $p_{\|}$. The corresponding transfer equation in this case has the following form:[18]

$$-\frac{m_e c}{\omega_B}\frac{\partial \omega_B}{\partial z}\frac{\partial I_\omega}{\partial p_{\|}} = \mu_{11}(p_{\|}, \alpha)[S - I_\omega], \tag{8.160}$$

where μ_{11} and S are given by the formulas (8.145) and (8.148), respectively, and the magnetic field $\boldsymbol{B}_0$ is directed along the z-axis which, in its turn, is perpendicular to the layer boundaries. Introducing the optical depth of the gyroresonant layer to cyclotron scattering of the extraordinary radiation at the first harmonic $s=1$ in the direction $\alpha=0$:

$$\tau_g = \frac{16}{3}\frac{\pi^2 e^2 N L_B}{m_e c \omega_B} = 2.6 \cdot 10^{-8}\frac{N L_B}{B_0}, \tag{8.161}$$

and using the new independent variable

$$y = (2\pi m_e \kappa T_{\|})^{-1/2}\int_{-\infty}^{p_{\|}} \exp(-p_{\|}^2/2m_e\kappa T_{\|})\, dp_{\|}, \tag{8.162}$$

it is possible to write the equation (8.160) in the mode compact form:

$$\frac{\partial I_\omega}{\partial y} = \frac{3}{8}\frac{1+\cos^2\alpha}{|\cos\alpha|}\tau_g\,[S - I_\omega]. \tag{8.163}$$

This equation in the Schwarzschild–Shuster approximation yields the following pair:

$$\begin{aligned} \frac{\partial I_1}{\partial y} &= \tau_g[-I_1 + \frac{1}{2}(I_1 + I_2)], \\ \frac{\partial I_2}{\partial y} &= \tau_g[-I_2 + \frac{1}{2}(I_1 + I_2)]. \end{aligned} \tag{8.164}$$

The boundary conditions at both sides of the gyroresonant layer are written as

$$I_1(y=0) = I^{\text{inc}}, \qquad I_2(y=0) = 0, \tag{8.165}$$

[18] Note that the spatial variation of gyrofrequency in an inhomogeneous magnetic field is analogous to the Doppler shift of the resonant frequency in a spatially inhomogeneous flow of a gas of two-level atoms. This analogy allows one to relate gyroresonant layers in plasma [298] and surfaces of constant velocity in moving stellar envelopes [434, 435]. Hence, the considered transfer equation is similar to the Sobolev equation studied in the theory of moving stellar envelopes.

where I^{inc} corresponds to the point z at which $\omega_B = \omega$. These boundary conditions correspond to the case when the gyroresonant layer is illuminated from one side by isotropic radiation with the intensity

$$I^{\text{inc}} = \int_0^{\pi/2} I_\omega^{\text{inc}} \sin \alpha d\alpha.$$

The solution of the system (8.164) under the conditions (8.165) has the following form:

$$I_1 = \frac{I^{\text{inc}}}{2} \left[1 + e^{-\tau_g y}\right], \qquad I_2 = \frac{I^{\text{inc}}}{2} \left[1 - e^{-\tau_g y}\right], \tag{8.166}$$

so that the source function

$$S = \frac{I^{\text{inc}}}{2} \tag{8.167}$$

is constant over the gyroresonant layer, contrary to the case of homogeneous magnetic field (8.154) in which S decreases with increasing z. This leads to the qualitative difference in the form of the solutions of the transfer equations in the cases of quasi-homogeneous and strongly inhomogeneous magnetic field (see below).

Using the obtained source function (8.167), it is possible to solve the equation (8.160) and determine the angle-dependent intensity of radiation escaping the upper boundary of the gyroresonant layer [433]:

$$I_\omega^{\text{out}}(y = 1, \cos \alpha > 0) = I^{\text{inc}} \exp \Phi + \frac{1}{2} I^{\text{inc}}(1 - \exp \Phi) \tag{8.168}$$

and reflected to the region below this layer:

$$I_\omega^{\text{ref}}(y = 1, \cos \alpha < 0) = \frac{1}{2} I^{\text{inc}}(1 - \exp \Phi) \tag{8.169}$$

The following notation is used in the above expressions:

$$\Phi = -\frac{3}{8} \frac{1 + \cos^2 \alpha}{|\cos \alpha|} \tau_g. \tag{8.170}$$

According to the above formulas, cyclotron scattering in an optically thick gyroresonant layer cannot extinguish more than half of the incoming radiation:

$$I_\omega^{\text{out}} = I_\omega^{\text{ref}} = \frac{1}{2} I^{\text{inc}} \qquad (\tau_g \gg 1). \tag{8.171}$$

This feature makes the process of cyclotron scattering in a strongly inhomogeneous magnetic field radically different from either the case of scattering

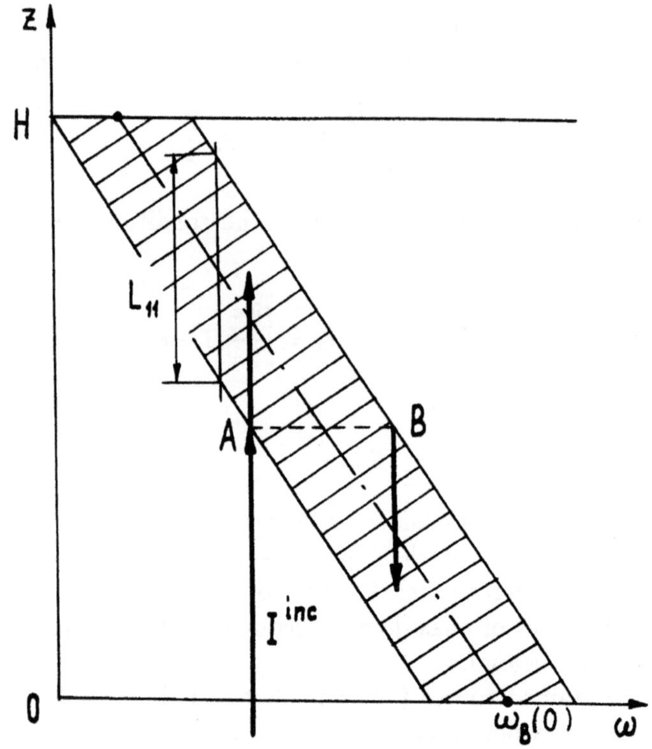

Figure 8.3. "Forward–backward" scattering of cyclotron radiation in an optically thick gyroresonant layer. The hatched region marks the location of the gyroresonant layer at the plane (ω, z). The geometrical size of this layer equals L_{11}. Backward scattering of the incoming radiation $I^{\rm inc}$ near the lower boundary of the gyroresonant layer leads to the frequency shift $\Delta\omega$.

in homogeneous field, in which $I^{\rm out} \sim \tau_{\rm c}^{-1}$ at $\tau_{\rm c} \gg 1$, or pure absorption by a gyroresonant layer in collisional plasma (see section 6.2), in which $I^{\rm out}$ decreases exponentially with increasing τ_g. The obtained result can be explained quite clearly using the "forward–backward" approximation in which it is assumed for simplicity that an electron may resonantly scatter cyclotron radiation in two distinct directions, forward and backward, with equal probabilities (see Fig. 8.3).

The first scattering of the radiation coming into the inner boundary of an optically thick ($\tau_g \gg 1$) gyroresonant layer occurs at the point A near this boundary. The cyclotron optical depth of the interval from this point to the lower boundary of the gyroresonant layer is about unity. The frequency of the forward-scattered radiation is unchanged, whereas the one for the backward-scattered radiation changes by the quantity $\Delta\omega \sim 2\beta_{T_\parallel}\omega_B |\cos\alpha|$. Hence, the backward scattering shifts radiation from the point A to the point B on the coordinate–frequency diagram in Fig. 8.3. Thus, just after the first scattering, the incoming radiation is separated into two components of equal intensity $I^{\rm inc}/2$. One of these components corresponds to the radiation coming into the inner boundary of the layer at the point A, while the other one, to the radiation which enters the outer boundary at the point B. If variations of the parameters of the gyroresonant layer over the frequency interval $\Delta\omega_B \sim \Delta\omega$ are negligible, further scattering of these components result in the formation of the outgoing radiation of the same intensity $I^{\rm inc}/2$ at both boundaries of the layer, in accord with the formula (8.171).

Contrary to the case of homogeneous magnetic field, an extended plasma layer in an inhomogeneous magnetic field cannot form a narrow cyclotron line centered at the gyrofrequency ω_B. The spectrum of radiation escaping such a plasma layer comprises the depression band at frequencies between the values of the gyrofrequency ω_B at the upper and lower boundaries of the plasma layer. If the optical depth τ_g of the gyroresonant layer $\omega \simeq \omega_B$ is much larger than unity, the intensity of the outgoing extraordinary radiation equals one half of the incoming one. If, moreover, the optical depth of this gyroresonant layer for the ordinary mode is much less than unity, the ordinary component is unchanged. Therefore, the outgoing intensity in both modes equals 3/4 of the intensity of the incoming unpolarized radiation. But if the optical depth of the gyroresonant layer for the ordinary mode exceeds unity, the observed intensity of this component equals the blackbody intensity at the temperature $T_\perp$.

The theory of cyclotron radiation transfer in a scattering plasma which is permeated by an inhomogeneous magnetic field is important for the analysis of the cyclotron scattering in extended plasma envelopes of magnetic white dwarfs.

The above analysis is valid if the effect of vacuum birefringence on the mode polarization in plasma is negligible. If, on the contrary, mode polarizations are dominated by the birefringent vacuum (see the criteria (2.187) and (2.189)), the probabilities of cyclotron transitions under the action of ordinary and extraordinary radiation are comparable: for example, the ratio of the cyclotron absorption coefficients (6.50) and (6.51) equals $\mu_{11}/\mu_{21} = \cos^2\alpha$. In this case the influence of both modes on the transverse electron temperature should be taken into account. For rough estimates it is possible to neglect the difference in transfer coefficients of different modes and put $\mu \simeq (\mu_{21} + \mu_{11})/2$ in the transfer equation (8.144).

More rigorous analysis [54,314] allows to find the transverse temperature $T_\perp$ of plasma electrons which is determined in this case by the formula (7.52). Substituting this value of the transverse temperature into the radiation transfer coefficients a_{j1} and μ_{j1}, we get the following system of the transfer equations for radiation in both modes ($j=1,2$) at the first harmonic $s=1$:

$$\frac{dN_{\mathbf{k}}^{(1)}(p_{\|},\alpha)}{dl} = -\mu_{11}\left[N_{\mathbf{k}}^{(1)}(p_{\|},\alpha) - x_1 - x_2\right],$$
$$\frac{dN_{\mathbf{k}}^{(2)}(p_{\|},\alpha)}{dl} = -\mu_{21}\left[N_{\mathbf{k}}^{(2)}(p_{\|},\alpha) - x_1 - x_2\right]. \qquad (8.172)$$

The absorption coefficients in the system "plasma + birefringent vacuum" are determined by the formulas (6.50)–(6.53) and (7.52), while the quantities x_1 and x_2 by the formulas (7.53). The solution of the system (8.172) allows one to determine the transverse temperatre $T_\perp$ and, therefore, the transfer coefficients a_{js}, μ_{js} at higher harmonics $s \geq 2$. The intensity of outgoing radiation at these harmonics can be found by solving the transfer equations (8.143). In a homogeneous plasma layer the solution looks like

$$I_{js}^{\text{out}} = I_{js}^{\text{inc}} e^{-\tau_{js}} + \frac{a_{js}}{\mu_{js}}\left(1 - e^{-\tau_{js}}\right), \qquad (8.173)$$

where the optical depth of the layer, $\tau_{js} = \mu_{js} L$ is determined by the formula (6.58) in which one should substitute the factor β_T^{2s-2} by $\beta_{T_\|}^{2s-2}$. To solve the system (8.172) is more difficult. Such a solution for a plasma layer in a homogeneous magnetic field has been obtained in [54, 314] by the Schwarzschild–Shuster method and used in these papers for the analysis of the origin of cyclotron lines in gamma-ray burst spectra.

The described approach to the description of radiation transfer at cyclotron harmonics in collisionless plasma which has been outlined above is based upon the simplifying approximation that transverse electron distribution over Landau levels is entirely determined by the radiation at the cyclotron fundamental. The established distribution, in its turn, settles the values of emissivities and absorption coefficients at all the higher harmonics. Such a multi-level transfer scheme involves the following elementary processes: at the first harmonic — resonant scattering and, only in the case of birefringent vacuum, mode conversion; at higher harmonics — emission of a photon after the absorption of several first harmonic ones and inverse process of true absorption of a photon with subsequent emission (spawning) of several first harmonic ones. Thus, our treatment has missed resonant scattering at higher harmonics and the influence of higher harmonic spawning on the intensity at the cyclotron fundamental. The account of these proceses, in the cases when they are important,[19] requires more complete

[19]Note that the effect of higher harmonic scattering is generally less important than the spawning of higher harmonic photons into several first harmonic ones. Indeed, since the probability of emission at the first harmonic is much greater than at a higher one, an upward transition of an electron due to extinction of a photon $s\hbar\omega_B$ ($s \geq 2$) will, in

analysis. In this respect we mention the numreical simulations of [436] which however do not include some of the effects considered in the framework of our approach.

8.2.4. TRANSFER OF RADIATION AT CYCLOTRON FREQUENCIES IN PLASMA WITH PARTICLE COLLISIONS

Collisions of plasma particles lead to the bremsstrahlung emission and absorption. The corresponding emissivities and absorption coefficients are given in sections 5.2, 6.2. In addition, collisions may lead to non-radiative transitions between the Landau levels. As a result, resonant cyclotron scattering is accompanied by two other processes: (i) collisional excitation (transition of an electron to a higher Landau level) followed by the cyclotron emission (spontaneous or induced); and (ii) cyclotron absorption followed by collisional de-excitation. Contrary to the above considered case of pure cyclotron scattering in which the photon number is conserved, these processes result in emission or true absorption of photons. We will demonstrate below that these processes can significantly affect the intensity of cyclotron radiation even in the limit of "collisionless" plasma at which the parameter ϵ, (7.57), (7.59), is much less than unity.

Electron distribution over Landau levels under the combined influence of cyclotron radiation and rare collisions $\epsilon \ll 1$ has been found in section 7.1 (see the formula (7.56) for the Landau level populations N_m.) If plasma parameters satisfy the criterion of rarefied plasma (2.144), this non-equilibrium, steady-state distribution is influenced by extraordinary radiation $j=1$ at the first cyclotron harmonic $s=1$. The emissivity (7.18) in this case takes the following form:

$$a_{11} = \hbar\omega \sum_{m=1}^{\infty} A_m^{m-1} f(p_{\|}) N_m \, dp_{\|}, \tag{8.174}$$

where $f(p_{\|})$ characterizes the distribution over longitudinal momenta which is assumed fixed in the following analysis,

$$A_m^{m-1} = \frac{e^2 \omega \omega_B}{4\pi m_e c^3} m(1+\cos^2\alpha)\delta\left\{\omega\left(1 - \frac{p_{\|}\cos\alpha}{m_e c}\right) - \omega_B\right\} \tag{8.175}$$

is the probability of cyclotron transition $(m) \to (m-1)$ [223], m is the Landau level number. Note that the recoil effect is neglected, and thus the corresponding term in the delta-function argument is dropped. For Maxwell

most cases, be followed by emission of a first harmonic photon $\hbar\omega_B$ rather than a higher harmonic one $s'\hbar\omega_B$ $(s' \geq 2)$.

distribution over the longitudinal momenta the emissivity (8.174) looks like:

$$a_{11} = \frac{\hbar\omega^3}{(2\pi)^3 c^2} F \sum_{m=1}^{\infty} m N_m, \tag{8.176}$$

$$F = \sqrt{\frac{\pi}{8}} \frac{\omega_L^2}{\omega_B c} \frac{\omega_B^2}{\omega^2 \beta_{T_\|}} \frac{1+\cos^2\alpha}{|\cos\alpha|} \frac{e^{-Z_1^2}}{N},$$

where Z_1 is determined by the formula (8.146) and N is electron density in plasma. Substituting (7.56) into (8.176), we find [41]:

$$a_{11} = \frac{\hbar\omega_B^3}{(2\pi)^3 c^2} FN \left[x_1 + \epsilon \left(\frac{\kappa T_\|}{\hbar\omega_B} - x_1 \right) \right]. \tag{8.177}$$

This formula is valid at arbitrary x_1, both in the quasi-calssical limit $x_1 \gg 1$ and in the quantum limit $x_1 \ll 1$.[20] Next, the absorption coefficient (7.17) can be written as

$$\mu_{11} = F \sum_{m=1}^{\infty} m(N_{m-1} - N_m) = FN. \tag{8.178}$$

Note that this quantity is independent on the form of the transverse electron distribution, a fact pointed out in [437]. In particular, the absorption coefficient μ_{11} has the same form both in classical and in quantized plasma.

The transfer equation of cyclotron radiation at the first harmonic in a plane-parallel plasma layer with $\epsilon \neq 1$ permeated by homogeneous magnetic field can be derived in the same way as the transfer equation (8.150) for the case $\epsilon = 0$ in which the effect of collisions is completely ignored. The resulting equation:[21]

$$\cos\alpha \frac{\partial I_\omega}{\partial z} = F(p_\|, \alpha) N [S - I_\omega] \tag{8.179}$$

is apparently identical to (8.150), though the form of the source function is different:

$$S(z, p_\|) = \tag{8.180}$$
$$\frac{1}{\epsilon+1} \int_0^\pi \frac{3}{8}(1+\cos^2\alpha) I_\omega(z, p_\|, \alpha) \sin\alpha \, d\alpha + \frac{\epsilon}{\epsilon+1} I_\omega^{(0)}(T_\|).$$

The above expression valid at arbitrary ϵ has been derived in [315]. In the collisionless limit $\epsilon \to 0$ the source function (8.181) is reduced to (8.148)

[20] In the absence of collisions $\epsilon = 0$ these limits correspond to the conditions $\kappa T_\perp \gg \hbar\omega_B$ and $\kappa T_\perp \ll \hbar\omega_B$, respectively (see section 7.1).

[21] The axis z, as before, is directed along the magnetic field $\mathbf{B}_0$.

which has been derived in [41], while the case of equilibrium plasma with isotropic distribution ($T_\perp = T_\| = T$) is realized in the opposite limiting case $\epsilon \to \infty$.

The equation (8.179) with the source function (8.181) is similar to the well-known Milne–Eddington equation which describes transfer of radiation in ion resonant lines [177]. On the other hand, the ansatz $\lambda \equiv \epsilon/(\epsilon+1) \to p \equiv \mu_b/\mu_{11}$, where μ_b is bremsstrahlung absorption coefficient, transforms (8.179), (8.181) into the transfer equation which accounts for resonant scattering and bremsstrahlung emission/absorption of radiation in the cyclotron line. This equation has been derived and analyzed in detail in [183]. The noted analogy allows us to immediately write the solution of the equation (8.179), (8.181) in a homogeneous plasma layer $0 \le z \le H$ with the boundary conditions (8.151) using the results of this paper (see also [315]). The desired solution:

$$\begin{aligned} I_1 + I_2 &= I_\omega^{(0)} \left(2 + C_1 e^{-2\sqrt{\lambda}\tau} + C_2 e^{2\sqrt{\lambda}\tau}\right), \\ I_1 - I_2 &= I_\omega^{(0)} \sqrt{\lambda} \left(-C_1 e^{-2\sqrt{\lambda}\tau} + C_2 e^{2\sqrt{\lambda}\tau}\right) \end{aligned} \tag{8.181}$$

for the angle-averaged intensities I_1 and I_2 (see (8.153)) has been found by the method of Schwarzschild–Shuster. The constants

$$\begin{aligned} C_1 &= -\Delta^{-1} \left[2(1+\sqrt{\lambda})e^{2\sqrt{\lambda}\tau_c} + (\gamma-2)(1-\sqrt{\lambda})\right], \\ C_2 &= \Delta^{-1} \left[2(1-\sqrt{\lambda})e^{-2\sqrt{\lambda}\tau_c} + (\gamma-2)(1+\sqrt{\lambda})\right], \end{aligned} \tag{8.182}$$

where

$$\Delta = (1+\sqrt{\lambda})^2 e^{2\sqrt{\lambda}\tau_c} - (1-\sqrt{\lambda})^2 e^{-2\sqrt{\lambda}\tau_c}$$

accounts for the adopted boundary conditions, τ_c and $I^{\rm inc}$ are determined by (8.154) and (8.156), respectively, $\gamma = 2I^{\rm inc}/I_\omega^{(0)}(p_\|)$, and

$$\tau = \mu_{11}(\alpha=0)\,[H - z] \tag{8.183}$$

stands for the cyclotron optical depth from the upper boundary of the layer $z = H$. The scattering coefficient in (8.183) is given by the formula (8.145). The intensity of radiation escaping the upper boundary of the layer equals

$$I^{\rm out} = \sqrt{\lambda} I_\omega^{(0)}(T_\|)(C_2 - C_1). \tag{8.184}$$

In the limiting case of collisionless plasma, $\epsilon \to 0$ (or, equivalently, $\lambda \to 0$), this solution is reduced to the one for pure scattering layer:

$$\begin{aligned} I_1 + I_2 &= I_\omega^{(0)}(T_\|) \frac{1+2\tau}{1+\tau_c} \frac{\gamma}{2}, \\ I_1 - I_2 &= I_\omega^{(0)}(T_\|) \frac{1}{1+\tau_c} \frac{\gamma}{2}, \end{aligned} \tag{8.185}$$

$$I^{\rm out} = I^{\rm inc}\frac{1}{1+\tau_{\rm c}}. \tag{8.186}$$

Comparison of the last formulas to the expression (8.158) obtained for the same case reveals that the latter provides a more accurate description of the radiation field, whereas (8.185), (8.186) account for the angular dependence of the intensity, $I_\omega(\alpha)$, in a rather crude way: by the difference of the quantities I_1 and I_2. Such a discrepancy results from the difference in the solution procedures. The formulas (8.185), (8.186) are the straightforward yield of the Schwarzschild–Shuster method. On the other hand, to arrive at (8.158), this method has been applied to find the approximate form of the source function S which has then been used to get the angle-dependent solution of the transfer equation.

In the opposite limit of collisional plasma at which $\epsilon \to \infty$ (and $\lambda \to 1$) the solution (8.181) transforms into

$$\begin{aligned} I_1 + I_2 &= I_\omega^{(0)}(T_\|)\left[2 - e^{-2\tau} + (\gamma - 2)e^{-2(\tau_{\rm c}-\tau)}\right],\\ I_1 - I_2 &= I_\omega^{(0)}(T_\|)\left[e^{-2\tau} + (\gamma - 2)e^{-2(\tau_{\rm c}-\tau)}\right], \end{aligned} \tag{8.187}$$

$$I^{\rm out} = I^{\rm inc}e^{-2\tau_{\rm c}} + I_\omega^{(0)}(T_\|)(1 - e^{-2\tau_{\rm c}}). \tag{8.188}$$

These formulas describe pure absorption and emission of cyclotron radiation in equilibrium plasma with $T_\perp = T_\|$.

Transfer of cyclotron radiation in plasma with arbitrary $\epsilon \equiv \lambda/(1-\lambda)$ depends strongly on the geometrical size of the layer H. Radiation inside a "thick" layer,

$$2\sqrt{\lambda}\tau_{\rm c} \gg 1, \tag{8.189}$$

at distances $\Delta z > (2\sqrt{\lambda}\mu_{11})^{-1}$ from its boundaries becomes isotropic, and its intensity gets close to the equilibrium one at the temperature $T_\|$:

$$I_1 \simeq I_2 \simeq I_\omega^{(0)}(T_\|) \tag{8.190}$$

In other words, cyclotron radiation in such a layer is completely thermalized. This means, in particular, that plasma in the central parts of a "thick" layer becomes thermally isotropic: $T_\perp = T_\|$. On the other hand, the outgoing intensity is entirely determined by the parameters of the plasma and does not depend on $I^{\rm inc}$:

$$I^{\rm out} = \frac{2\sqrt{\lambda}}{1+\sqrt{\lambda}} I_\omega^{(0)}(T_\|). \tag{8.191}$$

In the case of a "thin" layer,

$$2\sqrt{\lambda}\tau_{\rm c} \ll 1, \tag{8.192}$$

thermalization is negligible, and the outgoing intensity is given by

$$I^{\text{out}} = \frac{I^{\text{inc}}}{1+(1+\lambda)\tau_{\text{c}}} + 2\lambda\tau_{\text{c}}I_{\omega}^{(0)}(T_{\parallel}). \tag{8.193}$$

The outgoing intensity can be dominated by either proper emission of the layer or the scattered incoming radiation. If

$$I^{\text{inc}}\left[1+(1+\lambda)\tau_{\text{c}}\right]^{-1} \ll 2\lambda\tau_{\text{c}}I_{\omega}^{(0)}(T_{\parallel}), \tag{8.194}$$

the outgoing intensity is mainly contributed to by the proper emission of the layer:

$$I^{\text{out}} = 2\lambda\tau_{\text{c}}I_{\omega}^{(0)}(T_{\parallel}). \tag{8.195}$$

If the condition opposite to (8.194) is met for the plasma parameters:

$$I^{\text{inc}}\left[1+(1+\lambda)\tau_{\text{c}}\right]^{-1} \gg 2\lambda\tau_{\text{c}}I_{\omega}^{(0)}(T_{\parallel}), \tag{8.196}$$

the outgoing intensity is nothing but the scattered incoming one:

$$I^{\text{out}} = \frac{I^{\text{inc}}}{1+(1+\lambda)\tau_{\text{c}}}. \tag{8.197}$$

If $\tau_{\text{c}} \ll 1$, the influence of the layer on the incoming radiation is insignificant. In the line core this case can be realized at sufficiently small plasma density N, whereas in the Doppler wings, due to the exponential decrease of the optical depth τ_{c} with increasing $|\omega - \omega_B|$ (see (8.138) and (8.155)).

If the condition (8.196) is met at the center of the cyclotron line, then it holds for the entire line profile. Then pure scattering takes place at the first cyclotron harmonic. It is clear from (8.197) that a line in absorption appears in the spectrum of the outgoing radiation at frequencies near ω_B. The depth of this line will be rather strong if the optical depth at the line center $\tau_{\text{c}} \lesssim 1$. As τ_{c} and λ increase, the inequality (8.194) opposite to (8.196) becomes valid at the line center. The intensity of the outgoing radiation in this case depends on the plasma parameters (see (8.195)). If

$$I^{\text{inc}} < 2\lambda\tau_{\text{c}}I_{\omega}^{(0)}(T_{\parallel}) \tag{8.198}$$

the cyclotron line in emission is formed in the spectrum of the outgoing radiation.[22]

The above consideration concerned the case of the "thin" layer (8.192). But if the layer is "thick" (8.189) and the effect of thermalization strongly

[22] Note that the intermediate, more complicated, case of the cyclotron line with emission core and absorption wings can be realized if the inequality (8.198) is violated in the wings of the line due to the exponential decrease of the optical depth τ_{c}.

influences the outgoing radiation, the line profile depends on the ratio $I^{\text{inc}}/I_\omega^{(0)}(T_\parallel)$ and the value of the parameter λ. The cyclotron line at $\omega \simeq \omega_B$ is seen in absorption if

$$I^{\text{inc}} > \frac{2\sqrt{\lambda}}{1+\sqrt{\lambda}}\, I_\omega^{(0)}(T_\parallel) \tag{8.199}$$

and in emission otherwise. Finally, in the collisional case $\epsilon \gg 1$ the cyclotron line from the hot layer with $T_\parallel > T_{\text{inc}}$, where T_{inc} is the brightness temperature of the incoming radiation, is seen in emission at arbitrary τ_c.

Transfer of radiation at higher cyclotron harmonics can be considered using the equation (8.143) in which a_{js} and μ_{js} are found in the approximation of the fixed electron distribution. The latter is determined by the intensity of extraordinary radiation at the first harmonic. The expressions for the transfer coefficients a_{js} and μ_{js} valid at arbitrary ϵ have been derived in [315]. As $\epsilon \to 0$, these general formulas are reduced to the corresponding transfer coefficients (5.145) and (6.49) for plasma with temperature anisotropy. The solution of the transfer equation (8.143) for a homogeneous plasma layer of the size H has the following form (cf. (4.33)):

$$I_{js}^{\text{out}} = I_{js}^{\text{inc}} e^{-\tau_{js}/\cos\alpha} + \frac{1}{\cos\alpha}\int_0^{\tau_{js}} \frac{a_{js}}{\mu_{js}}\, e^{-\xi/\cos\alpha} d\xi, \tag{8.200}$$

where $\tau_{js} = \int_0^H \mu_{js} dz$. Radiation escaping an optically thick layer $\tau_{js} \gg 1$ is determined by the source function $S = a_{js}/\mu_{js}$ in the region of the size Δh near the upper boundary of the layer $z = H$. The size Δh is determined by the condition

$$\int_{H-\Delta h}^{H} \mu_{js} \simeq 1. \tag{8.201}$$

The analysis performed in [315] has revealed that in a pure scattering layer (8.192), (8.196) the spectrum of the outgoing radiation comprises the sequence of lines in absorption at cyclotron harmonics. The depths of these lines decrease with increasing harmonic number s. A "thin" layer with proper emission (8.192), (8.194) forms lines in emission if the criterion (8.198) is satisfied or lines with emission core and absorption wings otherwise. Finally, all the higher harmonic lines emitted by the "thick" layer (8.189) are seen in emission, and the intensity in the line centers equals $I_\omega^{(0)}(T_\parallel)$. Thus, line profiles at any harmonic and for both modes are generally similar.

Concluding this section, consider the transfer of cyclotron radiation in a plasma atmosphere with homogeneous magnetic field and inhomogeneous density profile. Let plasma density vary with height according to the barometric formula

$$N = N_0 \exp(-z/H) \tag{8.202}$$

where

$$H = 2\kappa T_{\parallel}/m_{\rm p}g \tag{8.203}$$

is the scale height of pure hydrogen plasma in a uniform gravitational field, and g is the free-fall acceleration.

Transfer of radiation in continuum outside the cyclotron line is affected by the non-resonant Thomson scattering as well as bremsstrahlung emission and absorption. The relevant transfer equation has the form (4.64), and its approximate solution which ignores the angular dependence of scattering and absorption coefficients is given by (4.73). As a result, the outgoing intensity can be written as [184]

$$I_1^{\rm out} = I_\omega^{(0)}(T_{\parallel})\left(\frac{2}{3}\right)^{2/3}\frac{\Gamma(2/3)}{\Gamma(4/3)}M_{\rm T}^{1/3}, \tag{8.204}$$

where $M_{\rm T} = \mu_{\rm b}/H(\sigma_{\rm T}N)^2$ (cf. with the formula (4.74)).

Transfer of cyclotron radiation at frequencies $\omega \simeq \omega_B$ is described by the equation (8.179), (8.181). The solution of this equation in the Schwarzschild-Shuster approximation in the collisionless limit $\epsilon \ll 1$ has the following form [438]

$$\begin{gathered} I_1 + I_2 = I_\omega^{(0)}(\omega_B)\left[2 - \frac{2\,{\rm Ai}\,(\Phi^{2/3}M_{\rm c}^{1/3}\tau_{\rm c})}{{\rm Ai}(0) - {\rm Ai}'(0)(M_{\rm c}/\Phi)^{1/3}}\right], \\ I_1 - I_2 = \Phi^{-1}\frac{\partial}{\partial\tau_{\rm c}}(I_1 + I_2), \end{gathered} \tag{8.205}$$

where the angle-averaged intensities I_1 and I_2 are defined by (8.153),

$$\tau_{\rm c}(z) \equiv \frac{4\sqrt{2}\pi^{3/2}e^2}{m_{\rm e}c\omega_B\beta_{T_{\parallel}}}N(z)H \tag{8.206}$$

is the line center optical depth from the upper boundary of the atmosphere,

$$\Phi(p_{\parallel}) = \exp(-p_{\parallel}^2/2m_{\rm e}\kappa T_{\parallel}) \tag{8.207}$$

is the Doppler profile of the cyclotron line, and the parameter

$$M_{\rm c} = \epsilon/\tau_{\rm c}. \tag{8.208}$$

Since, according to (7.59) and (8.206), $\epsilon \propto N(z)$ and $\tau_{\rm c} \propto N(z)$, the ratio (8.208) is independent on the coordinate z. The solution (8.205) matches the same boundary conditions:

$$I_2|_{\tau_{\rm c}=0} = 0, \qquad I_{1,2}|_{\tau_{\rm c}\to\infty} \to {\rm const},$$

as the continuum radiation. These conditions mean that no radiation enters the atmosphere from the outside, while the intensity remains finite deep in the atmosphere. According to (8.205), the outgoing intensity at cyclotron frequencies is equal to

$$I^{\mathrm{out}} \equiv I_1(\tau_{\mathrm{c}}=0) = \frac{-2\mathrm{Ai}'(0)(M_{\mathrm{c}}/\Phi)^{1/3} I_\omega^{(0)}(\omega_B)}{\mathrm{Ai}(0) - \mathrm{Ai}'(0)(M_{\mathrm{c}}/\Phi)^{1/3}}. \tag{8.209}$$

If mode polarization is determined by the effect of vacuum birefringence (see the criterion (2.187) which can be easily met for plasmas on neutron stars), the coefficients of cyclotron absorption (scattering) are comparable: $\mu_{11} = \mu_{21}\cos^2\alpha$. To get a simple approximate solution of the transfer equation, one may ignore the difference between them and put $\mu_{\mathrm{c}} = (\mu_{11} + \mu_{21})/2$. The resulting transfer equation of the radiation intensity in both modes is similar to (8.179), (8.181) in which the equilibrium intensity should be multiplied by the factor 2 and the scattering coefficient should be divided by this factor. Therefore, the following formula for the outgoing intensity should be used instead of (8.209):

$$I^{\mathrm{out}} = \frac{-4\mathrm{Ai}'(0)(M_{\mathrm{c}}/\Phi)^{1/3} I_\omega^{(0)}(\omega_B)}{\mathrm{Ai}(0) - \mathrm{Ai}'(0)(M_{\mathrm{c}}/\Phi)^{1/3}}. \tag{8.210}$$

Here the quantities τ_{c} and M_{c} are determined by (8.206) and (8.208), respectively.

8.2.5. CYCLOTRON FEATURES IN THE SPECTRA OF MAGNETIC DEGENERATE STARS

Cyclotron frequencies in the magnetic fields of white dwarfs ($B_0 \sim 10^8$ G) and neutron stars ($B_0 \sim 10^{12}$ G) correspond to optical (ultra-violet) and X-ray bands, respectively. It seems thus quite natural to explain the unidentified absorption features in the spectra of some magnetic white dwarfs [439–441] by the cyclotron interaction of radiation with magnetized plasmas on these stars. Fig. 8.4 shows a couple of examples of such spectral features. Discovery of X-ray emitting coronas with temperature $T_{\mathrm{c}} \sim 10^7$ K around non-magnetic dwarfs has stimulated the investigations of the interaction of hot plasmas with radiation on magnetic ones in order to explain the origin of absorption features in the spectra of these objects. Note that, up to now, there are no reliable detections of X-rays from magnetic white dwarfs. As an example, we mention the case of the magnetic white dwarf GR 290. The search in the database of the *Einstein* observatory have yielded the X-ray flux from this star at better than 99 % confidence [443]. However, the forthcoming observations with more sensitive facility onboard the *ROSAT*

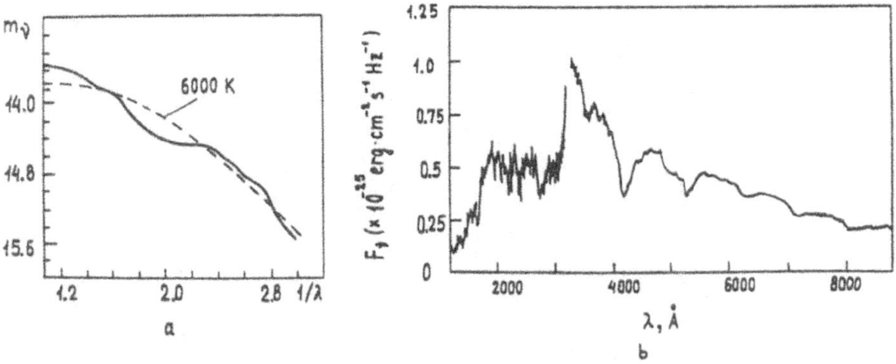

Figure 8.4. Frequency spectra of magnetic white dwarfs: *a*, the spectrum (visual spectral magnitude m_V versus inverse wavelength in microns) of "cool" dwarf G 240-72 with the photospheric temperature $T_{ph} \simeq 6 \cdot 10^3$ K [442]; *b*, the spectrum of "hot" dwarf GD 229 with $T_{ph} \simeq (1.5 - 2) \cdot 10^4$ K [440]. The dotted line is the blackbody spectrum at the photospheric temperature.

satellite have shown no detectable X-rays and imposed only an upper limit which is 40 times less than the flux deduced from the *Einstein* archive [40]. Such a discrepancy means either that the *Einstein* observation is nothing but a statistical fluctuation or that the coronal density has decreased by a factor of 6 during the time between two observations.[23] As we see below, pronounced cyclotron depressions can be formed at much lower electron densities.

Assume that the temperatures of white dwarf coronas are similar to the values typical for the majority of stellar ones: $T_c \sim 10^6 - 10^7$ K. Since the surface gravity of a white dwarf is rather strong, such a hot plasma surrounds the star as a thin envelope with the scale height $H \sim (10^{-2} - 10^{-3})R_*$, where $R_* \sim 10^9$ cm is white dwarf radius. Transfer of radiation in such a corona can be analyzed with the equation (8.179), (8.181) for a homogeneous magnetic field which accounts for cyclotron and collisional transitions between the Landau levels. Solutions of this equation have been analyzed in detail in the previous section. The above analysis allows us to point out the regions of coronal parameters N, B_0 and $T_\parallel$ at which the process of radiation transfer and the resulting cyclotron lines in the optical spectrum of the star are qualitatively different. Such parameter regions are shown in Fig. 8.5.

[23] If a magnetic white dwarf accretes matter from the binary companion, the intense X-rays may be emitted by the accreting flow even in the case when no hot corona surrounds the star.

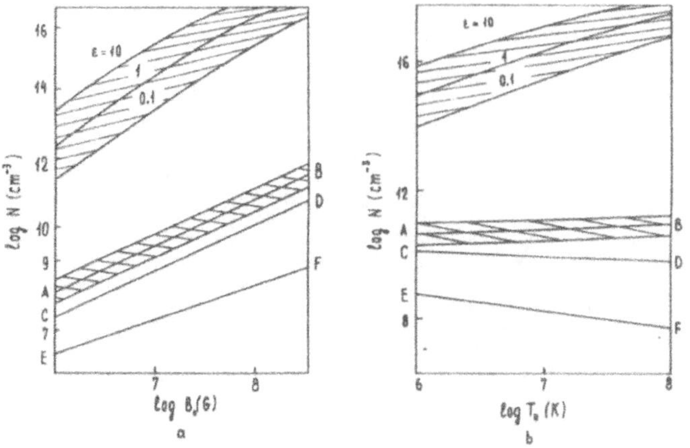

Figure 8.5. The regions of the parameters N, B_0 and $T_\parallel$ which correspond to qualitatively different cases of cyclotron radiation transfer in the magnetic white dwarf corona [41]: *a*, the plane lg $N -$lg B_0 at $T_\parallel = 10^7$ K; *b*, the plane lg $N -$lg $T_\parallel$ at $B_0 = 10^8$ G.

The hatched region at the diagrams of Fig. 8.5 centered at the line $\epsilon = 1$ separates the regions of collisional ($\epsilon > 10$) and collisionless ($\epsilon < 0.1$) plasma. At high densities N relevant to collisional limit the cyclotron lines in emission are formed in the corona. Brightness temperature in optically thick lines, $\tau_{js} \gg 1$, reaches the coronal value: $T_{\rm b} \simeq T_{\rm c}$. The strip AB at the diagrams separate the regions corresponding to the cases of "thick" and "thin" corona determined by the conditions (8.189) and (8.192), respectively. [24] The former case at which the outgoing radiation is strongly affected by thermalization is realized at higher plasma densities. The line CD defined by the equality

$$I^{\rm inc} = \frac{2\sqrt{\lambda}}{1+\sqrt{\lambda}} I_\omega^{(0)}(T_\parallel) \tag{8.211}$$

(see the formula (8.199)) separates the regions of parameter values at which cyclotron lines are seen in emission or in absorption. Location of this line depends on the ratio $T_{\rm ph}/T_\parallel$. In particular, the diagrams in Fig. 8.5 are plotted for $T_{\rm ph}/T_\parallel = 3 \cdot 10^{-3}$. Coronal plasma with parameters below the line CD can only scatter the photospheric radiation at gyrofrequency. The line EF at the diagrams corresponds to the unit optical depth for the extraordinary mode at the first harmonic: $\tau_{\rm c} = \mu_{11} H = 1$. Therefore, cyclotron scattering in the region below this line is insignificant, and the resulting absorption feature is rather weak.

[24] The parameter $2\sqrt{\lambda}\tau_{\rm c}$ ranges from 0.3 to 3 within the strip AB.

It is clear from the above discussion that a rather strong cyclotron line in absorption can be formed by a plasma layer with the parameters in the region between EF and CD. At $B_0 \simeq 10^8$ G and $T_{\parallel} \simeq 10^7$ K this corresponds to electron density

$$2 \cdot 10^8 < N < 6 \cdot 10^{10} \mathrm{cm}^{-3}.$$

The inhomogeneity of the magnetic field over the white dwarf surface sweeps the narrow cyclotron line into a broad, though shallower, depression band in the integral radiation flux from the star. An example of such a depression feature is presented in Fig. 8.4, *a*.

Thus, the appearance of strong absorption features in the spectra of magnetic white dwarfs evidences the existence of plasma coronas with densities within the quite narrow interval indicated above. Investigation of cyclotron features in magnetic white dwarf spectra is a unique source of the information on the plasma envelopes of these objects. Indeed, soft X-ray flux from a thin plasma corona with density $N < 6 \cdot 10^8 \mathrm{cm}^{-3}$ surrounding such a small star as a white dwarf is too weak to be detected with modern X-ray facilities. On the contrary, spectra of white dwarfs at cyclotron frequencies are very sensitive to the electron densities which are one-three orders of magnitude less than the ones corresponding to detection limits of X-ray telescopes.

It will be shown in the next section 8.3 that the structure of plasma envelopes of hot dwarfs with $T_{\mathrm{ph}} \gtrsim 2 \cdot 10^4$ K is drastically different. The pressure force by the photospheric radiation at cyclotron frequencies may exceed the gravitational force on a proton in the vicinity of such stars. This leads to ejection of plasma from the photosphere and the formation of an extended envelope in the star magnetosphere as well as the disk near the magnetic equator [39, 444]. If the plasma density of such an envelope is large enough, it may strongly distort the photospheric spectrum and give rise to the broad and deep depression bands in the observed radiation from such an object called radiation-driven diskon. The magnetic white dwarf GD 229 gives an example of such a spectrum (Fig. 8.5 *b*). This star exhibits a strong depression band in ultraviolet at wavelengths 2000-3000 Å.

In an extended plasma envelope of the size of several white dwarf radii the criterion of a strongly inhomogeneous magnetic field $H \gg \beta_{T_{\parallel}} L_B$ is satisfied. In this case, as has already been noted above, cyclotron scattering at frequency ω takes place in narrow gyroresonant layers $\omega \simeq \omega_B$ (see (6.60)). The optical depth τ_g of a gyroresonant layer to cyclotron scattering of the extraordinary mode is given by the formula (8.161), and the outgoing intensity at $\tau_g > 1$ is described by (8.171). Since the influence of the layer $\omega \sim \omega_B$ on the ordinary radiation is negligible compared to that on the extraordinary one, the intensity in the depression band equals 3/4 of the photospheric level. Using the formula (8.161) it is easy to see that a strong

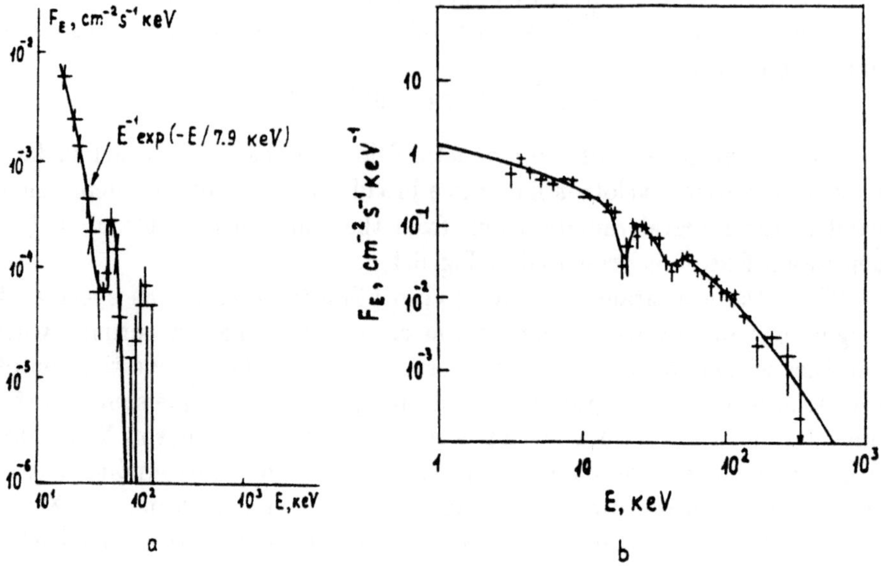

Figure 8.6. X-ray spectra (spectral flux of photons F_E versus photon energy E) of the X-ray pulsar Her X-1 (a) [44] and the gamma-ray burst GB 870303 (b) [445].

depression is formed in the plasma envelope around the white dwarf GD 229 with $B_0 \sim 5 \cdot 10^8$ G and $L_B \simeq R_*/3 \sim 3 \cdot 10^8$ cm if electron density $N > 10^8$ cm^{-3}.

Strong magnetic fields of neutron stars may also influence the observed radiation spectra. Cyclotron effects in magnetic fields $B_0 \sim 10^{12}$ G appear in the X-ray band and may be important for X-ray pulsars and gamma-ray bursts. Examples of the spectra observed from these objects are given in Fig. 8.6.

The possibility of the formation of cyclotron features in the spectra of X-ray pulsars was predicted in [446] before the discovery of cyclotron lines in spectra of X-ray sources. X-ray pulsars are neutron stars in binary systems. The second component in such a system supplies the matter to be accreted onto magnetic pole regions of the neutron star. The observed X-rays are emitted by plasma with temperature $T \sim 10^8$ K (deduced from the continuum spectra of such sources) in polar hot spots. Rotation of the the neutron star leads to the pulsations of the radiation intensity. Since hot plasma is localized near the star surface at magnetic pole region, cyclotron radiation transfer can be considered in the approximation of a quasi-homogeneous magnetic field. Cyclotron lines formed in this case are rather narrow. This allows one to determine the magnetic field strengths of neutron stars — X-ray pulsars using the positions of cyclotron lines in their spectra. For example, $B_0 \simeq 4 \cdot 10^{12}$ G for Her X-1 and $B_0 \simeq 10^{12}$ G for 4U 0115-69.

Since the transfer of cyclotron radiation in plasmas on magnetized neutron stars is dominated by scattering, it has been assumed [45] that cyclotron lines in the spectra of X-ray pulsars are formed in a way similar to Fraunhofer lines in ordinary stars. We repeat that the majority of the latter occur due to resonant scattering caused by electron transitions between discrete energy levels of atoms and ions. This idea has been developed in [184,447]. It has been assumed that the source of X-ray emission in a continuum and cyclotron line is an isothermal plasma atmosphere with the barometric density profile (8.202). The temperature in the source $T \sim 10^8$ K is estimated using the observed shape of the continuum spectrum. Despite the high temperature, the scale height of the plasma in a strong gravitational field of a neutron star is rather small: $H \sim 10^2$ cm. According to the analysis given above (see the formula (8.204)), the intensity of radiation in a continuum equals

$$I^{\text{out}} \simeq I_\omega^{(0)} M_{\text{T}}^{1/3}. \tag{8.212}$$

For neutron star conditions the parameter $M_{\text{T}} \sim 10^{-6}$, so the intensity of radiation in a continuum is approximately two orders of magnitude less than the blackbody one for the adopted plasma temperature. This fact should be accounted for when the distance to an X-ray pulsar is estimated using the observed radiation flux [104]. Respectively, radiation intensity in the center of the cyclotron line $\omega = \omega_B$ can be estimated using (8.209):

$$I^{\text{out}} \simeq I_\omega^{(0)} M_{\text{c}}^{1/3}. \tag{8.213}$$

Here $M_{\text{c}} = \epsilon / \tau_{\text{c}} \sim 10^{-11}$, the parameter ϵ is given by (7.57), (7.59), and τ_{c} is the optical depth of the atmosphere for cyclotron scattering in magnetized vacuum. The intensity in the line center is only about $2 \cdot 10^{-4}$ of the blackbody level and also much lower than the continuum one. As a result, a line in absorption is formed in the spectrum of an X-ray pulsar due to strong cyclotron scattering with $\mu_{\text{c}} \gg \mu_{\text{T}}$. Such a line appears in a continuum which is also depressed far below the blackbody level due to the effect of Thomson scattering.

Note that the approximate formulas (8.212), (8.213) are given to illustrate the role of cyclotron effects on neutron stars. These formulas, as well as more complicated ones (8.204) and (8.209), neglect the angular dependence of scattering coefficients, mode conversion and the recoil effect. Moreover, the model of an isothermal hot spot with the barometric density profile is not the only one discussed in the theory of X-ray pulsar radiation. For example, the case of strong accretion, when the density of the accretion flow is large enough to ensure its effective cyclotron interaction with radiation, is better described by the alternative model of an accretion column extended along the magnetic field above the poles.

Consider now the formation of cyclotron lines in the spectra of cosmic gamma-ray bursts. Such lines have been discovered by *KONUS* detectors onboard the Soviet "*Venera-11*" and "*Venera-12*" probes [47]. Absorption features detected in gamma-burst spectra indicate that magnetic fields in the burst sources are similar to X-ray pulsar ones. Cyclotron lines in gamma-burst spectra have been also seen in many observations thereafter. The latest results of this kind are the *PHOBOS-2* [49] and *Ginga* [445] detections of absorption features at multiple frequencies. An example of cyclotron features in a gamma-burst spectrum is shown in Fig. 8.6, *b*.

Cyclotron harmonic lines in gamma-ray burst spectra is one of the most convincing arguments in favour of the neutron star model of their sources. However, copious observations of gamma-bursts by the *GRO* facility have yielded no example of line features in their spectra. In addition, distribution of gamma-burst sources over the celestial sphere has been found to be highly isotropic and shown no correlation with the position of the Galactic Center or disk. These results gave an impact to the investigations of other possible scenarios, including extra-galactic ones, for the gamma-ray burst phenomenon (see section 1.1).

Bearing in mind the above remarks, consider nevertheless the strongly magnetized neutron star model of gamma-burster. The above model of an isothermal hot spot cannot explain the appearance of relatively narrow cyclotron lines on a hard X-ray continuum. Such a continuum with characteristic photon energy $\hbar\omega \sim 100\,\mathrm{keV}$ can only be emitted by particles of similar energy. However, the Doppler width of a cyclotron line formed in such a hot plasma would be much larger than the observed ones. Explanation of narrow cyclotron features requires the presence of relatively cold plasma with mean particle energy of about several tens keV above the hot spot. Such a two-layer model has been proposed in [448] and considered in detail in [54, 314, 436]. It is quite probable that the colder upper layer comprises electron-positron plasma. Electron-positron pairs can be created in the strong magnetic field of the neutron star by the photons with energy in excess of 1 MeV from the hard tail of the hot spot spectrum. The formation of spectral lines at cyclotron harmonics has been studied analytically in [54] using the transfer equation (8.172). The Monte-Carlo calculations of cyclotron radiation transfer has been performed in [436]. Both methods yield the parameters of the line-forming region. In addition, upper limits for the distances to the gamma-bursters have been obtained in former work which reveals that the bursts considered are Galactic in origin. Unfortunately, both papers have missed some essential processes which strongly influence the formation of cyclotron lines in gamma-burst spectra.

8.3. Radiation pressure force in plasma

Calculation of the radiation pressure force is one of the traditional problems in astrophysics, and the Eddington limit is an important and useful characteristic of an astrophysical object. Investigations of radiation pressure have recently received an additional impact due to the development of the theory of accreting compact objects. In particular, radiation pressure due to Thomson scattering by free electrons is analyzed in [449–453]. If the scattering cross-section exceeds the Thomson one, e.g. due to resonant scattering in spectral lines, the radiation pressure force can greatly increase. This fact is well known in the theory of stellar atmospheres [454–457]. Ion resonance line radiation pressure plays the leading role in driving winds of early-type stars [177,455,458,459]. In strong magnetic fields of degenerate stars electron gyrofrequency ω_B may shift to the optical or X-ray band. Resonant cyclotron scattering of radiation by a degenerate star in surrounding plasma may considerably enhance the radiation pressure force. This effect and its astrophysical applications are considered in detail later on.

8.3.1. RADIATION PRESSURE DUE TO THOMSON SCATTERING

It is known that the momentum of radiation with specific intensity I_ω equals I_ω/c. In case of scattering this momentum is gained by electrons. As a result, the radiation pressure force on the electron in a plasma with $n_j \simeq 1$ appears which equals the decrease of radiation momentum per unit time:

$$\boldsymbol{f} = \int_\omega \oint_{4\pi} \frac{\sigma_{\rm sc} I_\omega}{c} \boldsymbol{l}\, d\Omega d\omega. \tag{8.214}$$

Here $\sigma_{\rm sc}$ is the scattering cross section, $\boldsymbol{l} = \boldsymbol{k}/k$ is unit vector along the wave vector $\boldsymbol{k}$, and $d\Omega$ is the solid angle element. Note that the formula (8.214) is valid if the total momentum of the scattered radiation is equal to zero.

If an electron is located near the surface of a star which emits isotropic radiation into the upper hemisphere, the radiation pressure force is directed radially outward and is equal to

$$f = \frac{\sigma_{\rm T}}{c} \int_\omega \pi I_\omega \, d\omega = \frac{\sigma_{\rm T} \mathcal{L}}{4\pi R_*^2 c}. \tag{8.215}$$

Here

$$\mathcal{L} = 4\pi R_*^2 \int F_\omega \, d\omega \tag{8.216}$$

is the luminosity of a star of radius R_*, $F_\omega = \pi I_\omega$ is the spectral flux of radiation through the star surface, and the scattering cross-section is equal to the Thomson one (8.62) which characterizes scattering of radiation by a free, non-relativistic electron in the absence of a magnetic field. At the same time, gravitational force on a proton near the stellar surface is given by

$$f_g = GM_* m_p / R_*^2, \tag{8.217}$$

where G is the gravitational constant and M_* is the mass of the star. The forces (8.215) and (8.217) become equal if the star luminosity equals the critical, or Eddington, value

$$\mathcal{L}_{cr} = \frac{4\pi GM_* m_p c}{\sigma_T}. \tag{8.218}$$

If the radiation pressure force f on an electron exceeds the gravitational force f_g on a proton with the mass $m_p \gg m_e$, a quasi-neutral plasma is ejected from the star surface by radiation pressure. The inequality

$$\mathcal{L} > \mathcal{L}_{cr} \simeq 10^{38} \frac{M_*}{M_\odot}, \tag{8.219}$$

where $M_\odot = 2 \cdot 10^{33}$ g is the solar mass, constitutes the criterion for the appearance of a stellar wind driven by Thomson scattering. The same criterion determines the luminosity at which the radiation pressure force on free electrons prevents accretion of optically thin plasma onto a compact object. Thus, $\mathcal{L}$ is an upper limit for the luminosity of an accreting X-ray source (white dwarf or neutron star).

Let us now find the radiation pressure force in a scattering plasma layer of a non-vanishing optical depth. Distribution of radiation intensity in such a layer of the size Δh is described by the radiation transfer equation (4.13) in which $a_\omega = 0$ and $\mu_j = 0$. Using the approximate method of Schwarzschild-Shuster, one can find the intensities I_1, I_2 in this layer (see (4.55)) and get the value

$$I_1 = \frac{I^{\rm inc}}{1 + \tau_T} \tag{8.220}$$

for the outgoing intensity (4.59). Here $I^{\rm inc}$ is the incoming intensity, or the value of I_1 at the lower boundary of the layer, $\mu_T = \sigma_T N$ is the Thomson scattering coefficient in plasma with electron density N, and $\tau_T = \mu_T \Delta h$ is the corresponding optical depth of the layer. To obtain the approximate value for the radiation pressure force (8.214) on an electron,

$$f_T = \int_\omega \int_0^{2\pi} I_\omega \cos\theta \sin\theta \, d\theta d\omega, \tag{8.221}$$

one should, according to the Schwarzschild-Schuster method, substitute $\cos\theta$ with 1/2 at the interval $0 \le \theta \le \pi/2$ and with -1/2 for $\pi/2 \le \theta \le \pi$.

Then (8.221) becomes

$$f_{\mathrm{T}} \simeq \frac{\pi \sigma_{\mathrm{T}}}{c} \int_{\omega} (I_1 - I_2)\, d\omega .$$

According to (4.57), the difference I_1–I_2 does not depend on the coordinate h inside a pure scattering layer with $p = 0$. Since $I_2 = 0$ at the upper boundary of the layer, this difference equals the outgoing intensity (8.220) at any point inside. Therefore,

$$f_{\mathrm{T}} \simeq \frac{\pi \sigma_{\mathrm{T}}}{c} \frac{1}{1+\tau_{\mathrm{T}}} \int_0^{\infty} I^{\mathrm{inc}}\, d\omega . \tag{8.222}$$

If plasma layer is illuminated by the equilibrium radiation with the Planck spectrum (1.44), the integral over frequency in (8.222) yields $(\sigma/\pi)T^4$, where $\sigma = \pi\kappa^4/60\hbar^3c^2$ is the Stefan-Boltzmann constant. In this case the force (8.222) equals

$$f_{\mathrm{T}} = \frac{\sigma_{\mathrm{T}}}{c} \frac{\sigma T^4}{1+\tau_{\mathrm{T}}} . \tag{8.223}$$

According to (8.222) and (8.223), the radiation pressure force f_{T} does not depend on the location of an electron in a scattering layer and decreases with increasing optical depth.

8.3.2. PRESSURE FORCE BY RADIATION AT CYCLOTRON FREQUENCIES

Pressure tensor for radiation from a unit frequency range can be introduced as

$$\mathcal{P}_{ik} = \frac{1}{c} \oint_{4\pi} I_{\omega}\, l_i l_k \, d\Omega , \tag{8.224}$$

where l_i, l_j are the projection of the vector $\boldsymbol{l} \parallel \boldsymbol{k}$ onto i-th and j-th coordinate axes, respectively. Each component of the tensor (8.214) equals the j-th projection of the momentum flux of radiation from unit frequency range through a unit surface perpendicular to the i-th axis. To calculate the tensor (8.224) for radiation at cyclotron frequencies $\omega \simeq \omega_B$ one should take into account that such a radiation from unit frequency range can only interact with particles from the momentum interval

$$\left| \frac{\partial p_{\parallel}}{\partial \omega} \right| = \frac{m_e c}{\omega_B |\cos\alpha|} \tag{8.225}$$

(see (8.138)). Radiation pressure on particles from the unit interval of momenta is then determined by the tensor

$$\mathcal{P}_{ij} = \frac{1}{c} \oint_{4\pi} I_{\omega} l_i l_j \left| \frac{\partial \omega}{\partial p_{\parallel}} \right| d\Omega , \tag{8.226}$$

so that $-\mathrm{div}\mathcal{P}_{ij}$ equals the force on particles from the unit interval of longitudinal momenta $p_{\|}$. Finally, dividing this value by $f(p_{\|})$, the density of particles from unit momentum range, we arrive at the expression for the cyclotron radiation pressure force per particle:

$$\boldsymbol{f}_B = -f^{-1}(p_{\|})(\boldsymbol{l}\nabla)\left(\oint_{4\pi} \frac{I_\omega(p_{\|},\alpha)}{c}\boldsymbol{l}\left|\frac{\partial\omega}{\partial p_{\|}}\right| d\Omega\right). \tag{8.227}$$

Let a plane-parallel plasma layer be permeated by a homogeneous magnetic field $\boldsymbol{B}_0$ perpendicular to its boundaries. Then the symmetry of the configuration makes the tensor $\mathcal{P}_{ij}$ diagonal. The force $\boldsymbol{f}_B$ is directed vertically, its value is given by

$$f_B = \frac{2\pi\omega_B}{m_e c^2 f(p_{\|})}\left[-\frac{\partial}{\partial z}\int_0^\pi I_\omega(p_{\|},\alpha)\cos^2\alpha|\cos\alpha|\sin\alpha\, d\alpha\right]. \tag{8.228}$$

To derive this formula, multiply the radiation transfer equation (8.150) by $|\cos\alpha|\cos\alpha\sin\alpha$ and integrate over α from 0 to π. The resulting equation clearly shows that the expression in square brackets in (8.228) equals

$$\frac{\sqrt{2}\pi^{3/2}e^2 N}{(m_e\kappa T)^{1/2}\omega_B}\, e^{-p_{\|}^2/2m_e\kappa T_{\|}} \int_0^\pi (1+\cos^2\alpha)\cos\alpha I_\omega(z,p_{\|},\alpha)\sin\alpha\, d\alpha.$$

Assuming that distribution of plasma particles $f(p_{\|})$ is Maxwellian, we finally obtain the following expression for cyclotron radiation pressure force per particle [460]:

$$f_B(p_{\|}) = \frac{4\pi^3 e^2}{m_e c^2}\int_0^\pi (1+\cos^2\alpha)\cos\alpha I_\omega(z,p_{\|},\alpha)\sin\alpha\, d\alpha. \tag{8.229}$$

The integral in the expression (8.229) can be found using the solution of the equation (8.150) in the Schwarzschild–Shuster approximation. As follows from the solution of the system (8.152),

$$f_B(p_{\|}) = \frac{3\pi^3 e^2}{m_e c^2}(I_1 - I_2), \tag{8.230}$$

where I_1 and I_2 are given by the formulas (8.153). Note that the difference

$$I_1 - I_2 = \frac{I^{\mathrm{inc}}(p_{\|})}{1 + \int_0^H \mu_{11}(z,p_{\|},\alpha=0)\, dz}, \tag{8.231}$$

where

$$I^{\mathrm{inc}}(p_{\|}) = \int_0^{\pi/2} I_\omega^{\mathrm{inc}}\left(\omega_B(1 - p_{\|}\cos\alpha/m_e c)^{-1}, \alpha\right)\sin\alpha\, d\alpha \tag{8.232}$$

$I_\omega^{\rm inc}$ is the intensity of the incoming radiation, and H is the height of the layer, is independent on the coordinate z. If $I_\omega^{\rm inc}$ is isotropic over the interval $0 \le \alpha \le \pi/2$ and slowly varies over the width of the cyclotron line, the intensity (8.232) can be approximated as

$$I^{\rm inc}(p_\|) \simeq \int_0^{\pi/2} I_\omega^{\rm inc}(\omega_B) \sin\alpha \, d\alpha = I^{\rm inc}. \tag{8.233}$$

The obtained value is independent on $p_\|$. Then

$$f_B(p_\|) = \frac{3\pi^3 e^2}{m_e c^2} \frac{I^{\rm inc}}{1 + \tau_c \exp(-p_\|^2/2m_e\kappa T_\|)}, \tag{8.234}$$

where

$$\tau_c = \frac{(2\pi)^{3/2} e^2}{m_e c \beta_{T_\|} \omega_B} N_H \tag{8.235}$$

is the cyclotron scattering optical depth in the center of the line, and

$$N_H = \int_0^H N(z)\, dz$$

is the column density of the layer.

According to (8.234), the radiation pressure force depends on the longitudinal momentum of a particle. It increases with increasing modulus $|p_\||$, i.e. with increasing detuning from the line center, due to the decrease of the corresponding optical depth.

Averaging of the force (8.234) over the Maxwell distribution leads to the expression for the average force per particle [460]:

$$f_B = f_B^{\max} \Psi(\tau_c). \tag{8.236}$$

Here

$$f_B^{\max} = \frac{3\pi^3 e^2}{m_e c^2} I^{\rm inc} \tag{8.237}$$

is the maximal average force per particle, and the integral factor

$$\Psi(\tau_c) = \frac{1}{\sqrt{\pi}} \int_{-\infty}^{\infty} \frac{d\xi}{e^{\xi^2} + \tau_c} \tag{8.238}$$

is close to unity at $\tau_c \ll 1$ and obeys the asymptotic expansion $\Psi \sim \tau_c^{-1}\sqrt{\ln \tau_c}$ at large τ_c. The following approximation for the factor $\Psi(\tau_c)$ is frequently used in the theory of line radiation transfer:

$$\Psi(\tau_c) \simeq \min\{1; \tau_c^{-1}\}. \tag{8.239}$$

Magnetic field leads to the considerable increase of the radiation pressure force if

$$\Gamma = f_B/f_{\rm T} \gg 1, \tag{8.240}$$

where f_B and $f_{\rm T}$ are given by the formulas (8.236) and (8.222), respectively. The function $\Gamma(N_H)$ is shown in Fig. 8.7. It is clearly seen that Γ decreases monotonically with increasing N_H from the maximal value at $N_H \to 0$,

$$\Gamma_{\max} = \frac{9\pi}{8}\frac{\hbar c}{e^2}\frac{m_e c^2}{\hbar}\frac{I^{\rm inc}(\omega_B)}{\int_0^\infty I_1^{\rm inc}\,d\omega} \tag{8.241}$$

down to the minimal one,

$$\Gamma_{\min} = \frac{3}{2}\sqrt{\frac{\pi}{2}}\beta_{T_\parallel}\omega_B\frac{I^{\rm inc}(\omega_B)}{\int_0^\infty I_1^{\rm inc}\,d\omega}, \tag{8.242}$$

which corresponds to the opposite limit $N_H \to \infty$. If the incoming radiation has the blackbody spectrum characterized by the temperature T, then the formula (8.243) is transformed into

$$\Gamma_{\max} = \frac{135}{16\pi^3}\frac{\hbar c}{e^2}\frac{m_e c^2}{\kappa T}\frac{(\hbar\omega_B/\kappa T)^3}{\exp(\hbar\omega_B/\kappa T) - 1} \tag{8.243}$$

Of course, this expression is identical to the one obtained in [461] in which the pressure force on a particle in the radiation field with a blackbody spectrum has been calculated without taking into account the effects of the non-vanishing optical depth $\tau_c \neq 0$.

If $\Gamma_{\max} \ll 1$, the contribution of cyclotron radiation to the total radiation pressure force does not depend on the optical depth τ_c (on the column density N_H). The influence of cyclotron radiation pressure may be strong if $\Gamma_{\max} \gg 1$. In this case the inequality $\Gamma > 1$ is realized under the condition [460]

$$\tau_c < \Gamma_{\max} \tag{8.244}$$

(see (20.28), (8.239), (8.222)). The ratio of cyclotron radiation pressure force f_B to the gravitational force on proton $f_g = m_p g$ equals

$$\Delta = \frac{3\pi^3 e^2}{m_e m_p c^2} I^{\rm inc}(\omega_B)\min\{1; \tau_c^{-1}\}\frac{R_*^2}{GM_*}. \tag{8.245}$$

If the incoming radiation has the blackbody spectrum with temperature T, the quantity

$$I^{\rm inc}(\omega_B) = \frac{\hbar\omega_B^3}{(2\pi)^3 c^2}\frac{1}{\exp(\hbar\omega_B/\kappa T) - 1} \tag{8.246}$$

(see (1.44) (8.233)).

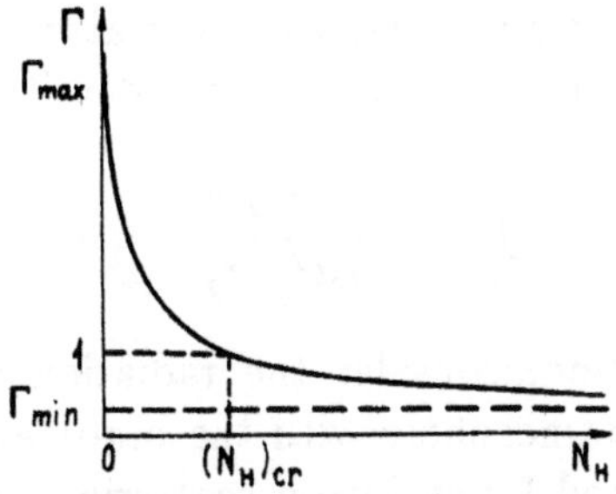

Figure 8.7. The ratio Γ of the radiation pressure forces in cyclotron line and in a continuum as a function of the optical depth τ_c determined by (8.235) [460].

The above formulas are related to the case when the mode polarization is determined by plasma. If, on the contrary, the criterion (2.187) is met and the mode polarization is dominated by the effect of vacuum birefringence, one may ignore the difference between the cyclotron absorption (scattering) coefficients of extraordinary and ordinary radiation, μ_{11} and μ_{21}, and put $\mu \simeq (\mu_{11} + \mu_{21})/2$. The resulting optical depth $\tau_c \simeq (\mu_{11} + \mu_{21})N_H/2$ is two times less than (8.235), while the radiation intensity used in calculations should comprise both components.

Transfer of radiation in extended plasma envelopes in a strongly inhomogeneous magnetic field is drastically different from the case of a quasi-homogeneous field (see 8.2). This also changes the value of the cyclotron radiation pressure force. General expression for this force (8.229) is the same for both cases. However, the intensity of cyclotron radiation $I_\omega(z, p_{||}, \alpha)$ in a strongly inhomogeneous magnetic field is determined by the formulas (8.168), (8.169). This leads to the following expression for the radiation pressure force per particle with the longitudinal momentum $p_{||}$:

$$f_B(p_{||}) = \frac{3\pi^3 e^2}{m_e c^2} I^{\rm inc}(\omega_B) e^{-\tau_g y}, \tag{8.247}$$

where

$$I_{\rm inc} = \int_0^{\pi/2} I_\omega^{\rm inc}(\omega_B, \alpha) \sin\alpha \, d\alpha,$$

$I_\omega^{\rm inc}(\omega_B, \alpha)$ is specific intensity of radiation incident on the gyroresonant layer $\omega \simeq \omega_B$, τ_g is the optical depth of the gyroresonant layer (8.161), and the variable y is introduced by (8.162). Averaging of (8.247) over the Maxwell distribution $f(p_{||})$ yields the mean force per particle [460]:

$$f_B = \frac{3\pi^3 e^2}{m_e c^2} I^{\rm inc}(\omega_B) \frac{1 - \exp(-\tau_g)}{\tau_g}. \tag{8.248}$$

Respectively, the force on particles in the unite volume equals $f_B N$. The cyclotron radiation force in an optically thick, scattering gyroresonant layer $\tau_g \gg 1$ is inversely proportional to τ_g or plasma density at the point $\omega \simeq \omega_B$:

$$f_B \simeq \frac{3\pi^3 e^2}{m_e c^2} \frac{I^{\rm inc}}{\tau_g}. \tag{8.249}$$

Concluding this section, consider the radiation pressure force in an inhomogeneous plasma atmosphere with the barometric density profile in which radiation is affected by cyclotron scattering and particle collisions [39,438]. The forthcoming analysis will allow us to find the stability criterion of such an atmosphere against radiation pressure. Violation of this criterion inevitably leads to the radiation-driven mass loss from the surface of hot, strongly magnetized stars. Moreover, it will be shown below that cyclotron radiation pressure may strongly influence the structure and dynamics of plasma envelopes formed by the ejected matter in the magnetospheres of these stars (see the following section).

Consider an isothermal plasma atmosphere with non-relativistic longitudinal temperature $T_{\|} =$ const and the barometric density profile (8.202). The magnetic field is considered to be homogeneous and directed along the z-axis.[25] Using again the expression (8.229) for the cyclotron radiation force per particle and taking into account the formulas (8.205) which describe the radiation intensity in the Schwarzschild–Shuster approximation, we obtain the following approximate formula [438]:

$$f_B = \frac{6\pi^3 e^2}{m_e c^2 \sqrt{\pi}} M_{\rm c}^{1/3} I_\omega^{(0)}(\omega_B) \int_{-\infty}^{+\infty} \frac{-{\rm Ai}'(\Phi^{2/3} M_{\rm c}^{1/3} \tau_{\rm c}) \Phi \, d\xi}{{\rm Ai}(0)\Phi^{1/3} - {\rm Ai}'(0) M_{\rm c}^{1/3}}, \tag{8.250}$$

where Ai and Ai′ denote the Airy function and its derivative, respectively, $\Phi(\xi) = \exp(-\xi^2)$, the parameter $M_{\rm c} = \epsilon/\tau_{\rm c}$, and the optical depth $\tau_{\rm c}$ is determined by the formula (8.206).

The integral in (8.250) can be easily estimated in the approximation of the rectangular line profile. Since this integral is contributed to mainly by the interval $|\xi| \lesssim 1$, we put $\Phi = \Phi(0) = 1$ if $|\xi| \le \sqrt{\pi}/2$ and $\Phi = 0$ otherwise. Then (8.250) is reduced to the following expression:

$$f_B \simeq \frac{6\pi^3 e^2}{m_e c^2} M_{\rm c}^{1/3} I_\omega^{(0)}(\omega_B) \frac{-{\rm Ai}'(M_{\rm c}^{1/3} \tau_{\rm c})}{{\rm Ai}(0)}. \tag{8.251}$$

To get the last expression, the second term in the denominator of the integrand in (8.250) has been neglected since it is small compared to the first one. According to (8.251), the force f_B is maximal at the upper layers

[25] The general case of an oblique magnetic field $\mathbf{B}_0$ has been considered in [462].

of the atmosphere where $\tau_c \to 0$ and decreases monotonically with increasing τ_c. The latter is explained by the effect of thermalization which makes the radiation field isotropic and thus nullifies the force f_B.

The hydrostatic equilibrium of the atmosphere is disrupted if the radiation pressure force per unit volume, $f_B N$, exceeds the gravitational force $m_p g N$. This condition may be written in the form

$$\Delta = f_B / m_p g > 1. \tag{8.252}$$

According to this criterion, hydrogen plasma can be ejected from the photosphere if the criterion (8.252) is satisfied at its upper boundary where the radiation pressure force equals

$$f_B \simeq \frac{6\pi^3 e^2}{m_e c^2} M_c^{1/3} I_\omega^{(0)}(\omega_B) \frac{-\mathrm{Ai}'(0)}{\mathrm{Ai}(0)}. \tag{8.253}$$

8.3.3. RADIATION PRESSURE EFFECTS ON MAGNETIC DEGENERATE STARS

In section 8.2 the formation of cyclotron depression bands due to cyclotron scattering of radiation in geometrically thin plasma coronas of magnetic white dwarfs was discussed. The existence of such envelopes with a scale height much less than the star radius is only possible if the radiation pressure force is negligible as compared to the force of gravity. The estimates show that for typical white dwarf parameters the component of radiation pressure force due to Thomson scattering is much less than the one caused by resonant cyclotron scattering. The influence of the latter component can be estimated using Fig. 8.7 and the relevant formulas. For example, at $B_0 \simeq 2 \cdot 10^8$ G and $T \simeq 6 \cdot 10^3$ K the ratio $\Gamma_{max} \simeq 4 \cdot 10^7$ is considerably larger than unity. According to Fig. 8.7, radiation pressure in the cyclotron line is much greater that the force by continuum radiation if $\tau_c \ll \Gamma_{max}$. To form a pronounced depression band, the optical depth to cyclotron scattering should be large enough: $\tau_c > 1$. In the case $\tau_c \simeq 1$ the ratio (8.245) is less than unity: $\Delta \simeq 0.3$ if $R_* \simeq 10^9$ cm, $M_* \simeq M_\odot \simeq 2 \cdot 10^{33}$ g, and the intensity $I_{inc}(\omega_B)$ is given by (8.246). Therefore, plasma coronas of white dwarfs are stable, and the radiation pressure is generally unimportant. However, at higher plasma temperatures the existence of a thin corona around such stars is impossible.

If hydrostatic equilibrium of plasma on magnetic white dwarfs can be disrupted by large radiation pressure and the radiation-driven ejection from the star photosphere is possible, the more rigorous analysis of plasma stability is necessary. In this case one should take into account that

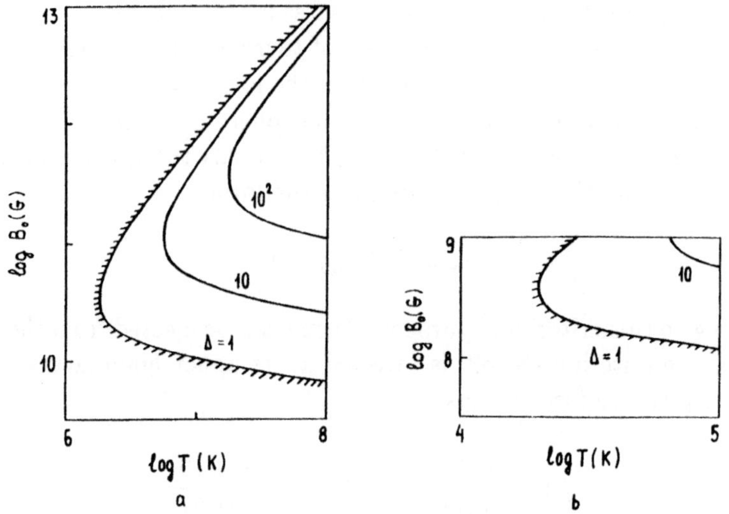

Figure 8.8. The isolines of the parameter Δ, the ratio of radiation pressure force to the force of gravity in the upper layers of the photosphere, plotted for: (*a*) magnetic white dwarf with free-fall acceleration $g \simeq 10^8$ cm s^{-2}; (*b*) magnetized neutron star with $g \simeq 10^{14}$ cm s^{-2} [39, 438, 462]. The region to the right from the curve $\Delta = 1$ corresponds to the parameters at which radiation-driven ejection of plasma from the photosphere is possible.

radiation flux in the cyclotron line is attenuated[26] and use the formulas (8.251) and (8.253) to estimate cyclotron radiation force inside such a photosphere at its upper boundary, respectively. Hydrostatic equilibrium of the photosphere is impossible if the condition (8.252) is met at its upper boundary. Radiation-driven ejection of plasma may occur in this case. On the contrary, the photosphere is stable if $\Delta < 1$. The line $\Delta = 1$ separating these cases is drawn in Fig. 8.8 on the plane $\lg B_0 - \lg T_{||}$. The diagrams in this figure correspond to the case when the magnetic field is directed radially from the star. This case is realized at the magnetic poles. However, the analysis [462] of the general case of an oblique magnetic field have shown that positions and shapes of isolines $\Delta =$ const are practically insensitive to deviation of the magnetic field from the radial direction.

According to Fig. 8.8, *a*, radiation-driven ejection of plasma from the surface of an isolated magnetic white dwarf is possible at $T_{||} \gtrsim 2 \cdot 10^4$ K and $B_0 \sim 10^8 - 10^9$ G. Two of the known white dwarfs, GD 229 ($T \simeq (1.5-2) \cdot 10^4$ K and $B_0 \gtrsim 5 \cdot 10^8$ G) and PG 1031+234 ($T \simeq 2.5 \cdot 10^4$ K and $B_0 \gtrsim 5 \cdot 10^8$ G) can be mentioned as examples of such objects. An intense plasma ejection is also

[26]Such an effect can be ignored when calculating the radiation pressure in hot white dwarf coronas illuminated from below by the photospheric radiation. Indeed, due to the dramatic difference in temperatures of photosphere and corona, the Doppler width (8.142) of the cyclotron line in the former is much less than in the latter.

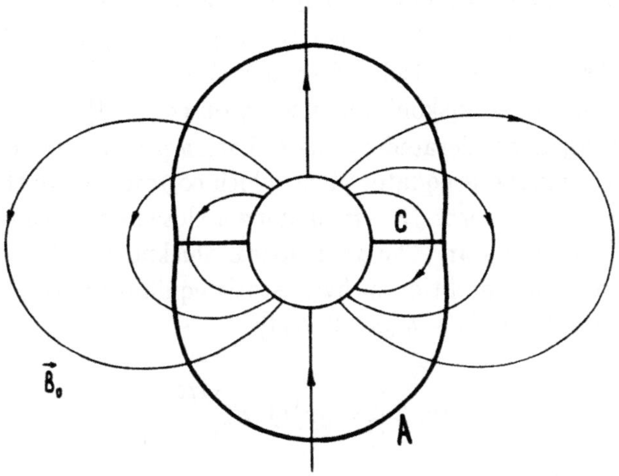

Figure 8.9. **The structure of plasma envelope formed by radiation pressure in the magnetosphere of a white dwarf [39]: A – closed shell; C – equatorial disk.**

possible in young white dwarfs with extremely high temperatures $T \gtrsim 10^5$ K born from planetary nebula nuclei in the course of their evolution. As far as magnetized neutron stars are concerned, radiation-driven plasma ejection may occur in young neutron stars with $T \gtrsim 2 \cdot 10^6$ K and $B_0 \simeq 10^{10} - 10^{11}$ G and also during the bursts of X-ray bursters – accreting neutron stars in binary systems.

The role of the radiation pressure force due to Thomson scattering in a continuum can be judged using the value of the critical temperature T_{cr} which equals $6 \cdot 10^5$ K and $2 \cdot 10^7$ K for white dwarfs and neutron stars, respectively. If $T > T_{cr}$, this force exceeds the force of gravity and leads to the plasma ejection even in the absence of a strong magnetic field. It is clear from the above estimates of T_{cr} and Fig. 8.8 that the magnetic field strongly reduces the maximal value of temperature at which the hot photosphere of a degenerate star remains stable against a radiation pressure force. Moreover, the presence of magnetic field may lead to the plasma ejection even at those values of the photospheric temperature for which the mass loss from non-magnetic stars is absolutely impossible.

Plasma ejected from the photosphere of a magnetic degenerate star can freely move along the field lines under the combined action of radiation pressure and gravity. At the same time, its transverse motion is suppressed by the strong magnetic field of a star. Thus, the transverse components of the above forces can only drive the drift currents in the plasma envelope formed by the ejected matter in the star magnetosphere.

Plasma ejected along the opened field lines in the magnetic pole regions forms a pair of counter-streaming jets along the magnetic axis of the star. Besides, an extended envelope and disk near the magnetic equator are formed in the closed part of the magnetosphere (see Fig. 8.9). The presence of a plasma disk around an isolated (non-accreting) star gave birth to the name "radiation-driven diskon" for such an object [444].

Contrary to jets accelerated by the radiation pressure force, the disk is formed near the magnetic equator plane which corresponds to the minimum of the potential of the force $f_B - m_p g$ along a closed field line of the dipole magnetic field of the star. The geometrical thickness of the disk can be determined from the condition of hydrostatic equilibrium of plasma under the action of the force $f_B - m_p g > 0$ [444]:

$$\mathcal{D} \simeq 2 \left[2\kappa T_d r^3 / f_0 R_*^2 \right]^{1/2} . \tag{8.254}$$

Here T_d is the plasma temperature in the disk at a distance r from the star center. This formula is valid if the dynamic pressure of the plasma flow feeding the disk, $\rho v^2/2$, is much less than the gas pressure in the disk, $P = 2N\kappa T$. Ordinary barometric distribution of plasma density along the closed magnetic field lines is maintained in this case. It should be mentioned that the formula (8.254) has been derived assuming that the difference between the force of radiation pressure and gravity is inversely proportional to the distance from the star:

$$f_B - m_p g = f_0 (R_*/r)^2 . \tag{8.255}$$

This formula accounts for the spatial variation of the force f_B due to the spherical dilution of the star radiation. However, the resonant frequency $\omega \simeq \omega_B$ of radiation providing this force decreases rapidly with r due to the decrease of the field strength B_0 in the dipole magnetosphere. As a result, the difference $f_B - m_p g$ is positive in the vicinity of the star where the radiation pressure is large and decreases to negative values at sufficiently large r. Therefore, this difference is nullified at some closed surface around the star (see the surface A in Fig. 8.9). The shape of this surface at which $f_B = m_p g$ has been investigated in [463, 464]. According to the later estimates [39, 465], such an equilibrium surface A in the dipole magnetosphere of the white dwarf GD 229 crosses the magnetic equator at the distance $r \simeq 2.2\, R_*$ from the star center and the magnetic axis at $r \simeq 3\, R_*$. The following parameters have been used to get these estimates: $T \simeq 2 \cdot 10^4$ K, $B_0 \simeq 10^9$ G, and $g \simeq 10^8$ cm s^{-2}. The hydrostatic profile of the plasma density near this surface is described by the formula [39]

$$N(\xi) = N_0 \exp(-\xi^2/\mathcal{D}^2), \tag{8.256}$$

in which $\xi = (r - r_0)/R_*$, r_0 is the distance from the star center to the equilibrium surface, and

$$\mathcal{D} = \frac{r_0}{R_*}\sqrt{\frac{2\kappa T_{\|}}{m_{\rm p} g R_*}} \tag{8.257}$$

is the characteristic width of the plasma layer.

The above analysis of the structure of the plasma envelope of the radiation-driven diskon is relevant to the case when the optical depth of gyroresonant layers in the magnetospheric plasma is negligibly small: $\tau_g \ll 1$. Since the optical depth $\tau_g \propto N$ (see (8.161)), this condition may fail if the magnetosphere of a hot white dwarf is steadily filled by the plasma of the photospheric wind. As a result, the new, optically thick, equilibrium configuration can be maintained. This new envelope with $\tau_g \gg 1$ extends from the photosphere to the equilibrium surface A. Since, according to (8.248), the radiation pressure force at large τ_g scales as $f_B \propto \tau_g^{-1}$, the density in an optically thick plasma envelope is determined by the following equilibrium condition [39,465]:

$$f_B^{\|} \simeq m_{\rm p} g \, \frac{R_*^2}{r^2} \, \cos\delta, \tag{8.258}$$

where δ is the angle between the radius-vector $\boldsymbol{r}$ and the magnetic field $\boldsymbol{B}_0$, and $f_B^{\|}$ denotes the projection of the radiation pressure force onto $\boldsymbol{B}_0$.

An optically thick plasma envelope of the radiation-driven diskon enclosing the whole star may strongly influence the spectrum of the white dwarf. The magnetic white dwarfs GD 229 and PG 1031+234 give examples of such spectra with strong and broad depression bands in the ultraviolet. The influence of plasma envelopes on the observed spectra of white dwarfs has been already discussed in the previous section 8.2. In this section estimates have also been given for the electron density of plasma at which the observed depressions can be formed in the spectra of these radiation-driven diskon candidates.

8.4. Raman scattering in plasma

Scattering of radiation by plasma particles has been considered in sections 8.1, 8.2. This section deals with effects called Raman (combinational) scattering which appear as coupling between different waves due to their nonlinear interaction. In plasma physics this process is known as coalescences and decays of waves in plasma. The most important process among these is the so-called three-wave interaction when the energy of a wave (or wave packet) changes due to the energy of two other waves.[27] The investigation of Raman scattering in plasma was started in [466] in which the scattering of electromagnetic waves on plasma ones was considered, and in [283] in which conversion of plasma waves into electromagnetic radiation was considered and the role of this process in explaining some components of solar sporadic radio emission analyzed. Today one can find a lot of publications on Raman scattering in plasma. We especially refer the reader to the monographs and reviews [72,89,381–384,467].

8.4.1. CONSERVATION LAWS FOR ELEMENTARY SCATTERING EVENTS

As a rule, wave phases in natural astrophysical plasmas are random. This favors the use of an elementary quantum approach instead of the classical one to describe scattering processes. According to this approach, an elementary act of Raman scattering of some waves (or wave packets) by the others can be treated as the coalescence of two photons (or plasmons) with the appearance of a single one [283]. Energy and momentum conservation in in this case demand that

$$\hbar\omega_1 + \hbar\omega_2 = \hbar\omega, \qquad \hbar\boldsymbol{k}_1 + \hbar\boldsymbol{k}_2 = \hbar\boldsymbol{k}. \tag{8.259}$$

In addition to the coalescence process $\boldsymbol{k}_1 + \boldsymbol{k}_2 \rightarrow \boldsymbol{k}$, the inverse one, a decay of a single plasmon into a pair: $\boldsymbol{k} \rightarrow \boldsymbol{k}_1 + \boldsymbol{k}_2$, is also possible under the same conditions (8.259). The conservation laws (8.259) called synchronism conditions appear in classical treatment as the necessary conditions ensuring three-wave interaction.

According to the dispersion relations for normal waves in plasma, the frequencies entering the first synchronism condition are the functions of the wave vector, the possibility to satisfy both of the conditions (8.259) is not *a priori* clear. If interacting waves are of the same type, they obey the same dispersion equation $\omega = \omega(\boldsymbol{k})$. The synchronism conditions can therefore be met if

$$\omega(\boldsymbol{k}_1) + \omega(\boldsymbol{k}_2) = \omega(\boldsymbol{k}_1 + \boldsymbol{k}_2) \tag{8.260}$$

The spectrum $\omega(\boldsymbol{k})$ is called decaying in this case and non-decaying otherwise. Examples of decaying and non-decaying spectra are shown in Fig. 8.10.

There is a rule allowing to judge whether a spectrum $\omega(\boldsymbol{k})$ is decaying or not [309]. According to (8.259), the vectors $\boldsymbol{k}_1$, $\boldsymbol{k}_2$ and $\boldsymbol{k}$ lie in the same plane, each having two components, k_x and k_y. With ω as the third coordinate, we get the three-dimensional space in which a point corresponds

[27] It will be shown further that a three-wave interaction is the second-order effect in radiation energy density $W_{\mathbf{k}}$. Higher-order interactions can be neglected for the moderate values of $W_{\mathbf{k}}$ considered below.

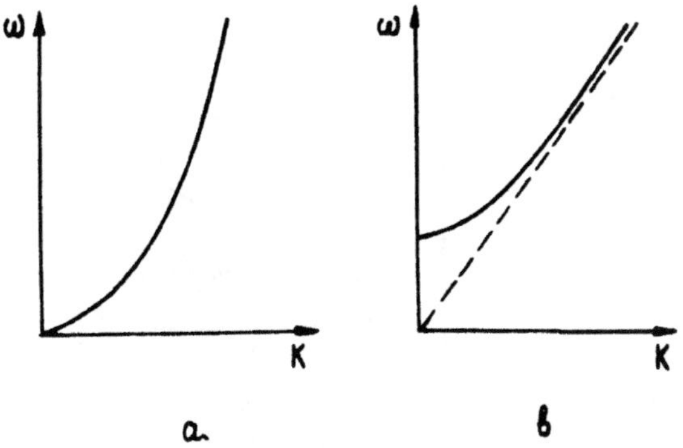

Figure 8.10. Examples of decaying (a) and non-decaying (b) spectra.

to a wave with parameters ω and $\boldsymbol{k}$ (see Fig. 8.11). Then, according to (8.259), the coalescence of waves is nothing but the vector summation in this space. For example, coalescence of waves corresponding to the vectors OA and OO' in Fig. 8.11 gives rise to the wave determined by OB. On the other hand, the spectrum $\omega = \omega(k_x, k_y)$ determines the "rotation" surface which is axisymmetric with respect to ω-axis. The one shown in Fig. 8.11 corresponds to the spectrum Fig. 8.10, a. A spectrum is decaying if one can draw a parallelogram with one corner at the origin at three other ones on the surface $\omega = \omega(k_x, k_y)$. One can easily see from Fig. 8.11 that such a procedure is possible if this surface crosses the similar one drawn from the origin O' lying on the first surface $\omega = \omega(k_x, k_y)$. Thus, the spectrum $\omega = \omega(k)$ is decaying if these two surfaces cross each other; otherwise the spectrum is non-decaying.

The spectrum Fig. 8.11 is of special interest to us, since it has the same form as the dispersion curves for electromagnetic and plasma waves (see (2.103) and (2.115)). This spectrum is non-decaying, so the coalescences and decays of the type

$$t + t \leftrightarrow t, \qquad l + l \leftrightarrow l, \tag{8.261}$$

where t and l stand for transverse and longitudinal waves, are forbidden in isotropic plasma by the conservation laws (8.259). At the same time, these relations allow the three-wave interactions in which one wave is of another type than the others[28]

$$l + t \leftrightarrow t, \qquad l + l \leftrightarrow t, \tag{8.262}$$

[28]In isotropic, non-isothermal plasmas three-wave interactions $l + s \leftrightarrow t$ and $t + s \leftrightarrow l$ are possible where the symbol s labels ion-sound waves. These processes, though probably playing a role in cosmic conditions, are not considered in this book. They are analyzed, for example, in [10, 89].

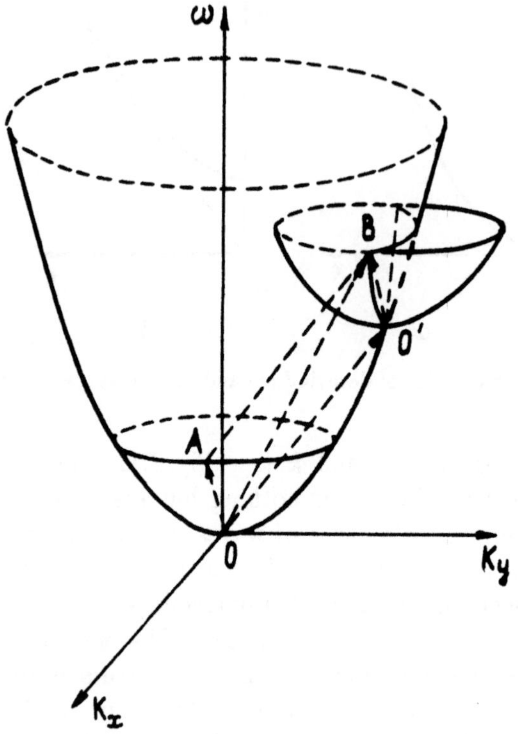

Figure 8.11. Criterion of decaying spectrum.

The process $t + t \leftrightarrow l$ is forbidden by conservation laws. The fact is the frequencies of plasma waves ω_{pl} are close to ω_{L}, so the frequencies ω_{em} of the coalescing electromagnetic waves should be less than ω_{L}. But the latter is possible only for pure imaginary $\boldsymbol{k}_{\mathrm{em}}$ which do not satisfy the second relation (8.259) for real $\boldsymbol{k}_{\mathrm{pl}}$.

In cosmic plasma conditions a special role is played by the coalescence of plasma waves into an electromagnetic one: $l + l \leftrightarrow t$. Since plasma wave frequencies are close to ω_{L}, the frequency of the electromagnetic wave

$$\omega = \omega_{\mathrm{em}} \approx 2\omega_{\mathrm{L}}, \tag{8.263}$$

the fact resulting from the first relation in (8.259). In the second relation describing the momentum conservation one should account that

$$k_{\mathrm{pl}} = \omega_{\mathrm{pl}} / v_{\mathrm{ph}} \approx \omega_{\mathrm{L}} / v_{\mathrm{ph}},$$

where v_{ph} is the phase velocity of plasma waves, and

$$k_{\mathrm{em}} = (\omega_{\mathrm{em}} / c) n_{\mathrm{em}}(\omega_{\mathrm{em}}) \approx (2\omega_{\mathrm{L}} / c) n_{\mathrm{em}}(2\omega_{\mathrm{L}}) = \sqrt{3}\omega_{\mathrm{L}} / c.$$

Then, if the phase velocity of one of the coalescing waves is

$$v_{ph} \ll \frac{c}{\sqrt{3}}, \qquad (8.264)$$

the wave vector $k_{em} \ll k_{pl}$ and the momentum conservation in (8.259) is only provided under the condition

$$\boldsymbol{k}_1 \approx -\boldsymbol{k}_2. \qquad (8.265)$$

Thus, under the condition (8.264) the Raman scattering can only occur for counter plasma waves with nearly equal phase velocities. The electromagnetic radiation at the double frequency appears as a result of this process [283]. The efficiency of plasma to electromagnetic waves conversion depends essentially on the intensity of plasma waves satisfying the relation (8.265).

Note that plasma waves with higher phase velocities

$$v_{ph} > \frac{c}{\sqrt{3}}, \qquad (8.266)$$

cannot convert to electromagnetic ones. Indeed, under the condition (8.266) the wave vector $k_{em} > 2k_{pl}$, so the parameters of such plasma waves cannot satisfy the momentum conservation law in (8.259). Plasma waves with such parameters appear, for example, due to spectral redistribution to the small k_{pl} range caused by the effective induced scattering by plasma particles (see section 8.1).

The appearance of electromagnetic radiation at frequencies $\omega \approx \omega_L$ and $\omega \approx 2\omega_L$ due to Rayleigh and Raman scattering of intense plasma waves provides the possible explanation for two harmonic bands in dynamical spectra of drifting type II and III bursts of solar radio emission [283]. These bands correspond to "fundamental" and "second harmonic", their frequency ratio to be 1:2. The close connection of this features of dynamical spectra with plasma wave conversion serves as one of the major arguments in favor of the plasma mechanism of generation of type II and III bursts.

At the same time, the third harmonic $\omega \approx 3\omega_L$ is observed in some rare events of solar radio emission. This radiation may be the result of either four-wave interaction $l + l + l \to t$ leading to the coalescence of three plasma waves into one electromagnetic or the process $l + t \to t$ (see (8.262)) of coalescence of plasma waves with second harmonic electromagnetic waves. The latter process is more efficient, but the estimates for the specific solar plasma conditions show that the valuable level of the third harmonic in solar radio emission can only appear at rather high intensity of plasma waves (i.e., the observed radio emission) [468]. The frequency of the third harmonic is close to $3\omega_L$. The corresponding phase velocities of plasma waves should lie in the interval $c/(2\sqrt{2}+\sqrt{3}) < v_{\rm pl} < c/(2\sqrt{2}-\sqrt{3})$ which may be determined by the conservation laws (8.259) applied to the process $l+t \to t$.

8.4.2. KINETIC EQUATION FOR PHOTONS AND THE RADIATION TRANSFER EQUATION

Coalescence and decay of photons can be accounted for in kinetic equations for photons (plasmons) in the similar way as in section 8.1 where the kinetic equation for scattering process has been derived. Let $w(\boldsymbol{k}, \boldsymbol{k}_1, \boldsymbol{k}_2)$ be the probability per unit time of the spontaneous splitting of a photon $\hbar\omega$ into a pair $\hbar\omega_1$ and $\hbar\omega_2$. Then $(2\pi)^{-9} w N_{\mathbf{k}} d^3k_1 d^3k_2$, where $N_{\mathbf{k}}$ is the number of quanta per mode in the unit phase volume,[29] equals the number of decays $\boldsymbol{k} \to \boldsymbol{k}_1 + \boldsymbol{k}_2$ per unit time of photons from unit $\boldsymbol{k}$-space volume into the ones in d^3k_1 and d^3k_2. In addition to spontaneous decays, the corresponding processes with probability proportional to $N_{\mathbf{k}_1} + N_{\mathbf{k}_2}$, the numbers of plasmons which the initial photon is splitted into, are also possible. These processes are of induced (stimulated) type, and they are called stimulated Raman scattering in quantum radiophysics.[30] The total rate of decays, spontaneous or induced, per unit time and unit phase volumes near $\boldsymbol{k}$, $\boldsymbol{k}_1$, and $\boldsymbol{k}_2$ is equal to

$$w N_{\mathbf{k}} (2\pi)^{-9} + \tilde{w} N_{\mathbf{k}} (N_{\mathbf{k}_1} + N_{\mathbf{k}_2})(2\pi)^{-9}. \tag{8.267}$$

In addition to the process $\boldsymbol{k} \to \boldsymbol{k}_1 + \boldsymbol{k}_2$ there is also the inverse one of coalescence $\boldsymbol{k}_1 + \boldsymbol{k}_2 \to \boldsymbol{k}$, the efficiency of the latter being proportional to $N_{\mathbf{k}_1} N_{\mathbf{k}_2}$ or, respectively, $W_{\mathbf{k}_1} W_{\mathbf{k}_2}$. The corresponding rate of coalescences per unit phase volume is written as

$$\tilde{\tilde{w}} N_{\mathbf{k}_1} N_{\mathbf{k}_2} (2\pi)^{-9}. \tag{8.268}$$

In the classical approach the coalescence (Raman scattering) process is the growth of a weak wave $\boldsymbol{k}$ due to the non-linear interaction of two intense ones $\boldsymbol{k}_1$ and $\boldsymbol{k}_2$. This process vanished if one of the coalescing wave amplitudes is zero.

The detailed balance conditions applied to the state of complete thermodynamic equilibrium yield the relations between scattering coefficients w, $\tilde{w}$ and $\tilde{\tilde{w}}$. Indeed, let the direct and inverse processes $\boldsymbol{k}_1 + \boldsymbol{k}_2 \leftrightarrow \boldsymbol{k}$ maintain the photon numbers $N_{\mathbf{k}}$, $N_{\mathbf{k}_1}$, $N_{\mathbf{k}_2}$ equal to the equilibrium values determined

[29] The value $N_{\mathbf{k}}$ is related to energy density $W_{\mathbf{k}}$ of decaying waves as $W_{\mathbf{k}} = (2\pi)^{-3} \hbar\omega N_{\mathbf{k}}$ (see (1.35) and (7.29)).

[30] As far as terminology is concerned, we notice here that a decay process resulting in the decrease of frequency: $\omega_1 - \omega_2 = \omega$, is called Stokes scattering, contrary to anti-Stokes scattering in which frequency is increased due to coalescence $\omega_1 + \omega_2 = \omega$.

by the Planck formula (8.264). The rates of decays (8.267) and coalescences (8.268) should then be equal. In the region $\hbar\omega \ll \kappa T$ the equilibrium photon numbers are given by $N_{\mathbf{k}} = \kappa T/\hbar\omega$, $N_{\mathbf{k}_1} = \kappa T/\hbar\omega_1$, $N_{\mathbf{k}_2} = \kappa T/\hbar\omega_2$. Here the balance between coalescences and decays requires that

$$\tilde{w} = \tilde{\tilde{w}}. \tag{8.269}$$

On the other hand, in the low temperature region $\hbar\omega \gg \kappa T$ the equilibrium values $N_{\mathbf{k}} = e^{-\hbar\omega/\kappa T}$, $N_{\mathbf{k}_1} = e^{-\hbar\omega_1/\kappa T}$,$N_{\mathbf{k}_2} = e^{-\hbar\omega_2/\kappa T}$, so the balance of the processes leads to the relation

$$w = \tilde{\tilde{w}}. \tag{8.270}$$

Thus, the coefficients w, $\tilde{w}$ and $\tilde{\tilde{w}}$ must be equal. Since they are entirely determined by the type of a single coalescence or decay event, their values are set by the non-linear properties of the medium and do not depend on radiation energy density. The latter justifies the validity of the relations (8.269), (8.270) for non-equilibrium radiation.

However, in the absence of thermodynamic equilibrium the detailed balance between the processes (8.267) and (8.268) is violated, and the number densities of interacting photons vary in time. The variation of $N_{\mathbf{k}}$ is evidently described by the following equation:

$$\frac{\partial N_{\mathbf{k}}}{\partial t} = \int w(N_{\mathbf{k}_1} N_{\mathbf{k}_2} - N_{\mathbf{k}_1} N_{\mathbf{k}} - N_{\mathbf{k}_2} N_{\mathbf{k}}) \frac{d^3k_1 d^3k_2}{(2\pi)^6}. \tag{8.271}$$

The corresponding equations for $N_{\mathbf{k}_1}$ and $N_{\mathbf{k}_2}$ look like[31]

$$\frac{\partial N_{\mathbf{k}_1}}{\partial t} = \int w(N_{\mathbf{k}_2} N_{\mathbf{k}} + N_{\mathbf{k}_1} N_{\mathbf{k}} - N_{\mathbf{k}_1} N_{\mathbf{k}_2}) \frac{d^3k_1 d^3k_2}{(2\pi)^6}, \tag{8.272}$$

$$\frac{\partial N_{\mathbf{k}_2}}{\partial t} = \int w(N_{\mathbf{k}_1} N_{\mathbf{k}} + N_{\mathbf{k}_2} N_{\mathbf{k}} - N_{\mathbf{k}_1} N_{\mathbf{k}_2}) \frac{d^3k_1 d^3k_2}{(2\pi)^6}. \tag{8.273}$$

The term $wN_{\mathbf{k}}(2\pi)^{-6}$ describing spontaneous decays is dropped in the above equations. This is quite possible under the conditions $N_{\mathbf{k}_2} \gg 1$, $N_{\mathbf{k}_2} \gg 1$ which hold in the Rayleigh-Jeans limit $\hbar\omega \ll \kappa T$ even for thermal fluctuations.

The above kinetic equations for decay processes may be presented in the form of radiation transfer equations. With (7.29) taken into account,

[31] First, in [469,470] the method of kinetic equation has been applied to describe decays in plasma. The investigations were then continued in [309, 467, 471–474]. In [309] the kinetic equations are presented in the illustrative "quantum" form (8.271)-(8.273).

the equation (8.271) is reduced to

$$\frac{\partial W_{\mathbf{k}}}{\partial t} + v_{\text{gr}} \frac{\partial W_{\mathbf{k}}}{\partial l} =$$

$$\int w\omega \left(\frac{W_{\mathbf{k}_1} W_{\mathbf{k}_2}}{\omega_1 \omega_2} - \frac{W_{\mathbf{k}_1} W_{\mathbf{k}}}{\omega_1 \omega} - \frac{W_{\mathbf{k}_2} W_{\mathbf{k}}}{\omega_2 \omega} \right) \frac{d^3k_1 d^3k_2}{\hbar (2\pi)^3}. \quad (8.274)$$

The term $v_{\text{gr}} \partial W_{\mathbf{k}} / \partial l$ describing the possible variation of energy density along a ray is added to the left-hand side of this equation (see for comparison (4.28)). The similar equations describing the space-time behavior of $W_{\mathbf{k}_1}$ and $W_{\mathbf{k}_2}$ may be easily obtained from (8.272) and (8.273).

The term

$$\alpha_{\mathbf{k}} = \int w W_{\mathbf{k}_1} W_{\mathbf{k}_2} \frac{\omega}{\omega_1 \omega_2} \frac{d^3k_1 d^3k_2}{\hbar (2\pi)^3} \quad (8.275)$$

in the equation (8.274) does not depend on $W_{\mathbf{k}}$. It plays a role of emissivity for the wave $\boldsymbol{k}$. The other terms are proportional to $W_{\mathbf{k}}$ with the coefficient

$$\beta_{\mathbf{k}} = 2\gamma_{\mathbf{k}} = \int w \left(\frac{W_{\mathbf{k}_1}}{\omega_1} + \frac{W_{\mathbf{k}_2}}{\omega_2} \right) \frac{d^3k_1 d^3k_2}{\hbar (2\pi)^3}, \quad (8.276)$$

which characterizes the relative decrease of the energy density $W_{\mathbf{k}}$ per unit time. In other words, $\gamma_{\mathbf{k}}$ is the decrement for the radiation $\boldsymbol{k}$. The value of $\gamma_{\mathbf{k}}$ is non-negative which seems to be quite natural for the decaying process $\boldsymbol{k} \rightarrow \boldsymbol{k}_1 + \boldsymbol{k}_2$ removing energy from the waves $\boldsymbol{k}$. Using (8.272) and (8.273) it is not difficult to get the decrements for coalescing waves $\boldsymbol{k}_1$ and $\boldsymbol{k}_2$:

$$\gamma_{\mathbf{k}_1} = \frac{1}{2} \int w \left(\frac{W_{\mathbf{k}_2}}{\omega_2} - \frac{W_{\mathbf{k}}}{\omega} \right) \frac{d^3k d^3k_2}{\hbar (2\pi)^3}, \quad (8.277)$$

$$\gamma_{\mathbf{k}_2} = \frac{1}{2} \int w \left(\frac{W_{\mathbf{k}_1}}{\omega_1} - \frac{W_{\mathbf{k}}}{\omega} \right) \frac{d^3k d^3k_1}{\hbar (2\pi)^3}. \quad (8.278)$$

These decrements are negative for sufficiently large $W_{\mathbf{k}}$. The energy gain to the waves $\boldsymbol{k}_1$ and $\boldsymbol{k}_2$ may, due to this fact, exceed the losses due to their coalescence into the waves $\boldsymbol{k}$. Thus, the processes $\boldsymbol{k} \leftrightarrow \boldsymbol{k}_1 + \boldsymbol{k}_2$ allow for instability and amplification of the waves $\boldsymbol{k}_1$ and $\boldsymbol{k}_2$.

The solution of the system of equations (8.271)-(8.273) describes the temporal evolution of energy densities $W_{\mathbf{k}}, W_{\mathbf{k}_1}, W_{\mathbf{k}_2}$. Since these equations describe the processes conserving energies and momenta of interacting photons (8.259), the total energy and momentum densities are conserved, contrary to the processes of interaction of radiation with plasma particles: pure absorption, spontaneous and induced emision, and scattering. To account for these interactions, the terms with $\alpha_{\mathbf{k}}$ and β_j from (7.23), (7.24), (8.18)

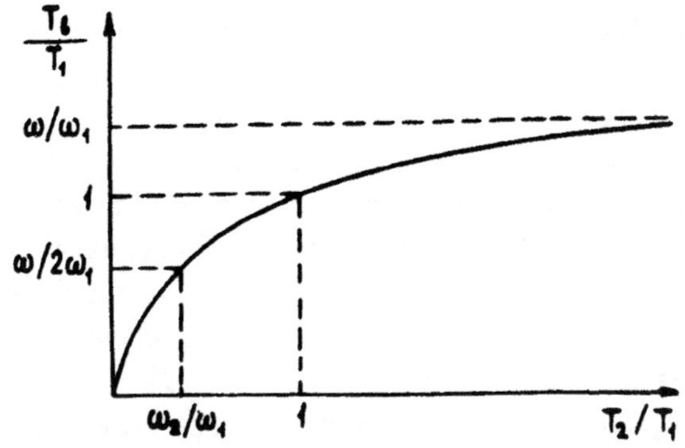

Figure 8.12. The dependence of the ratio T_b/T_1 on T_2/T_1 in the case $\omega_2/\omega_1 < 1$.

and (8.19) should be introduced in the equation of the type (8.274). This, of course, makes the solution of kinetic equations even more complicated.

On the other hand, analysis of the equation (8.274) is drastically simplified if the energy densities of the coalescing waves, $W_{\mathbf{k}_1}$ and $W_{\mathbf{k}_2}$, may be considered given. For example, for the process $l + l \leftrightarrow t$ the energy densities of plasma waves $\mathbf{k}_1$ and $\mathbf{k}_2$ may be maintained due to their effective generation by a charged particle beam.

In a quasi-stationary, extended source the value of $W_{\mathbf{k}}$ can be estimated from the following considerations. If variations of energy densities $W_{\mathbf{k}_1}$ and $W_{\mathbf{k}_2}$ are small over time $\Delta t \sim 1/\gamma_{\mathbf{k}}$ and distance $\Delta l \sim 1/\mu_j$, where $\mu_j = 2\gamma_{\mathbf{k}} v_{\mathrm{gr}}^{-1}$ is the absorption coefficient along a ray due to decay processes, then their established values should satisfy the condition $dW_{\mathbf{k}}/dt + v_{\mathrm{gr}} dW_{\mathbf{k}}/dl \approx 0$. According to (8.274), this occurs if

$$\frac{T_1 T_2}{\omega_1 \omega_2} - \frac{T_1 T_b}{\omega_1 \omega} - \frac{T_2 T_b}{\omega_2 \omega} = 0, \tag{8.279}$$

where T_1, T_2 and T_b are brightnessradiation temperatures related to the corresponding energy densities as

$$W_{\mathbf{k}_1} = \frac{\kappa T_1}{(2\pi)^3}, \qquad W_{\mathbf{k}_2} = \frac{\kappa T_2}{(2\pi)^3}, \qquad W_{\mathbf{k}} = \frac{\kappa T_b}{(2\pi)^3} \tag{8.280}$$

(the definition of the brightness temperature is given in section 1.2). The equation (8.279) yields [475]

$$T_b = \frac{\omega T_1 T_2}{\omega_1 T_2 + \omega_2 T_1}. \tag{8.281}$$

Evidently, this is the maximal temperature which can be achieved for a given level of coalescing waves corresponding to the temperatures T_1 and T_2.

The ratio T_b/T_1 as a function of T_2/T_1 is sketched in Fig. 8.12. Note that for $T_1 = T_2$ the brightness temperature of the resulting radiation is the same. For example, the brightness temperature T_b of electromagnetic waves at frequency $\omega \approx 2\omega_L$ generated due to coalescence $l + l \to t$ of two plasma waves with frequencies $\omega_1 \approx \omega_2 \approx \omega_L$, can not exceed the plasma wave temperature $T_1 = T_2$. If coalescing wave frequencies are greatly different, $\omega_1 \gg \omega_2$ (as an example, we mention here the coalescence of plasma waves with ion-cyclotron waves, Bernstein modes or whistlers), the maximal brightness temperature of electromagnetic radiation T_b at frequencies $\omega = \omega_1 + \omega_2 \approx \omega_1$ corresponds to the level of high-frequency waves ($T_b \sim T_1$) over the wide range of $T_2/T_1 \gtrsim \omega_2/\omega_1$ and cannot exceed the value T_1, ever large the ratio T_2/T_1 is.

Note, finally, that in the framework of three-wave interaction the same wave can participate in several interaction processes at a time. For example, a wave $\boldsymbol{k}$ may enter the processes $\boldsymbol{k} \leftrightarrow \boldsymbol{k}_1 + \boldsymbol{k}_2$ and $\boldsymbol{k} + \boldsymbol{k}_4 \leftrightarrow \boldsymbol{k}_5$. This circumstance was not taken into account above. If another interaction process is also imporatnt, the right-hand side of the equation (8.271) should be supplemented with the expression

$$\int w'(N_{\boldsymbol{k}_4} N_{\boldsymbol{k}_5} + N_{\mathbf{k}} N_{\boldsymbol{k}_5} - N_{\mathbf{k}} N_{\boldsymbol{k}_4}) \frac{d^3k_4 d^3k_5}{(2\pi)^6}, \tag{8.282}$$

where w' is the probability of the spontaneous decay $\boldsymbol{k}_5 \to \boldsymbol{k} + \boldsymbol{k}_4$. The corresponding changes should also be made both in the transfer equation (8.273) and in the coefficients (8.275), (8.276).

The kinetic equations (8.271)-(8.273) are written in terms of spontaneous decay probability $w(\boldsymbol{k}, \boldsymbol{k}_1, \boldsymbol{k}_2)$ which depends on plasma properties, type of interacting waves etc. The methods of calculation of this value are given in [89,384], see also [476–478]. We do not give these methods here but rather restrict ourselves to general remarks on the method of calculating the probability w and show, as an example, the probabilities of the processes of the type (8.262) in isotropic plasma.

The probability w can be found by calculation of $\partial W_{\mathbf{k}}/\partial t$ in a homogeneous system at vanishing $W_{\mathbf{k}}$, when $\partial W_{\mathbf{k}}/\partial t = \alpha_{\mathbf{k}}$ (see (8.275)). Such a calculation can be done using a quantum-mechanical approach. The coefficient $\alpha_{\mathbf{k}}$ can be found in the quasi-classical limit $\hbar\omega \ll \kappa T$ using classical electrodynamics. According to the correspondence principle, both results should be the same. Thus, the probability w can be determined by comparison of the quantum formula (8.275) with the classical expression for $\alpha_{\mathbf{k}}$. The calculation is done in two stages: first, the non-linear current induced by the waves 1 and 2 is found; second, the emission by this current is calculated.

To elucidate this in more detail, the total current j_{tot} induced by the waves 1 and 2 in a medium can be written in the hydrodynamic approximation (2.75) as

$$j_{tot'} = j'_o + j'_{nl}, \qquad j_{nl} = -eN'_1 v'_2 - eN'_2 v'_1. \tag{8.283}$$

Here N'_1 and N'_2 are perturbations of electron density from the unperturbed value N_0; v'_1 and v'_2 are electron velocities in waves 1 and 2, respectively; j'_o is the linear part of the total current (2.75), and j'_{nl} is the non-linear component due to the interaction of two waves. The term j'_{nl} can be found with the perturbation technique by substituting for N'_1 and N'_2 the corresponding values matching the linearized equations (2.74). They are proportional to the electric field components E_1 and E_2 of the waves. The desired term j'_{nl} thus is expressed in terms of the product of the components of the vectors E_1 and E_2. Next, the obtained j'_{nl} is treated as an external current induced in plasma by the fields E_1 and E_2 and the spectral power of radiation by this current into the third wave, $\alpha_{\mathbf{k}} = \partial W_{\mathbf{k}}/\partial t$, is found. The resulting expression is non-zero under the synchronism conditions (8.259) only. The obtained radiation power is proportional to the product $W_{\mathbf{k}_1} W_{\mathbf{k}_2}$ in the case of interaction of random-phase waves. Finally, the probability w is found by the comparison of the obtained expression to the coefficient $\alpha_{\mathbf{k}}$ in the form (8.275).

8.4.3. COALESCENCES AND DECAYS IN ISOTROPIC PLASMA

The probability of scattering of the type $l + l \leftrightarrow t$ in isotropic plasma is given by [472,473]

$$w = w_t^{ll} = \tag{8.284}$$
$$2\pi^5 \frac{\hbar e^2}{m_e^2 \omega_L} \frac{(k_1^2 - k_2^2)^2}{k^2} \frac{[\mathbf{k}_1 \times \mathbf{k}_2]^2}{k_1^2 k_2^2} \delta(\mathbf{k} - \mathbf{k}_1 - \mathbf{k}_2)\delta(\omega - \omega_1 - \omega_2).$$

The difference $k_1^2 - k_2^2$ appears in this formula due to the presence of two terms in the expression (8.283) for the non-linear current. One of these terms is for scattering of the wave 2 on density perturbations N'_1 caused by the wave 1, and another one is for scattering of the wave 1 on similar perturbations N'_2 caused by the wave 2. These contributions are exactly balanced for equal absoulte values of wave vectors of coalescing waves, $k_1 = k_2$, so the resulting probability w_t^{ll} is zero in this case [479,480].

Let us find the emissivity $\alpha_{\mathbf{k}}$ and decay decrement $\gamma_{\mathbf{k}}$ determined by (8.275), (8.276) for electromagnetic radiation at frequency $\omega \approx 2\omega_L$ as a result of the coalescence of plasma waves with $\omega_1 \approx \omega_2 \approx \omega_L$. According to

(8.285), these coefficients look like the following:

$$\alpha_{\mathbf{k}} = 2\pi^2 \frac{e^2 k^2}{m_e^2 \omega_L^2} \int \frac{k_1^2}{k_2^2} \sin^2\theta \left(\cos\theta - \frac{k}{2k_1}\right)^2 \times W_{\mathbf{k}_1} W_{\mathbf{k}_2} \delta(\boldsymbol{k} - \boldsymbol{k}_1 - \boldsymbol{k}_2)\delta(\omega - \omega_1 - \omega_2) d^3k_1 d^3k_2, \quad (8.285)$$

$$\gamma_{\mathbf{k}} = \frac{\pi^2}{2} \frac{e^2 k^2}{m_e^2 \omega_L^2} \int \sin^2\theta \left(\cos\theta - \frac{k}{2k_1}\right)^2 (W_{\mathbf{k}_1} + W_{\mathbf{k}_2}) \times \delta(\boldsymbol{k} - \boldsymbol{k}_1 - \boldsymbol{k}_2)\delta(\omega - \omega_1 - \omega_2) d^3k_1 d^3k_2. \quad (8.286)$$

It is taken into account here that, according to (8.259), $(k_1^2 - k_2^2)^2 = 4k_1^2 k^2(\cos\theta - k/2k_1)^2$ and $[\boldsymbol{k}_1 \times \boldsymbol{k}_2]^2 = [\boldsymbol{k}_1 \times \boldsymbol{k}]^2 = k_1^2 k^2 \sin^2\theta$, where θ is the angle between $\boldsymbol{k}$ and $\boldsymbol{k}_1$.

Assume the plasma wave spectrum to be isotropic ($W_{\mathbf{k}_1}$ and $W_{\mathbf{k}_2}$ do not depend on the directions of $\boldsymbol{k}_1$ and $\boldsymbol{k}_2$) and plasma waves in this spectrum are concentrated within the region (8.264) where $k \ll k_1, k_2$ and $\boldsymbol{k}_1 \approx -\boldsymbol{k}_2$. Then integrating over d^3k_2 and taking into account the delta function over wave vectors, we find that

$$\alpha_{\mathbf{k}} = 6\pi^2 \frac{e^2}{m_e^2 c^2} \int \sin^3\theta \cos^2\theta W_{\mathbf{k}_1} W_{-\boldsymbol{k}_1} \delta(\omega - \omega_1 - \omega_2) k_1^2 d\theta d\phi dk_1 \quad (8.287)$$

Here the relation $d^3k_1 = k_1^2 \sin\theta d\theta d\phi dk_1$, where ϕ is the angle describing the position of the projection of the vector $-\boldsymbol{k}_1$ onto the plane perpendicular to $\boldsymbol{k}$. Since $k_1^2 W_{\mathbf{k}_1} W_{-\boldsymbol{k}_1}$ does not depend on the angles θ and ϕ, one can easily integrate (8.287) over θ from 0 to π and over ϕ from 0 to 2π. Then the integration over k_1 may be performed taking into account that the frequencies $\omega_1(\boldsymbol{k}_1)$ and $\omega_2(-\boldsymbol{k}_1)$ in delta-function argument are related to the wave vector $\boldsymbol{k}_1$ by the dispersion equation (2.115). As a result of these calculations, the following expression for emissivity is derived:

$$\alpha_{\mathbf{k}} = \frac{(2\pi)^3}{15} \frac{e^2 \omega_L k_1}{m_e c^2 v_T^2} W_{\mathbf{k}_1}^2. \quad (8.288)$$

The similar procedure for (8.286) yields

$$\gamma_{\mathbf{k}} = \frac{(2\pi)^3}{30} \frac{e^2 \omega_L k_1}{m_e^2 c^2 v_T^2} W_{\mathbf{k}_1}. \quad (8.289)$$

In the transfer equation (4.30) for radiation intensity the emissivity $a_\omega = k^2 \alpha_{\mathbf{k}} (|\cos\vartheta| v_{gr})^{-1}$ and the absorption coefficient $\mu = 2\gamma_{\mathbf{k}} v_{gr}^{-1}$ are used instead of $\alpha_{\mathbf{k}}$ and $\gamma_{\mathbf{k}}$. These values for the process considered are given by the following expressions:

$$a_\omega = \frac{2(2\pi)^3}{5\sqrt{3}} \frac{e^2\omega_L^4}{m_e^2 c^5 v_T^2 v_{ph}} W_{\mathbf{k}_1}^2, \tag{8.290}$$

$$\mu = \frac{2(2\pi)^3}{15\sqrt{3}} \frac{e^2\omega_L^2}{m_e^2 c^3 v_T^2 v_{ph}} W_{\mathbf{k}_1}, \tag{8.291}$$

where it has been taken into account that $k_1 \approx \omega_L/v_{ph}$ for plasma waves; $k = \sqrt{3}\omega_L/c$ and $v_{gr} = \sqrt{3}c/2$ for electromagnetic radiation at frequency $\omega \approx 2\omega_L$; and, finally, $\cos\vartheta = 1$ in isotropic plasma.

The brightness temperature T_b of electromagnetic radiation in a homogeneous, steady-state source containing plasma waves with isotropic spectrum $W_{\mathbf{k}_1}$ reaches the value $T_l = (2\pi)^3 W_{\mathbf{k}_1}/\kappa$ if $\mu L \gg 1$, where L is the linear size of the source, and the absorption coefficient μ is given by (8.281). In other words, if a source is optically thick with respect to the process of decay of electromagnetic waves into a couple of plasma ones, the brightness temperature T_b can rise to the maximal value set by the brightness temperature of plasma turbulence, T_l (see also the notes concerning the formula (8.281)). On the contrary, the intensity of electromagnetic radiaton at frequency $\omega \approx 2\omega_L$ escaping an optically thin source

$$\mu L = \frac{2}{15\sqrt{3}} \frac{e^2\omega_L^2 \kappa T_l}{m_e^2 c^3 v_T^2 v_{ph}} L \ll 1, \tag{8.292}$$

is much smaller: $I_\omega \approx a_\omega L$. Being expressed in terms of brightness temperature (1.39), it corresponds to

$$T_b \approx \frac{2}{15\sqrt{3}} \frac{e^2\omega_L^2 \kappa T_l^2}{m_e^2 c^3 v_T^2 v_{ph}} L \ll T_l \tag{8.293}$$

(see (8.280) and (8.290)).

The following parameter values may be adopted for solar corona conditions: $v_T \sim 4 \cdot 10^8\ \mathrm{cm\,s^{-1}}$, $\omega_L \sim 2\pi \cdot 10^8\ \mathrm{s^{-1}}$, and $v_{ph} \sim c/5$. Then (8.292) yields the inequality $T_l L \ll 5 \cdot 10^{22}$ K·cm. Assuming that decay absorption of electromagnetic radiation occurs in a layer of the size $L \sim 10^8$ cm (see [468] on the choice of this estimate), we get the criterion $T_l \ll 5 \cdot 10^{14}$ K allowing (8.293) to be used for calculation of the brightness temperature of radio emission at the second harmonic of plasma frequency. At $T_l > 5 \cdot 10^{14}$ K the source becomes optically thick and the brightness temperature T_b is comparable to T_l. According to the observational data, however, such a high values of T_b are not observed or are observed very rarely. Therefore, the process of conversion of plasma into electromagnetic waves due to coalescence processes $l + l \to t$ typically results in the significant drop of the brightness temperature $T_b/T_l \ll 1$.

The situation is radically different in radio pulsars. If the plasma mechanism of radio emission is realized here, the conversion of plasma into electromagnetic waves due to Raman scattering may become quite effective since the radiation intensity in the source is rather high. This fact has been discussed in [345,346] for the case of induced conversion of plasma waves due to scattering by plasma particles. The similar conclusion for Raman scattering may be done by estimating the parameters in neutron star magnetospheres by similar values as in the solar corona. Of course, this yields the same criterion $T_1 \gtrsim 5 \cdot 10^{14}\,\mathrm{K}$ of strong conversion $T_\mathrm{b} \sim T_1$. Since brightness temperatures of pulsar radiation reach $10^{30}\,\mathrm{K}$, this criterion is well satisfied for these objects.

Note that the initial assumptions on the isotropy of an excited plasma wave spectrum, which the formulas (8.290), (8.291) for the coefficients α_ω and μ rely upon, pose severe limitations on the applicability of these expressions. The plasma wave spectrum may be quite anisotropic in astrophysical plasmas. The example may be the source region of fast-drifting type III bursts in the solar corona. The spectrum of plasma waves excited in this region by electron beams is most likely one-dimensional (see the section 7.2).

According to a model of type III burst source, plasma waves with wave vectors $\boldsymbol{k}_1$ directed along the bulk velocity of the beam $\boldsymbol{v}_\mathrm{S}$ are excited in this region. Radio emission at the second harmonic of plasma frequency $\omega \approx 2\omega_\mathrm{L}$ is generated due to coalescence of the above plasma waves with the ones having a wide angular spectrum $W_{\mathbf{k}_2}$. In the initial theory of type III bursts [283] only thermal (fluctuating) plasma waves have been accounted in the spectrum $W_{\mathbf{k}_2}$. In further developments of the theory [336,337] it has been taken into account that Rayleigh scattering of plasma waves excited by the beam (see section 8.1) makes more a significant contribution to $W_{\mathbf{k}_2}$. The angular pattern of radio emission at the second harmonic of plasma frequency has been calculated in [481,482].

In such a model the emissivity for second-harmonic radio emission can be easily found from (8.285) assuming that the spectrum of the excited plasma waves is described by the expression of the type (8.26) comprising the delta-function over directions of the wave vector $\boldsymbol{k}_1$. Substituting such a spectrum in (8.285) and integrating the resulting expression, we get

$$\alpha_\mathbf{k} = \frac{\pi^2 e^2 \omega_L}{m_\mathrm{e}^2 c^2 v_T^2 k_1} \Psi(\theta_0) W(k_1) W_{\mathbf{k}_2}(k_1, \Omega_0). \tag{8.294}$$

The corresponding emissivity is determined by the expression

$$a_\omega = 2\sqrt{3}\pi^2 \frac{e^2 \omega_L v_\mathrm{ph}}{m_\mathrm{e}^2 c^5 v_T^2} \Psi(\theta_0) W(k_1) W_{\mathbf{k}_2}(k_1, \Omega_0). \tag{8.295}$$

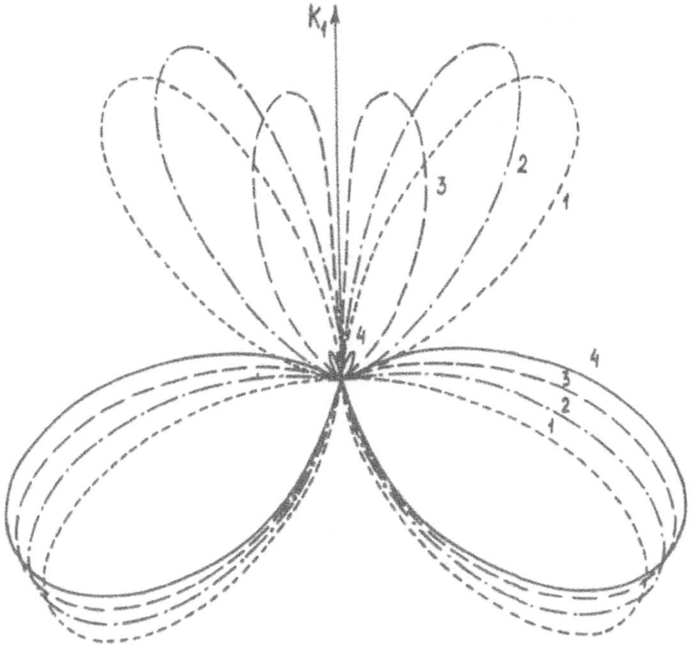

Figure 8.13. Radiation pattern $\Psi(\theta_0)$ of second-harmonic radio emission in isotropic plasma: 1 – $v_{ph}/c=0.2$, 2 – $v_{ph}/c=0.4$, 3 – $v_{ph}/c=0.6$, 4 – $v_{ph}/c=0.8$.

Here

$$\Psi(\theta_0) = \frac{\sin^2\theta_0(\cos\theta_0 - k/2k_1)^2}{1 + k^2/k_1^2 - 2(k/k_1)\cos\theta_0}, \tag{8.296}$$

v_{ph} is phase velocity of plasma waves with the one-dimentional spectrum, θ_0 is the angle between $\boldsymbol{k}$ and $\boldsymbol{\Omega}_0$, the direction of wave vectors $\boldsymbol{k}_1$ in the one-dimentional spectrum. The energy density $W_{\mathbf{k}_2}$ depends on k_1 and θ_0 even for isotropic angular spectrum. This fact results from the conservation laws (8.259) which yield $k_2^2 = k_1^2 + k^2 - 2k_1k\cos\theta_0$. The dependence on θ_0 may be neglected for $k/k_1 \equiv \sqrt{3}v_{ph}/c \ll 1$, i.e., under the condition (8.264). In this case the radiation pattern of radio emission at frequency $\omega \approx 2\omega_L$ has four identical lobes:

$$a_\omega(\theta_0) \propto \Psi(\theta_0) \approx \sin^2\theta_0\cos^2\theta_0. \tag{8.297}$$

The dependence of $W_{\boldsymbol{k}_2}$ on the angle θ_0 can be neglected for the excited plasma waves with higher phase velocities V_{ph} if this spectrum is weakly dependent on k_2. The radiation pattern in this case is given by the function $\Psi(\theta_0)$ again. From Fig. 8.13 it is clear that the symmetry of the radiation pattern breaks at $v_{ph}/c \gtrsim 1/\sqrt{3}$. The bakward lobes directed oppositely with respect to the wave vectors $\boldsymbol{k}_1$ of excited plasma waves become larger than the forward ones. This effect allows to observe the second harmonic of solar radio bursts in radiation reflected by deeper coronal layers [30].

References

1. Ginzburg, V.L., and Syrovatskii, S.I. (1964) *The Origin of Cosmic Rays.* Pergamon Press, Oxford.
2. Kaplan, S.A., and Pikel'ner, S.B. (1970) *The Interstellar Medium.* Harvard Univ. Press, Cambridge.
3. Zheleznyakov, V.V. (1970) *Radio Emission of the Sun and Planets.* Pergamon Press, Oxford.
4. Kundu, M.R. (1965) *Solar Radio Astronomy.* Interscience Publ., New York
5. Shklovskii I.S. (1968) *Supernovae.* Wiley, London.
6. Pacholczyk, A.G. (1970) *Radio Astrophysics.* Freeman, San Francisco.
7. Krüger, A. (1979) *Introduction to Solar Radio Astronomy and Radio Physics.* D. Reidel Publ., Dordrecht.
8. Benz, A.O. (1993) *Plasma Astrophysics.* Kluwer Academic Publ., Dordrecht.
9. Beskin, V., Gurevich, A., and Istomin, Ya. (1993) *Physics of the Pulsar Magnetosphere.* Cambridge Univ. Press
10. Kaplan, S.A., and Tsytovich, V.N. (1973) *Plasma Astrophysics.* Pergamon Press, Oxford.
11. Melrose, D.B. (1980) *Plasma Astrophysics*, Vol. I, II. Gordon and Breach, New York.
12. Zheleznyakov, V.V. (1977) *Electromagnetic Waves in Cosmic Plasma.* Nauka, Moscow.
13. Langmuir, I. (1929) Phys. Rev. **33**, 954
14. Goldreich, P., and Julian, W.H. (1969) *Astrophys. J.*, **157**, p. 869
15. Mathewson, D.S., and Milne, D.K. (1964) *Nature*, **203**, p. 1273
16. Mathewson, D.S., and Milne, D.K. (1968) *Astrophys. J. Lett.*, **153**, p. 47
17. Razin, V.A., and Khizhnyakova, I.P. (1969) *Izv. VUZ Radiofizika*, **12**, p. 379
18. Dalgarno, A., and McCray, R.A. (1972) *Ann. Rev. Astron. Astrophys.*, **10**, p. 375
19. Mezger, P.G. (1974) in: *Proc. ESO/SRC/CERN Conf. on Research Programmes for the New Large Telescopes* Geneva. p. 79
20. Woltjer, L. (1972) *Ann. Rev. Astron. Astrophys.*, **10**, p. 129
21. Field, G.B. (1972) *Ann. Rev. AstronȦstrophys.*, **10**, p. 227
22. Ness, N.F. (1968) *Ann. Rev. Astron. Astrophys.*, **6**, p. 79
23. Carr, T.D., and Gulkis, S. (1969) *Ann. Rev. Astron. Astrophys.*, **7**, p. 577
24. McCulloch , P.M., and Komesaroff, M.M. (1973) *Icarus*, **19**, p. 83
25. Mohajan, K.K., and Kar, J. (1988) *Space Sci. Rev.*, **47**, p. 303
26. Bagenal, F. (1992) *Ann. Rev. Earth Planet. Sci.*, **20**, p. 289
27. Belcher, J.W., McNutt, R.L. (Jr.), and Richardson, J.D. (1990) *Adv. Space Res.*, **10**, p. 5
28. Dessler, A.J. (Ed.) (1983) *Physics of Jovian Magnetosphere.* Cambridge Univ. Press.
29. Newkirk, G. (1961) *Astrophys. J.*, **133**, p. 983
30. Wild, J.P., Smerd, S.F., and Weiss, A.A. (1963) *Ann. Rev. Astron. Astrophys.*, **1**, p. 291
31. Linsky, J.F., and Serio, S. (Eds.) (1993) *Physics of Stellar Coronae: G. S. Vaiana Memorial Symposium. Proc. Conference. Palermo, Italy.* Kluwer Acad. Publ.
32. Newkirk, G., Altschuler, M.D., and Harvey, J. (1969) *Solar Phys.*, **9**, p. 131
33. Eddy, J.A. (1973) *Solar Phys.*, **30**, p. 385
34. Michaud, G., Megessier, C., and Charland, Y. (1981) *Astron. Astrophys.*, **103**, p. 244
35. Drake, S.A., Abbot, D.C., Bastian, T.S., Bieging, J.H., Churchwell, E., Dulk, G., and Linsky, J.L. (1987) *Astrophys. J.*, **322**, p. 902
36. Willson, R.F., Lang, K.R., and Foster, P. (1988) *Astron. Astrophys.*, **199**, p. 255
37. Zheleznyakov, V.V., and Tikhomirov, Yu.V. (1984) *Astrophys. Space Sci.*, **102**, p. 189

38. Schmidt, G.D. (1989) in: G.Wegner (Ed.) *White Dwarfs. Proc. IAU Colloquium No. 114* Springer-Verlag. p. 305
39. Zheleznyakov, V.V., and Serber, A.V. (1994) *Space Sci. Rev.*, **68**, p. 275
40. Cavallo, R., Arnaud, K.A., and Trimble, V. (1993) *J. Astrophys. Astron.*, **14**, p. 141
41. Zheleznyakov, V.V., and Litvinchuk, A.A. (1984) *Astrophys. Space Sci.*, **105**, p. 73
42. Manchester, R.N., and Taylor, J.H. (1977) *Pulsars*. Freeman, San Francisco.
43. Ginzburg, V.L. (1964) *Soviet Phys. Doklady*, **9**, p. 329
44. Trumper, J., et.al. (1976) *Ann. N.-Y. Acad. Sci.*, **302**, p. 538
45. Trumper, J., et.al. (1978) *Astrophys. J.*, **219**, p. L105
46. Wheaton, W.A., et.al. (1979) *Nature*, **282**, p. 240
47. Mazets, E.P., et.al. (1980) *Sov. Astron. Lett.*, **6**, p. 372
48. Murakami, T., Fujii, M., Hayashida, K., Itoh, M., Nishimura, J., Yamagami, T., Conner, J.P., Evans, W.D., Fenimore, E.E., Klebesadel, R.W., Yoshida, A., Kondo, I., and Kawai, N. (1988) *Nature*, **335**, p. 234
49. Barat, C., Atteia, J.L., Jourdain, E., Vedrenne, G., Sunyaev, R., Blinov, N., Kuznetsov, A., Lvov, M., and Terekhov, O. (1991) *Planet. Space Sci.*, **39**, p. 67
50. Paciesas, W., and Fishman, G. (Eds.) *Gamma-Ray Bursts. AIP Conf. Proc. No. 265* New York, AIP.
51. Hurley, K., (1994) *Astrophys. J. Suppl.*, **90**, p. 857
52. Sturrock, P.A. (1970) *Nature*, **227**, p. 465
53. Sturrock, P.A. (1971) *Astrophys. J.*, **164**, p. 529
54. Zheleznyakov, V.V., and Serber, A.V. (1993) *Astron. Rep.* **37**, p. 507
55. Landau, L.D., Lifshits, Ye.M., and Pitayevskii, L.P. (1980) *Statistical Physics*. Pergamon Press, Oxford.
56. Landau, L.D., and Lifshits, Ye.M. (1975) *Classical Theory of Fields*. Pergamon Press, Oxford.
57. Zheleznyakov, V.V. (1968) *Astrophys. Space Sci.*, **2**, p. 403
58. Cohen, M.H. (1958) *Proc. IRE*, **46**, p. 183
59. Zheleznyakov, V.V. (1971) *Astrophys. Space Sci.* **13**, p. 74
60. Sorokin, Yu. M. (1974) *Radiophys. Quantum Electronics*, **17**, p. 258
61. Mirabel, I.F., and Rodriguez, L.F. (1994) *Nature*, **371**, p. 46
62. Rees, M.J. (1966) *Nature*, **211**, p. 468
63. Ginzburg, V.L., and Zheleznyakov, V.V. (1975) *Ann. Rev. Astron. Astrophys.*, **13**, p. 511
64. Smith, F.G. (1970) *MNRAS*, **149**, p. 1
65. Epstein, R.I., and Feldman, P.A. (1967) *Astrophys. J. Lett.*, **150**, p. 109
66. Ginzburg, V.L., Sazonov, V.N., and Syrovatsky, S.I. (1968) *Sov. Phys. Uspekhi*, **11**, p. 34
67. Takakura, T., and Uchida, Y. (1968) *Astrophys. Lett.*, **1**, p. 147
68. Zheleznyakov, V.V., and Shaposhnikov, V.E. (1972) *Astrophys. Space Sci.*, **18**, p. 141
69. Ginzburg, V.L., and Zheleznyakov, V.V. (1961) *Sov. Astron.*, **5**, p. 1
70. Ginzburg, V.L., and Zheleznyakov, V.V. (1970) *Comments Astrophys. Space Sci.*, **2**, 167, 197
71. Ginzburg, V.L. (1961) *Propagation of Electromagnetic Waves in Plasma*. Gordon and Breech. New York/ North Holland Publ. Co., Amsterdam.
72. Akhiezer, A.I., ed. (1974) *Plasma Electrodynamics*. Nauka, Moscow.
73. Mikhailovskii, A.B. (1970) *Theory of Plasma Instabilities*. Vol. I. Atomizdat, Moscow.
74. Mandelschtam, L.I. (1941) *J. Phys. USSR*, **4**, p. 9
75. Ginzburg, V.L. (1949) *Theory of Radio Wave Propagation in Ionosphere*. Gostekhizdat, Moscow.
76. Kadomtsev, B.B. (1957) *Sov. Phys. JETP*, **6**, p. 117

77. Gertsenshtein, M.E. (1952) *Zh. Eksp. Teor. Fiz.*, **22**, p. 303
78. Silin, V.P., Rukhadze, A.A. (1961) *Electromagnetic Properties of Plasma and Plasma-Like Media*. Atomizdat, Moscow.
79. Zheleznyakov, V.V. (1959) *Izv. VUZ Radiofizika* **2**, p. 14
80. Shafranov, V.D. (1963) in: M.A.Leontovich (ed.) *Reviews of Plasma Physics*. Consultants Bureau, New York. **3**, p. 1
81. Ginzburg, V.L., and Rukhadze, A.A., (1975) *Waves in Magnatoactive Plasma*. Nauka, Moscow.
82. Shafranov, V.D. (1958) *Plasma Physics and the Problem of Controlled Fusion*, **4**, p. 416
83. Landau, L.D. (1946) *Zh. Eksp. Teor. Fiz.*, **16**, p. 574
84. Andronov, A.A., Zheleznyakov, V.V., and Petelin, M.I. (1964) *Izv. VUZ Radiofizika*, **7**, p. 251
85. Bekefi, J. (1966) *Radiation Processes in Plasma*. Wiley, New York.
86. Mikhailovskii, A.B. (1963) in: *Reviews of Plasma Physics*, **3**, p. 159.
87. Sizonenko, V.L., and Stepanov, K.N. (1971) *Ukr. Fiz. Zh.*, **15**, p. 438
88. Bernstein, I.B. (1958) *Phys. Rev.*, **109**, p. 10
89. Tsytovich, V.N. (1970) *Non-Linear Effects in Plasma*. Plenum Press, New York.
90. Poeverlein, H. (1961) *Phys. Fluids*, **4**, p. 397
91. Gershman, B.N. (1964) *Zh. Tekhn. Fiz.*, **34**, p. 1351
92. Storey, L.R.O. (1953) *Phil. Trans. Roy. Soc.*, **A246**, p. 113
93. Alfven, H., and Fälthammar, C.-G. (1963) *Cosmic Electrodynamics, Fundamental Principles*. Clarendon Press, Oxford.
94. Gershman, B.N. (1960) *Sov. Phys. JETP*, **11**, p. 657
95. Tataronis, J.A., and Crawford, F.W. (1970) *J. Plasma Phys.*, **4**, p. 231
96. Zheleznyakov, V.V., and Zlotnik, E.Ya. (1975) *Solar Phys.*, **43**, p. 431
97. Zheleznyakov, V.V., and Zlotnik, E.Ya. (1975) *Solar Phys.*, **44**, p. 447
98. Zheleznyakov, V.V., and Zlotnik, E.Ya. (1975) *Solar Phys.*, **44**, p. 461
99. Lominadze, D.G. (1975) *Cyclotron Waves in Plasma*. Metsniereba.
100. Suvorov, Ye.V., and Chugunov, Yu.V. (1975) *Astrophysics* , **11**, p. 203
101. Adler, S.L. (1971) *Ann. Phys., N.Y.*, **67**, p. 599
102. Novick, R., Weisskopf, M.C., Angel, J.R.P., and Sutherland P.G. (1977) *Astrophys. J.*, **215**, L117
103. Gnedin, Yu.N., Pavlov, G.G., and Shibanov, Yu.A. (1978) *JETP Lett.*, **27**, p. 305
104. Zheleznyakov, V.V. (1980) *Astrophysics*, **16**, p. 316
105. Pavlov, G.G., and Shibanov, Yu.A., (1979) *Sov. Phys. JETP*, **49**, p. 741
106. Barabanenkov, Yu.N., Kravtsov, Yu.A., Rytov, V.I., and Tatarskii, V.I. (1971) *Sov. Phys. Uspekhi*, **13**, p. 551
107. Lotova, N.A. (1968) *Sov. Phys. Uspekhi*, **11**, p. 424
108. Scheuer, P.A. (1968) *Nature*, **218**, p. 920
109. Salpeter, E.E. (1969) *Nature*, **221**, p. 31
110. Erukhimov, L.M., and Pisareva, V.V. (1969) *Radiophys. Quantum Electronics*, **12**, p. 716
111. Erukhimov, L.M. (1972) *Radiophys. Quantum Electronics*, **15**, p. 625
112. Rankin, J.M., and Counselman, C.C. (1973) *Astrophys. J.*, **181**, p. 875
113. Shishov, V.I. (1974) *Sov. Astron*, **17**, p. 598
114. Zheleznyakov, V.V., and Zlotnik, E.Ya (1962) *Izv. VUZ Radiofizika*, **5**, p. 644
115. Zheleznyakov, V.V., and Zlotnik, E.Ya (1963) *Izv. VUZ Radiofizika*, **6**, p. 634
116. Ginzburg, V.L., and Zheleznyakov, V.V. (1959) *Sov. Astron*, **3**, p. 235
117. Tatarskii, V.I. (1967) *The Effects of the Turbulent Atmosphere on Wave Propagation*. Keter Press, Jerusalem.
118. Landau, L.D., and Lifshits, E.M. (1984) *Electrodymanics of Continuous Media*. Pergamon Press, Oxford.
119. Reule, A. (1952) *Naturforsch.*, **A7**, p. 234
120. Hewish, A., Bell, S.J., Pilkington, D.H., Scott, P.F., and Collins, R.A. (1968)

Nature, **217**, p. 709
121. Lyne, A.G., and Rickett, B.J. (1968) *Nature*, **218**, p. 326
122. Manchester, R.N., and Taylor, J.H. (1972) *Astrophys. Lett.*, **10**, p. 67
123. Gershman, B.N., and Ugarov, V.A. (1960) *Sov. Phys. Uspekhi*, **3**, p. 743
124. Helliwell, R.A., Crary, J.H., Pope, H., and Smith, J. (1956) *J. Geophys. Res.*, **61**, p. 139
125. Ginzburg, V.L. (1943) *J. Phys. USSR*, **7**, p. 289
126. Ginzburg, V.L. (1948) *Zh. Eksp. Teor. Fiz.*, **18**, p. 487
127. Golant, V.E., and Piliya, A.D. (1972) *Sov. Phys. Uspekhi*, **14**, p. 413
128. Yerokhin, N.S., and Moiseev, S.S. (1973) in: *Reviews of Plasma Physics*, **7**. Atomizdat, Moscow.
129. Kravtsov, Yu.A. (1969) *Sov. Phys. Doklady*, **13**, p. 1125
130. Zheleznyakov, V.V., Kocharovskii, V V., and Kocharovskii, Vl.V. (1979) *Sov. Phys. JETP*, **50**, p. 51
131. Cohen, M.H. (1960) Astrophys. J., **131**, p. 664
132. Suvorov, Ye.V. (1972) *Izv. VUZ Radiofizika*, **15**, p. 1320
133. Kocharovskii, V.V., and Kocharovskii, Vl.V. (1980) *Fizika Plazmy*, **6**, p. 565
134. Kravtsov, Yu.A., and Naida, O.N. (1976) *Sov. Phys. JETP*, **44**, p. 122
135. Budden, K.G. (1952) *Proc. Roy. Soc.*, **215**, p. 215
136. Budden, K.G. (1961) *Radio Waves in the Ionosphere*. Cambridge Univ. Press.
137. Zheleznyakov, V.V., Kocharovskii, V.V., and Kocharovskii, Vl.V. (1983) *Sov. Phys. Uspekhi*, **26**, p. 877
138. Zheleznyakov, V.V. (1977) *Sov. Phys. JETP*, **46**, p. 292
139. Melrose, D.B. (1974) *Australian J. Phys.*, **27**, p. 31
140. Melrose, D.B. (1974) *Australian J. Phys.*, **27**, p. 43
141. Zheleznyakov, V.V., and Zlotnik, E.Ya. (1963) *Sov. Astron.*, **7**, p. 485
142. Brekhovskikh, L.M. (1973) *Waves in Stratified Media*. Nauka, Moscow.
143. Heding, J. (1965) *Introduction to the Method of Phase Integrals*. Mir, Moscow.
144. Stückelberg, E.C. (1932) *Helv. Phys. Acta*, **5**, p. 369
145. Denisov, N.G. (1956) *Sov. Phys. JETP*, **2**, p. 342
146. Denisov, N.G. (1957) *Sci. Notic. Gorky University. Fizika*, **35**, p. 3
147. Zheleznyakov, V.V., and Zlotnik, E.Ya. (1978) *Radiophys. Quantum Electronics*, **20**, p. 997
148. Zheleznyakov, V.V., Kocharovskii, V.V., and Kocharovskii, Vl.V. (1995) *Astron. Astrophys.* (in press)
149. Harris, E.G. (1962) *Nuovo Cim.*, **23**, p. 115
150. Rankin, J.M., Campbell, D.B., and Backer, D.C. (1974) *Astrophys. J.*, **188**, p. 609
151. Tanaka, H., and Kakinuma, T. (1959) in: *Paris Symp. on Radio Astronomy*. Stanford Univ. Press. p. 215
152. Cohen, M.H. (1961) *Astrophys. J.*, **133**, p. 978
153. Sheridan, K.V., McLean, D.J., and Smerd, S.F. (1973) *Astrophys. Lett.*, **15**, p. 139
154. White, S.M., Thejappa, G., and Kundu, M.R. (1992) *Solar Phys.*, **138**, p. 163
155. Peterova, N.G., and Akhmedov, Sh.B. (1974) *Sov. Astron.*, **17**, p. 768
156. Melrose, D.B. (1973) *Proc. Astron. Soc. Australia*, **2**, p. 208
157. Schindler, K.J. (1974) *J. Geophys. Res.*, **79**, p. 2803
158. Galeev, A.A. (1083) in: A.Nishina (Ed.) *Magnetosphjeric Plasma Physics*. Center for Acad. Publ. Japan, Tokyo. Ch.3, p. 143
159. Gopalswamy, N., Zheleznyakov, V.V., White, S.M., and Kundu, M. (1994) *Solar Phys.*, **155**, p. 339
160. Shklovskii, I.S. (1946) *Astron. Zh.*, **23**, p. 233
161. Martyn, D.F. (1947) *Nature*, **159**, p. 26
162. Zheleznyakov, V.V. (1958) *Sov. Astron.*, **2**, p. 206
163. Zheleznyakov, V.V. (1965) *Sov. Astron.*, **9**, p. 617

164. Ginzburg, V.L., and Ozernoy, L.M. (1966) *Izv. VUZ Radiofizika*, **9** p. 221
165. Zheleznyakov, V.V. (1958) *Izv. VUZ Radiofizika*, **1** p. 32
166. Zheleznyakov, V.V. (1959) *Izv. VUZ Radiofizika*, **2** p. 858
167. Zheleznyakov, V.V., Kocharovskii, V.V., and Kocharovskii, Vl.V. (1980) *Sov. Phys. JETP*, **52**, p. 877
168. Pavlov, G.G. , and Shibanov, Yu.A. (1979) *Sov. Phys. JETP*, **49**, p. 741
169. Zheleznyakov, V.V., and Litvinchuk, A.A. (1983) *Fizika Plazmy*, **9**, p. 553
170. Harris, E.G. (1965) *Phys. Rev.*, **138**, p. 479
171. Zheleznyakov, V.V. (1966) *Izv. VUZ Radiofizika*, **9**, p. 1057
172. Smerd, S.F., and Westfold, K.C. (1949) *Phil. Mag.*, **40**, p. 831
173. Pomraning, G.C. (1968) *Astrophys. J.*, **153**, p. 321
174. Sobolev, V.V. (1969) *Course in Theoretical Astrophysics*. NASA, Washington.
175. Sobolev, V.V. (1963) *A Treatise on Radiative Transfer*. Van Nostrand, Princeton.
176. Sobolev, V.V. (1975) *Light Scattering in Planetary Atmospheres*. Pergamon Press, Oxford.
177. Mihalas, D. (1977) *Stellar Atmospheres*. 2nd Ed. Freeman, San Francisco.
178. Dicke, R.H., Peebles, P.J.E., Roll, P.G., and Wilkinson, D.T. (1965) *Astrophys. J.*, **142**, p. 414
179. Zel'dovich, Ya.B., and Novikov, I.D. (1967) *Relativistic Astrophysics*. Vol. 1. Univ. of Chicago Press, Chicago.
180. Zheleznyakov, V.V. (1969) *Astrophys. J.*, **155**, p. 1129
181. Mercier, R.P. (1964) *Proc. Phys. Soc.*, **83**, p. 811
182. Longair, M.S., and Sunyaev, R.A. (1971) *Uspekhi Fiz. Nauk*, **105**, p. 41
183. Zheleznyakov, V.V. (1983) *Astrophys. Space Sci.*, **97**, p. 229
184. Zheleznyakov, V.V. (1981) *Astrophys. Space Sci.*, **77**, p. 279
185. Zel'dovich, Ya.B., and Shakura, N.N (1969) *Sov. Astron.*, **13**, p. 175
186. Hatanaka, T. (1956) *Publ. Astron. Soc. Japan*, **8**, p. 73
187. Akabane, K., and Cohen, M.H. (1961) *Astrophys. J.*, **33**, p. 258
188. Alissandrakis, C.E., and Chuideri-Drago, F. (1994) *Astrophys. J.*, **428**, p. L73
189. Razin, V.A. (1958) *Sov. Astron.*, **2**, p. 216
190. Getmantsev, G.G., and Razin, V.A. (1956) in: *Proc. 5-th Meeting on the Problems of Cosmogony*. USSR Acad. Sci. Publ. p. 496
191. Kawabata, K. (1964) *Publ. Astron. Soc. Japan*, **16**, p. 30
192. Zheleznyakov, V.V., Suvorov, E.V., and Shaposhnikov, V.E. (1974) *Sov. Astron.*, **18**, p. 142
193. Pacholczyk, A.G., and Swihart, T.L. (1970) *Astrophys. J.*, **161**, p. 415
194. Pacholczyk, A.G., and Swihart, T.L. (1975) *Astrophys. J.*, **196**, p. 125
195. Sazonov, V.N. (1969) *Sov. Phys. JETP*, **29**, p. 578
196. Sazonov, V.N. (1974) *Astrophysics*, **10**, p. 251
197. Sazonov, V.N., and Tsytovich, V.N. (1968) *Izv. VUZ Radiofizika*, **11**, p. 1287
198. Shaposhnikov, V.E. (1977) *Astrophysics*, **13**, p. 209
199. Erukhimov, L.M., and Kirsh, P.I. (1973) *Radiophys. Quantum Electronics*, **16**, p. 1379
200. Gel'freikh, G.B. (1964), *Izv. GAO*, **23**, No. 5, p. 59
201. Ginzburg, V.L., and Syrovatskii, S.I. (1969) *Ann. Rev. Astron. Astrophys.*, **7**, p. 375
202. Ramaty, R. (1969) *Astrophys. J.*, **158**, p. 753
203. Gnedin, Yu.N., and Pavlov, G.G. (1974) *Sov. Phys. JETP*, **38**, p. 903
204. Sazonov, V.N. (1975) *Sov. Astron.*, **19**, p. 42
205. Dolginov, A.Z., Gnedin, Yu.N., and Silant'ev, N.A. (1979) *Propagation and Polarization of Radiation in Cosmic Media*. Nauka, Moscow.
206. Zel'dovich, Ya.B., Levich, E.V., and Sunyaev, R.A. (1972) *Sov. Phys. . JETP*, **35**, p. 733
207. Galeev, A.A., and Sunyaev, R.A. (1973) *Sov. Phys. JETP*, **36**, p. 669
208. Ginzburg, V.L. (1989) *Applications of Electrodynamics in Theoretical Physics and*

Astrophysics. Gordon and Breach, New York.
209. Ginzburg, V.L. (1993) in: E.Wolf (ed.) *Progress in Optics*. Elsevier Science Publ. **32**, p. 267
210. Berestetskii, V.B., Lifshits, E.M., and Pitayevskii, L.P. (1973) *Relativistic Quantum Theory*. Pergamon Press, Oxford
211. Heitler, W. (1954) *Quantum Theory of Radiation*. Clarendon Press, Oxford.
212. Melrose, D.B. (1972) *Astrophys. Space Sci.*, **18**, p. 267
213. Tsytovich, V.N. (1973) *Fizika Plazmy. Trudy FIAN*, **66**.
214. Eidmann, K. (1975) *Plasma Phys.*, **17**, p. 121
215. Akopian, A.V., and Tsytovich, V.N. (1975) *Fizika Plasmy*, **1**, p. 673
216. Ginzburg, V.L. (1940) *Zh. Eksp. Teor. Fiz.*, **10**, p. 589
217. Andronov, A.A. (1961) *Izv. VUZ Radiofizika*, **4**, p. 861
218. Cohen, M.H. (1961) *Phys. Rev.*, **123**, p. 711
219. Sitenko, A.G., and Kolomenskii, A.A. (1956) *Sov. Phys. JETP*, **3**, p. 410
220. Eidman, V.Ia. (1962) *Sov. Phys. JETP*, **14**, p. 1401
221. Zheleznyakov, V.V. (1984) *Sov. Sci. Rev. E Astrophys. Space Phys.*, **3**, p. 157
222. Landau, L.D., and Lifshits, E.M. (1981) *Quantum Mechanics: Non-Relativistic Theory*. Pergamon Press, Oxford.
223. Melrose, D.B., and Zheleznyakov, V.V. (1981) *Astron. Astrophys.*, **95**, p. 86
224. Tsytovich, V.N. (1951) *Vestnik MGU*, No. 11, p. 27
225. Ginzburg, V.L. and Frank, I.M. (1947) *Doklady Academii Nauk SSSR*, **56**, p. 583
226. Ginzburg, V.L., and Eidman, V.Ya. (1959) *Izv. VUZ Radiofizika*, **2**, p. 331
227. Suvorov, E.V., and Chugunov, Yu.V. (1973) *Astrophys. Space. Sci.*, **23**, p. 189
228. Eidman, V.Ia. (1958) *Sov. Phys. JETP*, **7**, p. 91
229. Eidman, V.Ia. (1959) *Sov. Phys. JETP*, **9**, p. 947
230. Zheleznyakov, V.V. (1964) *Izv. VUZ Radiofizika*, **7**, p. 67
231. Ginzburg, V.L. and Zheleznyakov, V.V. (1959) *Izv. VUZ Radiofizika*, **1**, p. 59
232. Stepanov, K.N., and Pakhomov, V.I. (1960) *Sov. Phys. JETP*, **11**, p. 1126
233. Kippenheuer, K.O. (1950) *Phys. Rev.*, **79**, p. 738
234. Ginzburg, V.L. (1951) *Doklady Akademii Nauk SSSR*, **76**, p. 377
235. Gordon, I.M. (1954) *Doklady Akademii Nauk SSSR*, **94**, p. 813
236. Shklovskii, I.S. (1953) *Doklady Akademii Nauk SSSR*, 90, p. 983
237. Sokolov, A.A., and Ternov, I.M. (Eds.) (1968) *Synchrotron Radiation*. Akademie-verlag, Berlin.
238. Ivanenko, D.D., and Sokolov, A.A. (1951) *Classical Field Theory*. Nauka, Moscow.
239. Schwinger, J. (1949) *Phys. Rev.*, **75**, p. 1912
240. Westfold, K.C. (1959) *Astrophys. J.*, **130**, p. 241
241. Jackson, J. (1962) *Classical Electrodynamics*. Wiley, New York.
242. Ginzburg, V.L.(1953) *Uspekhi Fiz. Nauk*, **51**, p. 343
243. Ginzburg, V.L.(1954) *Forschrifte d. Physik*, **1**, p. 659
244. Ginzburg, V.L. (1959) *Sov. Phys. Uspekhi*, **2**, p. 874
245. Razin, V.A. (1960) *Izv. VUZ Radiofizika*, **3**, p. 584
246. Razin, V.A. (1960) *Izv. VUZ Radiofizika*, **3**, p. 594
247. Ginzburg, V.L., and Syrovatskii, S.I. (1966) *Sov. Phys. Uspekhi*, **8**, p. 674
248. Zheleznyakov, V.V., and Trakhtengerts, V. Yu. (1965) *Sov. Astron.*, **9**, p. 775
249. Kaplan, S.A., and Trakhtengerts, V. Yu. (1967), *Radiophys. Quantum Electronics*, **10**, p. 5
250. Feenberg, E., and Primakoff, H. (1948) *Phys. Rev.*, **73**, p. 449
251. Peterson, A.M., and Hower, G.L. (1966) in: B.M.McCormac (Ed.) *Radiation Trapped in the Earth Magnetic Field*. D.Reidel Publ. Co., p. 714
252. Chugunov, Yu.V., Eidman, V.A., and Suvorov, E.V. (1975) *Astrophys. Space. Sci.*, **32**, p. 7
253. Goldreich, P., Pacini, F., and Rees, M.J. (1972) *Comments Astrophys. Space Sci.*, **4**, p. 23
254. Greene, J. (1959) *Astrophys. J.*, **130**, p. 693

255. Ginzburg, V.L. and Syrovatskii, S.I. (1965) *Sov. Phys. Uspekhi*, **7**, p. 696
256. Trubnikov, B.A., (1958) *Soviet Phys. Doklady*, **3**, p. 136
257. Hirshfield, J.L., Baldwin, D.E., and Brown, S.C. (1961) *Phys. Fluids*, **4**, p. 198
258. Gnedin, Yu.N., and Sunyaev, R.A. (1973) *Astron. Astrophys.*, **25**, p. 233
259. Petrosian, V. (1981) *Astrophys. J.*, **251**, p. 727
260. Abramovitz, M., and Stegun, I.A. (1970) *Handbook of Mathematical Functions*. Dover, New York. p. 365
261. Ramaty, R., and Lingenfelter, R.E. (1967) *J. Geophys. Res.*, **72**, p. 879
262. Ramaty, R., and Petrosian, V. (1972) *Astrophys. J.*, **178**, p. 241
263. Lamb, D.Q., and Masters, A.R. (1979) *Astrophys. J.*, **234**, p. L117
264. Trubnikov, B.A., (1961) *Phys. Fluids*, **4**, p. 195
265. Petrosian, V., and McTiernan, J.M. (1983) *Phys. Fluids*, **26**, p. 3023
266. Dulk, G.A., and Marsh, K.A. (1982) *Astrophys. J.*, **259**, p. 350
267. Robinson, P.A., and Melrose, D.B. (1984) *Austral. J. Phys.*, **37**, p. 675
268. Wild, J.P., and Hill, E.R. (1971) *Austral. J. Phys.*, **24**, p. 43
269. Endean, V.G., and Allen, J.E. (1970) *Nature*, **228**, p. 348
270. Garibyan, G.M., and Gol'dman, I.I. (1954) *Izv. Armenian Acad. Sci.*, **7**, p. 31
271. Getmantsev, G.G. (1952) *Doklady Akademii Nauk SSSR*, **83**, p. 557
272. Daniel, R.R., and Stephens, S.A. (1970) *Space Sci. Rev.*, **10**, p. 599
273. Zheleznyakov, V.V., and Belyanin, A.A. (1994) *Astron. Astrophys.*, **287**, p. 782
274. Shklovsky, I.S. (1953) *Doklady Akademii Nauk SSSR*, **90**, p. 983
275. Ginzburg, V.L. (1954) in: *Proc. 3-rd Meeting on the Problems of Cosmogony*. USSR Acad. Sci., Moscow.
276. Ginzburg, V.L. (1946) *Doklady Akademii Nauk SSSR*, **52**, p. 491
277. Martyn, D.F. (1946) *Nature*, **158**, p. 632
278. Martyn, D.F. (1948) *Nature*, **193**, p. 44
279. Smerd, S.F. (1950) *Austral. J. Sci. Res.*, **A3**, p. 34
280. Bohm, D., and Gross, E.P. (1949) *Phys. Rev.*, **75**, p. 1851
281. Bohm, D., and Gross, E.P. (1949) *Phys. Rev.*, **75**, p. 1864
282. Shafranov, V.D. (1958) *Sov. Phys. JETP*, **7**, p. 1019
283. Ginzburg, V.L., and Zheleznyakov, V.V. (1958) *Sov. Astron.*, **2**, p. 653
284. Zlotnik, E.Ya. (1968) *Astron. Zh.*, **45**, p. 310
285. Zlotnik, E.Ya. (1968) *Sov. Astron.*, **12**, p. 464
286. Gershman, B.N. (1953) *Zh. Eksp. Teor. Fiz.*, **24**, p. 659
287. Stepanov, K.N. (1959) *Sov. Phys. JETP*, **8**, p. 195
288. Akhiezer, A.I., Akhiezer, I.A., Polovin, R.V., Sitenko, A.G., and Stepanov, K.N. (1975) *Plasma Electrodynamics*. Vol. 1, 2. Pergamon Press, Oxford.
289. Suvorov, E.V., and Fraiman, A.A. (1977) *Izv. VUZ Radiofizika*, **20**, No. 1
290. Pavlov, G.G., Shibanov, Yu.A., and Gnedin, Yu.N. (1979) *JETP Lett.*, **30**, p. 125
291. Zheleznyakov, V.V., and Zlotnik, E.Ya. (1971) *Solar Phys.*, **20**, p. 85
292. Lamb, F.K., and Sutherland, P.G. (1974) in: C.T.Hansen (Ed.) *Physics of Dense Matter*. D. Reidel Publ. Comp., p. 265
293. Field, G.B. (1959) *J. Geophys. Res.*, **61**, p. 1169
294. Hewitt, R.G., Melrose, D.B., and Rönnmark, K.G. (1982) *Austral. J. Phys.*, **35**, p. 447
295. Wu, C.S. (1985) *Space Sci. Rev.*, **41**, p. 215
296. Topka, K., and Marsh, K.A. (1982) *Astrophys. J.*, **254**, p. 641
297. Gary, D.E., and Lynsky, J.L. (1981) *Astrophys. J.*, **250**, p. 284
298. Zheleznyakov, V.V. (1963) *Sov. Astron.*, **7**, p. 630
299. Zheleznyakov, V.V. (1962) *Sov. Astron.*, **6**, p. 3
300. Kakinuma, T., and Swarup, G. (1962) *Astrophys. J.*, **136**, p. 975
301. Zheleznyakov, V.V., and Zlotnik, E.Ya. (1980) *Solar. Phys.*, **68**, p. 317
302. Zheleznyakov, V.V., and Zlotnik, E.Ya. (1980) *Astron. Zh.*, **57**, p. 778
303. Zheleznyakov, V.V., and Tikhomirov, Yu.V. (1981) *Solar Phys.*, **81**, p. 121
304. Willson, R.F. (1989) *Astrophys. J.*, **298**, p. 911

305. Zheleznyakov, V.V., and Zlotnik, E.Ya. (1989) *Solar Phys.*, **121**, p. 449
306. Titchmarsch, E. (1967) *Introduction to the Theory of Fourier Integrals.* Oxford.
307. Twiss, R.Q. (1951) *Proc. Phys. Soc.*, **B64**, p. 654
308. Sturrock, P.A. (1959) *Phys. Rev.*, **112**, p. 1488
309. Vedenov, A.A. (1963) in: *Reviews of Plasma Phys.*, **3**, p. 229
310. Twiss, R.Q. (1958) *Austral. J. Phys.*, **11**, p. 564
311. Zheleznyakov, V.V. (1967) *Sov. Phys. JETP*, **24**, p. 381
312. McCray, R. (1966) *Science*, **154**, p. 1320
313. Litvak, A.G., Permitin, G.V., Suvorov, E.V., and Frajman, A.A. (1977) *Nucl. Fusion*, **17**, p. 659
314. Zheleznyakov, V.V., and Serber, A.V. (1992) in: W.S.Paciesas and G.J.Fishman (Eds.) *Gamma-Ray Bursts. AIP Conf. Proc. No. 265.* AIP, New York. p. 262
315. Serber, A.V. (1990) *Astron. Zh.*, **67**, p. 582
316. Romanov, Yu.A., and Filippov, G.F. (1961) *Sov. Phys. JETP*, **13**, p. 87
317. Vedenov, A.A., Velikhov, E.P., and Sagdeev, R.Z. (1961) *Sov. Phys. Uspekhi*, **4**, p. 332
318. Drummond, W.E., and Pines, D. (1962) *Nucl. Fusion Suppl.*, **3**, p. 1049
319. Akhiezer, A.I., and Fainberg, Ya.B. (1949) *Doklady Akademii Nauk SSSR*, **69**, p. 555
320. Akhiezer, A.I., and Fainberg, Ya.B. (1949) *Zh. Eksp. Teor. Fiz.*, **21**, p. 1262
321. Ginzburg, V.L., and Zheleznyakov, V.V. (1965) *Phyl. Mag.*, **11**, p. 197
322. Buneman O. (1959) *Phys. Rev.*, **115**, p. 603
323. Sagdeev, R.Z. (1966) *Reviews of Plasma Phys.*, **4**, p. 23
324. Zaitsev, V.V. (1969) *Radiophys. Quantum Electronics*, **12**, p. 622
325. Rosenbluth, M.N., and Post, R.F. (1965) *Phys. Fluids*, **8**, p. 547
326. Trakhtengerts, V.Yu. (1968) *Geomagn. and Aeronomy*, **8**, p. 332
327. Zaitsev, V.V., and Stepanov, A.V. (1975) *Astron. Astrophys.*, **45**, p. 135
328. Galeev, A.A. (1967) *J. Plasma Phys.*, **1**, p. 105
329. Ivanov, A.A., and Rudakov, L.I. (1967) *Sov. Phys. JETP*, **24**, p. 1027
330. Vedenov, A.A., and Ryutov, D.D. (1972) in: *Reviews of Plasma Phys.*, **6**. Atomizdat, Moscow.
331. Sizonenko, V.L., and Stepanov, A.V. (1966) *Sov. Phys. JETP*, **22**, p. 832
332. Ivanov, A.A., Soboleva, T.K., and Yushmanov, P.N. (1976) *Sov. Phys. JETP*, **42**, p. 1027
333. Fainberg, Ya.B., and Shapiro, V.D. (1965) in: *Physics of Plasmas.* Kiev. p. 92
334. Fainberg, Ya.B., Shapiro, V.D., and Shevchenko, V.I. (1970) *Sov. Phys. JETP*, **30**, p. 528
335. Fainberg, Ya.B., and Shapiro, V.D. (1965) *Sov. Phys. JETP*, **20**, p. 937
336. Zheleznyakov, V.V., and Zaitsev, V.V. (1970) *Sov. Astron.*, **14**, p. 47
337. Zheleznyakov, V.V., and Zaitsev, V.V. (1970) *Sov. Astron.*, **14**, p. 250
338. Ryutov, D.D., and Sagdeev, R.Z. (1970) *Sov. Phys. JETP*, **31**, p. 396
339. Zaitsev, V.V., Mityakov, N.A., and Rapoport, V.O. (1972) *Solar Phys.*, **24**, p. 444
340. Zaitsev, V.V., Kunilov, M.V., Mityakov, N.A., and Rapoport, V.O. (1974) *Sov. Astron.*, **18**, p. 147
341. Zaitsev, V.V. (1975) *Sov. Astron.*, **18**, p. 475
342. Pikel'ner, S.B., and Gintsburg, M.A. (1964) *Sov. Astron*, **7**, p. 639
343. Zaitsev, V.V. (1966) *Sov. Astron.*, **9**, p. 572
344. Stepanov, A.V. (1974) *Sov. Astron.*, **17**, p. 781
345. Ginzburg, V.L., Zheleznyakov, V.V., and Zaitsev, V.V. (1969) *Sov. Phys. Uspekhi*, **12**, p. 378
346. Ginzburg, V.L., Zheleznyakov, V.V., and Zaitsev, V.V. (1969) *Astrophys. Space Sci.*, **4**, p. 464
347. Sen, H.K. (1952) *Phys. Rev.*, **88**, p. 816
348. Pearlstein, L.D., Rosenbluth, M.N., and Chang, D.B. (1966) *Phys. Fluids*, **9**, p.

953
349. Slottje, C. (1972) *Solar Phys.*, **25**, p. 210
350. Kuijpers, J. (1975) *Thesis*. Utrecht.
351. Harris, E.G. (1959) *Phys. Rev. Lett.*, **2**, p. 34
352. Zlotnik, E.Ya. (1973) *Radiophys. Quantum Electronics*, **16**, p. 1278
353. Zlotnik, E.Ya. (1974) *Radiophys. Quantum Electronics*, **17**, p. 11
354. Hirschfield, J., and Bekefi, G. (1963) *Nature*, **198**, p. 20
355. Schneider, J. (1959) *Phys. Rev. Lett.*, **2**, p. 504
356. Gaponov, A.V. (1959) *Izv. VUZ Radiofizika*, **2**, p. 450
357. Gaponov, A.V. (1959) *Izv. VUZ Radiofizika*, **2**, p. 836
358. Sagdeev, R.Z., and Shafranov, V.D. (1961) *Sov. Phys JETP*, **12**, p. 130
359. Zheleznyakov, V.V. (1960) *Izv. VUZ Radiofizika*, **3**, p. 57
360. Zheleznyakov, V.V. (1960) *Izv. VUZ Radiofizika*, **3**, p. 180
361. Zheleznyakov, V.V. (1961) *Izv. VUZ Radiofizika*, **4**, p. 619
362. Zheleznyakov, V.V. (1961) *Izv. VUZ Radiofizika*, **4**, p. 849
363. Kovner, M.S. (1960) *Izv. VUZ Radiofizika*, **3**, p. 631
364. Kovner, M.S. (1960) *Izv. VUZ Radiofizika*, **3**, p. 746
365. Petelin, M.I. (1961) *Izv. VUZ Radiofizika*, **4**, p. 455
366. Zayed, K.E., and Kitsenko, A.B. (1968) *Fizika Plazmy*, **10**, p. 147
367. Mikhailovskii, A.B. (1972) in: *Reviews of Plasma Phys.*, **6**. Atomizdat, Moscow.
368. Bratman, V.L., and Suvorov, E.V. (1968) *Sov. Phys JETP*, **28**, p. 740
369. Shklovsky, I.S. (1969) *Astron. Circ. (Russ.)*, No. 527, p. 1
370. Zheleznyakov, V.V. (1969) *Sov. Phys Uspekhi*, **12**, p. 807
371. Bekefi, G., Hirshfield, J.L., and Brown, S.C. (1961) *Phys. Rev.*, **122**, p. 1037
372. Zheleznyakov, V.V. (1967) *Sov. Astron.*, **11**, p. 33
373. Zheleznyakov, V.V., and Suvorov, E.V. (1968) *Sov. Phys JETP*, **27**, p. 335
374. Sazonov, V.N. (1970) *Sov. Astron.*, **13**, p. 797
375. Heyvaerts, M.J. (1968) *Ann. d'Astrophys.*, **31**, p. 129
376. Zheleznyakov, V.V., and Suvorov, E.V. (1972) *Astrophys. Space Sci.*, **15**, p. 3
377. Goreva, T.A., and Suvorov, E.V. (1973) *Sov. Phys. JETP*, **35**, p. 1122
378. Bratman, V.L., and Bychkov, V.V. (1970) *Sci. Notic. Gorky University. Radiofizika*, No. 124, p. 9
379. Wu, C.S. (1985) *Space Sci. Rev.*, **41**, p. 215
380. Charikov, Yu.E, Mosunov, A.N., and Prokopjev, A.V. (1993) *Solar Phys.*, **147**, 157
381. Kadomtsev, B.B. (1964) *Reviews oF Plasma Phys.*, **4**. Atomizdat, Moscow.
382. Kovrizhnykh, L.M. (1966) *Plasma Phys. Trudy FIAN*, **32**, p. 173
383. Tsytovich, V.N. (1977) *Theory of Turnulent Plasma*. Consultants Bureau, New York.
384. Pustovalov, V.V., and Silin, V.P. (1972) in: D.V.Skobel'tsyn (ed.) *Theory of Plasmas. Proc. (Trudy) of the P.N.Lebedev Physical Inst.*, **61**, p. 37
385. Pauli, W. (1923) *Z. Phys.*, **18**, p. 272
386. Plachek, G. (1935) *Rayleigh Scattering and Raman Effect*. Ukraine Gostekhizdat.
387. Gailitis, A., and Tsytovich, V.N. (1964) *Sov. Phys. JETP*, **19**, p. 1165
388. Gailitis, A., and Tsytovich, V.N. (1965) *Sov. Phys. JETP*, **20**, p. 987
389. Melrose, D.B., and Sy, W.N. (1972) *Astrophys. Space Sci.*, **17**, p. 343
390. Melrose, D.B., and Sy, W.N. (1972) *Australan . J. Phys.*, **25**, p. 387
391. Gorbunov, L.M., and Silin, V.N. (1964) *Sov. Phys. JETP*, **20**, p. 135
392. Ginzburg, V.L., and Tsytovich, V.N. (1990) *Transition Radiation and Transition Scattering*. Higler, Bristol.
393. Kompaneets, A.S. (1956) *Zh. Eksp. Teor. Fiz.*, **31**, p. 876
394. Tsytovich, V.N., and Shapiro, V.D. (1965) *Nuclear Fusion*, **5**, p. 228
395. Andronov, A.A. (1968) *Ph.D. Thesis*, Gorky University
396. Trakhtengerts, V.Yu. (1968) *Geomagn. i Aeronom.*, **8**, p. 966
397. Zaitsev, V.V. (1971) *Solar Phys.*, **20**, p. 95

398. Zaitsev, V.V. (1974) *Izv. VUZ Radiofizika*, **14**, 1438
399. Zaitsev, V.V. (1970) *Radiophys. Quantum Electronics*, **13**, p. 661
400. Zaitsev, V.V., and Stepanov, A.V. (1975) *Studies on Geomagn., Aeronom. and Solar Phys.*, **37**, p. 11
401. Smith, D.F., Fung, P.C.W. (1971) *J. Plasma Phys.*, **5**, p. 1
402. Smith, D.F. (1973) *Solar Phys.*, **33**, p. 213
403. Bardwell, S., and Goldman, M.V. (1976) *Astrophys. J.*, **209**, p. 912
404. Ginzburg, V.L., and Tsytovich, V.N. (1974) *Sov. Phys. JETP*, **38**, p. 909
405. Andronov, A.A., and Gorodinsky, G.V. (1962) *Izv. VUZ Radiofizika*, **5**, p. 234
406. Gertsenshtein, M.E. (1952) *Zh. Eksp. Teor. Fiz.*, **23**, p. 678
407. Ginzburg, V.L., and Syrovatsky, S.I. (1964) *Sov. Phys. JETP*, **19**, p. 1255
408. Felten, J.E., and Morrison, P. (1966) *Astrophys. J.*, **146**, p. 686
409. Blumenthal, G.R., and Gould, R.J. (1970) *Rev. Mod. Phys.*, **42**, p. 237
410. Kaplan, S.A., Tsytovich, V.N., and Chikhachev, A.S. (1970) *Astrophysics*, **6**, p. 253
411. Melrose, D.B. (1971) *Astrophys. Space Sci.*, **13**, p. 56
412. Zel'dovich, Ya.B. (1975) *Sov. Phys. Uspekhi*, **18**, p. 79
413. Getmantsev, G.G., and Tokarev, Yu.V. (1972) *Astrophys. Space Sci.*, **18**, p. 135
414. Gunn, J.E., and Ostriker, J.P. (1971) *Astrophys. J.*, **165**, p. 523
415. Rees, M.J. (1971) *Nature*, **229**, p. 312
416. Rees, M.J. (1971) *Nature Phys. Sci.*, **230**, p. 55
417. Weyman. R. (1965) *Phys. Fluids*, **8**, p. 2112
418. Weyman, R. (1966) *Astrophys. J.*, **145**, p. 560
419. Sunyaev, R.A., and Titarchuk, L.G. (1980) *Astron. Astrophys.*, **86**, p. 121
420. Pozdnyakov, L.A., Sobol', I.M., and Sunyaev, R.A. (1982) in: *Sov. Sci. Rev. E. Astrophys. Space Phys.*, **2**, p. 189
421. Sunyaev, R.A., and Zel'dovich, Ya.B. (1980) *Ann. Rev. Astron. Astrophys.*, **18**, p. 537
422. Felten, J.E., and Rees, M.J. (1972) *Astron. Astrophys.*, **17**, p. 226
423. Illarionov, A.F., and Sunyaev, R.A. (1972) *Sov. Astron.*, **16**, p. 45
424. Arons, J. (1972) *Astrophys. J.*, **164**, p. 437
425. Zel'dovich, Ya.B., and Shakura, N.N. (1969) *Sov. Astron.*, **13**, p. 175
426. Zel'dovich, Ya.B., and Sunyaev, R.A. (1969) *Astrophys. Space Sci.*, **4**, p. 301
427. Chaplin, G., and Stevens, J. (1973) *Astrophys. J.*, **184**, p. 1041
428. Shapiro, S., Lightman, A., and Eardley, D. (1976) *Astrophys. J.*, **204**, p. 187
429. Sunyaev, R.A., and Trümper, I. (1979) *Nature*, **279**, p. 506
430. Sobol', I.M. (1973) *Monte-Carlo Numerical Methods*. Nauka, Moscow.
431. Canuto, V., Lodenquai, J., and Ruderman, M. (1971) *Phys. Rev. D*, *3*, p. 2303
432. Chandrasekhar, S. (1945) *Rev. Mod. Phys.*, **17**, p. 138
433. Zheleznyakov, V.V., and Litvinchuk, A.A. (1986) in: *Plasma Astrophysics. Porc. Joint Varenna-Abastumani Int. School and Workshop*. ESA Publ. Division, Noordwijk. p. 375
434. Sobolev, V.V. (1955) *Moving Envelopes of Stars*. Harvard College Observatory, Cambridge.
435. Castor, J. (1970) *MNRAS*, **149**, p. 111
436. Wang, J.C.L., Lamb, D.Q., Loredo, T.J., et al. (1989) *Phys. Rev. Lett.*, **63**, p. 1550
437. Rowlands, J., Sizonenko, V.L., and Stepanov, K.N. (1966) *Sov. Phys. JETP*, **23**, p. 661
438. Zheleznyakov, V.V., and Serber, A.V. (1991) *Sov. Astron. Lett.*, **17**, p. 419
439. Green, R.F., and Liebert, J. (1981) *Publ. Astron. Soc. Pacific*, **93**, p. 105
440. Schmidt, G.D., Latter, W.B., and Foltz, C.B. (1990) *Astrophys. J.*, **350**, p. 758
441. Angel, J.P.P. (1974) *Ann. Rev. Astron. Astrophys.*, **16**, p. 487
442. Liebert J. (1976) *Publ. Astron. Soc. Pacific*, **88**, p. 490
443. Arnaud, K.A., Zheleznyakov, V.V., and Trimble, V. (1992) *Publ. Astron. Soc.*

Pacific, **104**, p. 239
444. Bespalov, P.A., and Zheleznyakov, V.V. (1990) *Sov. Astron. Lett.*, **16**, p. 442
445. Murakami, T., (1989) in: *Proc. 23rd ESLAB Symp. on Two Topics in X-ray Astron.* ESA SP-296. p. 173
446. Gnedin, Yu.N., and Sunyaev, R.A. (1974) *Astron. Astrophys.*, **36**, p. 379
447. Zheleznyakov, V.V. (1987) *Radiophys. Quantum Electronics*, **30**, p. 102
448. Zheleznyakov. V.V. (1982) *Astrophys. Space Sci.*, **83**, 81
449. Basko, M.M., and Sunyaev, R.A. (1976) *MNRAS*, **175**, p. 395
450. Klein, R.I., Stockman, H.S., and Shevalier, R.A. (1980) *Astrophys. J.*, **237**, p. 912
451. Stellingwerf, R.F., and Buff, J. (1982) *Astrophys. J.*, **260**, p. 755
452. Wang, Y.-M. (1982) *Astron. Astrophys.*, **112**, p. 24
453. Burger, H.L., and Katz, J.I. (1983) *Astrophys. J.*, **265**, p. 393
454. Sobolev, V.V. (1957) Sov. Astron., **1**, p. 678
455. Lucy, L.B., and Solomon, P.M. (1970) *Astrophys. J.*, **159**, p. 879
456. Lucy, L.B. (1971) *Astrophys. J.*, **163**, p. 95
457. Castor, J.I. (1974) *MNRAS*, **169**, p. 279
458. Castor, J.I., Abbott, D.C., and Klein, R.I. (1975) *Astrophys. J.*, **195**, p. 157
459. Cassinelli, J.R. (1979) *Ann. Rev. Astron. Astrophys.*, **17**, p. 273
460. Zheleznyakov, V.V., and Litvinchuk, A.A. (1987) *Sov. Astron.*, **31**, p. 159
461. Mitrofanov, I.G., and Pavlov, G.G. (1982) *MNRAS*, **200**, p. 1033
462. Zheleznyakov, V.V., and Serber, A.V. (1994) *Astrophys. J. Suppl.*, **90**, p. 783
463. Dermer, C.D., and Sturner, S.J. (1991) *Astrophys. J.*, **382**, p. L23
464. Sturner, S.J., and Dermer, C.D. (1992) in: W.S.Paciesas and G.J.Fishman (Eds.) *Gamma-Ray Bursts. AIP Conf. Proc. No. 265.* AIP, New York. p. 277
465. Zheleznyakov. V.V., Serber, A.V., and Kuijpers, J. (1995) *Astron. Astrophys.* (in press)
466. Akhiezer, A.I., Prokhoda, I.G., and Sitenko, A.G. (1958) *Sov. Phys. JETP*, **6**, p. 576
467. Galeev, A.A., Karpman, V.I., and Sagdeev, R.Z. (1965) *Nuclear Fusion*, **5**, p. 20
468. Zheleznyakov, V.V., and Zlotnik, E.Ya. (1974) *Solar. Phys.*, **36**, p. 443
469. Kadomtsev, V.V., and Petviashvili, V.I. (1963) *Sov. Phys. JETP*, **16**, p. 1578
470. Camac, M., Kantrowitz, A.R., Litvak, M.M., Patrick, R.M., and Petchek, H.E. (1962) *Nucl. Fusion Suppl.*, Part 2, p. 423
471. Gorbunov, L.M., Pustovalov, V.V., and Silin, V.P. (1965) *Sov. Phys. JETP*, **20**, p. 967
472. Al'tshul', I.A., and Karpman, V.I. (1965) *Sov. Phys. JETP*, **20**, p. 1043
473. Akhiezer, I.A., Daneliya, N.L., and Tsintsadze, N.N. (1964) *Sov. Phys. JETP*, **19**, p. 208
474. Kovrizhnykh, L.M., and Tsytovich, V.N. (1964) *Sov. Phys. JETP*, **20**, p. 978
475. Smith, D.F., and Sturock, P.A. (1971) *Astrophys. Space Sci.*, **12**, p. 411
476. Melrose, D.B. (1975) *Austral. J. Phys.*, **28**, p. 101
477. Zlotnik, E.Ya. (1976) *Radiophys. Quantum Electronics*, **19**, p. 337
478. Trakhtengerts, V.Yu. (1970) *Radiophys. Quantum Electronics*, **13**, p. 697
479. Silin, V.P. (1964) *Prikl. Mat. Tech. Fiz.*, **1**, p. 31
480. Terashima, Y., and Yajima, N. (1964) *Progr. Theor. Phys.*, **30**, p. 443
481. Zaitsev. V.V. (1970) *Izv. VUZ Radiofizika*, **13**, p. 837
482. Zlotnik, E.Ya. (1981) *Astron. Astrophys.*, **101**, p. 250

Index

Absolute instability, 289-292
Absorption index, 53
Ap stars, 14, 16
Bremsstrahlung from H II regions, 264, 265
Brightness temperature, 25, 26
  under the steady-state three-wave interaction, 441, 442
Comptonization of radiation in astrophysical objects, 389, 392
Convective instability, 289-292
Cotton–Mouton (QT Faraday) effect, 179
Critical value of magnetic field, 91
Cyclotron frequency, 7,8
Cyclotron radiation
  spectral power of, 216, 217, 220, 221
  depression in magnetoactive plasma, 218
Decaying and non-decaying spectra of normal waves, 434, 435
Decrement of normal wave, 54
Degree of circular polarization, 32
Degree of linear polarization, 32
Dispersion measure, 109
Doppler effect
  formula of, 34
  normal and anomalous, 210
Eikonal equation, 102
Einstein relations, 293
Energy losses of particle moving in a medium, 202, 203
Equilibrium radiation, 26
Faraday rotation (QL Faraday effect), 178, 179
  depolarization under, 182-183
Free-free absorption in the solar corona, 265
Galaxy
  interstellar medium, 9
  H I and H II regions 9, 11
Gyrofrequency, *see* Cyclotron frequency
Gyroresonant layer
  linear size, 279
  optical depth, 279
  scattering of radiation in, 403-405
Interplanetary medium, 11
Inverse Compton scattering in cosmic conditions, 382
Kirchhoff law, 257, 258
Linear mode coupling
  characteristic function for, 116
  coupling parameter, 118
  in neutral current sheet, 138-140
  with transverse magnetic field, 140-141
  in transverse magnetic field, critical frequency for, 124, 135
  transformation coefficient, 114, 115, 117, 118
Magnetic bremsstrahlung
  Doppler formula, 210
  recoil effect, 209

spectral power of, 213-215
Magnetic white dwarfs, 16, 17
radiation-driven ejection from atmospheres of, 430-431
Milne–Eddington equation, 409
Neutron stars
magnetic fields, 17
plasma envelopes, 18
Phase and group velocities, 67, 68, 70
Phase integral method, 126-134
Planets, plasma environments of, 11
Plasma dielectric tensor
elementary approach, 55-56
in magnetized vacuum, 91-92
kinetic approach, 59-66
Plasma properties
criterion of quantized plasma, 19
Debye radius, 7
degree of ionization, 6
electron density, 6
Langmuir (plasma) frequency, 7
quasi-neutrality, 6
Plateau formation, 318-323
Polarization coefficients of modes, 77-79, 81
Pulsation measure, 180
Pulse compression effect, 41
Pulsed regime of plasma instability, 370-371
Quasi-linear equations for plasma waves, 318
Quasi-longitudinal and quasi-trans- verse propagation limits, 75, 81
Radiation-driven diskon, 432-433
Radiation mechanisms
antenna and maser, 43
coherent and incoherent, 41, 42
Radiation transfer
in expanding Universe, 160-163, 170, 171
intensity invariant in an inhomogeneous medium, 158, 165
in scattering medium, equation, 160
solution of, 171-175
Reflection point in an inhomogeneous plasma, 105, 107
Refraction, 157
Refractive indices of ordinary and extraordinary modes, 72, 92, 93
Relativistic beaming, 35, 37, 39
Rotation measure, 178
Schwarzschild–Shuster method, 171
Solar radio emission
cyclotron lines in the spectrum of, 284-285
harmonics in spectra of drifting types of solar bursts, 437, 446-448
limb brightening effect, 267-268
origin of fine structures in, 331, 334-336
radio emission of "quiet" Sun, 265-268
thermal cyclotron mechanism of, 280
Solar corona
active regions, 12, 14
physical conditions in, 12-13
sunspots, 14
Stokes lines, 132
Superluminal radio sources, 36,37

Synchrotron mechanism of cosmic radio emission, 254-256
Synchrotron radiation
in a medium
depression of, 226, 227
spectral power of, 227-229
in vacuum, spectral power of, 214-226
Synchrotron reabsorption in astrophysical plasmas 345
Thermal and non-thermal radiation, 41
Vacuum birefringence, 20
Vavilov–Čerenkov effect
criterion of, 204, 205, 209, 211
spectral power, 206

GPSR Compliance

The European Union's (EU) General Product Safety Regulation (GPSR) is a set of rules that requires consumer products to be safe and our obligations to ensure this.

If you have any concerns about our products, you can contact us on ProductSafety@springernature.com

In case Publisher is established outside the EU, the EU authorized representative is:

Springer Nature Customer Service Center GmbH
Europaplatz 3
69115 Heidelberg, Germany

Batch number: 10755284

Printed by Printforce, the Netherlands